W0105986

COMPUTATIONAL METHODS
IN WATER RESOURCES X

Water Science and Technology Library

VOLUME 12

Series Editor:
V. P. Singh, *Louisiana State University,*
Baton Rouge, U.S.A.

Editorial Advisory Board:

S. Chandra, *Roorkee (U.P.), India*
J. C. van Dam, *Delft, The Netherlands*
M. Fiorentino, *Potenza, Italy*
W. H. Hager, *Zürich, Switzerland*
N. Harmancioglu, *Izmir, Turkey*
V. V. N. Murty, *Bangkok, Thailand*
J. Nemec, *Genthod/Geneva, Switzerland*
A. R. Rao, *West Lafayette, Ind., U.S.A.*
Shan Xu Wang, *Wuhan, Hubei, P.R. China*

The titles published in this series are listed at the end of this volume.

COMPUTATIONAL METHODS IN WATER RESOURCES X

Volume 1

edited by

ALEXANDER PETERS
IBM Heidelberg, Germany

GABRIEL WITTUM
Universität Stuttgart, Germany

BRUNO HERRLING
Universität Karlsruhe, Germany

UDO MEISSNER
Technische Hochschule Darmstadt, Germany

CARLOS A. BREBBIA
Wessex Institute of Technology, UK

WILLIAM G. GRAY
University of Notre Dame, USA

and

GEORGE F. PINDER
University of Vermont, USA

SPRINGER-SCIENCE+BUSINESS MEDIA, B.V.

Library of Congress Cataloging-in-Publication Data

Computational methods in water resources X / editors, Alexander Peters
 ... [et al.].
      p.   cm. -- (Water science and technology library ; v. 12)
  Edited proceedings of the Tenth International Conference on
Computational methods in Water Resources, held at Universität
Heidelberg, Germany, Jul. 1994.

   1. Hydrology--Data processing--Congresses. 2. Hydrology-
-Mathematical models--Congresses.   I. Peter, A. (Alexander), 1956-
. II. International Conference on Computational Methods in Water
Resources (10th : 1994 : Universität Heidelberg) III. Series.
GB656.2.E42C653   1994
551.49'0285--dc20   94-17902

ISBN 978-94-010-9206-7   ISBN 978-94-010-9204-3 (eBook)
DOI 10.1007/978-94-010-9204-3

Printed on acid-free paper

All Rights Reserved
© 1994 Springer Science+Business Media Dordrecht
Originally published by Kluwer Academic Publishers in 1994

No part of the material protected by this copyright notice may be reproduced or
utilized in any form or by any means, electronic or mechanical,
including photocopying, recording or by any information storage and
retrieval system, without written permission from the copyright owner.

EDITORS:

A. Peters
IBM Heidelberg STSS
Vangerowstr. 18
69020 Heidelberg
Germany

G. Wittum
ICA / Numerik
Universität Stuttgart
Pfaffenwaldring 27
70550 Stuttgart
Germany

B. Herrling
Institut für Hydromechanik
Universität Karlsruhe
Kaiserstr. 12
76131 Karlsruhe
Germany

U. Meissner
Technische Hochschule Darmstadt
Institut für Numerische Methoden und Informatik im Bauwesen
Petersenstr. 13
64287 Darmstadt
Germany

C.A. Brebbia
Wessex Institute of Technology
University of Portsmouth
Ashurst Lodge
Southhampton SO4 2AA
UK

W.G. Gray
University of Notre Dame
Department of Civil Engineering and Geological Sciences
Notre Dame, IN 46566-0767
USA

G.F. Pinder
College of Engineering and Mathematics
University of Vermont
101 Votey Building
Burlington, VT 05405
USA

PREFACE

These volumes constitute the edited proceedings of the Tenth International Conference on Computational Methods in Water Resources (formerly Finite Elements in Water Resources), held at Universität Heidelberg, Germany in July 1994. The biennial series began in 1976 at Princeton University, U.S.A., as a forum for researchers in the expanding field of applications of finite element methods to problems in water resources. Alternating between the U.S.A. and Europe, meetings have been held at Imperial College, U.K. (1978); the University of Mississippi (1980); Hannover University, Germany (1982); the University of Vermont (1984); the Laboratorio Nacional de Engenharia Civil, Portugal (1986); the Massachusetts Institute of Technology (1988); the Giorgio Cini Foundation, Italy (1990); and the University of Colorado at Denver (1992). The Heidelberg conference is organized jointly by Interdisziplinäres Zentrum für Wissenschaftliches Rechnen (Interdisciplinary Center for Scientific Computing) and Sonderforschungsbereich 359 of Universität Heidelberg and Institute of Supercomputing and Applied Mathematics of IBM Heidelberg.

The 1994 proceedings present the work of authors from 23 countries. Numerical methods, mathematical modeling and applications to subsurface and surface hydrology are covered by a wide variety of papers. Issues of formation description and modeling, including parameter estimation, heterogeneity, and scaling up continue to attract the attention of a large number of researchers. It is significant to mention that several papers edited in this book concern the solution of the Navier-Stokes equations.

The organizers of the Heidelberg meeting greatly appreciate the efforts of featured lecturers A.J. Baker, M.A. Celia, W. Jäger, and W. Kinzelbach. We wish to thank the invited speakers P. Ackerer, H.G. Bock, J. Carrera, G. Dagan, H. Daniels, R.E. Ewing, E.O. Frind, P. Gresho, W. Hackbusch, I. Herrera, G. Gambolati, P. Knabner, H. Kobus, M. Kawahara, S.P. Neuman, K. Pruess, R. Rannacher, J. Troesch, T. van Genuchten, P. Wesseling, J.J. Westerink, and W.G. Yeh. We are also indebted to Anja McKellar, who did most of the secretarial work.

The papers appearing in this volume have been reproduced directly from material submitted by the authors, who are wholly responsible for their content.

The Editors

CONTENTS

VOLUME 1

2. SUBSURFACE TRANSPORT

xii

3. SCALING AND HETEROGENEITY

4. GEOSTATISTICS

5. REACTIVE FLOW

6. FRACTURED POROUS MEDIA

7. PARAMETER ESTIMATION

xviii

VOLUME 2

8. REMEDIATION OPTIMIZATION

9. SUBSURFACE MULTIPHASE FLOW

10. SALTWATER INTRUSION

11. SHALLOW WATER EQUATIONS

12. FLOW AND TRANSPORT IN RIVERS

13. NAVIER-STOKES EQUATIONS

14. COASTAL FLOW

15. SEDIMENT TRANSPORT

16. ALGEBRAIC METHODS

17. SOFTWARE DEVELOPMENT

18. PARALLEL METHODS

xxvi

3. SCALING AND HETEROGENEITY

UNSATURATED-SATURATED GROUNDWATER FLOW MODELLING BY MIXED HYBRID FINITE ELEMENT

Ph. Ackerer, B. Diaw, R. Mosé, P. Siegel
Institut de Mécanique des Fluides
Université Louis Pasteur, URA CNRS 854
2, rue Boussingault
F - 67000 STRASBOURG

INTRODUCTION

Mathematical simulation techniques are tools widely used in predicting flow and contaminant transport in groundwater. Conventionally, groundwater velocities are calculated by differentiation of a numerically obtained hydraulic potential field (Bear and Verruijt, 1987). Frind and Matanga (1985) showed that under certain conditions, such a procedure can have a dramatic effect on the resulting velocity field. A solution for the problem of differentiation of the numerically obtained hydraulic potential field is the mixed method. Meissner (1973) used this concept for the potential flow problem where potential function and velocity field are approximated simultaneously. But this method results in a large system of linear equation whose matrix is not positive definite. This drawback is circumvented by the hybridization (Arnold and Brezzi, 1985, Kaaschieter, 1990, Chavent and Roberts, 1991).

THE MIXED FORMULATION

The mixed approximation consists in approximating simultaneously the hydraulic head H (L) and the velocity field $\bar{q}$ (L/T) (Raviart and Thomas, 1977, Thomas, 1977).

On each element E, H and $\bar{q}$ are functions having following properties :

- $\nabla . \bar{q}$ is constant over E

- $\forall i = 1,...,nbe$, $\bar{q}.\bar{n}_{Ei}$ is constant over the edge A_i, $\bar{n}_{Ei}$ being the unit exterior normal vector of the edge A_i and nbe is the number of edges for element E

- $\bar{q}$ is perfectly determined by the knowledge of its flux Q_i through the edge A_i

These three very interesting properties of the vector $\bar{q}$ have motivated us in the choice of this numerical technique. The water balance is exact for each element. Moreover the normal component of the velocity is continuous from one element to the adjacent one

A. Peters et al. (eds.), Computational Methods in Water Resources X, 383–390.
© 1994 Kluwer Academic Publishers. Printed in the Netherlands.

and the velocity may be calculated over each element with the help of the appropriate interpolation function. In the same way as the hydraulic potential in the standard finite element method, the velocities can be calculated with the help of the basis functions. These vectorial functions $(\vec{w}_i)_{i=1,...,nbe}$ are defined by :

$$\int_{A_j} \vec{w}_i . \vec{n}_{Ej} = \delta_{ij} \text{ for } j=1,...nbe$$

where δ_{ij} is the Kronecker symbol.

The vector $\vec{w}_i$ corresponds to a vector having a flux of one through the edge A_j and a

zero flux through the others and has following properties : $\int_E \nabla . \vec{w}_i = \sum_{j=1}^{nbe} \int_{A_j} \vec{w}_i . \vec{n}_{Ej} = 1$

Numerical development of Darcy's law

We write the Darcy's law in the following form :

$\mathbf{K}^{-1} . \vec{q} = -\nabla H$ where K is the hydraulic conductivity tensor (L/T).

The Darcy law is written in a variational form using $\vec{w}_i$ as basis function and the Green formula is applied.

$$\int_E (\mathbf{K}_E^{-1} . \vec{q}_E) . \vec{w}_i = H_E \int_E \nabla . \vec{w}_i - \sum_{j=1}^{nbe} TH_{E,j} \int_{Aj} \vec{w}_i . \vec{n}_{E,j} \quad \forall i = 1,...,nbe$$

where H_E is the average head over the element E and $TH_{E,j}$ the average head over the edges of element E.

Using the properties of $\vec{w}_i$, the previous relation becomes :

$$\sum_{j=1}^{nbe} Q_{E,j} \int_E ((\mathbf{K}_E^{-1} . \vec{w}_j) . \vec{w}_i) = H_E - TH_{E,i} \quad \forall i = 1,..,nbe.$$

We now define the symetric matrix B_E associated to the element E as :

$$B_E(i,j) = \left[\int_E ((\mathbf{K}_E^{-1} . \vec{w}_j) . \vec{w}_i) \right]$$

The previous equation can be written in the following form, B_E^{-1} being the inverse of B_E :

$$Q_{E,i} = H_E \sum_{j=1}^{nbe} B_E^{-1}(i,j) - \sum_{j=1}^{nbe} B_E^{-1}(i,j) TH_{E,j} = H_E A_{E,i} - \sum_{j=1}^{nbe} B_E^{-1}(i,j) TH_{E,j}$$

With $A_{E,i} = \sum\limits_{j=1}^{nbe} B_E^{-1}(i,j)$

Numerical development of the flow equation

The second step is the integration of the flow equation over the control volume (or element) E :

$$\int_E S\frac{\partial H}{\partial t} + \int_E \nabla.\bar{q} = \int_E f$$

where S is the specific storativity of the porous material (L^{-1}) and f is the source or sink of the water term (T^{-1}).

Using a finite difference approximation and an implicit scheme for the time derivative, the equation becomes :

$$\int_E S_E \frac{H_E^n - H_E^{n-1}}{\Delta t} + \int_E \nabla.\bar{q}_E^n = \int_E f^n$$

where

$\quad$ H_E is a constant over E

$\quad$ $\nabla.\bar{q}_E^n$ is constant over E

$\quad$ S_E is an approximation of S over E, constant over E.

With the following relations resulting from the definition of $\bar{q}_E$,

$$\int_E \nabla.\bar{q}_E^n = \int_E \nabla.\left(\sum\limits_{i=1}^{nbe} Q_{E,i}^n \bar{w}_i\right) = \sum\limits_{i=1}^{nbe} Q_{E,i}^n \int_E \nabla.\bar{w}_i = \sum\limits_{i=1}^{nbe} Q_{E,i}^n,$$

and

$$\sum\limits_{i=1}^{nbe} Q_{E,i} = H_E \sum\limits_{i=1}^{nbe} A_{E,i} - \sum\limits_{i=1}^{nbe}\left(\sum\limits_{j=1}^{nbe} B_E^{-1}(i,j)TH_{E,j}\right) = H_E A_E - \sum\limits_{j=1}^{nbe} A_{E,j}TH_{E,j} \quad \text{with } A_E = \sum\limits_{i=1}^{nbe} A_{E,i}$$

$\quad$ the flow equation is written

$$|e|S_E \frac{H_E^n - H_E^{n-1}}{\Delta t} + H_E^n A_E - \sum\limits_{j=1}^{nbe} A_{E,j}TH_{E,j}^n = F_E^n$$

where F_E^n is an approximation of $\int_E f^n$ (constant over E) and $|e|$ is the area (2D) or the volume (3D) of the element E.

The mixed hybrid formulation

The unknowns are the mean head over each element H_E, the fluxes through each edges of the domain Q_i and the mean head over each edges TH_i. The formulation of Darcy's law, of the flow equation and the continuity of the fluxes from one element to the other (*i.e.* $Q_{E,i} + Q_{E',i} = 0$ for every edge i, E and E' being adjacent elements), are the three equations used to reduce the number of unknowns to the mean head over edges.

Dirichlet conditions are realized by an equality of the head over edges and Neumann conditions are realized by a flux equality. In the standard finite element method, Neumann conditions and sink/source terms are described by prescribed fluxes on specific nodes. The mixed approximation takes care of Neumann conditions in a very specific way by prescribed fluxes on element edges whereas the sink/source terms are averaged over the element. Initial conditions are given with the knowledge of $TH_{E_i}^0$. The calculations, which come into the system of TH_i unknowns, are fully presented in Chavent and Roberts (1991). Once the average piezometric head over the edges has been calculated, the average piezometric head over the elements and the flux through the edges are obtained by local equations.

COMPARISON WITH STANDARD FINITE ELEMENT

The first test case is based on an heterogeneous domain. The domain is a square of 100m by 100m. The hydraulic conductivities are 1 m/d (white), 0.1 m/d (medium hatches), 0.01 m/d (fine hatches) and 0.001 m/d (black). Boundary conditions, geometry and hydraulic conductivity distribution are presented in figure 1. The differences in head obtained by the mixed hybrid method (MHM) and the standard finite element (FEM) are less than 1%. The velocity distribution is compared on hand of pathlines. For the FEM, the pathlines calculation is improved by the postprocessing technique developped by Cordes and Kinzelbach (1992). Results obtained by different mesh sizes are presented in figure 2, 3 and 4. The FEM tends towards the mixed method with increasing grid refinement. The direct streamline computation as suggested by Frind and Matanga (1985) gives the same results as those given by the mixed approximation. By increasing the mesh size, the three approximations converge to the solution presented in figure 2

Another comparison is done with an unsaturated flow domain. The vertical cross section is 110m long and 100m deep. Boundary conditions are presecribed pressure of 1cm on the top and impervious limits elsewhere. The initial pression is equal to -100cm at the bottom of the domain and hydrostatic distributed. The standard finite element program used is SWMS_2D (Simunek *et al.*, 1992). The Mualem Van Genuchten model is used to describe the retention curve and the variation of the hydraulic conductivity with saturation (Van Genuchten, 1980). The heterogeneities are presented in figure 5. The hydrodynamic parameters are θ_r=0.02, θ_s=0.35, α=0.0410, n=1,964 and Ksat=7.2 10^{-3} m/d (white zones) and θ_r=0.02, θ_s=0.29, α=0.0410, n=1,964 and Ksat=7.2 10^{-7} m/d (grey zones). θ_r is the residual water content, θ_s the water content at

Prescribed head : 100m

No flux

No flux

Prescribed head : 60 m

Figure 1 : Boundary conditions and
hydraulic conductivities

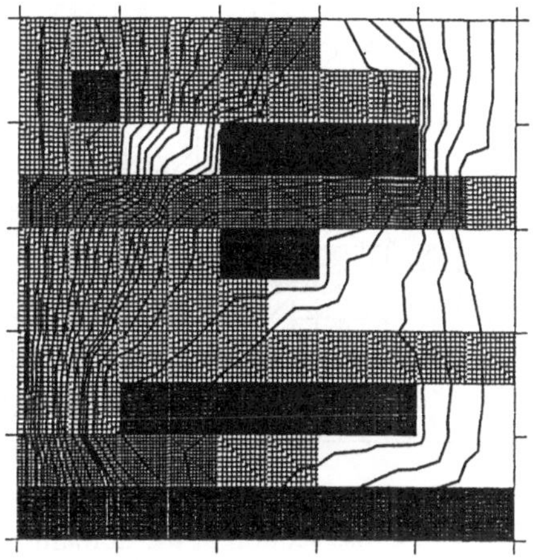

Figure 2 : MHM approximation
(mesh size : 10x10)

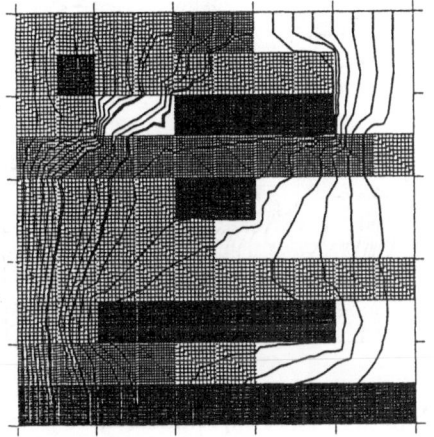

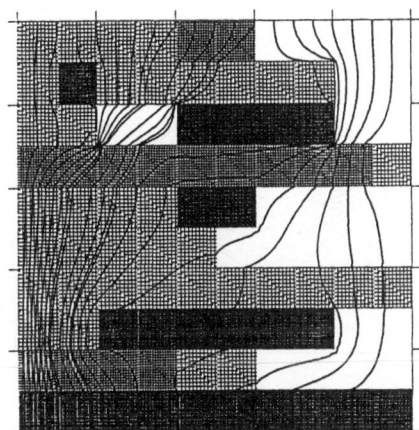

Figure 3 : FEM approximation
(mesh size : 10x10)

Figure 4 : FEM approximation
(mesh size : 40x40)

Figure 5 : Cross section of the domain

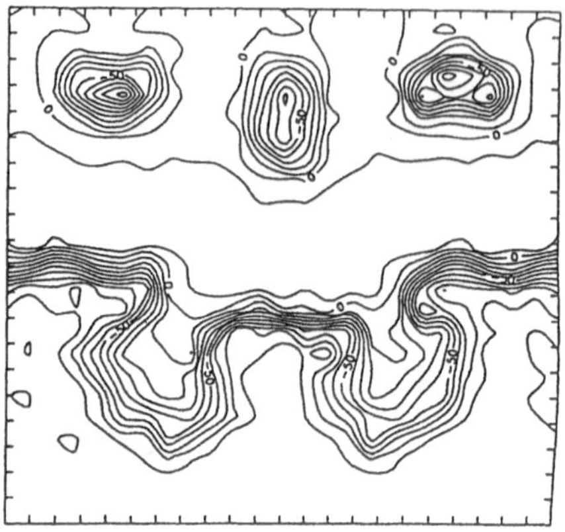

Figure 6 : MHM approximation
(mesh size : 11x10)

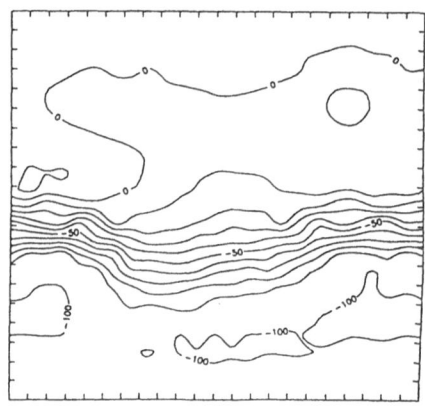

Figure 7 : FEM approximation
(mesh size : 11x10)

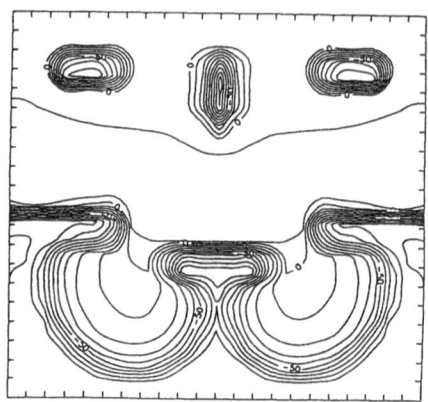

Figure 8 : FEM approximation
(mesh size : 44x40)

saturation and α and n describe the shape of the water retention curve. Comparisons of the pressure distribution are done at time 3600 days with different mesh sizes for SWMS_2D (fig. 6, 7 and 8).

CONCLUSION

For both saturated and unsaturated flow modeling, the MHM gives better results for heterogeneous formations.The differences between both methods are linked to hydraulic conductivity contrasts and to the distribution of the hydraulic conductivity within the medium. Drastic differences between MHM and FEM have been shown by Mosé *et al.* (1993) for heterogeneous saturated domains.

In the context of contaminant transport, the problem of accuracy of the velocities is of great importance. The errors in the velocities can be small enough to pass unnoticed when only the potential point of view is considered, but large enough when introduced in the transport model.

Velocities obtained by the mixed hybrid finite element method can be used for any solute transport numerical model. Moreover this method is very convenient for particle tracking technique (as the random walk). The velocity is obtained everywhere in the modelled domain. Since this velocity field is more accurate, its derivative (required by the random walk) will also not lead to erratic results as it is commonly observed using velocities calculated by the standard method.

Same conclusions were obtained by comparisons between MHM and finite differences (Ackerer *et al.*, 1991). Russel and Wheeler (1983) and Chavent and Roberts (1991) show that under certain quadrature rules used in the calculation of matrix B, the mixed method is equivalent to the block centered finite difference method. These quadrature rules are not employed in our numerical developments.

AKNWOLEDGMENTS

B. Diaw has been supported by the Ecole Inter-Etats de l'Equipement Rural de Ouagadougou (Burkina Faso) and P. Siegel by the Région Alsace (France).

REFERENCES

Ackerer P., Mosé R., Semra K., (1990)."Natural tracer test simulation by stochastic particle tracking method". Proc. Intern. Conf. on "Transport and Mass Exchange Processes in Sand and Gravel Aquifers", Ottawa, October 1-4 1990,G. Moltyaner (ed.), 595-604.

Arnold D.N., Brezzi F., (1985). "Mixed and non conforming finite element methods : implementation, postprocessing and error estimates". Mathematical Modelling and Numerical Analysis 19, 7-32.

Bear J., Verruijt A., (1987). Modelling groundwater flow and pollution, Reidel Publishing Company, Dordrecht.

Chavent G., Roberts J.E., (1991). "A unified physical presentation of mixed, mixed hybrid finite elements and standard finite difference approximations for the determination of velocities in waterflow problems", Adv. in Water Resources, Vol. 14, n°6, 329-348.

Cordes C., Kinzelbach W., (1992). "Continuous groundwater velocity field and pathlines in linear, bilinear, and trilinear finite elements", Water Resour. Res., Vol.28, n°11, pp. 2903-2911.

Frind E.O., Matanga G.O., (1985). "The dual formulation of flow for contaminant transport modelling. 1. Review of theory and accuracy aspects", Water Resour. Res., Vol.21, n°2,.159-169.

Kaaschieter E.F., (1990). "Mixed hybrid finite elements for saturated groundwater flow", Proc. Intern. Conf. on "Computer Method in Water Resources", Venice, Italy, june 11-15 1990, Computational Mechanics Publications, 17-22.

Meissner U.,(1973). "A mixed finite element model for use in potential flow problem", Int. J. Numer. Methods Eng., 6,.467-473.

Mosé R., Siegel P., Ackerer Ph., (1993). "Simulation des écoulements en milieu poreux par éléments finis mixtes hybrides". Hydrogéologie, 4.

Raviart P.A., Thomas J.M., (1977). "A mixed finite method for the second order elliptic problems", Mathematical Aspects of the Finite Element Method, Lecture Notes in Mathematics, Springer-Verlag, New York.

Russel T.F., Wheeler M.F., (1983). "Finite element and finite difference methods for continuous flows in porous media", The Mathematics of Reservoir Simulation, Society of Industrial and Applied Mathematics, R.E. Ewing Ed., 35-106.

Simunek J., Vogel T., Van Genuchten M. Th., (1992). The SWMS_2D Code for Simulating Water Flow and Solute Transport in Two-Dimensional Variably Saturated Media. U.S. Salinity Lab., Riverside, CA. Research Report n°126, 169p.

Thomas J.M., (1977). Sur l'analyse numérique des méthodes d'éléments finis hybrides et mixtes, Thèse de doctorat d'état, Université Pierre et Marie Curie.

Van Genuchten, M. Th. (1980). " A closed form equation for predicting the hydraulic conductivity of unsaturated soils", Soil Sci. Soc. Am. J., 44, 892-898.

STOCHASTIC FLOW ANALYSIS OF FLUX VARIANCES IN HETEROGENEOUS POROUS MEDIA

G. A. AKPOJI*, R. ABABOU** and F. DE SMEDT*
*Laboratory of Hydrology, Free University Brussels
Pleinlaan 2, B1050 Brussels, Belgium.
**Commissariat a l'Energie Atomique, Saclay,
DRN/DMT/SEMT/TTMF, 91191 Gif sur Yvette Cedex, France.

Three-dimensional (3-D) groundwater flow is simulated numerically in a porous medium characterized by an isotropic or anisotropic stationary conductivity field. The solution is obtained by the finite element technique by discretizing the medium into 60x60x60 elements. Various simulations are performed for different variances of lnK and anisotropy ratios. Analytical expressions for second-order moments of the flux field, are compared with numerical results. The kinetic energy of the velocity fluctuations is also studied.

INTRODUCTION

The natural variability of aquifer hydraulic properties is difficult to simulate using deterministic models. Alternatively, stochastic groundwater flow modelling takes into consideration the variability of material properties as random space functions (see e.g., Gelhar, 1986; Ababou, 1988; and Akpoji 1993). Gelhar and Axness (1983) developed a head-based solution for obtaining expressions for the variance of the random flux vector in anisotropic media. On the other hand, Ababou (1988) proposed a method in which the dependent variable was the flux vector and obtained new expressions for the flux variances. Comparisons between theoretical and direct flow simulations by Ababou (1988) showed that the flux variances obtained by the flux-based theory were more accurate than those obtained by the head-based method. These findings have been supported by the numerical results of Akpoji (1993), who also found that for high variability, both the standard spectral theory and the flux-based solutions deviate from the numerical results. In this study, new expressions are presented for flux variances of anisotropic media, and based on the numerical results, conjectures are proposed which fit the numerical results better than the previous solutions.

THEORY

Isotropic case

Consider a porous medium with an autocorrelated conductivity field that is lognormal, stationary and statistically isotropic. The flow through this medium is driven by a fixed hydraulic gradient, with the regional scale much larger than the integral scale (or range) of the conductivity field. Under these conditions, Ababou (1988) used the first-order spectral solutions (head-based approach) of Gelhar and Axness (1983) to derive the following expression for the flux standard deviations:

A. Peters et al. (eds.), Computational Methods in Water Resources X, 391–398.
© 1994 Kluwer Academic Publishers. Printed in the Netherlands.

$$\sigma_i = \sqrt{m_i}\, \sigma K_g J \tag{1}$$

where σ_i is the standard deviation of the flux component in direction i, (i = 1, 2, 3) with i = 1 as the mean flow direction, K_g is the geometric mean hydraulic conductivity, σ is the standard deviation of lnK, J is the hydraulic gradient, m = 8/15 for i = 1 and m = 1/15 for i = 2 or 3. The flux-based spectral solutions of Ababou (1988) give,

$$\sigma_i = \sqrt{m_i}\, \sigma \langle q_1 \rangle \tag{2}$$

where $\langle q_1 \rangle$ is the average flux in the mean flow direction. This flux can be estimated from the numerical mean flux value or inferred from the effective K and mean gradient. Gelhar and Axness (1983) and Matheron (1967) express the mean flux as a function of the mean regional gradient and an effective conductivity, K_e,

$$\langle q_1 \rangle = K_e J \tag{3}$$

where $K_e = K_g \exp(\sigma^2/6)$, for lnK normal, such that the σ_i of Eq. (2) become,

$$\sigma_i = \sqrt{m_i}\, \sigma K_g J \exp(\sigma^2/6) \tag{4}$$

Ababou (1988) compared both Eqs. (1) and (4) to numerical simulations and concluded in favour of Eq. (4).

Consider now the volumetric density of the kinetic energy of the fluctuating motion,

$$E = \frac{1}{2} \rho_w \left\langle \sum_{i=1}^{D} \left(v_i - \bar{v}_i \right)^2 \right\rangle \tag{5}$$

where ρ_w is the density of water, D is the number of space dimensions, v_i are the velocities, and $\bar{v}_i$ are the mean velocities. Assuming negligible porosity variations about their mean, $\langle n \rangle$, Eq. (5) can be rewritten as,

$$E = \frac{1}{2} (\rho_w / \langle n \rangle^2) \sum_{i=1}^{D} \sigma_i^2 \tag{6}$$

An associated variable is the pseudo kinetic energy, E_0, that characterizes the spectral content of fluctuations of the flux magnitude, defined here as,

$$E_0 = \sum_{i=1}^{D} \sigma_i^2 \tag{7}$$

Thus, E is directly related to E_0 by a multiplicative factor, and we shall for simplicity, refer to E_0 as the kinetic energy. Using previous equations for the flux variances, different expressions for E_0 can be obtained, for instance, the flux-based theory gives:

$$E_o = (1 - 1/D) \ \sigma^2 \langle q_1 \rangle^2 \tag{8}$$

Anisotropic case

Expressions for anisotropic media will now be presented. It is assumed that the covariance structure of ellipsoidal-anisotropic media is entirely determined by a set of fluctuation length scales ℓ_i (i = 1, 2, 3). The medium is considered to be horizontally stratified, two fluctuation scales (ℓ_1 and ℓ_2) are equal, while the third fluctuation scale (ℓ_3) is allowed to be either larger (case of columnar media) or smaller (stratified formations) than ℓ_1. The flux variance is therefore only a function of the anisotropy ratio, $\rho = \ell_1/\ell_3$. The anisotropic flux variances can be described in terms of normalized flux variances,

$$\tilde{\sigma}_i^2 = \sigma_i^2/(\sigma^2 \langle q_1 \rangle^2) \tag{9}$$

and the kinetic energy, E_0, may be represented in a scaled form as

$$\tilde{E}_o = \sum_{i=1}^{D} \tilde{\sigma}_i^2 \tag{10}$$

The closed-form results for the normalized flux variances and the kinetic energy given below were obtained by integration of the first-order flux-based solutions of Ababou (1988) for the case of ellipsoidal anisotropy. The results are:

$$\tilde{E}_o = 1 - (\rho^2 A - 1)/[2(\rho^2 - 1)] \tag{11}$$

$$\tilde{\sigma}_2^2 = [2 + \rho^2 - (4 - \rho^2)\rho^2 A]/[16(\rho^2 - 1)^2] \tag{12}$$

$$\tilde{\sigma}_3^2 = \rho^2 [-3 + (\rho^2 + 2)A]/[4(\rho^2 - 1)^2] \tag{13}$$

where $A(\rho)$ is defined as,

$$A(\rho) = \tan^{-1}\left(\sqrt{\rho^2 - 1}\right)/\sqrt{\rho^2 - 1} \qquad \text{for } \rho \geq 1 \tag{14}$$

$$A(\rho) = \tanh^{-1}\left(\sqrt{1 - \rho^2}\right)/\sqrt{1 - \rho^2} \qquad \text{for } \rho \leq 1 \tag{15}$$

The normalized longitudinal flux variance for i = 1 can be inferred from Eq. (10) in an obvious way. The mean flux vector is assumed to be horizontal, that is, parallel to direction 1. The special cases of a perfectly stratified medium and perfectly columnar medium can be obtained from the corresponding limits as ρ tends to infinity or to zero. As will be seen, the first limit yields a 1-D flow while the second limit yields a 2-D isotropic flow. Finally, the case $\rho = 1$ corresponds to 3-D isotropic flow, and by Taylor expansions of the functions involved, it can be shown that the corresponding normalized flux variances equal m_i.

EMPIRICAL CORRECTIONS AND CONJECTURES

For isotropic conditions, the following conjecture was made by Ababou (1988), in order to improve Eq. (4),

$$\sigma_i = \sqrt{\overline{m_i}}\,\sigma K_g J \exp(\xi_i \sigma^2) \tag{16}$$

where $\xi_i = 1/4$ for $i = 1$ and $\xi_i = 1/3$ for $i = 2, 3$. In consideration of the expressions in Eqs. (1) - (4), and Eq. (16), and taking into account the numerical simulation results discussed below, a different form of expressions is conjectured for σ_i. It is suggested that the normalization in Eq. (9) should take the form,

$$\tilde{\sigma}_i^2 = \sigma_i^2 / \left[\sigma^2 \left(q_1\right)^2 \exp(\sigma^2/6)\right] \tag{17}$$

The kinetic energy is scaled by a similar factor such that Eq. (10) remains valid. The analytical results derived in Eqs. (11) - (13) will be compared with the numerical results using Eq. (9) or Eq. (17).

RESULTS AND DISCUSSION

The global flow domain is discretized into 60x60x60 cubic finite elements. The up and down gradient sections have fixed hydraulic heads such that the mean gradient is unity. The top, bottom and lateral sides constitute no flow boundaries. The conductivity field is generated by the turning bands algorithm (Tompson et al., 1989). The K-field is assumed to be stationary, lognormally distributed and autocorrelated, with a spherical autocovariance function of range 4 m and a geometric mean of 1 m/d. Anisotropy is created by scaling of the vertical dimension by a factor, ρ, hence, the global flow domain becomes 60 m by 60 m horizontally and $(60/\rho)$ m vertically. The solution is obtained with a finite element model, with compact matrix storage and a preconditioned conjugate gradient solver specially adapted to enhance vectorization when run on supercomputers.

In order to reduce boundary effects, statistical analysis of the flux field is restricted to the middle portion of the domain of size, 20x20x20 elements, with the flux components calculated at the centre of each finite element. Variances of the numerically calculated flux components are estimated and compared with the analytical predictions. When the analytical results depend on the mean longitudinal flux component, the latter is estimated from the numerical simulation results.

First, the isotropic results are presented. Figure 1 plots E_0 as a function of σ^2 for the range, 0 to 5. The gradient and flux-based expressions agree with the numerical results for moderate values of σ^2, but the predictive ability for larger σ^2 is poor. The conjectural expressions fit excellently over the entire range, and therefore can be considered adequate for predictive purposes. Figure 2a illustrates the dependence of the standard deviation of q_1 as a function of σ, while Figure 2b gives a similar relationship for the standard deviations of q_2 and q_3. Both figures show again that, the standard spectral and flux-based theories tend to underestimate the results for large σ. The conjectural expressions perform well, but clearly, Ababou's (1988) proposal is better than the present conjecture. However, Ababou's (1988) conjecture implies that different exponential terms are used

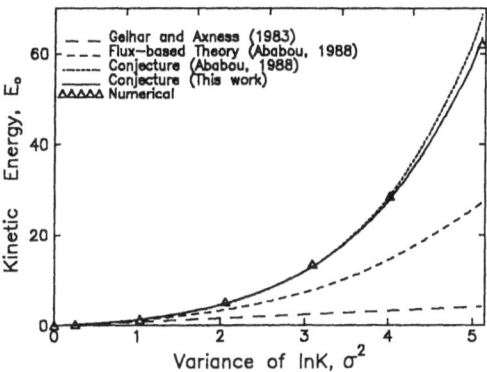

Figure 1. Comparison of kinetic energy from numerical simulations results, spectral theory, flux-based theory and conjectural expressions.

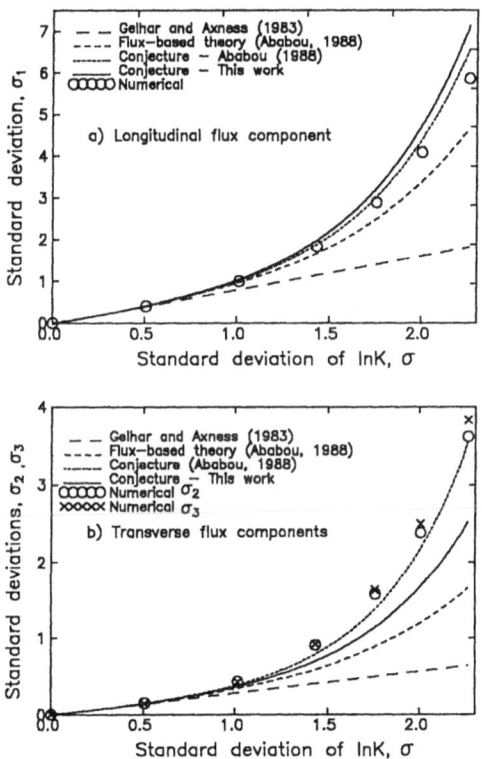

Figure 2. Comparison of standard deviations of flux components from numerical simulations, standard spectral theory, modified flux-based theory and conjectural expressions, a) Longitudinal flux component and b) Transverse flux components.

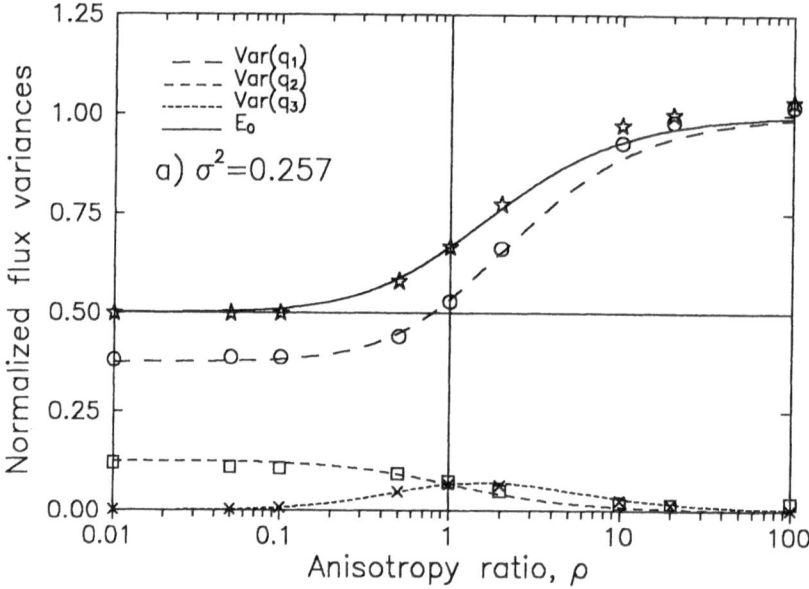

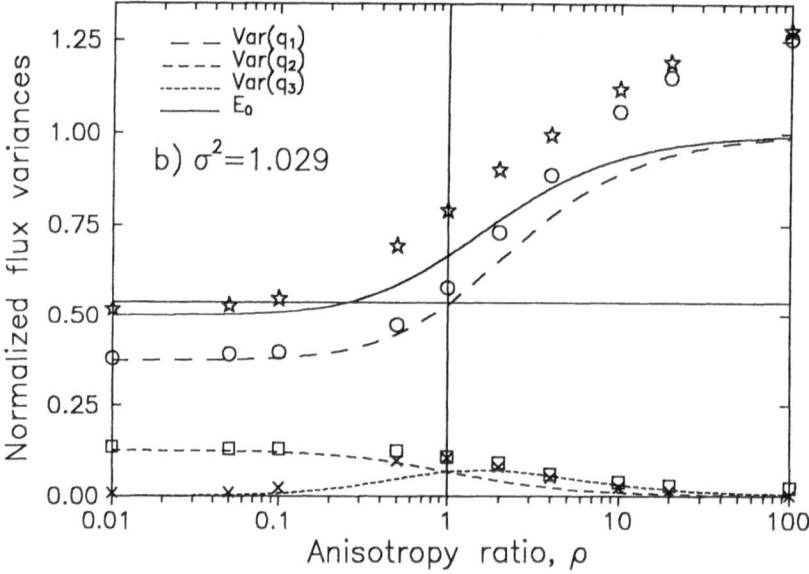

Figure 3. Comparison of results of numerical (in symbols) and analytical normalized flux variances and kinetic energy of anisotropic simulations, a) $\sigma = 0.5$ and b) $\sigma = 1$. Normalization by Eq. (9)

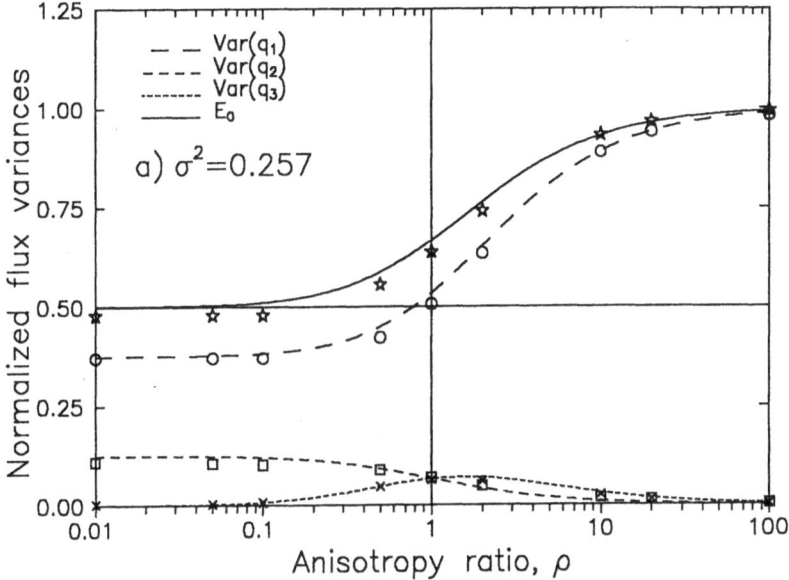

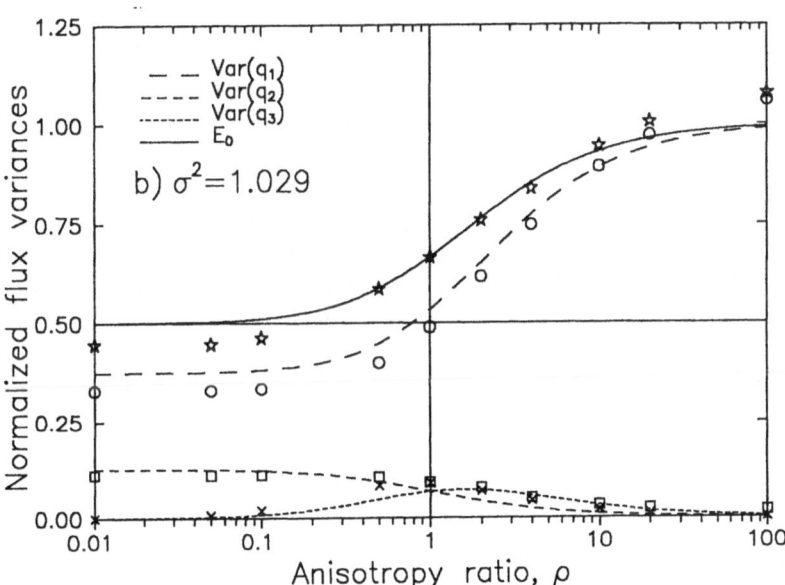

Figure 4. Comparison of results of numerical (in symbols) and analytical normalized flux variances and kinetic energy of anisotropic simulations, a) $\sigma = 0.5$ and b) $\sigma = 1$. Normalization by (17).

for different σ_i and the corresponding expression for E_0 becomes a mixture of these terms. On the contrary, the new conjectural expressions are based on one exponent, which uniquely applies to all σ_i as well as E_0. It will be shown shortly that this exponent is also useful for improving predictions of the anisotropic cases.

Anisotropic conductivity fields were generated ranging from $\rho = 0.01$ (columnar media, with mean flow perpendicular to the columns) to $\rho = 100$ (stratified formations, with mean flow parallel to the layers). Two illustrative cases are presented here, $\sigma = 0.5$ (exact value of the generated field being 0.507) and $\sigma = 1$ (actual value 1.015), corresponding to the lower range of the isotropic cases studied. Note that isotropic results ($\rho = 1$) are included in the plots depicting the anisotropic results.

Figure 3a compares the analytical expressions, Eqs. (11) - (13) with the numerical simulation results for $\sigma = 0.5$. The numerical results are obtained by normalizing σ_i and E_0 as in Eq. (9). The results are exceptionally good, in view of the broad range of anisotropy ratios considered. However, similar results for $\sigma = 1$, shown in Figure 3b, indicate a poorer fit, especially for high values of ρ. The deviations are very obvious for the normalized variance in the mean flow direction as well as for the kinetic energy.

Better estimates of the flux variances and the kinetic energy than the case considered above are obtained when the conjectural expression, Eq. (17), is used. Figure 4 shows the results, with a distinct improvement in the predictions. The results are encouraging. The proposed equations predict flux variances and kinetic energy very well under both isotropic and anisotropic conditions.

SUMMARY AND CONCLUSIONS

Three-dimensional groundwater flow was simulated in heterogeneous porous media with isotropic or anisotropic correlated hydraulic conductivity input. Statistical analysis of the flux components was performed to check analytical perturbation solutions of the head-based (standard) spectral theory and the flux-based theory. The results show that the flux-based theory agrees better with the numerical results than the standard spectral theory, though both tend to underestimates the flux variances for higher variances of lnK. Empirical corrections are proposed that fit the numerical results better than the theoretical expressions. The new expressions for the flux variances and the energy spectrum agree closely with the numerical results for a broad range of lnK variability and anisotropy.

REFERENCES

Ababou, R. (1988) "Three-dimensional flow in random porous media", Ph.D. Thesis, Parsons Laboratory, MIT, Cambridge, 833 pp.
Akpoji, G.A. (1993) "Groundwater modelling under uncertainty of the hydraulic conductivity", Ph.D. Dissertation, Free University Brussels (VUB), 282 pp.
Gelhar, L.W.(1986) "Stochastic subsurface hydrology: from theory to applications", Water Resour. Res., Vol. 22(9), pp. 135S-145S.
Gelhar, L.W., and Axness, C.L. (1983) "Three-dimensional stochastic analysis of macrodispersion in aquifers", Water Resour. Res., Vol. 19(1), pp. 161-180.
Matheron, G. (1967) Eléments pour une théorie des milieux poreux, Masson et Cie Paris.
Tompson, A.F.B., Ababou, R. and Gelhar, L.W (1989) "Implementation of the three-dimensional turning bands algorithm", Water Resour. Res., Vol. 25(10), 2227-2243.

NUMERICAL INVESTIGATION OF THE COUPLING BETWEEN THE MICROGEOMETRY AND THE PERMEABILITY TENSOR OF NATURAL POROUS MEDIA THROUGH THE VOLUME AVERAGING THEORY.

Y. ANGUY[*], D. BERNARD[*] and R. EHRLICH[**]
[*] Laboratoire Energétique et Phénomènes de Transfert, L.E.P.T.-ENSAM
Esplanade des Arts et Métiers, 33405 Talence, cédex, FRANCE
[**] Department of Geological Sciences, University of South Carolina.
Columbia, South Carolina, 29208, U.S.A.

ABSTRACT

The 'local change of scale method' (L.C.S.M.) carries the promise for a significant advance in the study of the instantaneous coupling, at the local scale, between the micro-geometry and the permeability tensor of natural porous media. Theoretical developments in L.C.S.M. permit explicit calculation of the permeability tensor as an implicit function of the micro-geometry without having recourse to the use of any constitutive assumptions. In a previous report, we demonstrated that the 'Whitaker / Quintard' formalism of the L.C.S.M. can be expressed numerically suitable for finite element / difference implementation and generate results in close agreement with analytical solutions. Several fundamental problems have to be solved however before this procedure can be used in practical applications : 1) a method to generate micro-structure similar to real media and of a large enough size to constitute a local Representative Elmentary Volume (R.E.V.), 2) to explore the relationship between this local micro-structural R.E.V. and the more classical R.E.V. associated with Darcy's law.
The local geometrical R.E.V. can be built by convolving a density power spectrum derived from a real medium with a random field. The absence of significant 'power' at low wave-numbers ensures 'weak' ergodicity meaning that all local heterogeneities are included many times within the R.E.V. Preliminary results involving the use multiple isotropic stochastic realizations of a local geometrical R.E.V. gives insights into the scale of the Darcy's R.E.V. Results indicate that a relatively small number of local geometrical R.E.Vs. (30 or fewer) may be required. Successful calculation of the permeability tensor constitute is a successful demonstration of the necessary condition for the L.C.S.M. that both micro-geometry and permeability are implicitly linked. Data from natural sandstones indicate that the size of the R.E.V. of Darcy's and micro-structural types are much larger than is commonly believed. This might pose computational problems for calculating the permeability tensor.

INTRODUCTION

This paper represents a progress report of an ongoing effort to build physically and structurally realistic models of flow through porous media using the local change of scale method (L.C.S.M.). In our previous reports, we showed that the 'Whitaker / Quintard' L.C.S.M. formalism can be successfully expressed in terms of two-dimensional and three-dimensional numericals models which produce a permeability tensor on closely periodic media close to the analytical solutions. That exercice provided realistic microscopic fields related to Stokes pressure and velocity in complex media.

A. Peters et al. (eds.), Computational Methods in Water Resources X, 399–406.
© 1994 Kluwer Academic Publishers. Printed in the Netherlands.

The results described below represent two more steps in our ongoing effort to bring to a practical level L.C.S.M. Firstly, we discuss a procedure to construct a realistic local micro-geometrical R.E.V. and, secondly, we discuss relationships between the local micro-geometrical R.E.V. and the larger local R.E.V. required by the L.C.S.M. ; e.g. the tensorial representation of Darcy's law.

The L.C.S.M. leads, from first principles, to spatially smoothed transport equations and to a closure problem, set at the microscopic scale, over a local R.E.V. of Darcy's type ($'V_\beta(r_0)'$). V_β (r_0) must be larger than the micro-structural R.E.V.($'V_\beta(\nu_0)'$) -- (minimal volume of length-scale of order ν_0, describing the geometrical variability over some scale of order L ; $\nu_0 \ll L$).

Development of a valid geometrical R.E.V. coupled with its relationship to the Darcy's R.E.V. will allow explicit calculation of the permeability tensor as an implicit function of the micro-geometry. In accordance with existence of a local closure form, *the detailed micro-structure must define the permeability tensor*. However, realistically characterizing the three-dimensional micro-geometry is a complex problem since current direct observations are of porosity exposed in planes intersecting the medium. Nonetheless, characteristics observed in section impose limits on the structural variations possible in the three-dimensional medium. If in addition, the sectioned plane is taken from the end of a cylinder used to make physical measurements, further restrictions on structural possibilities are imposed. Work by Ehrlich et al.[1,2] suggest that a finite number of parameters $\{P^i\}_{i=1,N}$, measurable through observations from petrographic thin sections are adequate to characterize the micro-porous structure. Anguy et al.[3] and Prince et al.[4] linked the structural parameters of the aforementioned authors to characteristics of the density power spectrum. Therefore, image analysis procedures can be used to assess the measure, $\{P^i\}_{i=1,N}$, of the micro-geometry.

Numerical experiments were performed on a totally synthetic medium to determine the maximum size of the local R.E.V. of Darcy's type relative to the local R.E.V. of micro-structural type.

GENERAL CONTEXT

It can be taken for granted that the permeability tensor **K** is not solely function of the porosity value ε_β ("It is obvious that no simple correlation between porosity and permeability can exist[5]"). The relationship between porosity and permeability can be strong or non-existant depending on the sample set evaluated. Even when the relationship exists, the fiunctional nature of this relationship varies widely. Accross all sandstones, samples with identical porosities values can vary in permeability by four orders of magnitude. Obviously the configuration of the porosity, the micro-structure, must have a stong control on permeability. In other words, in order for a strong and consistent relationship between porosity and permeability to exist, micro-structural diffferences between the studied sample-volumes must be related to the porosity ε_β :

$$P^1 = \varepsilon_\beta \; ; \; \{dP^i = F_i(P^1)\}_{i=2,N} \qquad (1)$$

Previous work[3,6,7] shows that **K** is closely related to the whole micro-structure <u>in sandstones</u> as an inescapable consequence of sedimentation processes[8]. The micro-structure must be characterized by a set of N statistical and / or deterministic parameters, $\{P^i\}_{i=1,N}$.

Diagenesis can affect sandstones without producing much changes in porosity but can induce large changes in permeability. This renders impossible a single simple relationship between porosity and permeability and indicates that changes in permeability may come about to the changes in the porous micro-structure. Therefore, one must determine the coupling between the permeability tensor and the micro-structure involving aspects independent of porosity :

- a measure of the micro-geometry is required in the form of N parameters $\{P^i\}_{i=1,N}$:

$$\text{micro-geometry} = f(\{P^i\}_{i=1..N}) \tag{2}$$

- in accordance with the existence of the local closure form of the change of scale, the permeability tensor **K** is taken as an implicit function of a measure :

$$\mathbf{K} = g(\{P^i\}_{i=1..N}) \tag{3}$$

- variations **dK**, due to geometrical differences are expressed in terms of the variations of the $\{P^i\}_{i=1,N}$ fully characterizing the micro-geometry alterations :

$$\mathbf{dK} = \sum_{i=1}^{N} \frac{\partial g}{\partial P^i} \, dP^i \tag{4}$$

where $\left(\dfrac{\partial g}{\partial P^i}\right)_{i=1,N}$ is an intrinsic function of the change of scale[9].

CHARACTERIZATION OF THE MICRO-GEOMETRY.

A direct three-dimensional assessment of the $\{P^i\}_{i=1,N}$ is non-practicable in that, current means of observation are two-dimensional. Image analysis is the method used by the authors to obtain micro-structural information from sections through the porous medium. In the field of stereology, the two-dimensional relevance to the three-dimensional system only holds under the assumption of one or another of two conditions :

- isotropy of the micro-geometry,
- existence of a single identified direction of anisotropy, i.e. statistical cylindricity of the micro-geometry.

To a first approximation, cylindrical symetry can be assumed in many sandstones in that permeabilities measured in direction parallel to bedding are much more similar than measurements made at high angles to bedding. This has been verified by the existence of strong statistical correlations between geometrical characteristics measured in sections and physical properties measured on associated samples including: permeability, formation factor and mercury porosimetry[1,2,7].

Image analysis coupled with pattern recognition procedures[1,2] has revealed that isotropy in sandstones is never present; a consequence of the dynamics of sedimentary processes[4,8]. This may explain the discrepancies between modeled and measured data in studies[10] that assume isotropy.

Intuitively, one will require the $\{P^i\}_{i=1,N}$ to be assessed at some local scale, v_0, over a local R.E.V. of geometrical type, large enough to avoid any interactions between sampling and micro-structure and small enough to have bearing on the micro-geometry of a larger volume of rock of practical interest[9]. The structure of most natural porous media is far too complicated to permit a rigorous analytical description[11,12], and so the description of the micro-structure has to contain statistical elements. It is unlikely that a small R.E.V. (a few cubic centimeters) can represent an entire sandstone body (tens of meters thick). However, a local R.E.V. of relatively small volume, within the sandstones may represent individual beds within the whole sandstone body The verification of the existence and scale of a local R.E.V. can be obtained using the concepts of *local homogeneity* of the micro-geometry or, more generally speaking, *local stationarity of order 2* (ergodicity in the weak sense of the term).

Provided that the local homogeneity condition is met, the two valued stationary random variable ('time series'[11]) describing the micro-structure may be adequately characterized by low moments such as *the bi-dimensional Fourier power spectrum*, noted $A^2(v_x, v_y))$ *or its equivalent the autocorrelation function* and its *probability density function* $(\varepsilon_\beta)^{[11,13]}$.

$$\{P^i\}_{i=1,,2} = (\varepsilon_\beta \; ; A^2(v_x, v_y)) \tag{5}$$

On these grounds, the micro-structure has been quantified by the authors[3,4,9] from thin sections cut perpendicular to bedding (plane enclosing most of the variance of the micro-geometry) for many sandstones using binary images (porosity equal 1, matrix equal 0) obtained by image analysis procedures combined with Fourier transforms. Areas covered by conventional high resolution imagery (pixel edge ≤ 10 microns) being too small to evaluate the micro-structure[3], seamless mosaics of conventional views have been constructed[3,4,9], producing images combining high resolution with large area (Fig. 1, for example).

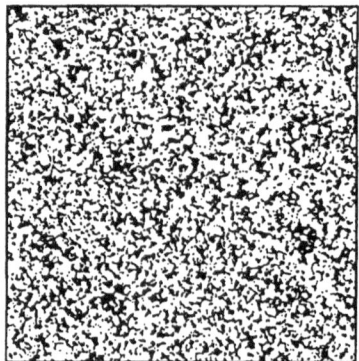

Figure 1 : 7.9 mm x 7.9 mm binary mosaic of a North Sea Jurassic sandstone produced by merging 20 conventional high resolution overlapping images, each measuring 480 x 640 pixels (pixel edge : 3.861 μm), digitized using a Symbiotic Concepts MIS-386 video digitizing system attached to a petrographic microscope (black represents porosity).

The Fourier transform permits quantitive assessment of the characteristic length-scales (Figs. 2 and 3) and of the nature of spatial complexity (Figs. 4a to 4c). By removing all spatial wave-numbers tied to length-scales equal to or less than a specified wave-length and inverting the remainder to generate a 'filtered' image, it is possible to visualize the structure at various scale represented by peaks in the power spectrum. Such visualization verifies the conjecture of Graton and Fraser[8] that first order structure in all sandstones consist of the juxtaposition of close-packed domains with loose-packed domains (Fig. 4a)[4]. The loose-packed domains are arrranged as cicuits with large scale continuity[3,4] (Fig. 4b), containing large pores and large throats, and representing a higher level of spatial organization. The third and final level in this example (Fig. 4c) has been shown to consist of different modes of clustering of the loose-packed circuits. A local R.E.V. of geometrical type, $V_\beta(v_0)$, must not be sensitive to an interaction between the sample-volume and structure heterogeneities ; the definition of such a volume can be attacked as a sample-support problem : $V_\beta(v_0)$ must achieve local homogeneity in that no significant power on the bi-dimensional power spectrum should occur at spatial wave-lengths approaching the size of the sample ('power' in Fourier parlance is equivalent to the amount to total variance tied to features expressed at any wave-length).

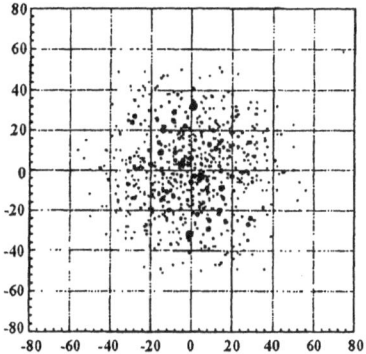

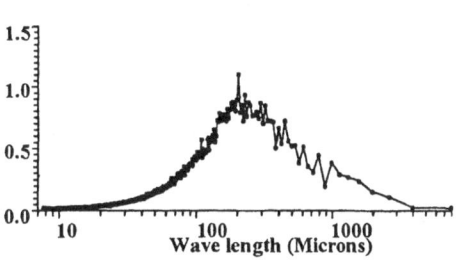

Figure 2 : 2-D power spectrum of Fig. 1. Numbers on axes are wave-numbers (v_x, v_v) (proportional to the inverse of wave-lengths). Circle size proportional to power. Note NW-SE anisotropy (concentration of large circles along that diagonal). No power near (0,0) suggests local homogeneity.

Figure 3 : radial power spectrum of Fig. 1. Horizontal axis is calibrated in wave-lengths (λ). Grain spike[3,4] at $\lambda = 208$ μm and adjacent plateau $(\lambda=208$ to $\lambda=309$ μm) associated to loose-packed domains[8,13]. Lack of power at long λ highlights that local homogeneity has been achieved .

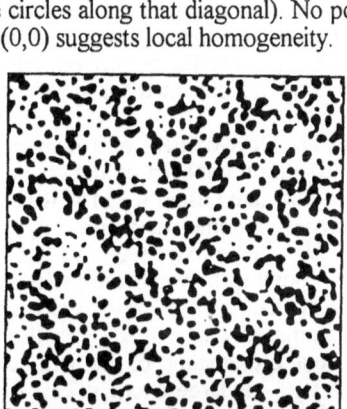

a)

b)

c)

Figure 4 :
a) North Sea sandstone image (7.9mm×7.9mm) after filtering all wave-lengths ≤ than grain size spike. The black pattern selectively overlays the oversized pores associated with packing flaws, the lowest level in the structural hierarchy. b) North Sea Jurassic sandstone image obtained after filtering all wave-lengths less than 659 μm, revealing circuits of packing flaws. c) North Sea Jurassic sandstone image, filtered at 1317 μm illustrating a higher level in the structural hierarchy.

In the example illustrated by Fig. 4, local homogeneity is achieved in that no significant power occurs for wave-numbers smaller than 4.

However, Fig. 4 represents an image of almost 0.64 squared centimeter ; containing many more pores than commonly used in many simulations[10]. Implicit assumption of a small R.E.V. makes explains the disparity[10] between modeled values and measurements and may also explain the high variance commonly associated with micro-permeabilities measurements.

The criterion $\{P^i\}_{i=1,2} = (\varepsilon_\beta \; ; \; A^2(v_x, v_y))$ may, in fact, not be totally justified in that, it is incomplete, lacking some phase information. Basic properties[14] of the Fourier transform permit use of a measure of the micro-geometry included in a more general ensemble :

$$\{\varepsilon_\beta \; ; \; A^2(v_x, v_y) \; ; \; \gamma(v_x, v_y)\} \tag{6}$$

where $\gamma(v_x, v_y)$ is a part of the phase of the bi-dimensional Fourier transform of the micro-geometry[9]. Some tests are in progress to investigate the nature of $\gamma(v_x, v_y)$.

RELATION BETWEEN THE LOCAL R.E.V. OF GEOMETRICAL TYPE AND THE LOCAL R.E.V. OF DARCY'S TYPE.

In previous work[9,15,16,17], the authors developed two-dimensional and three dimensional general numerical programs based on the local volume averaging technique. The bi-dimensional nature of the measure derived in the previous part requires a generator of three-dimensional numerical porous media, constrained by the $\{P^i\}_{i=1,,2}$, interfacing between the two-dimensional image-analysis-based characterization of natural porous media and the three-dimensional numerical codes.

The L.C.S.M. is not absolutely restricted to periodic media. Development of a numerically-tractable closure form requires only a weak[18] periodicity condition imposed on the boundaries of a local R.E.V of Darcy's type ($V_\beta(r_0)$). That is, $\mathbf{K}$ is not calculated using the local R.E.V. of geometrical type, $V_\beta(v_0)$, but a local periodic R.E.V., derived, in some way[9], from $V_\beta(v_0)$. In this, the numerical programs developed by the authors[9,15,16,17] have to be seen as a first step towards a complete understanding of the instantaneous coupling between the permeability tensor and the micro-geometry of natural porous media in that, *the sole complexity of the micro-geometry is rigorously effectively treated*. It is commonly asssumed[18,19,20] that the permeability tensor $\mathbf{K}$, derived this way, represents a close approximation of the permeability tensor $\mathbf{K}$ of natural random porous media, provided that $V_\beta(r_0)$ is large enough. That is, the effects of periodicity imposed at the area of entrances and exits for the fluid phase contained within $V_\beta(r_0)$ are negligible with regard to the no-slip condition imposed at the interfacial area of the solid-fluid interface contained within $V_\beta(r_0)$.

However, the random character of natural porous media requires assessment of the ratio "n" between the size v_0 of $V_\beta(v_0)$ and the size $r_0 = n_x v_0$ of $V_\beta(r_0)$. Because natural porous media are not totally periodic the local Darcy R.E.V., $V_\beta(r_0)$, must be larger than the micro-structural R.E.V., $V_\beta(v_0)$, in order for random effects correctly accounted for. On account of practical constraints, this problem has been so far investigated in two dimensions, however, principles determined from the two dimensional problem are entirely relevant for the three-dimensional case (at least as a necessary condition). A family of simulating media can be generated from a single set of parameters $(\varepsilon_\beta \; ; \; A^2(v))$ derived from a single medium by convolution of these empirical parameters with a gaussian field[21] using a stochastic simulation in a manner similat to that of Gujar[21], the pioneer in generating random sequences having desired probability density function and power density spectrum

A first rough estimate of n can be derived using synthetic bi-dimensional isotropic media. Two examples of such synthetic media, of size $50 x v_0 \times 50 x v_0$, are displayed of Figs. 5a and 5b.

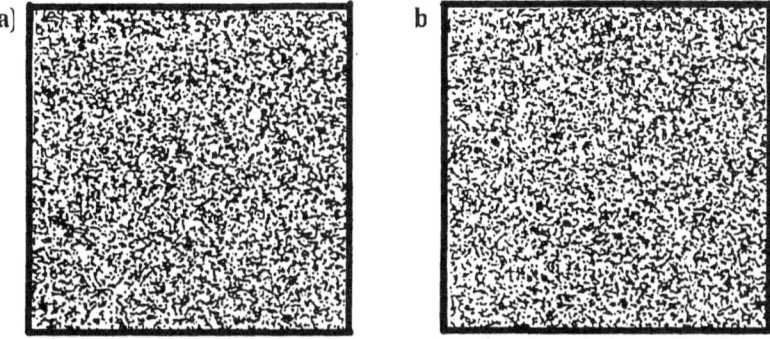

Figure 5 : a) and b) : examples of artificial isotropic media of size ($50 \times \nu_0$ x $50 \times \nu_0$) generated through simulation constrained by the same probability density and power density functions.

The curves on Figs. 6a and 6b display results obtained by calculating $\mathbf{K}$ for unit cells, $V_\beta(r_0)$, of increasing sizes and of common centers. Examination of Fig. 6 verifies that the local R.E.V. of Darcy's type, $V_\beta(r_0)$, is larger than the local R.E.V. of geometrical type, $V_\beta(\nu_0)$, ($r_0 \approx 30_x \nu_0$). Happily the number is not larger and, using natural anisotropic media as a template, it could be significantly smaller. These results are also encouraging in that they suggest that both realizations of the synthetic medium are similar, converging at about the same values for $\mathbf{K}$ (Fig. 6). In this respect, these results can be taken as a necessary condition in view of basing that $\mathbf{K}$ *is solely a function of both probability density function and power density spectrum of the micro-geometry* :

$$\mathbf{K} = g(\varepsilon_\beta \; ; \; A^2(\nu_x, \nu_y)) \tag{7}$$

An open question of prime importance is whether natural media and periodic media are equivalent from the transport point of view ? Some authors[14] suggest this but, this problem is more complex than commonly appreciated. Space does not permit a fuller discussion of this topic which will be presented in details in a subsequent contribution.

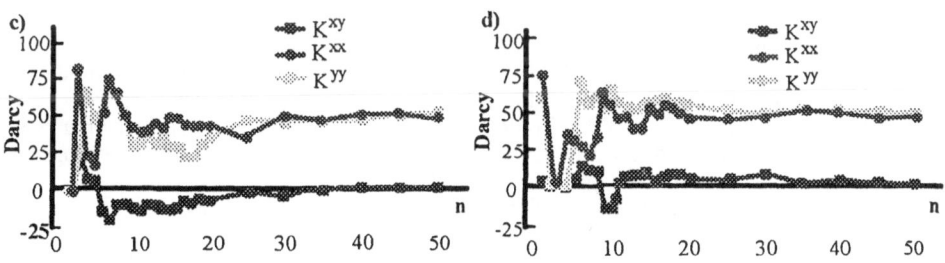

Figure 6 :
c) and d) Evolution of the permeability tensor $\mathbf{K}$ as a function of the size of the local R.E.V. of Darcy's type for the media displayed on Figs. 5a and 5-b respectively.

REFERENCES

1. Ehrlich, R., Crabtree, S.J., Horkowitz, K.O. and Horkowitz, J.P. (1991) "Petrography and reservoir physics I : objective classification of reservoir porosity", *A.A.P.G. Bulletin* 75, No 10, 1547-1562.
2. McCreesh, C.A., Ehrlich, R. and Crabtree, S.J. (1991) "Petrography and reservoir physics II : relating thin section porosity to capillary pressure, the association between pore types and throat sizes", *A.A.P.G. Bulletin* 75, No 10, 1563-1578.
3. Anguy, Y., Ehrlich, R., Prince, C.M., Riggert, V. and Bernard, D. (1994) "The sample support problem for permeability assessment in reservoir sandstones", *A.A.P.G. Spec. Publ. : Stochastic Modeling and Geostatistics, Practical Applications and Case Histories,* Eds. Yarus J. M. and Chambers R. L., in press.
4. Prince, C.M., Ehrlich, R. and Anguy, Y., "Analysis of saptial order in sandstones II : grain clusters, packing flaws, and the small-scale structure of sandstones", *Jour. of Sed. Pet.*, sub.
5. Scheidegger, A.E. (1974), *The physic of flow through porous media*, U.. Toronto P., 353 p.
6. Coskun, S.B. and Wardlaw, N.C. (1993) "Estimation of permeability from image anlysis of reservoir sandstones", *Journal of Petroleum Sciences and Engineering* 10, 1-16.
7. Ehrlich, R., Etris, E.L., Brumfield, D., Yan, L.P. and Crabtree, S.J. (1991) "Petrography and reservoir physics III : physical models for permeability and formation factor", *A.A.P.G. Bulletin* 75, No 10, 1579-1592.
8. Graton, L.C. and Fraser, H.J. (1935) "Systematic packing of spheres with particular relation to porosity and permeabilty" *Jour. Geology* 43(8), 785-909.
9. Anguy, Y. (1993), "Application de la prise de moyenne volumique à l'étude de la relation entre le tenseur de perméabilité et la micro-géométrie ds milieux poreux naturels", *thèse de doctorat,* mécanique, Univ. Bordeaux I, 170 p. .
10. Adler, P.M., Jacquin, C.G. and Quiblier, J.A. (1990) "Flow in simulated porous media", *Int. J. Multiphase Flow* 16, No 4, 691-712.
11. Joshi, M.Y. (1974) "A class of stochastic models for porous media", U. Kansas, Ph.D., Engng. Petroleum, 151 p.
12. Matheron, G. (1965) "Les variables régionalisées et leur estimation", thèse, Paris.
13. Quiblier, J.A. (1984) "A new three-dimensional modeling technique for studying porous media", *J. of Colloid and Interfacial Sci.* 98, No. 1, 84-102.
14. Ventzel, H. (1973) *Théorie des Probabilités,* Eds. Mir Moscou, 584 p.
15. Anguy, Y. and Bernard, D. (1992), "Numerical computation of the permeability tensor evolution versus microscopic geometry changes", v. 2 : *Mathematical Modeling in Water Resour., Russel et al. Eds., Comp. Mech. Pub., Southampton Boston, Elsevier Applied Sci.,* Computational Methods in Water Resources IX, London New York, pp. 385-392.
16. Anguy, Y., Bernard, D. and Ehrlich, R. "The local change of scale method for modeling flow in natural porous media (I) : numerical tools", submitted to *Adv. in Water Resour.*
17. Anguy, Y. and Bernard, D. (1991) "Etude Numérique des effets de la diagénèse minérale sur le tenseur de perméabilité des roches", *Journée SGF-SIS Objets Géologiques : Description Quantitative et Modélisation,* Paris, pp. 5-8.
18. Brenner, H. (1980) "Dispersion resulting from flow through spatially periodic porous media", *Trans. Roy. Soc. (London)* 297, 81-133.
19. Whitaker, S. (1986) "Flow in porous media I : a theoretical derivation of Darcy's law", *Transport in Porous Media* 1, 3-25.
20. Barrère, J., Gipouloux, O. and Whitaker, S. (1992) "On the closure problem for Darcy's law", *Transport in Porous Media* 7, 209-222.
21. Gujar, U.G. (1967) "Generation of random signals with specified probability density functions and power density spectra", M.Sc.E. thesis, Dept. of Elec. Engng., Univ. of New Brunswick, Frediriction, N.B., Canada.

MICROSCOPIC SIMULATIONS OF SUBSURFACE PROCESSES

M. BABIĆ, E. G. SACK and S. JAZIĆ
Department of Civil Engineering and Geological Sciences
University of Notre Dame
Notre Dame, IN 46556
U.S.A.

Abstract
The prediction of the permeability and dispersion characteristics of porous media is a long-standing problem of very important theoretical and practical interest. In the present study these transport properties are determined from microscopic numerical experiments performed on simulated two-dimensional porous media. These numerical experiments are based on the homogenization theory which provides the link between the microscopic and macroscopic descriptions of the system.

INTRODUCTION
The homogenization theory has been devised for the purpose of asymptotic analysis of media with periodic microstructure (Bensoussan, et al. 1978). Applications to the flow and transport in porous media have been considered by Ene (1990) and Mei (1992). In the present study the key results of the homogenization theory are utilized to determine the transport properties of numerically simulated, spatially periodic porous media. The material in the following section is based mostly on the work of Mei (1992). Even though the resulting equations are not new, their development is presented here for completeness.

HOMOGENIZATION THEORY
Fluid Flow
The equations governing the flow of an incompressible Newtonian fluid at the microscale are the conservation equations for mass and linear momentum:

$$\nabla' \cdot \mathbf{u}' = 0 \quad ; \quad \rho\,(\partial_{t'}\mathbf{u}' + \mathbf{u}' \cdot \nabla'\mathbf{u}') = -\nabla'p' + \mu\nabla'^2\mathbf{u}', \tag{1}$$

where $\mathbf{u}'$ is the velocity, ∇' is the spatial gradient, $\partial_{t'}$ denotes differentiation with respect to time t', $p' = \pi' + \rho g x'_3$ is the effective pressure, π' is the pressure, g is the acceleration of gravity, x'_3 is the vertical coordinate, ρ is the fluid density and μ is the fluid viscosity. Primes are used to denote dimensional variables. These equations will be scaled utilizing the following quantities: λ is the characteristic micro-length scale (pore scale), L is the characteristic macro-length scale, U is the characteristic velocity and G is the characteristic macroscopic pressure gradient. A typical porous media flow is characterized by the balance of the pressure forces $|\nabla'p'| \sim G$ and the viscous forces $|\mu\nabla'^2\mathbf{u}'| \sim \mu U/\lambda^2$. Hence, the characteristic velocity $U = G\lambda^2/\mu$. In terms of dimensionless variables $t = t'U/\lambda$, $\mathbf{x} = \mathbf{x}'/\lambda$, $\mathbf{u} = \mathbf{u}'/U$ and $p = p'/(GL)$, equations (1) become:

$$\nabla \cdot \mathbf{u} = 0 \quad ; \quad \varepsilon \mathrm{Re}\,(\partial_t \mathbf{u} + \mathbf{u} \cdot \nabla \mathbf{u}) = -\nabla p + \varepsilon \nabla^2 \mathbf{u}, \tag{2}$$

407

A. Peters et al. (eds.), Computational Methods in Water Resources X, 407–414.
© 1994 Kluwer Academic Publishers. Printed in the Netherlands.

where $\varepsilon = \lambda/L$ is the length scale ratio (assumed to be very small) and $Re = \rho U \lambda/\mu$ is the Reynolds number based on the micro-length scale. For simplicity, the inertial terms on the left-hand side of the momentum equation will be neglected. As shown in Mei (1992), keeping the inertial terms while assuming that Re is of the order of ε does not change the resulting macroscopic theory. It should be noted that, despite their appearance, the terms on the right-hand side of (2) are actually of the same order of magnitude.

The homogenization theory is based on asymptotic expansions in the limit $\varepsilon \to 0$ using multiple scales. The asymptotic expansion of a function $f(x_i)$ is given by:

$$f(x_i) = \sum \varepsilon^m f^{(m)}(x_i, X_i),\qquad(3)$$

where $m \geq 0$, $x_i = x'_i/\lambda$ and $X_i = \varepsilon x_i = x'_i/L$ are the dimensionless micro-coordinates and macro-coordinates, respectively. Functions $f^{(m)}(x_i, X_i)$ are periodic in the x_i variable and vary independently with respect to the two sets of coordinates. Using the chain rule of differentiation, the spatial derivative of $f(x_i)$ can be evaluated as:

$$df/dx_i = \sum \varepsilon^m (\partial f^{(m)}/\partial x_i + \varepsilon \partial f^{(m)}/\partial X_i).\qquad(4)$$

Expanding $\mathbf{u}$ and p in asymptotic series according to (3) and their derivatives according to (4) in (2) and collecting the like powers of ε gives the following sequence of equations:

$$\frac{\partial u_i^{(m)}}{\partial x_i} + \frac{\partial u_i^{(m-1)}}{\partial X_i} = 0;\qquad(5)$$

$$-\left(\frac{\partial p^{(m)}}{\partial x_i} + \frac{\partial p^{(m-1)}}{\partial X_i}\right) + \left(\frac{\partial^2 u_i^{(m)}}{\partial x_j \partial x_j} + 2\frac{\partial^2 u_i^{(m-1)}}{\partial x_j \partial X_j} + \frac{\partial^2 u_i^{(m-2)}}{\partial X_j \partial X_j}\right) = 0,\qquad(6)$$

where $u_i^{(m)}$ and $p^{(m)}$ should be set to zero for $m < 0$. The boundary conditions are the no-slip condition $u_i^{(m)} = 0$ on the fluid-solid interface Γ and periodicity conditions for $u_i^{(m)}$ and $p^{(m)}$. The zeroth order problem (equation (6) with $m = 0$) gives $\partial p^{(0)}/\partial x_i = 0$, hence $p^{(0)} = p^{(0)}(X_i)$. The first order problem (equation (5) for $m = 0$ and equation (6) with $m = 1$) reads:

$$\partial u_i^{(0)}/\partial x_i = 0 \quad ; \quad -\partial p^{(1)}/\partial x_i + \nabla^2 u_i^{(0)} = \partial p^{(0)}/\partial X_i.\qquad(7)$$

Linearity of the problem implies that $u_i^{(0)}$ and $p^{(1)}$ can be expressed as:

$$u_i^{(0)} = -K_{ij}(\partial p^{(0)}/\partial X_j) \quad ; \quad p^{(1)} = -A_j(\partial p^{(0)}/\partial X_j) + \bar{p}^{(1)}(X_i).\qquad(8)$$

Using (8) in (7) yields:

$$\partial K_{ij}/\partial x_j = 0 \quad ; \quad -\partial A_j/\partial x_j + \nabla^2 K_{ij} = -\delta_{ij}.\qquad(9)$$

The boundary conditions are $K_{ij} = 0$ on Γ and periodicity conditions for K_{ij}, A_j. Averaging the equation (8) over the unit cell and returning to the original, dimensional coordinates gives the Darcy law:

$$\langle u'_i \rangle = -\frac{\rho g \lambda^2}{\mu} \langle K_{ij} \rangle \frac{\partial \langle p' \rangle}{\partial x'_j},\qquad(10)$$

where the averaging operator is defined as $\langle f \rangle = (1/\Omega) \int_\Omega f(x_i) \, d\Omega$, where Ω is the volume of the unit cell. For a specified geometry of the unit periodic cell it is now possible to determine the dimensionless permeability tensor $\langle K_{ij} \rangle$ by solving the boundary value problem specified by (9). Analysis of equations (9) indicates that (in 2D) the equations for K_{11}, K_{21} and A_1 are completely decoupled from equations for K_{12}, K_{22} and A_2. In fact, both of these sets of equations have a very simple physical interpretation. For example, in the flow driven by an unit macroscopic pressure gradient in the X_1 direction, A_1 is the pressure while K_{11} and K_{21} are the velocity components in the x_1 and x_2 directions, respectively. Equations (9) are in fact two sets of Stokes flow equations with driving forces acting in the two coordinate directions. Each of these sets of equations corresponds to a Darcy-type experiment used to measure the permeability of a porous medium.

Dispersion

The conservation equation for mass of a passive solute at the microscale is the advection-diffusion equation:

$$\partial_{t'}C + \mathbf{u}' \cdot \nabla'C = D\nabla'^2 C, \tag{11}$$

where C is the solute concentration and D is the molecular diffusion coefficient. In terms of the previously defined dimensionless variables, the above equation becomes:

$$\partial_t C + \mathbf{u} \cdot \nabla C = (1/\text{Pe})\nabla^2 C, \tag{12}$$

where $\text{Pe} = U\lambda/D$ is the Peclet number, which is assumed to be of the order of one. The requirement of zero mass flux across the fluid-solid interface implies $\mathbf{n} \cdot \nabla C = 0$, where $\mathbf{n}$ is the unit normal vector on the interface. Considering the fact that advection and diffusion are processes characterized by different time scales C is asymptotically expanded as:

$$C(t, x_i) = \sum \varepsilon^m C^{(m)}(t_1, t_2, x_i, X_i), \tag{13}$$

where $C^{(m)}$ depend not only on the two independent sets of spatial coordinates but also on two independent time coordinates, the "fast time" t_1 and the "slow time" t_2. The appropriate scales for t_1 and t_2 can be found by dimensional analysis. The advection time scale is $T_1 = L/U$ and the diffusion time scale is $T_2 = L^2/D$. Since the dimensionless time is defined as $t = t'/(\lambda/U)$, we can set $t_1 = t'/T_1 = \varepsilon t$ and $t_2 = t'/(T_2\text{Pe}) = \varepsilon^2 t$. Hence, the temporal derivative of $C(t, x_i)$ can be evaluated as:

$$\partial C/\partial t = \sum \varepsilon^m [\varepsilon(\partial C^{(m)}/\partial t_1) + \varepsilon^2(\partial C^{(m)}/\partial t_2)]. \tag{14}$$

Employing (13) and (14) in (12) and collecting the like powers of ε yields:

$$\frac{\partial C^{(m-1)}}{\partial t_1} + \frac{\partial C^{(m-2)}}{\partial t_2} + \sum_{k=0}^{m} \binom{m}{k} u_i^{(m-k)} \frac{\partial C^{(k)}}{\partial x_j} + \sum_{k=0}^{m-1} \binom{m-1}{k} u_i^{(m-1-k)} \frac{\partial C^{(k)}}{\partial x_j}$$

$$= \frac{1}{\text{Pe}} \left(\frac{\partial^2 C^{(m)}}{\partial x_j \partial x_j} + 2\frac{\partial^2 C^{(m-1)}}{\partial x_j \partial X_j} + \frac{\partial^2 C^{(m-2)}}{\partial X_j \partial X_j} \right), \tag{15}$$

where $u_i^{(m)}$ and $C^{(m)}$ should be set to zero for $m < 0$. The corresponding boundary conditions at the fluid-solid interface Γ are:

$$n_i(\partial C^{(m)}/\partial x_i + \partial C^{(m-1)}/\partial X_i) = 0. \tag{16}$$

The first order problem $(m = 0)$ is given by:

$$u_j^{(0)}(\partial C^{(0)}/\partial x_j) - (1/\text{Pe})\nabla^2 C^{(0)} = 0 \quad ; \quad n_i(\partial C^{(0)}/\partial x_i)\big|_\Gamma = 0. \tag{17}$$

The only possible solution to this problem is that $C^{(0)}$ is not a function of x_i, i.e. $C^{(0)} = C^{(0)}(t_1, t_2, X_i)$. The second-order problem $(m = 1)$ reduces to:

$$u_j^{(0)} \frac{\partial C^{(1)}}{\partial x_j} - \frac{1}{\text{Pe}}\nabla^2 C^{(1)} = -\tilde{u}_j^{(0)} \frac{\partial C^{(0)}}{\partial X_j} \quad ; \quad n_i \frac{\partial C^{(1)}}{\partial x_i}\bigg|_\Gamma = -\frac{\partial C^{(0)}}{\partial X_i}, \tag{18}$$

where $\tilde{u}_j^{(0)} = u_j^{(0)} - \langle u_j^{(0)} \rangle / n$ is the deviatoric flow velocity and n is the porosity.

Linearity of the problem implies that $C^{(1)}$ can be expressed as:

$$C^{(1)}(t_1, t_2, x_i, X_i) = -N_k(x_i, X_i)(\partial C^{(0)}/\partial X_k) + \bar{C}^{(1)}(t_1, t_2, X_i). \qquad (19)$$

Using (19) in (18) gives:

$$-u_j^{(0)}(\partial N_k/\partial x_j) + (1/Pe)\nabla^2 N_k = -\tilde{u}_k^{(0)} \qquad ; \qquad n_j(\partial N_k/\partial x_j)\big|_\Gamma = n_k. \qquad (20)$$

Equations (20) constitute a boundary value problem posed on the unit periodic cell. The solution for N_k can be made unique by specifying $\langle N_k \rangle = 0$. The significance of N_k is revealed by averaging the first three equations of the sequence (15) over the fluid volume in the unit cell. After lengthy manipulations, it can be shown that $\langle C \rangle \approx \langle C^{(0)} \rangle + \varepsilon \langle C^{(1)} \rangle$ satisfies the following equation in the original, dimensional, coordinates:

$$n\frac{\partial \langle C \rangle}{\partial t'} + \langle u'_k \rangle \frac{\partial \langle C \rangle}{\partial x'_k} = \frac{\partial}{\partial x_j'}(D_{jk}\frac{\partial \langle C \rangle}{\partial x'_k}), \qquad (21)$$

where D_{jk} is the effective dispersion tensor given by:

$$D_{jk}/D = Pe\langle \tilde{u}_j^0 N_k + \tilde{u}_k^0 N_j \rangle/2 - \langle \partial N_k/\partial x_j + \partial N_j/\partial x_k \rangle/2 + n\delta_{jk}. \qquad (22)$$

Therefore, the macroscopic governing equation for the mass balance of a passive solute is the well-known advection-dispersion equation (21). This equation can be obtained by simpler means (e.g. Bear, 1972). However, the homogenization theory is also able to provide an explicit expression for the dispersion tensor (22). The dispersion tensor can be evaluated if the solutions to the flow problem and the boundary-value problem (20) are available. In the present study, these solutions are obtained numerically for the simulated two-dimensional porous media described below.

GENERATION OF SIMULATED POROUS MEDIA

The idealized porous media are simulated as binary random fields. The nearest neighbor method (Smith and Freeze, 1979) is used to generate random fields of normally distributed, spatially correlated real numbers. These continuous random fields are subsequently transformed into binary random fields which take the value of zero if a given point is occupied by the fluid and one if it is occupied by the solid. The nearest neighbor method is a first-order autoregressive stochastic process model which does not allow generation of random fields with arbitrary correlation structures, but can produce simulated porous media with specified porosities and correlation length scales. The nearest neighbor model in 2D is defined by the following first-order difference relation:

$$Y[x] = \alpha(Y[x + \lambda e_1] + Y[x - \lambda e_1] + Y[x + \lambda e_2] + Y[x - \lambda e_2]) + G[x], \qquad (23)$$

where x is the position vector on a 2-dimensional rectilinear lattice with spacing λ, α is the autoregressive parameter, $G[x]$ is a normally distributed, uncorrelated random field (white noise), and $Y[x]$ is the simulated random field. The generation of individual samples of two-dimensional (M_x by M_y) random fields $Y[x]$ is performed in two steps: (1) generation of a sequence of $M_x \cdot M_y$ normally distributed random numbers representing the white noise field $G[x]$, and (2) solution of $M_x \cdot M_y$ linear algebraic equations (23). Periodicity conditions are employed at the edges of the lattice. The system of linear equations (23) is solved by standard methods for sparse systems. The resulting random fields are second-order stationary and their autocorrelation function (Vanmarcke, 1983) is given by: $R_{YY}(x_1, x_2) \equiv R_{YY}(u) = \beta U K_1(\beta U)$, where $u = |x_2 - x_1|$, $U = u/\lambda$, $\beta = [(1 - 2\alpha)/\alpha]^{1/2}$ and K_1 is the modified Bessel function of the second kind. Transformation of the continuous random field $Y[x]$ into the binary random field $Z[x]$ is given by $Z = g(Y)$, where $g(y)$ is the deterministic function defined by $g(y) = 0$

for $y < b$ and $g(y) = 1$ for $y > b$. The value of b is found such that the mean of $Z[\mathbf{x}]$ equals the solid fraction 1-n, where n is the specified porosity. Since $Y[\mathbf{x}]$ is normally distributed, $b = \sqrt{2}\,\mathrm{erf}^{-1}(2n-1)$, where erf^{-1} is the inverse error function. Finally, the autocorrelation function of the binary field, $R_{ZZ}(u)$, can be related to $R_{YY}(u)$ as shown in Adler, et al. (1990). This relationship depends on the porosity n. Since R_{YY} depends on the autoregressive parameter α, the resulting function $R_{ZZ}(u)$ α depends on n and α. Therefore, the relationship between the correlation length of the binary field $l_Z = \int_0^\infty R_{ZZ}(u)\,du$ and parameters n and α can be obtained. Inversely, for specified values of n and l_Z/λ the required value of the parameter α can be determined. This parameter controls the clustering of solid elements in the simulated porous media.

FLUID FLOW

For a given porous media sample, the fluid flow field is determined numerically by solving the Navier-Stokes equations (1) in the pore space along with the no-slip boundary conditions imposed at the complex fluid-solid interface. Considering the flow in the unit periodic cell induced by the macroscopic pressure gradient $\mathbf{G}$ and expressing the pressure gradient term as $\nabla'p' = \mathbf{G} + \nabla'P'$, dimensionless equations (1) can be written as:

$$\nabla \cdot \mathbf{u} = 0 \quad ; \quad \mathrm{Re}\,(\partial_t \mathbf{u} + \mathbf{u} \cdot \nabla \mathbf{u}) = -\nabla P + \nabla^2 \mathbf{u} - \mathbf{G}, \qquad (24)$$

where $t = t'U/\lambda$, $\mathbf{x} = \mathbf{x}'/\lambda$, $\mathbf{u} = \mathbf{u}'/U$, $U = G\lambda^2/\mu$, $\mathrm{Re} = \rho U \lambda/\mu$, $\mathbf{G} = \mathbf{G}'/G$ and $P = P'/G\lambda$. In the limit $\mathrm{Re} \to 0$ this equation becomes equivalent to the boundary value problem (9) formulated by the homogenization theory. Specifically, for $\mathbf{G} = -\mathbf{e}_1$, $\mathbf{u} = (K_{11}, K_{12})$ and $P = A_1$, while for $\mathbf{G} = -\mathbf{e}_2$, $\mathbf{u} = (K_{21}, K_{22})$ and $P = A_2$. For isotropic porous media it is sufficient to solve only one of these boundary value problems in order to determine the dimensionless permeability $K = \langle K_{11} \rangle = \langle K_{22} \rangle$.

A finite-difference numerical model for solution of equations (24) has been developed. The model is based on the Marker-And-Cell (MAC) scheme (Roache, 1976). The MAC scheme has four distinguishing features: (1) the form of the primitive-variable equations used, (2) the differencing scheme, (3) the cell structure, and (4) the use of "marker particles" to determine the streaklines of the flow. For a 2D flow with $\mathbf{G} = -\mathbf{e}_1$, the x and y components of the momentum equation (conservative forms) are:

$$\frac{\partial u}{\partial T} + \mathrm{Re}\left(\frac{\partial u^2}{\partial x} + \frac{\partial uv}{\partial y}\right) = -\frac{\partial P}{\partial x} + \nabla^2 u + 1; \qquad (25)$$

$$\frac{\partial v}{\partial T} + \mathrm{Re}\left(\frac{\partial uv}{\partial x} + \frac{\partial v^2}{\partial y}\right) = -\frac{\partial P}{\partial y} + \nabla^2 v, \qquad (26)$$

where the time was rescaled as $T = t/\mathrm{Re}$ so that the solution to the steady problem can be obtained as the asymptotic solution to the unsteady problem. The conservation of mass equation is $D = \partial u/\partial x + \partial v/\partial y = 0$. By differentiating (25) with respect to x and (26) with respect to y and adding the results, a Poisson equation for the pressure is obtained:

$$\nabla^2 P = -\mathrm{Re}\left(\frac{\partial^2 u^2}{\partial x^2} + 2\frac{\partial^2 uv}{\partial x \partial y} + \frac{\partial^2 v^2}{\partial y^2}\right) - \frac{\partial D}{\partial T} + \nabla^2 D, \qquad (27)$$

in which the terms involving D are retained in order to avoid numerical instabilities. The porous media sample consists of M_x by M_y fluid or solid elements. Each element of the sample is discretized into N_c by N_c computational cells. In the MAC scheme the pressure $p_{i,j}$ is defined at the center of the cell (i,j) and the velocities $u_{i+1/2, j}$ and $v_{i, j+1/2}$ are defined at the centers of the right and top edges of the cell, respectively. Using this cell structure, all terms in equations (25)-(27) can be approximated by central differences and the resulting scheme is second-order accurate. Furthermore, implementation of the no-slip boundary conditions is straightforward. Time derivatives in equations (25)-(26) are approximated by forward differences. At each time step, the updated values of u and v are calculated explicitly from discretized equations (25)-(26) and the updated values of p are then calculated from the discretized Poisson equation (27) by the SOR method. Calculations can be started from an arbitrary initial condition and the steady state solution is approached asymptotically. Once the steady solution is obtained, the dimensionless permeability is calculated as $K = \langle u \rangle = \Sigma u_{ij}/N$, where $N = M_x M_y N_c^2$ is the total number of computational cells and $u_{ij} = 0$ in solid cells. It should be noted that $\langle u \rangle$ is the superficial (Darcy) velocity rather than the average pore velocity $\bar{u} = \langle u \rangle/n$. A typical flow field resolved by the MAC method described above is shown in Figure 1.

Figure 1: Example of the flow field resolved by the MAC model. The size of the field is 40 by 40, porosity n=0.75, autoregression coefficient $\alpha = 0.1$.

The permeability vs. porosity relationship was determined by performing a large number of Darcy-type numerical experiments on 40 by 40 random fields. The results are shown in Figure 2. The variability in the calculated values of permeability for different random samples of equal porosity and correlation length increases with the porosity.

DISPERSION

Application of the homogenization theory to the problem of solute transport in porous media with a periodic microstructure resulted in the transformation of the governing microscopic equation (11) into the corresponding macroscopic equation (21). Solution of (20) can be obtained as the steady solution of the following unsteady problem:

$$\partial N/\partial T + \mathbf{u} \cdot \nabla N - (1/Pe)\, \nabla^2 N = (\mathbf{u} - \langle \mathbf{u} \rangle/n) \tag{28}$$

Equation (28) is equivalent (in 2D) to a set of two uncoupled advection-diffusion equations for N_1 and N_2. Equation (28) is solved using the simple FTCS (forward in time,

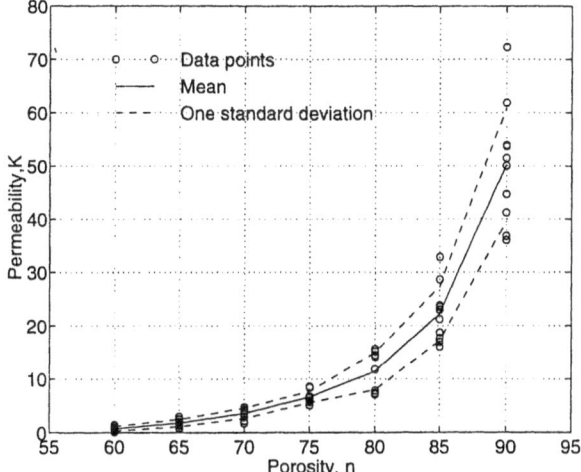

Figure 2: Permeability vs. porosity relation. The size of the fields is 40 by 40, autoregression coefficient $\alpha = 0.1$.

central in space) finite-difference scheme. From the resulting distribution of $N(x)$ the dispersion tensor is calculated according to (22). The relationship between the dispersion coefficient and the Peclet number was investigated. The results for a twenty by twenty field with the porosity of 0.75 is presented in Figure 3. These results are in a good qualitative agreement with results reported in the literature (e. g. Bear, 1972).

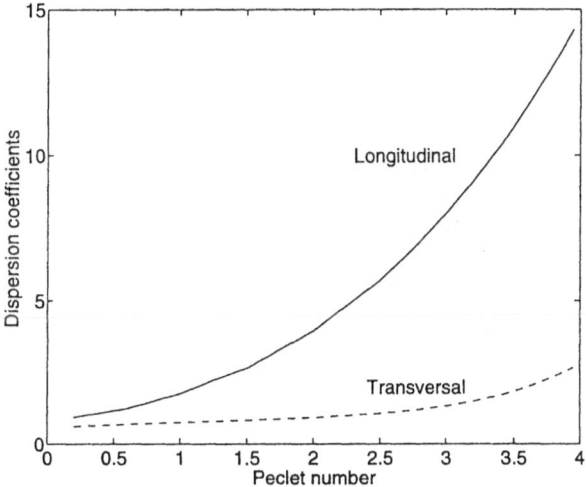

Figure 3: Longitudinal and transversal dispersion coefficients vs. Pe. The size of the field is 20 by 20, porosity $n = 0.75$, autoregression coefficient $\alpha = 0.1$.

CONCLUSIONS

The present study is concerned with the prediction of transport properties (permeability and dispersion coefficients) of numerically simulated, spatially periodic, two-dimensional porous media. The idealized porous media are simulated as binary random fields with specified porosities and correlation length scales using the nearest neighbor method. The required transport properties are obtained from numerical solutions of the appropriate microscopic boundary value problems posed on the unit periodic cell of the medium. These boundary value problems have been formulated by the homogenization theory, which provides a link between the distributions of microscopic variables and the corresponding macroscopic transport coefficients. Numerical solutions to these boundary value problems are obtained by the finite difference methods.

The results presented in this article include the following relationships: (1) permeability vs. porosity, and (2) dispersion coefficients vs. the Peclet number. Preliminary investigations on the effects of the field size and the autoregressive parameter α on these results have also been conducted. The results are qualitatively consistent with analogous relationships for real porous media. However, the 2D porous media studied here are characterized by much higher porosities than real, 3D, porous media. Extensions to 3D, which will lead to results that can be quantitatively compared with real porous media, are planned for future work. Conversion of numerical models developed in this study to 3D will be straightforward, but increased computational demands will eventually require modification of the explicit scheme into an implicit one.

The next step in microscopic modeling of subsurface processes will be to simulate the transport of discrete particles such as sediment, bubbles, colloids or microbes. Particle tracking will be performed by numerically integrating the equations of motion of the particles. Future studies will elucidate the effects of Brownian motion and interparticle interactions on the particle transport. Dispersion coefficients for the particulate suspensions will be compared to the dispersion coefficients for the passive solute. Microscopic simulations of subsurface processes will be used to establish fundamental relationships between the macroscopic transport coefficients and the microscopic parameters which characterize the properties of the solid matrix, the fluid, and the particles.

REFERENCES

1. Adler, P. M., Jacquin, C. G. and Quiblier, J. A. (1990) "Flow in simulated porous media", *Int. J. Multiphase Flow*, 16, 691-712.
2. Bear, J. (1972) *Dynamics of Fluids in Porous Media*, American Elsevier, New York.
3. Bensoussan, A., Lions, J. L. and Papanicolau, G. (1978) *Asymptotic Analysis for Periodic Structures*, North Holland, New York.
4. Ene, H. I. (1990) "Application of the homogenization method to transport in porous media", in J. Cushman (ed.), *Dynamics of fluids in hierarchical porous media*, Academic Press, New York, pp. 223-241.
5. Mei, C. C. (1992) "Method of homogenization applied to dispersion in porous media", *Transport in Porous Media*, 9, 261–274.
6. Roache, P. J. (1976) *Computational Fluid Dynamics*, Hermosa, Albuquerque, NM.
7. Smith, L. and Freeze, R. A. (1979) "Stochastic analysis of steady groundwater flow in bounded domain", *Water Resources Research*, 15, 1543–1559.
8. Vanmarcke, E. (1983) *Random Fields*, MIT Press, Cambridge, Ma.

ACKNOWLEDGEMENTS

Support by the Jesse H. Jones Fund, the Petroleum Research Fund and the U.S. Department of Energy is gratefully acknowledged.

UNSATURATED FLOW AND ADVECTION-DISPERSION IN THREE-DIMENSIONALLY HETEROGENEOUS GEOLOGIC MEDIA

Amvrossios C. Bagtzoglou and Vivek Kapoor
Center for Nuclear Waste Regulatory Analyses
Southwest Research Institute
San Antonio, TX 78238-5166
U.S.A.

ABSTRACT

Flow and transport in three-dimensional, heterogeneous porous media, with two statistically distinct layers, is numerically simulated, in a multiple realization mode, using a finite-difference method for the flow problem and a particle method for the transport problem. The coarser bottom layer acts as a capillary barrier to flow, which is however breached by localized zones of relatively rapidly moving water. The mean and variance of the concentrations of solute undergoing advection-dispersion in the flow fields are computed. The concentration coefficient of variation (CV) grows with time initially. On including local dispersion, its growth is dampened, and at large times the coefficient of variation decreases with time, unlike the zero local dispersion case.

INTRODUCTION

In assessing feasibility of waste disposal schemes in geologic media, of ultimate concern is the possible exposure of human beings to toxic substances. This requires assessing contaminant concentrations. The premise that the heterogeneity of the advective pathways will influence the assessment of the solute transport is unambiguously clear from results of large-scale solute transport experiments. In these experiments hundreds of point measurements of concentrations were taken to construct three-dimensional (3D) snapshots of contaminant plumes in the saturated zone at the Borden site in Canada, the Cape-Cod site in Massachusetts, and at the Columbus site in Mississippi, U.S.A. These large-scale experiments document the enhanced (compared to the laboratory scale) mass flux associated with transport in 3D varying hydraulic conductivity fields, and the concomitant complex 3D spatial distributions of solute concentration. The few large-scale experiments in unsaturated porous media show that the features of enhanced mass transport and rugged concentration distributions are common to both the saturated and unsaturated zones.

A. Peters et al. (eds.), Computational Methods in Water Resources X, 415–422.
© *1994 Kluwer Academic Publishers. Printed in the Netherlands.*

In this study, the detailed evaluation of flow and advection-dispersion of solute in unsaturated heterogeneous porous media is simulated to understand transport processes in the far-field of a two-layer system, analogous, in some respects, to Yucca Mountain, Nevada, the site of the proposed U.S. high level radioactive waste repository. Description of flow fields obeying basic constitutive and conservation laws in complex geologic media and an assessment of the impact of the flow field on solute transport is obtained with the executive numerical code Stochastic analysis of Unsaturated FLow And Transport (SUFLAT) (Bagtzoglou et al., 1994). The hydraulic parameters used to describe the relation between unsaturated conductivity and pressure, such as saturated conductivity (K_S) and slope of relative conductivity curve (Gardner α) are allowed to be cross-correlated.

FLOW AND TRANSPORT PROBLEM

A Topopah Spring welded/Calico Hills nonwelded-vitric (TSw/CHnv) flow system, of size 250×250×500m, is discretized in 18,081 computational cells each with a grid block size of $\Delta x=12.5m$. An exponential covariance spatial correlation is assumed for all properties, and the correlation length is assumed isotropic in the horizontal directions $(\lambda_X=\lambda_Y=37.5m)$. A correlation scale anisotropy of 3 is taken to represent a mild stratification, consistent with the depositional nature of tuffaceous rocks (i.e., $\lambda_Z=12.5m$). Data for tuffaceous materials were obtained from Peters et al. (1984).

The isotropic discretization interval of 12.5m yields 3 nodes and one node per correlation scale in the horizontal and vertical directions, respectively. This discretization imposes a high-wave number cutoff in the corresponding spectrum employed to generate the three-dimensional properties in the Turning Bands code (Tompson et al., 1987a). Therefore, the actual stochastic description employed here does not have the rich high-wave number characteristic of an exponential covariance function. The large range of hydraulic-conductivity values encountered in this problem can be seen in Figure 1. In this figure the symbols "*" and "o" correspond to the hydraulic conductivity function for the mean K_S and α parameters for the TSw and CHnv layers, respectively. The upper dashed line corresponds to a K_S one standard deviation larger than its mean, and α one standard deviation smaller than its mean, for the TSw unit. The lower dashed line corresponds to a K_S one standard deviation smaller than its mean, and α one standard deviation larger than its mean, for the TSw unit. Similarly, the range of hydraulic conductivity for the CHnv layer is "bounded" by the solid lines. The lateral boundaries of the system are no-flow, the bottom boundary is a water table condition, and the top boundary has a specified flux of 0.5 mm/year. The TSw and CHnv layers are 350 m and 150 m thick respectively. The computational domain spans 7 correlation lengths in the horizontal direction, and 40 in the vertical.

The unsaturated flow problem, represented by Richards equation, is solved with the BIGFLOW numerical code (Ababou and Gelhar, 1988; Ababou and Bagtzoglou, 1993).

In BIGFLOW, an implicit, low-order, seven-point centered finite difference scheme is implemented and steady-state solutions are attained by time-stepping to sufficiently long times. The unsaturated conductivity-pressure relation, K(h,**x**), is assumed to be a truncated exponential function. For the soil moisture retention curve, θ(h,**x**), BIGFLOW allows a choice between several functional forms, including a truncated exponential and the van Genuchten function (the one used here), among others. Both characteristic curves are allowed to include 3D spatially-variable parameters.

Conservative transport is simulated by particle tracking with the numerical code SLIM (Tompson et al., 1987b). A large number of particles (10,000) is released instantaneously from a single cell, located at the layer of cells adjacent to the top boundary, and approximately at the center of the domain. Steady-state flow results are used for particle transport for three values of dispersivity; zero for the purely advective case, 0.1m, and 1.0m. Within the context of this work, local dispersion is viewed as the manifestation of sub-continuum (grid block) variability, as it could be due to the presence of fractures in the rock.

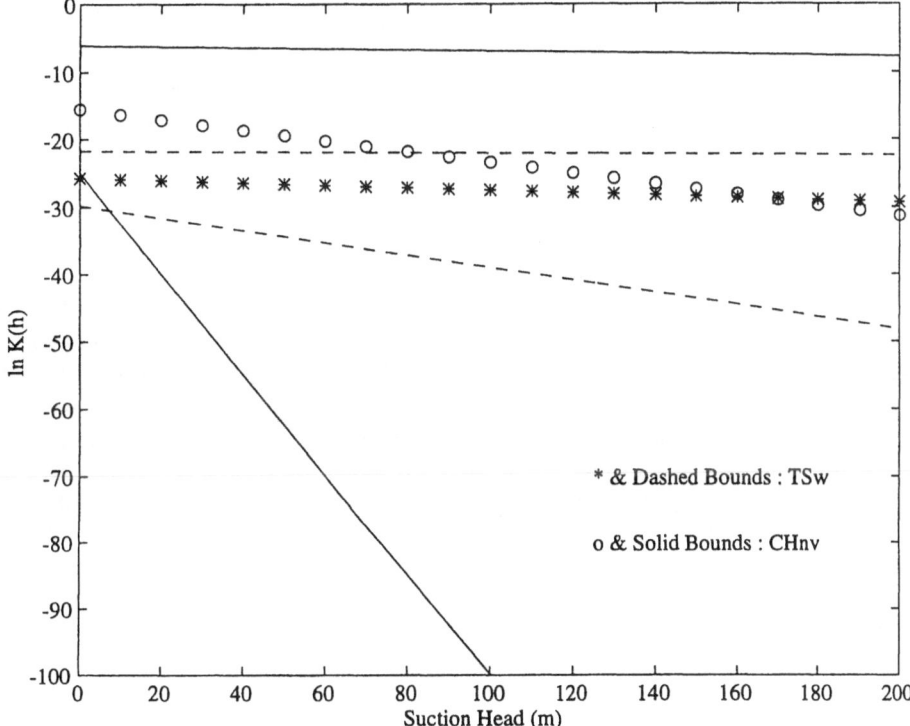

Figure 1: Range of unsaturated hydraulic conductivities.

It should be noted that transport is always advection dominated, insofar as the horizontal correlation scale-based Peclet number is 37.5, 375, and ∞ for dispersivity of 1.0m, 0.1m, and zero, respectively.

RESULTS

The suction field obtained from the numerical simulation shows significant spatial variation. The impact of such suction variation on the overall flux of moisture and solute transport are discussed by Gelhar (1993), with a summary of analytical results and an overview of different approaches in analyzing unsaturated flow and transport. A zone of high suction is formed in the vicinity of the interface. A wedge-like zone of almost zero suction is formed above the interface. This zone is characterized by very high saturations, in the range of 95 to 98%.

Figure 2 depicts the scalar magnitude of the water velocity. Notable is the overall contrast between the two distinct layers, with the lower layer acting as a capillary barrier (logarithm of velocity, in m/sec, is in the range of -45 to -39) as predicted by some U.S. Department of Energy researchers (Loeven, 1993) . However, in some distinct regions, high velocity zones bypass the barrier (velocity magnitude 9 orders magnitude greater than those of the surroundings). The locations of these regions are presumably determined by the nature of the heterogeneity of the porous medium and are related to the range of suction-based crossing points in the unsaturated conductivity curves (shown in Figure 1). Note that there exists a substantial range in suction head (greater than 80m) over which the TSw layer property outliers are more conductive than the CHnv layer average properties. Figure 3 shows the concentration distribution at time 250K years, for solute instantaneously introduced at the top of the domain, and the apparent breach in the capillary barrier. Figure 4 shows the spatial distribution of the concentration CV (the concentration standard deviation divided by the mean concentration), computed from 20 realizations. It is recognized that 20 different realizations is a small statistical sample. However, our flow analyses involve quite a substantial amount of variability (7×7×40 correlation lengths), and the transport analyses involved a very large number of particles (10,000) sampling numerous pathlines for every realization. The areas of the plot that are marked by light shades of gray to white correspond to relatively high degree of uncertainty, compared to the zones in gray or black. It is important to note that the persistence of the lateral movement across the interface is supported by the horizontal dark area above the 150m elevation. The relatively high degree of repeatability in the capillary barrier breaches is supported in the coefficient of variation plot. Finally, Figure 5 depicts the temporal evolution of the minimum (spatial) concentration CV for the pure advection case, and the two advection-dispersion cases.

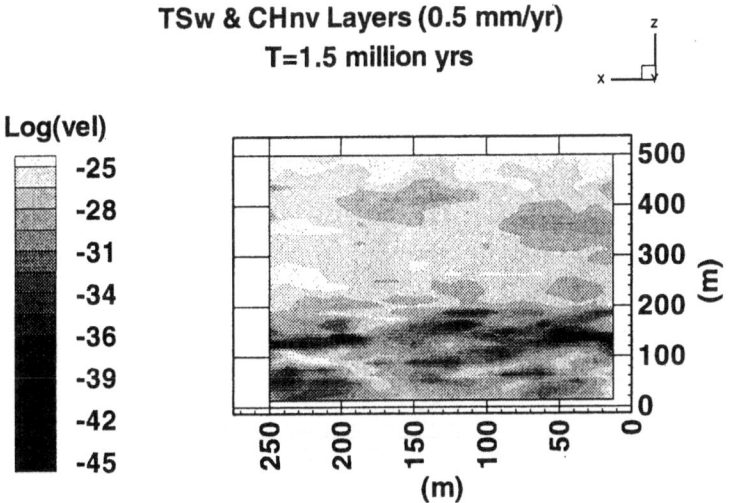

Figure 2: Magnitude of flow velocity, in ln(m/sec), for typical realization.

CONCLUSIONS

1. At early times the zero-local dispersion CV grows more rapidly than the advective-dispersive cases. From a time of 150K years, in both advective-dispersive cases, the CV decreases with time. This is especially pronounced in the case of the dispersivity being 0.1m. In contrast with this behavior, the zero-local dispersion case exhibits a mild increase in the CV with time from 150 to 300K years. This contrasting behavior was analytically shown by Kapoor (1993) and Kapoor and Gelhar (1994a,b) for transport in 3D heterogeneous saturated porous media. These numerical simulations support their conclusion that the singular capacity of dispersion to bring about a decrease in the concentration CV at large times, and the growth of the CV in the zero-local-dispersion transport model, render the zero-local-dispersion model to be of little value in predicting or interpreting the asymptotic behavior of contaminant concentration in heterogeneous porous media. The concentrations derived from a zero-local-dispersion transport model are subject to unrealistically large relative degrees of uncertainty.

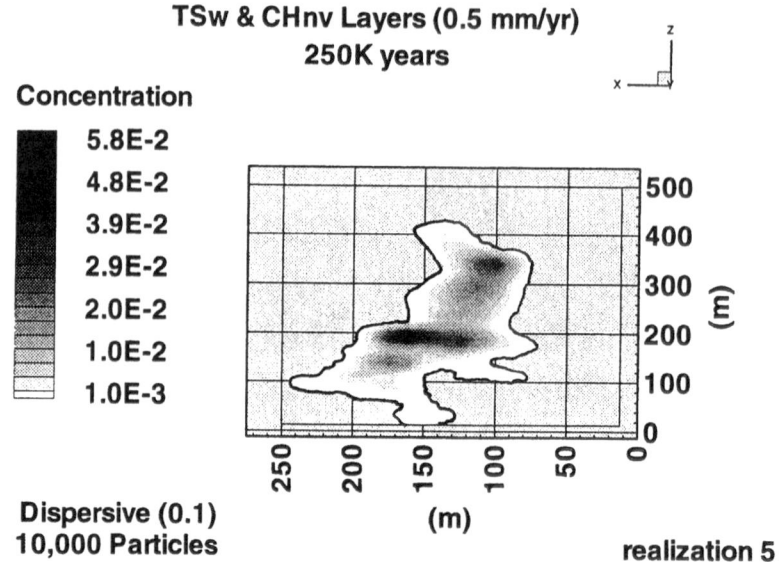

Figure 3: Single realization concentration field (Z_c=208.8m).

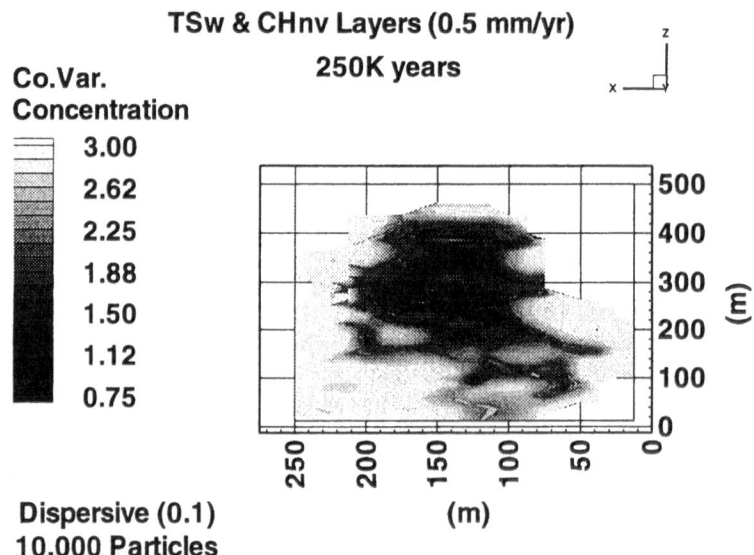

Figure 4: Concentration coefficient of variation field (Z_c=208.8m).

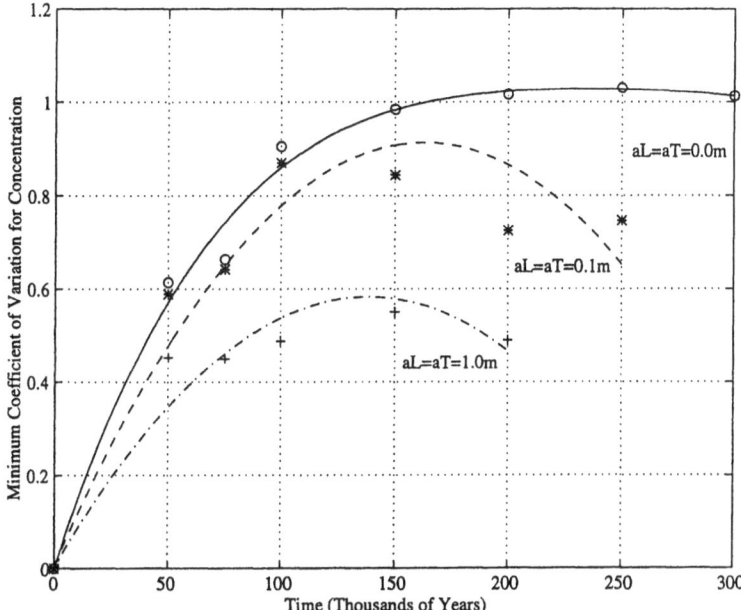

Figure 5: Temporal evolution of concentration CV for 3 levels of dispersion.

2. The variations in concentration due to the complexity of the flow fields, in unsaturated heterogeneous conditions, are large, and should be acknowledged in assessing contaminant concentrations. The concentration estimates calculated by any large-scale "mean transport" model, neglecting advective heterogeneity, should have large error bars. The magnitude of the error bars, as a multiple of the mean concentration, increases with a decrease in the local dispersion value. The concentration CV keeps increasing with distance from the center of mass (Figure 4), making the uncertainty around mean concentration predictions increasingly large multiples of the mean. The jaggedness of plumes as a consequence of porous media heterogeneity needs to be appreciated in assessing contaminant exposure levels. The acknowledgement of this variance is of key importance, especially since significant reduction, by conditioning on point measurements of hydraulic conductivity, may require an impractically large number of observations (Bagtzoglou et al., 1992).

3. The formation of capillary barriers when coarse "homogeneous" porous media is overlain by fine "homogeneous" porous media is well known, and so is the existence of the phenomenon of capillary barrier breaching by rapidly moving fingers of water (e.g., Baker and Hillel, 1990; and references therein). The work presented here demonstrated this phenomenon under the influence of high degrees of porous media heterogeneity (Figure 3).

ACKNOWLEDGEMENTS

This paper was prepared to document work performed by the Center for Nuclear Waste Regulatory Analyses (CNWRA) for the U.S. Nuclear Regulatory Commission (NRC) under Contract No. NRC-02-88-005. The activities reported here were performed on behalf of the NRC Office of Nuclear Regulatory Research, Division of Regulatory Applications (FIN B6664). The paper is an independent product of the CNWRA and does not necessarily reflect the views or regulatory position of the NRC. Technical assistance by Mike Muller and Rashid Islam, and critical reviews by Drs. Stu Stothoff and Budhi Sagar are gratefully acknowledged. The contributions of Tom Nicholson, the NRC Project Officer, are appreciated.

REFERENCES

Ababou, R., and L. W. Gelhar, 1988. A high-resolution finite difference simulator for 3D unsaturated flow in heterogeneous media. *Computational Methods in Water Resources.* Elsevier and Computational Mechanics Publications 1: 173-178.

Ababou, R., and A.C. Bagtzoglou. 1993. *BIGFLOW: A Numerical Code for Simulating Flow in Variably Saturated, Heterogeneous Geologic Media (Theory and User's Manual-Version 1.1).* NUREG/CR-6028. Washington, DC: U.S. Nuclear Regulatory Commission.

Bagtzoglou, A.C., D.E. Dougherty, and A.F.B. Tompson. 1992. Application of particle methods to reliable identification of groundwater pollution sources. *Water Resources Management* 6:15-23.

Bagtzoglou, A.C., M.R. Islam, and M. Muller. 1994. Stochastic analysis of unsaturated flow and transport with the SUFLAT executive numerical code. *Proceedings of 5th International High Level Radioactive Waste Management Conference* (in press).

Baker, R.S., and D. Hillel. 1990. Laboratory tests of a theory of fingering during infiltration into layered systems. *Soil Science Society of America Journal* 54:20-30.

Gelhar, L.W. 1993. *Stochastic Subsurface Hydrology*, Prentice-Hall, Inc., Engelwood Cliffs, NJ.

Kapoor, V. 1993. *Macrodispersion and Concentration Fluctuations in Three-Dimensionally Heterogeneous Aquifers.* Sc.D. Dissert., Massachusetts Institute of Technology, Cambridge, MA.

Kapoor, V., and L.W. Gelhar. 1994a. Transport in three-dimensionally heterogeneous aquifers:1. Dynamics of concentration fluctuations. *Water Resources Research* (in press).

Kapoor, V., and L.W. Gelhar. 1994b. Transport in three-dimensionally heterogeneous aquifers:2. Predictions and observations of concentration fluctuations. *Water Resources Research* (in press).

Loeven, C. 1993. *A Summary and Discussion of Hydrologic Data from the Calico Hills Nonwelded Hydrogeologic Unit at Yucca Mountain, Nevada.* Technical Report LA-12376-MS.

Peters, R.R, E.A. Klavetter, I.J. Hall, D.C. Blair, P.R. Heller, and G.W. Gee. 1984. *Fracture and Matrix Hydrologic Characteristics of Tuffaceous Materials from Yucca Mountain, Nye County, Nevada.* Report SAND84-1471.

Tompson, A.F.B., R. Ababou, and L.W. Gelhar. 1987a. *Application and Use of the Three-Dimensional Turning Bands Random Field Generator: Single Realization Problems.* Report 313. Ralph M. Parsons Laboratory. Cambridge. Massachusetts: Massachusetts Institute of Technology.

Tompson, A.F.B., E.G. Vomvoris, and L.W. Gelhar. 1987b. *Numerical Simulation of Solute Transport in Randomly Heterogeneous Porous Media: Motivation, Model Development, and Application.* UCID-21281. Livermore, CA: Lawrence Livermore National Laboratory.

CREEPING FLOW AND BROWNIAN PARTICLE DEPOSITION AROUND SPHEROIDAL OBJECTS

F.A. COUTELIERIS, V.N. BURGANOS and A.C. PAYATAKES
Institute of Chemical Engineering and High Temperature Chemical Processes, and
Department of Chemical Engineering, University of Patras
Patras 26500
Greece

The problem of creeping flow of suspensions in a swarm of spheroidal objects and of the concomitant deposition of suspended Brownian particles on the spheroidal surfaces is considered. The spheroid-in-cell model is used to represent the swarm. The full convective diffusion equation is used to describe Brownian particle transport and appropriate boundary conditions are employed to permit applicability over a broad range of Peclet number values. For fast flows the tangential diffusion terms are safely neglected and an initial value - two point boundary value problem obtains which is solved using cubic spline collocation. For slow flows the tangential diffusion terms are retained and a 2-D elliptic problem obtains which is solved using finite differences on a variable mesh. Numerical results show that the particle collection efficiency is severely overestimated if a Levich type of approximation is used for $Pe<10$ and that tangential diffusion is significant even for moderate Pe values.

INTRODUCTION

Suspension flow and particle deposition in unconsolidated porous media are encountered very frequently both during naturally occurring processes, such as river flow, subsurface transport, sedimentation, and in industrial fluid-solid separation processes. A swarm of solid particles is commonly used for the representation of an unconsolidated porous medium in theoretical models. Because of the complexity of flow and mass transport computation in swarms, use of "particle-in-cell" models has proved highly efficient. Happel (1958) and Kuwabara (1959) considered sphere- and cylinder-in-cell models to treat flow through packings of spheres and cylinders, respectively. In these models, the porous medium is represented by a rigid sphere or cylinder surrounded by a concentric spherical or cylindrical liquid envelope, the thickness of which is adjusted so that the solid volume fraction of the cell equals that of the medium. The sphere-in-cell model has also been widely used in mass transport problems. Tardos et al. (1976) have treated flow and adsorption in a sphere-in-cell using the analytical solution by Levich (1962) for convective diffusion around isolated spheres for low Peclet number values. Prieve and Ruckenstein (1974) and Spielman and Friedlander (1974) used the Levich approach for solving flow and adsorption problems in sphere-in-cell systems for moderate and high Pe values neglecting the complication caused by the development of thick concentration layers that cross the outer boundary of the liquid envelope. This possibility was explored by Song and Elimelech (1992) who modified the standard Levich approach and provided numerical solution to the problem of suspension flow and Brownian particle deposition in a sphere-in-cell.

However, in practice, the particles or grains that constitute the swarm are rarely

A. Peters et al. (eds.), Computational Methods in Water Resources X, 423–430.
© 1994 Kluwer Academic Publishers. Printed in the Netherlands.

spherical and even if they initially are, their shape is quickly altered by the physical or chemical processes that occur (deposit build-up, fluid-solid reaction, etc.). In fact, the experimental evidence suggests that the actual shape of the grains is closer to spheroidal than to spherical. Epstein and Masliyah (1972) computed numerically the flow field that develops among solid spheroids under creeping flow conditions whereas Dassios et al. (1994) proposed a complete "spheroidal-in-cell" model that is quite analogous to the Kuwabara (1959) model (except for the change in geometry) and formulated a series expansion solution to the creeping flow equation.

Recently, Coutelieris et al. (1993) solved analytically the problem of mass transfer in a swarm of adsorbing spheroidal particles for high Pe values using the "spheroidal-in-cell" model. Masliyah and Epstein (1972) solved numerically the convective diffusion equation for isolated spheroidal particles under low and moderate Pe conditions and produced an asymptotic solution for infinitely large Pe values.

In this work, the problem of suspension flow and Brownian particle deposition in a swarm of spheroidal particles is formulated using the spheroidal-in-cell model and numerical solutions to the full convection and diffusion formulation are provided for high, moderate, and small Pe values. A set of boundary conditions that respect flux and concentration continuity at the outer boundary of the cell is used to accompany the full convective diffusion equation. The role of the customarily ignored tangential diffusion is quantified and numerical estimates of the particle collection efficiency are provided over a broad range of Pe values and for prolate and oblate spheroidal geometries.

MATHEMATICAL FORMULATION

(I) Prolate-in-cell case

Consider a solid stationary spheroid with semiaxes $\tilde{a}_1 < \tilde{a}_3$. The semifocal distance is $\tilde{a} = \sqrt{\tilde{a}_3^2 - \tilde{a}_1^2}$ and the eccentricity $e = \tilde{a}/\tilde{a}_3$. Consider also an outer confocal prolate spheroidal surface with semiaxes $\tilde{b}_1 < \tilde{b}_3$, surrounding the solid surface. The space between the two spheroidal surfaces is filled with a Newtonian fluid of viscosity $\tilde{\mu}$ and density $\tilde{\rho}$. The relative sizes of the two surfaces are adjusted so that the solid volume fraction of this spheroid-in-cell is equal to that of the swarm, γ. It is assumed that the fluid approaches the cell in the direction parallel to the long axis and flows at a velocity with magnitude $\tilde{u}$ (Figure 1). The approaching stream is a dilute suspension of Brownian particles with concentration $\tilde{c}_\infty$, which diffuse in the space between the two spheroidal surfaces and, should they reach the solid surface, they deposit instantaneously. For convenience, we work with dimensionless variables using $\tilde{a}_1$ as the characteristic length, $\tilde{u}$ as the characteristic velocity, $\tilde{c}_\infty$ as the characteristic concentration, and dropping the tilde from the symbols.

Assuming pseudo-steady state for the mass transfer problem and that the diffusion coefficient $\tilde{D}$ is constant, the convective diffusion equation becomes

$$\underline{v} \cdot \nabla c = Pe^{-1} \nabla^2 c \qquad\qquad [1]$$

where $\underline{v}$ is the fluid velocity and Pe is the Peclet number defined by $Pe = \tilde{u}\,\tilde{a}_1 / \tilde{D}$ [2]

Using the orthogonal prolate spheroidal coordinates (η, θ), eq. [1] becomes

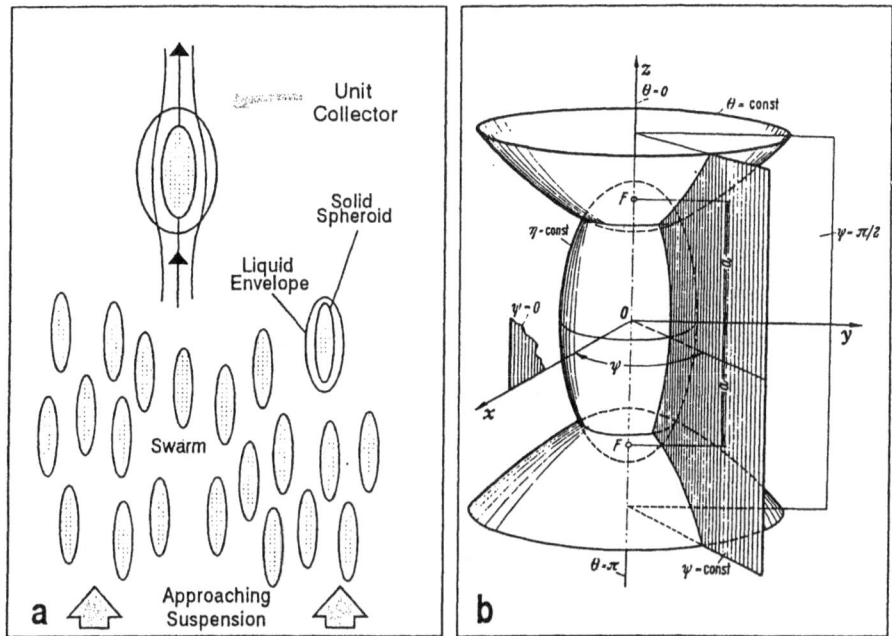

Figure 1. (a) Schematic representation of the prolate spheroid-in-cell model. (b) Prolate spheroidal coordinates.

$$v_\eta \frac{\partial c}{\partial \eta} + v_\theta \frac{\partial c}{\partial \theta} = \frac{Pe^{-1}}{a\sqrt{\sinh^2\eta + \sin^2\theta}} \left(\frac{\partial^2 c}{\partial \eta^2} + \coth\eta \frac{\partial c}{\partial \eta} + \frac{\partial^2 c}{\partial \theta^2} + \cot\theta \frac{\partial c}{\partial \theta} \right) \qquad [3]$$

where v_η and v_θ are the η- and θ-components, respectively, of the fluid velocity. These components can be obtained in a straightforward manner from the stream function solution derived by Dassios et al. (1994) for creeping flow in spheroids-in-cell.

For $Pe \geq 1000$ the θ-depedence of the concentration is negligible compared to the η-depedence. In this case, the appropriate boundary conditions are

c=0	on	$\eta = \eta_\alpha$	[4a]
c=1	on	$\eta \rightarrow \infty$	[4b]
c is finite	at	$\theta = \pi$ ($\eta \neq \eta_\alpha$)	[4c]

This problem formulation for spheroids-in-cell is analogous to the Levich (1962) formulation for isolated spheres and has been given an analytical solution by Coutelieris et al. (1993). For high Pe values, eq. [4b] is equivalent to

c=1	on	$\eta = \eta_\beta$	[4d]

since the concentration layer is much thinner than the fluid envelope.

However, for moderate and low Pe values eq. [4d] is not valid as the concentration layer may cross the outer boundary of the fluid envelope and concentration values lower than unity may obtain on $\eta=\eta_\beta$. To overcome this problem, we suggest the following boundary condition, which ensures flux continuity of the particle flux at the outer boundary:

$$c=1 \qquad \text{at} \qquad (\eta,\theta)=(\eta_\beta,\pi) \qquad\qquad\qquad [5a]$$

$$\partial c/\partial\eta = 0 \quad \text{on} \quad \eta=\eta_\beta \text{ for } 0\leq\theta<\pi \qquad\qquad [5b]$$

Note that the sole assumption made in [5] is that the thickness of the diffusion layer is smaller than that of the fluid envelope at the point of impact [5a]. It is evident that for extremely low Pe values eq. [5a] is violated. The full p.d.e. [3] should be used in this Pe value range complemented, in addition to eq. [5], by [4a] and

$$\partial c/\partial\theta = 0 \quad \text{on} \quad \theta=\pi \qquad\qquad\qquad [6]$$

$$\partial c/\partial\theta = 0 \quad \text{on} \quad \theta=0 \qquad\qquad\qquad [7]$$

The set of eqs [3], [4a], [5],[6],[7] was solved numerically using the techniques described in the section Computational Aspects (see below).

The removal efficiency of the spheroidal collector, η_0, is defined as the ratio of particle capture rate to upstream particle entrance rate.

It is straightforward to show that

$$\eta_0 = \frac{2a\sinh\eta_\alpha}{\text{Pe } b_1^2} \int_\pi^0 \left(\frac{\partial c}{\partial\eta}\right)_{\eta=\eta_\alpha} \sin\theta \; d\theta \qquad\qquad [8]$$

II) Oblate-in-cell case

The oblate-in-cell case $(\tilde{a}_1>\tilde{a}_3)$ can be treated in a similar manner using the oblate spheroidal coordinates $(\overline{\eta},\theta)$. The convective diffusion equation becomes

$$\overline{v}_\eta \frac{\partial \overline{c}}{\partial\eta} + \overline{v}_\theta \frac{\partial \overline{c}}{\partial\theta} = \frac{\text{Pe}^{-1}}{\overline{a}\sqrt{\sinh^2\overline{\eta}-\sin^2\theta}} \left(\frac{\partial^2 c}{\partial\overline{\eta}^2} + \tanh\overline{\eta} \frac{\partial\overline{c}}{\partial\overline{\eta}} + \frac{\partial^2\overline{c}}{\partial\theta^2} + \cot\theta \frac{\partial\overline{c}}{\partial\theta}\right) \qquad [9]$$

The boundary conditions are identical to those presented in the prolate case above. The expression for the removal efficiency becomes

$$\overline{\eta}_0 = \frac{2}{\text{Pe } b_1^2} \int_\pi^0 \left(\frac{\partial\overline{c}}{\partial\overline{\eta}}\right)_{\overline{\eta}=\overline{\eta}_\alpha} \sin\overline{\theta} \; d\overline{\theta} \qquad\qquad [10]$$

COMPUTATIONAL ASPECTS

The set of equations [3] (or [9]),[4a],[5],[6],[7] were discretized using finite differences on a variable mesh. The resulting set of linear algebraic equations was solved numerically using a modified elimination algorithm that proved to save ~ 25% on computer memory. For the sake of quantifying the effects of tangential diffusion, we repeated the calculations omitting the first and second θ-derivatives from eqs [3] and [9]. In this case, an initial value-two point boundary value problem obtains and eqs [3] and [9] reduce to o.d.e.'s along the line of impact $\theta=\pi$, $\eta_\alpha \leq \eta \leq \eta_\beta$ ($v_\theta=0$). Solution to these o.d.e.'s -accompanied by the Dirichlet conditions [4a] and [5a]- is straightforward and yields the concentration profile on $\theta=\pi$ which, in turn, serves as initial condition for the integration of the p.d.e.'s with respect to θ. Alternatively, we have used cubic spline collocation (de Boor, 1978) on logarithmically distributed elements of constant θ followed by θ-integration of the resulting o.d.e.'s using a stiff integration routine. In this way, we managed to get fine details on the particle concentration profile where needed.

Finally, the removal efficiency was calculated from eqs [8] and [10] through the use of a modified Newton-Cotes method with adjustable step size for the evaluation of the integral on the right-hand side of the equations. The need for adjusting the step size arose from the use of non-uniform mesh in the discretization of the p.d.e.'s (eqs [3],[9]).

RESULTS AND DISCUSSION

The comparison of the particle concentration profiles as calculated by the Levich approach to those obtained from the solution of the complete model (including tangential diffusion) for a moderate Pe value is presented in Figure 2. Note that the shape of the concentration profiles as obtained by the two methods is the same along the line of impact. However, on the equator ($\theta=\pi/2$) the Levich approach leads to a quite unrealistic profile in the vicinity of the outer boundary owing to the Dirichlet condition c=1 imposed there. It is quite unlike that the particle flux continuity remains in validity at that boundary. On the contrary, the concentration profile obtained in the present work appears quite realistic on the equator and acquires a maximal value which is considerably lower than the bulk value due to the strong diffusional limitations that develop at low Pe values.

The significance of the tangential diffusion in the overall transport problem can be quantified from Figure 3, which presents the particle concentration profiles as obtained with and without incorporation of the tangential diffusion terms in eqs [3] and [9]. In both types of spheroidal geometry, it is seen that omission of the tangential diffusion leads to overestimation of the particle concentration throughout the fluid envelope, the deviation becoming increasingly important as the exit region is approached.

Finally, estimated values of the particle removal efficiency for prolate and oblate spheroids-in-cell are plotted in Figures 4a and 4b, respectively, against the Peclet number. The removal efficiency as calculated from our model increases with decreasing Pe and remains below unity as it should. The Levich type of approach, on the other hand, overestimates the particle removal efficiency for Pe<20. For Pe<1, this approach leads to unrealistic values of the removal efficiency ($\eta_0 > 1$) for both prolate and oblate spheroids-in-cell. This is not surprising, however, since the Levich type of approach assumes convection-controlled transport and, consequently, its use should be restricted to mass transport problems in the high Pe value range. It is also notworthy that the aspect ratio effect on the removal efficiency is considerably less important in the low Pe range than it is in the convection-controlled case (Pe$\geq$100; see also Coutelieris et al., 1993).

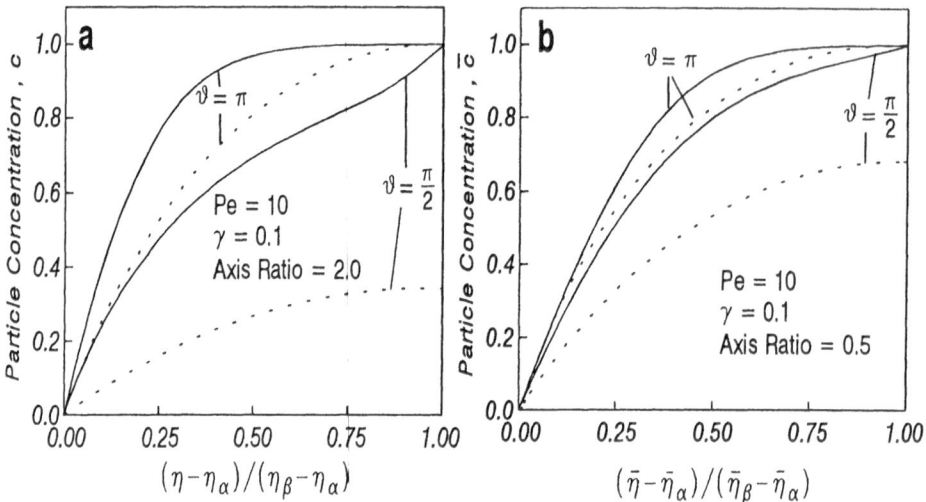

Figure 2. Concentration profiles for a moderate Pe value in (a) prolate and (b) oblate spheroids-in-cell. Comparison of the results from the complete model (---) to those from the Levich type of approach (———).

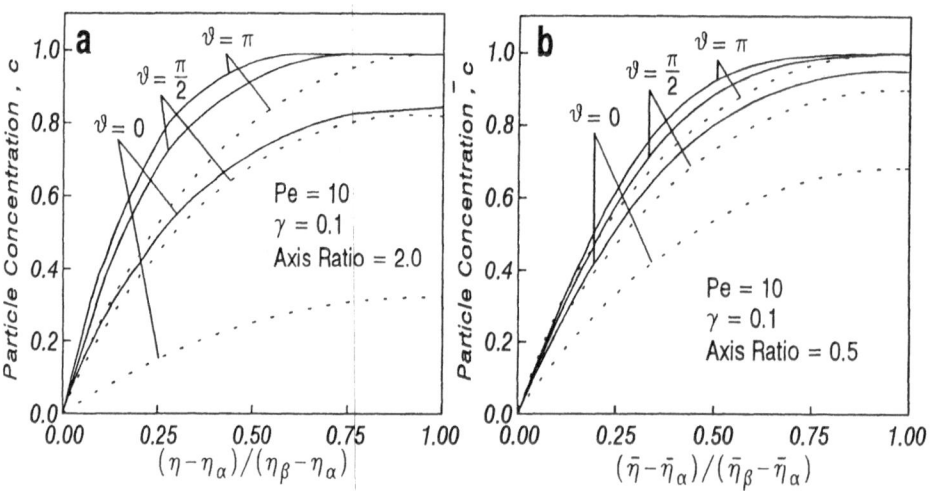

Figure 3. Effect of tangential diffusion on the concentration profile at three different angular positions for a moderate Pe value. (a) Prolate case. (b) Oblate case.
---: with,——— without tangential diffusion.

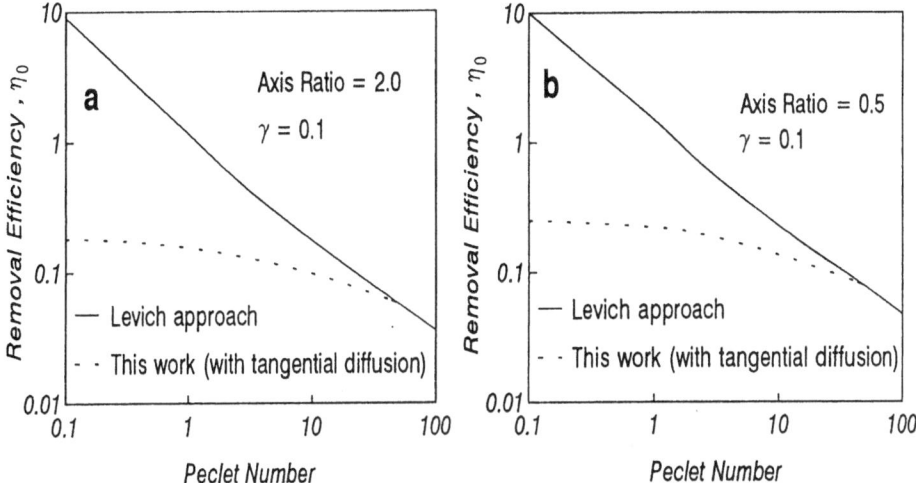

Figure 4. Dependence of the particle removal efficiency on the Pe value for (a) prolate and (b) oblate spheroids-in-cell. Comparison with the prediction of the Levich type of approach for spheroidal geometries.

CONCLUSIONS

The problem of suspension flow and Brownian particle deposition in a swarm of spheroidal grains was considered. The spheroid-in-cell model was used to represent the swarm. The suspension was assumed to behave as a Newtonian fluid under creeping conditions with approach velocity parallel to the axis of symmetry of the spheroid. The fluid velocity at any point within the fluid envelope was calculated from the stream function expression derived by Dassios et al. (1994). The complete convection diffusion equation in spheroidal coordinates was then used to describe Brownian particle transport within the envelope. Boundary conditions that describe instantaneous capture upon collision with the inner surface, and respect flux and concentration continuity at the outer boundary are employed. Numerical solution of the complete model was accomplished through the use of finite differences on a variable mesh. The role of tangential diffusion was quantified by simply eliminating the corresponding terms from the convective diffusion equation and solving the resulting initial value-two point boundary value problem with cubic spline collocation. It was found that tangential diffusion is significant for moderate and low Pe values and its omission leads to higher concentration estimates than the actual ones at any point within the liquid envelope. Comparison of our results with the predictions of a Levich type of approach for spheroidal geometries (Coutelieris et al., 1993) revealed that the simplifications involved in the Levich methodology are justifiable in the high Pe value range but cause severe overestimation of the particle removal efficiency for Pe<10 and lead to unrealistic predictions (η_0>1) for Pe<1.

NOTATION

$\tilde{a}_1$, $\tilde{a}_3$ semiaxes of the solid spheroid

$\tilde{b}_1$, $\tilde{b}_3$ semiaxes of the spheroidal envelope

$\tilde{a}$, $\tilde{\bar{a}}$	semifocal distance of the prolate, oblate spheroid
$\tilde{c}$, $\tilde{\bar{c}}$	concentration of the suspension in the prolate, oblate cell
$\tilde{c}_\infty$	concentration of the suspension for away from the solid spheroid
$\tilde{D}$	diffusion coefficient
e, $\bar{e}$	eccentricity of the prolate and oblate spheroid, respectively
Pe	Peclet number [$=\tilde{u}\tilde{a}_1/D$]
$\tilde{u}$	approach velocity
v	dimensionless fluid velocity

Greek Letters

γ	solid volume fraction
η, $\bar{\eta}$	prolate, oblate spheroidal coordinate
η_α, η_β	value of η on the inner, outer spheroidal surface of the prolate cell
$\bar{\eta}_\alpha$, $\bar{\eta}_\beta$	value of $\bar{\eta}$ on the inner, outer spheroidal surface of the oblate cell
η_0, $\bar{\eta}_0$	removal efficiency of the prolate and the oblate-in-cell, respectively
θ, $\bar{\theta}$	prolate, oblate spheroidal coordinate
$\tilde{\mu}$	fluid viscosity
$\tilde{\rho}$	fluid density

REFERENCES

de Boor, C. (1978) A Practical Guide to Splines, Springer, New York.

Coutelieris, F.A. Burganos, V.N. and Payatakes A.C. (1993) "On mass transfer from a Newtonian fluid to a swarm of adsorbing spheroidal particles for high Peclet numbers", J. Colloid Interface Sc., 153, 43-52

Dassios, G., Hadjinicolaou, M. and Payatakes A.C. (1994) "Generalized eigenfunctions and complete semiseparable solutions for stokes flow in spheroidal coordinates", Quart. Appl. Math., in press

Epstein, N. and Masliyah, J.H. (1972) "Creeping flow through clusters of spheroids and elliptical cylinders", Chem. Eng. J. 3, 169-175

Happel, J. (1958) "Viscous flow in multiparticle systems: Slow motion of fluids relative to beds of spherical particles", AIChE J. 4, 197-201

Kuwabara, S. (1959) "The forces experienced by randomly distributed parallel circular cylinders or spheres in a viscous flow at small Reynolds numbers", J. Phys. Soc., Japan 14, 527-532

Levich, V.G. (1962) Physicochemical Hydrodynamics, Englewood Cliffs, N.J.

Masliyah, J.H. and Epstein, N. (1972) "Numerical solution of heat and mass transfer from spheroids in steady axisymmetric flow", Prog. Heat Mass Transfer, 6, 613-632.

Prieve, D.C. and Ruckenstein, E. (1974) "Effect of London forces upon the rate of deposition of Brownian particles", AIChE J. 20, 1178-1187

Spielman, L.A. and Friendlander, S.K. (1974) "Role of the electrical double-layer in particle deposition by convective-diffusion", J. Colloid Interface Sci., 44, 22-28

Song, L. and Elimelech, M. (1992) "Deposition of Brownian particles in porous media: modified boundary conditions for a sphere-in-cell model" J. Colloid Interface Sci., 153, 294-297

Tardos, G.I., Cutfinger, C. and Abuaf, N. (1976) "High Peclet number mass transfer to a sphere in a fixed or fluidized bed" AIChE J., 22, 1147-1150

UPSCALING OF DISPERSION COEFFICIENTS IN TRANSPORT THROUGH HETEROGENEOUS FORMATIONS

GEDEON DAGAN
Faculty of Engineering
Dept. of Fluid Mechanics and Heat Transfer
Tel Aviv University
Ramat Aviv, 69978
Tel Aviv, Israel

An inert solute is transported in a heterogeneous formation of spatially variable transmissivity, which is modeled as a two-dimensional random space function of given statistical structure. In Monte-Carlo simulations, the flow (Darcy's law and continuity) and transport (advection - dispersion) equations are solved numerically in different realization of the transmissivity field. To reproduce accurately the spatial variability of the velocity field, a dense numerical grid is needed, of blocks size much smaller than the heterogeneity scale. It is customary to use coarser grids, leading to a loss in the velocity variability and in the spread of the solute. The study provides a method to compensate for this loss by introducing enhanced upscaled dispersion coefficients in the transport equations. The upscaled coefficients are effectively determined under a few simplifying assumptions: uniform mean flow in an unbounded formation, lognormal and stationary transmissivity of finite integral scale and of isotropic exponential or Gaussian covariance, first-order solution in the logtransmissivity variance. The upscaled dispersion coefficients (longitudinal and transverse) are derived as functions of the ratio between the numerical block size and the logtransmissivity integral scale on one hand, and the travel time from the source, on the other.

INTRODUCTION

Numerous studies carried out in the last decade indicate that the spreading of plumes transported through heterogeneous formations is due to the spatial variability of formation properties and mainly to that of hydraulic conductivity. Thus, due to the contrast between the large velocities in zones of high permeability and the low ones in less conductive areas, plumes disperse at a much higher rate than in laboratory experiments, which capture the pore-scale dispersion only. To model the advective effect of heterogeneity, numerical models are used in order to solve the equations of flow and subsequently those of transport. To obtain an accurate solution for the velocity variations, such models need to use a fine discretization of the domain with numerical blocks of size smaller than say 1/10 of the heterogeneity scale. Such a requirement may pose a heavy computational burden upon numerical schemes and it is common to discretize the domain in larger elements. As a result, sub-grid variations of the velocity are lost and the simulated plumes are "smoother" than those derived with a fine grid. To compensate for this undesirable effect, one has to "upscale" the transport parameters. Only recently the problem of upscaling of permeability has been investigated in the hydrological literature (e.g. Rubin and Gomez-Hernandez, 1990, Durlofsky, 1991, Indelman and Dagan, 1993, Indelman, 1993). We do not know about any

431

A. Peters et al. (eds.), Computational Methods in Water Resources X, 431–439.
© 1994 Kluwer Academic Publishers. Printed in the Netherlands.

article on upscaling of transport parameters and this is the objective of the present study. In Sects.2 and 3, we review a few results of interest to the present developments. In Sect. 4 we suggest the general methodology, while Sect. 5 presents illustrative results.

BACKGROUND

We consider here flow and transport through formations of spatially variable transmissivity T, i.e. T is a function of $x(x_1, x_2)$. Following a common procedure, the seemingly erratic variation of T and the lack of data are accounted for by modeling T as a random space function (RSF). Though we restrict the discussion to 2D flows, the methodology presented here can be extended to three-dimensional flows and to spatial variability of other properties (for details see e.g. Dagan, 1989).

We cast the transport problem in a Lagrangian framework by regarding the plume as made up from solute particles of random trajectories $x = X_t(t,a) = X + X_d$, where X satisfies the kinematical equations

$$\frac{dX}{dt} = V(X_t) \quad ; \quad X(0,a) \equiv a \tag{1}$$

while dX_d is a "brownian motion" type of elementary displacement such that $\langle dX_{di}(t) \, dX_{dj}(t') \rangle = D_{dij} \, \delta(t-t') \, dt \, dt'$. Then the concentration field associated with a plume of initial concentration C_0 is given by

$$C(x, t) = \int_{A_0} C_0(a) \, \delta(x - X_t) \, da \tag{2}$$

The Lagrangian approach is the one we followed in the past (e.g. Dagan, 1984) and it is adopted here as well. In a numerical context the solution of (1) by discrete time steps is known as "particle tracking". It is easy to express the spatial moments of the plume by using (7) as follows

$$M = \int_{A_0} n \, C_0(a) \, da \quad ; \quad R = \frac{1}{M} \int_{A_0} n \, X_t \, C_0(a) \, da$$

$$\tag{3}$$

$$S_{ij} = \frac{1}{M} \int_{A_0} n \, (X_{ti} - R_i)(X_{tj} - R_j) \, C_0(a) \, da, \, ...$$

where M is the mass, R is the centroid coordinate and S_{ij} are second spatial moemnts with respect to the centroid.

Then, the transport problem becomes: for given V and X_d, derive the statistical moments of the random variables (8) by solving Eqs. (5). In the "particle tracking" procedure, the computation of the spatial moments (8) is based on evaluating the statical moments of the mass distributed among particles at each time step and in each realization of V.

A spatial moment is assumed to be *ergodic* if its value in each realization is equal to its ensemble mean, or more precisely if the variance tends to zero. Thus, for a conservative solute M (8) is ergodic, while ergodicity is attained for R and S_{ij} if the transverse scale of the initial plume is large compared to the heterogeneity scale (Dagan 1990, 1991). Conversely, ergodicity can be reached after such a long transport time that the diffusive mechanism represented by $\mathbf{D}_d$ causes the plume to spread over a sufficiently large volume. However, since for natural formation this mechanism is slow compared to that associated with advection, its effect may be felt only at very large distances from the input zone, beyond the range of interest in many applications.

For ergodic moments, for a stationary velocity field and for C_0=const Eqs. (3) yield $\langle \mathbf{X} \rangle = \mathbf{a} + \mathbf{U}t$, $\langle \mathbf{R} \rangle = \mathbf{R}(0) + \mathbf{U}t$ and $\langle S_{ij}(t) \rangle = S_{ij}(0) + X_{ij}(t) + X_{dij}$ (Dagan, 1989). The important quantity $X_{ij} = \langle X_i'(t,\mathbf{a})\ X_j'(t,\mathbf{a}) \rangle$, the one particle trajectory covariance, is related to the velocity covariance through (5). It has been evaluated explicitly under a few simplifying conditions in an analytical form (Dagan, 1984) or numerically (Rubin 1990, Bellin et al, 1992, Chin and Wang, 1992). One can define effective dispersion coefficients by the equation

$$D_{ij} = \frac{1}{2}\frac{d\langle S_{ij} \rangle}{dt} = \frac{1}{2}\frac{dX_{ij}}{dt} + \mathbf{D}_{dij} \tag{4}$$

If $\mathbf{X}$ are normal it can be shown that $\langle C \rangle$ satisfies a transport equation similar to (2)

$$\frac{\partial \langle C \rangle}{\partial t} + \mathbf{U}.\nabla \langle C \rangle = \sum_i \sum_j D_{ij}\frac{\partial^2 \langle C \rangle}{\partial x_i \partial x_j} \tag{5}$$

It is emphasized that generally D_{ij} is nonlocal since it depends on the travel time or, equivalently, on the travel distance from the input zone. If it tends to constant values after a sufficiently long time, transport is coined as "Fickian". D_{ij} at this limit has been evaluated under different conditions by Dagan (1984, 1988), Gelhar and Axness (1983) and Neuman et al (1987).

UPSCALING OF DISPERSION COEFFICIENTS

We assume that the equations of flow are solved numerically, by using a grid with numerical blocks ω of length scale ℓ. The resulting velocity field $\tilde{\mathbf{V}}$ is used subsequently in order to solve the transport problem in order to obtain the concentration $\tilde{C}$ and/or the spatial moments $\tilde{M}$, $\tilde{R}$ and $\tilde{S}_{ij}$. This can be achieved either within the Eulerian framework by solving (2) for $\tilde{C}$, with $\tilde{\mathbf{V}}$ replacing $\mathbf{V}$ and an upscaled dispersion coefficients tensor $\tilde{\mathbf{D}}_d$ replacing $\mathbf{D}_d$. Alternatively, in the Lagrangian context one has to solve (5) with $\tilde{\mathbf{V}}$ replacing $\mathbf{V}$ to derive the trajectories $\tilde{\mathbf{X}}_t = \tilde{\mathbf{X}} + \tilde{\mathbf{X}}_d$, where the latter represents an upscaled local dispersion mechanism.

The velocity field $\mathbf{V}$, which we shall denote as the actual or the pointwise one, has statistical moments different from those of the upscaled one $\tilde{\mathbf{V}}$, since subgrid variability is smoothed out in the latter and the fluctuations $\tilde{\mathbf{u}}$ are fully or strongly correlated over elements ω. To simplify the discussion, we focus it on representing the solute plume by its spatial moments (8), derived for instance by "particle tracking". Two extreme cases of numerical solution of transport are simple to grasp. First, for a fine grid $\tilde{\mathbf{V}} \simeq \mathbf{V}$ and $\tilde{\mathbf{D}}_d = \mathbf{D}_d$ and the numerical solution reproduces the actual moments, numerical errors related to truncation and to the particular scheme, notwithstanding (we disregard such errors here). At the other limit, of an ergodic plume and $\ell \gg I$, transport is modeled as of advection with $\mathbf{U}$ and effective dispersion coefficients D_{ij} (9), and again the spatial moments are recovered. The salient question, to be answered here, is how to model transport in an intermediate case? The same basic problem is faced in upscaling of flow, when one is using the actual logconductivity statistics for a fine grid or the effective conductivity for large numerical elements.

The basic requirement we propose for upscaling is that the statistics of the actual spatial moments and of the upscaled ones should be equal, i.e.

$$\tilde{M} = M, \langle R \rangle = \langle \tilde{R} \rangle \ , \ \tilde{R}_{ij} = R_{ij} \ , \ \langle \tilde{S}_{ij} \rangle = \langle S_{ij} \rangle \ , \ ... \tag{6}$$

There are two major difficulties with satisfying (6) : (i) we do not know the actual moments and the purpose of the entire development is to avoid their derivation and (ii) since V and $\tilde{V}$ are given as solutions of the flow problem in the actual and the upscaled domain, the only freedom left in order to satisfy (6) is in selecting an appropriate upscaled dispersion coefficients tensor $\tilde{D}_{ij}$. Then, the question is whether such a $\tilde{D}_{ij}$ exists.

The first difficulty is circumvented like in the case of effective conductivity : we derive the actual moments under simple flow conditions (uniform in the average, unbounded domain) and assume that the results are applicable to complex flows if the latter have slowly varying means.

The second question is answered by splitting the *actual* solute plume into subplumes which originate from numerical blocks ω within A_0 (Fig. 1a). Each such subplume is characterized by its spatial moments M^ω, R^ω and S_{ij}^ω. It is easy to ascertain from their definition that

$$M = \sum_\omega M^\omega \ , \ R = \frac{1}{M} \sum_\omega R^\omega M^\omega$$

$$\tag{7}$$

$$S_{ij} = \frac{1}{M} [\sum_\omega \int_\omega n(X_i - R_i^\omega)(X_j - R_j^\omega) \, C_0 \, da + \sum_\omega n \, (R_i^\omega - R_i)(R_j^\omega - R_j) \, M^\omega] + X_{dij}$$

By the same token we split the plume in the upscaled medium into the same subplumes (Fig. 1b) originating from ω but advected by $\tilde{V}$ rather than V. For a conservative solute we have exactly $M^\omega = \tilde{M}^\omega$ and approximately, at the numerical error, $R^\omega = \hat{R}^\omega$ (Fig. 3). Furthermore, by the nature of the numerical approximation $\tilde{V}$ is practically constant within ω and the solute particles originating from ω do not disperse in absence of $\tilde{X}_d$. Hence, the first three relationships in (17) are obeyed whereas (18) becomes in the upscaled medium

$$\tilde{S}_{ij} = \frac{1}{M} \sum_\omega (\tilde{R}_i^\omega - R_i)(\tilde{R}_j^\omega - R_j) \, M^\omega] + \tilde{X}_{dij} \tag{8}$$

It is seen that in order to obey the last relationship in (7) and with S_{ij} (7, 8) a sufficient condition is

$$< (X_i - R_i^\omega)(X_j - R_j^\omega) > = \tilde{X}_{dij} \quad \text{i.e.} \quad D_{ij}(t,\omega) = \tilde{D}_{ij} \tag{9}$$

which is the main result of the present analysis. The term on the left-hand side of (9) represents the second spatial moment of particles originating from an element ω around the centroid of the subplume in the actual medium (Fig. 1a) and half of its rate of change is $D_{ij}(t,\omega)$, the effective dispersion tensor for such an element. The term on the right-hand side of (9) is a fictitious, upscaled, dispersion term to be introduced in the solution of the transport in order to satisfy (7). Hence, Eq. (9) solves the upscaling problem.

DERIVATION OF UPSCALED DISPERSION TENSOR FOR UNIFORM AVERAGE FLOW AND FIRST-ORDER APPROXIMATION IN LOGTRANSMISSIVITY VARIANCE

The dispersion tensor $D_{ij}(t,\omega)$ in (9) is precisely the nonergodic one analyzed in Sect. 3. If ω

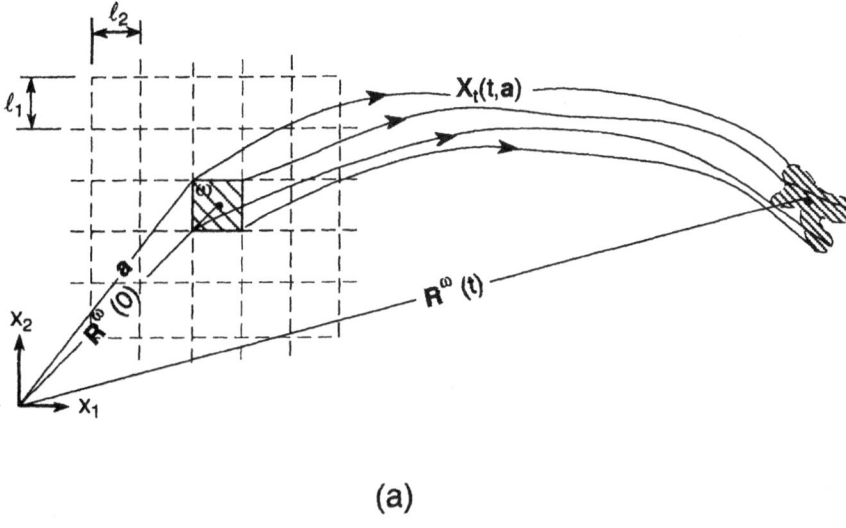

(a)

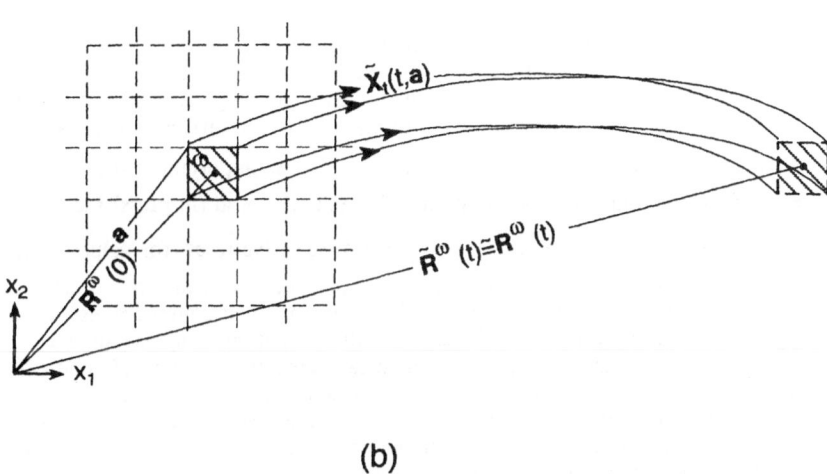

(b)

Fig. 1 Schematic representation of the advective transport of a solute subplume, initially in a numerical block ω : (a) by the actual velocity field $\mathbf{V}$ and (b) by the numerical (upscaled) velocity field $\hat{\mathbf{V}}$.

is selected as a rectangle $\ell_1 \times \ell_2$, the expressions of D_{ij} are given in Eq. 15 of Dagan, 1991 under the assumptions: average uniform flow $U(U, 0)$, stationary velocity covariance u_{ij}, unbounded domain and first-order approximation in σ_Y^2. Calculating $D_{ij}(t, \ell_1, \ell_2)$ involves three integrations and to simplify matters we take advantage of the result (Dagan, 1991) that the streamwise dimension ℓ_1 has little effect upon D_{ij}, and carry out the calculations for $\ell_1 = 0$. Under these circumstances $D_{ij}(t, \ell_2)$ is given by $\tilde{D}_{ij}(t, \ell_2) = D_{ij}(t) - r_{ij}(t, \ell_2)$ with

$$D_{ij} = r_{ij}(t, 0) = \int_0^t u_{ij}(Ut', 0) \, dt' \quad \text{and} \quad r_{ij} = \frac{2}{\ell_2^2} \int_0^{\ell_2} \int_0^t (\ell_2 - b_2) \, u_{ij}(Ut', b_2) \, dt' \, db_2 \quad (10)$$

D_{ij} (10) were calculated for the two logtransmissivity covariances, exponential and Gaussian. We present and discuss the final results in the next section.

UPSCALED DISPERSION COEFFICIENTS FOR EXPONENTIAL AND GAUSSIAN LOGTRANSMISSIVITY AUTOCORRELATIONS (16)

The dimensionless longitudinal upscaled dispersion coefficient $D_{11}/\sigma_Y^2 UI$ is represented as function of the dimensionless distance from the input line Ut/I in Fig. 2 for different values of the ratio ℓ_2/I (it is reminded that I is the linear integral scale of Y and ℓ_2 is the transverse dimension of ω). Fig. 2 shows how D_{11} tends to its asymptotic limit as Ut/I increases and this tendency is more rapid for small ℓ_2/I ($Ut/I \simeq 1$ for $\ell_2/I = 1$) than for large ones ($Ut/I \simeq 20$ for $\ell_2/I \to \infty$). It is emphasized that the asymptotic limits for large t were given in a close analytical form by Dagan (1991, Eqs. 28, 29). The ergodic limit $D_{11}(t) = D_{11}(t, \infty)$ was derived for the exponential ρ_Y in the past (Dagan, 1984).

The transverse dimensionless dispersion coefficients D_{22} are represented in Fig. 3. Unlike D_{11}, D_{22} tends to zero for large Ut/I for reasons discussed in previous studies, namely because u_{22} (14) is a "hole" covariance. Hence, in this case, the neglected local dispersion term $\mathbf{D}_d$ and its interaction with advection starts to play a role for sufficiently large t.

A few issues deserve further elucidation in the future: (i) how important is the impact of nonlinear terms in σ_Y^2? Preliminary results obtained by Salandin et al (in press) indicate that these effects are quite small and the ratio D_{11}/D_{11} is not sensitive to σ_Y^2 for values as large as 1.5; (ii) a subject of interest is the extension of the present results to three-dimensional flows. The process is straightforward, but it implies more involved computations; (iii) an additional topic is the impact of conditional simulations, i.e. making use of the logtransmissivity-head covariances conditioned on measurements of Y and H, upon the upscaled dispersion coefficients. Since conditional covariances are non-stationary, this topic is related also to the possible trends of $\langle H \rangle$ or m_Y; (iv) in the case of transport of reactive solutes, one has also to derive upscaling rules for the coefficients characterizing the reactive properties and their interaction with upscaled dispersion coefficients.

BIBLIOGRAPHY

Bellin, A. Salandin, P. and Rinaldo, A. (1992) "Simulation of dispersion in heterogeneous porous formations: statistics, first-order theories, convergence of computations", Water Resour. Res., 28, 2211-2228.

Chin, D.A. and Wang, T. (1992) "An investigation of the validity of first-order stochastic dispersion theories in isotropic porous media", Water Resour. Res., 28, 1531-1542.

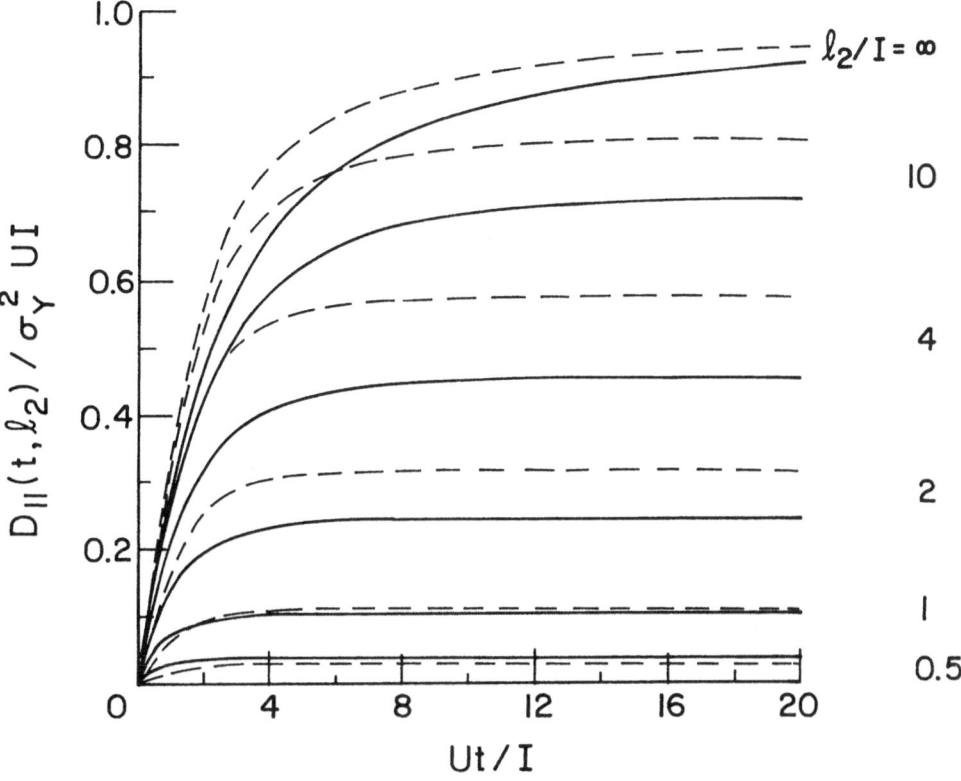

Fig. 2 The dependence of the dimensionless effective longitudinal dispersion coefficient $D_{11}(t,\ell_2)$ (Eq. 10) on the dimensionless travel time from the source for a few values of the ratio between the transverse dimension of the element ℓ_2 and the integral scale of the logtransmissivity I. Full line (exponential covariance), dashed line (Gaussian covariance)

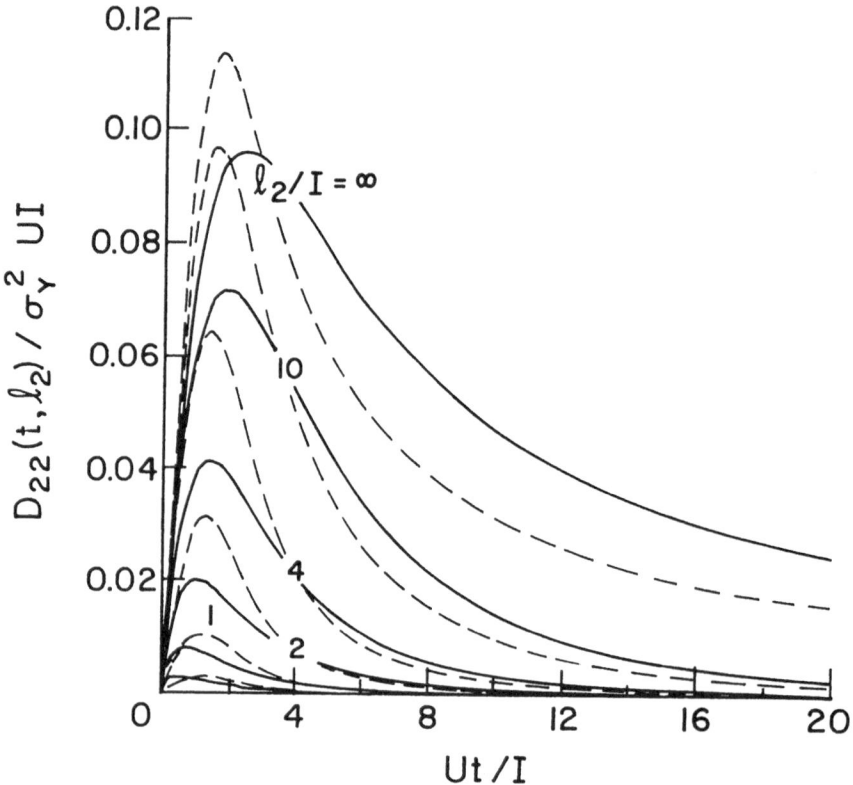

Fig. 3 The dependence of the dimensionless effective transverse dispersion coefficient $D_{22}(t,\ell_2)$ (Eq. 10) on the dimensionless travel time from the source for a few values of the ratio between the transverse dimension of the element ℓ_2 and the integral scale of the logtransmissivity I. Full line: exponential covariance. Dashed line: Gaussian covariance (Eqs. 16)

Dagan, G. (1984) "Solute transport in heterogeneous porous formations", J. Fluid Mech., 145, 151-177.

Dagan, G. and Rubin, Y. (1988) "Stochastic identification of recharge, transmissivity and storativity in aquifer unsteady flow: a quasi-steady approach", Water Resour. Res., 24, 1698-1710.

Dagan, G. (1989) Flow and Transport in Porous Formations, Springer-Verlag, 465 p..

Dagan, G. (1990) "Transport in heterogeneous formations: spatial moments, ergodicity and effective dispersion", Water Resour. Res., 26(6), 1281-1290.

Dagan, G. (1991) "Dispersion of a passive solute in non-ergodic transport by steady velocity fields in heterogeneous formations", Journ. Fluid. Mech., 233, 197-210.

Dagan, G. Cvetkovic V. and Shapiro, A. (1992) "A solute-flux approach to transport in heterogeneous formations 1. The general framework", Water. Resour. Res.,28, 1369-1376.

Durlofsky, L. J. (1991) "Numerical calculation of equivalent grid block permeability tensors for heterogeneous porous media", Water Resour. Res.,27, 699-708.

Gelhar, L.J. and Axness, C.L. (1983) "Three-dimensional stochastic analysis of macrodispersion in aquifers", Water Resour. Res., 19, 161-180.

Graham, W. and McLaughlin, D. (1989a) "Stochastic analysis of nonstationary subsurface solute transport 1. Unconditional moments", Water Resour. Res., 25(2), 215-232.

Hoeksema, R.J. and Kitanidis, P.K. (1985) "Analysis of spatial structure of properties of selected aquifers", Water Resour. Res., 21, 563-572.

Indelman, P. and Dagan, G. (1993) "Upscaling of permeability of anisotropic heterogeneous formations. Part 1: The general framework", Water Resour. Res., 29, 917-923.

Indelman, P. and Dagan, G. (1993) "Upscaling of permeability of anisotropic heterogeneous formations. Part 2: General structure and small perturbation analysis", Water Resour. Res., 29, 923-933.

Indelman,P. (1993) "Upscaling of permeability of anisotropic heterogeneous formations. Part 3: Applications", Water Resour. Res., 29, 935-943.

Kitanidis, P.K. (1988) "Prediction by the method of moments of transport in a heterogeneous formation", Journ. Hydrology, 102, 453-473.

Neuman, S.P. Winter, C.L. and Newman, C.M. (1987) "Stochastic theory of field-scale Fickian dispersion in anisotropic porous media", Water Resour. Res., 23, 453-466.

Rubin, Y. and Gomez-Hernandez, J. J. (1990) "A stochastic approach to the problem of upscaling of conductivity in disordered media: Theory and unconditional numerical simulations", Water Resour. Res., 26, 691--701.

Rubin, Y. (1990) "Stochastic modeling of macrodispersion in heterogeneous porous media", Water Resour. Res., 26, 133-142.

Rubin, Y. and Dagan, G. (1992) "A note on head and velocity covariances in three-dimensional flow through heterogeneous anisotropic porous media", Water Resour. Res., 28, 1463-1470.

Salandin, P. and Fiorotto, V. (1993) "Numerical simulations of non-ergodic transport in natural formations, Proc. XXV IAHR Congress, Tokyo (submitted)

A MICROSCALE MODEL OF MICROBIAL TRANSPORT IN POROUS MEDIA

ROBERT DILLON[§], LISA FAUCI[§] and DONALD GAVER III[‡]

[§] Department of Mathematics
Tulane University
New Orleans, LA 70118

[‡] Department of Biomedical Engineering
Tulane University
New Orleans, LA 70118

Abstract

In this paper, we introduce a mathematical and computational model at the microscale level of the chemotaxis and transport of motile bacteria. We use the *immersed boundary method*, originally introduced by C. Peskin, to couple microbial motion and the convection-diffusion of contaminant with the full incompressible Navier-Stokes equations. Simulations using a preliminary two-dimensional model are presented.

BACKGROUND

The development of effective strategies for *in situ* bioremediation depends upon understanding the detailed pore-level behavior of contaminants and microorganisms within porous media. This is due to the fact that bioavailability of microorganisms to the toxin site depends upon the local geometry and physicochemical conditions (e.g. pH, temperature, concentrations of dissolved gasses). First, these conditions are primary determinants of bioavailability because they influence the propensity of microbes to form aggregates and adhere to the local pore structure. Increased flocculation hinders microbial migration by lessening forced convection and diffusive transport of the colloidal mixture through small pores. The local physicochemical conditions influence bioavailabilty because microbes swim preferentially by chemotaxis, the directional motion induced by variations of chemical concentrations. Thus, if concentration gradients are appropriate, microbes may more readily swim towards contaminated regions and aid in the elimination of toxic waste. Once the microbes are

A. Peters et al. (eds.), Computational Methods in Water Resources X, 441–448.
© 1994 *Kluwer Academic Publishers. Printed in the Netherlands.*

at the contamination site, restoration will be governed by metabolic kinetics, which in turn are functions of the local physicochemical state. All of the aforementioned processes occur in a moving viscous fluid, and therefore the fluid dynamical events must be included in any realistic model. As is evident by the above description, factors controlling the local environments of microbial communities in the subsurface interstices are critical for any *in situ* remediation technology. Unfortunately, knowledge of this small-scale system has yet to be fully investigated, and is extremely complex due to the many components that govern the physicochemical and flow conditions.

Efficient penetration of microbes into a porous media requires motility and chemotaxis [18]. Motile bacteria have been identified that degrade a variety of toxins. For example, *Pseudomonas putida*, found widely distributed in soil and freshwater environments, exhibits positive chemotaxis toward aromatic acids and chlorinated benzoates. Moreover, these compounds are also metabolized by the organisms [11, 12]. Strains of motile *Pseudomonas* have been identified that are capable of using benzene, chlorobenzene, or toluene as the sole source of carbon and energy [1].

Current analysis of bacteria transport through porous media relies primarily on continuum modelling resulting in a macroscopic mass balance equation for the microbes (for example, see [2, 5]). The resulting macroscopic mass balance equations contain several parameters which are difficult to evaluate on theoretical grounds [10]. Bacterial chemotaxis has been studied experimentally and theoretically (for example see [3, 9, 13, 14, 17]). A common feature of the theoretical models is that explicit dependence upon fluid dynamics is ignored. Recent studies of bacterial movement in microchannels suggest that surface interaction and hydrodynamic forces must be included in models at the micropore scale [4, 10].

Our studies will focus on micropores with diameters ranging from 1-100 cell-diameters. It is appropriate to model this system in a manner such that the fluid is regarded as a continuum while the bacteria are defined as discrete objects of finite size. In this paper we present a mathematical model and numerical method coupling fluid dynamics, contaminant transport and microbial motility. We hope the results from this model will provide detailed information concerning transport that is not available from purely continuum representations.

MATHEMATICAL MODEL

The dynamic evolution of a single contaminant that is initially deposited within a pore filled with a viscous fluid depends upon the fluid motion induced by motile bacteria, background flow, diffusion and microbial uptake. Moreover, microbes move in direct response to the surrounding contaminant field (chemotaxis). This nonlinear coupled system of equations is described below. The governing equations describing the fluid dynamics are:

$$\rho(\mathbf{u}_t + (\mathbf{u} \cdot \nabla)\mathbf{u}) = -\nabla p + \mu \nabla^2 \mathbf{u} + \mathbf{F}_{\text{external}}, \tag{1}$$

and

$$\nabla \cdot \mathbf{u} = 0, \tag{2}$$

which are the Navier-Stokes and continuity equations for the incompressible fluid. These equations represent the balance of momentum and conservation of mass, and hold within the fluid domain. Here, ρ is the fluid density, $\mathbf{u}$ is the fluid velocity vector, p is pressure, μ is the fluid viscosity and $\mathbf{F}_{\text{external}}$ are forces on the fluid due to suspended neutrally-buoyant microorganisms and the pore walls (described below). These forces are localized in a thin layer surrounding the immersed microorganisms and elastic pore walls.

The equation describing the convection and diffusion of the contaminant species within the fluid-filled pore is

$$c_t + (\mathbf{u} \cdot \nabla)c = D\nabla^2 c - R(c)c, \tag{3}$$

where c is the concentration, D is the molecular diffusivity and $R(c)$ is a concentration-dependent consumption rate that is nonzero only near the site of a microbe.

The presence of microorganisms influences both the flow dynamics and the contaminant field. In turn, the microorganisms respond to the fluid and contaminant fields. For this reason, we incorporate discrete representations of microorganisms that are mechanically coupled to the fluid-contaminant system described above. These organisms have finite volume and exert stress on the fluid and thus alter convection and contaminant transport. Moreover, the swimming orientation of these organisms is influenced by chemical concentration gradients. Contaminant concentrations are modified locally due to consumption by these organisms.

The microorganisms influence the fluid through $\mathbf{F}_{\text{external}}$ in Equation 1. This function represents the force created by the N microorganisms and the walls on the fluid and has the following components:

$$\mathbf{F}_{\text{external}} = \sum_{i=1}^{N} \left[\mathbf{F}_{\text{microbe}(i)} + \mathbf{F}_{\text{swim}(i)} \right] + \mathbf{F}_{\text{walls}}. \tag{4}$$

A single microorganism in this two-dimensional model is modelled as a neutrally buoyant elastic ring, whose configuration is defined by $\mathbf{X}_i(s, t)$, where s is a Lagrangian label, t is time and i denotes the ith microbe. The boundary force per unit length $\mathbf{f}_i(s, t)$ at each point on the ring consists of an elastic spring force which maintains the integrity of the membrane and a bending-resistant force which resists deformation. This elastic force is transmitted directly to the fluid through

$$\mathbf{F}_{\text{microbe}(i)}(\mathbf{x}, t) = \int \mathbf{f}_i(s, t)\delta(\mathbf{x} - \mathbf{X}_i(s, t))ds. \tag{5}$$

Here, the integration is over the ring structure and δ is the two-dimensional Dirac delta function. This force gives the microbe its material integrity. Note that each organism contributes such forces to the flow field and therefore their interactions, mediated through the fluid, are included in this model. The walls are modelled in the same manner as the microbe rings, that is as neutrally buoyant elastic filaments immersed within the fluid domain. However, these walls are not free to move throughout the fluid, since they are elastically tethered to fixed points in space. $\mathbf{F}_{\text{walls}}$ is thus expressed in a manner analogous to Equation 5. We choose this representation of the walls so that the geometry of the pore can be easily changed.

The swimming of bacteria is achieved through the action of one or more flagella. In this preliminary model of bioremediation the flagella are not explicitly represented as immersed elastic structures. However, idealized forces of the microbes locomotory movements are applied to the fluid. Since inertial forces are negligible for bacterial swimming (Reynolds numbers are small), the swimming forces induced by the microbe's on the fluid sum to zero. These forces do, however, result in a swimming velocity relative to the fluid [15], and are incorporated into our model in the following manner:

$$\mathbf{F}_{\text{swim}(i)}(\mathbf{x}, t) = \int \mathbf{f}_{\text{swim}(i)}(s, t)\delta(\mathbf{x} - \mathbf{X}_i(s, t))ds. \qquad (6)$$

where the direction of the swimming force is given by $\nabla c(\mathbf{X}_i(s, t), t)$ rotated by a random angle, θ_i. Here $\nabla c(\mathbf{X}_i(s, t), t)$ denotes the principal direction of chemotaxis as determined by conditions near the microbe, and θ_i represents a random variable that reflects a bacteria's inability to exactly align its swimming direction towards the concentration gradient. Note that in order to conserve momentum, the sum of $\mathbf{f}_{\text{swim}(i)}$ around each organism equals zero.

Finally, the system is closed by requiring microbes to move at the local fluid velocity using

$$\frac{d\mathbf{X}_i(s, t)}{dt} = \mathbf{u}(\mathbf{X}_i(s, t), t). \qquad (7)$$

The salient feature of this representation is that suspended organisms are replaced by suitable contributions to a force density term in the fluid dynamics equations. A single set of fluid equations holds in the entire domain and there are no internal boundary conditions. Consequently, the fluid dynamics equations may be solved efficiently using finite-difference methods on a uniform computational grid. We are able to model the interaction of more than one organism in the same domain of fluid. It is not assumed that the motion is steady-state, and therefore, transient effects can be modelled.

NUMERICAL METHOD

The numerical method that we use couples microbial motion with fluid dynamics, and is known as the *immersed boundary method*. This method was introduced by Peskin [16] to model blood flow in the heart. Subsequently, this approach has been used to simulate the swimming of microorganisms [7] and to model platelet aggregation in the blood's clotting response [8]. The full incompressible Navier-Stokes equations are solved in a domain of fluid within which neutrally buoyant elastic objects (i.e.

microorganisms) undergoing time-dependent movements are immersed. Fluid quantities are represented on a grid (Eulerian description), and the swimming organism is modelled by a discrete collection of moving points (Lagrangian description) connected by elastic links. The external force of an organism on the fluid is represented as a delta-function layer of force supported only by the region of fluid which coincides with material points of the organism as described in Equations 4-6; away from these points the external force is zero. The strength of this delta-function force is determined at each instant by the local configuration of the organism and the local contaminant field. For a detailed description of these forces and numerical implementation see [6]. The solution of the velocity field from this stage of the computation is then used to update the contaminant field.

The algorithm for the numerical solution of Equations 1-7 may be summarized as follows: at the beginning of each time step n, we have the fluid velocity field $\mathbf{u}^n$, the configuration of the elastic boundaries, and the chemical concentration field c^n. In order to update these values to those occurring at the next time step we: (1) calculate the force density $\mathbf{f}_i^n$ from the boundary configuration defined by the elastic boundaries; (2) calculate the swimming forces $\mathbf{f}_{\text{swim}(i)}^n$ imposed on the fluid by each organism; (3) spread the force densities to the grid to determine the $\mathbf{F}_{\text{external}}$ on the fluid; (4) solve the Navier-Stokes equations (Equations 1-2) for $\mathbf{u}^{n+1}$; (5) solve the convection-reaction equation for c^{n+1} (Equation 3) and (6) interpolate the fluid velocity field at each immersed boundary point and convect that point at the local fluid velocity (Equation 7).

Figure 1 shows four 'snapshots' of a simulation intended to demonstrate the behavior of the coupled fluid/contaminant/microbe system. In this computational experiment, 18 microorganisms were initially placed randomly within a pore with a small bolus of contaminant. The contaminant diffuses and convects due to fluid motion induced by the microbes, which swim preferentially towards the region of higher contaminant concentration. In this simulation $\theta_i = 0$ so there is no random component to the swimming force vectors. The microbes simultaneously ingest the contaminant, which in turn modifies the concentration field. In these figures the vectors represent the fluid velocity field. Contour levels of contaminant concentration are also indicated with the darkest regions corresponding to the highest level. Note the net migration of cells is towards the contaminant site. However, at any instant any individual cell may be moving away from the contaminant due to the convection induced by other cells. Also, the consumption and the microbe swimming clearly affect the contaminant distribution.

DISCUSSION

Our preliminary simulations, as shown above, suggest the potential importance of fluid dynamical effects on microbial transport at the microscale. In our two-dimensional model we have the capability of examining the effects of microbe size, swimming

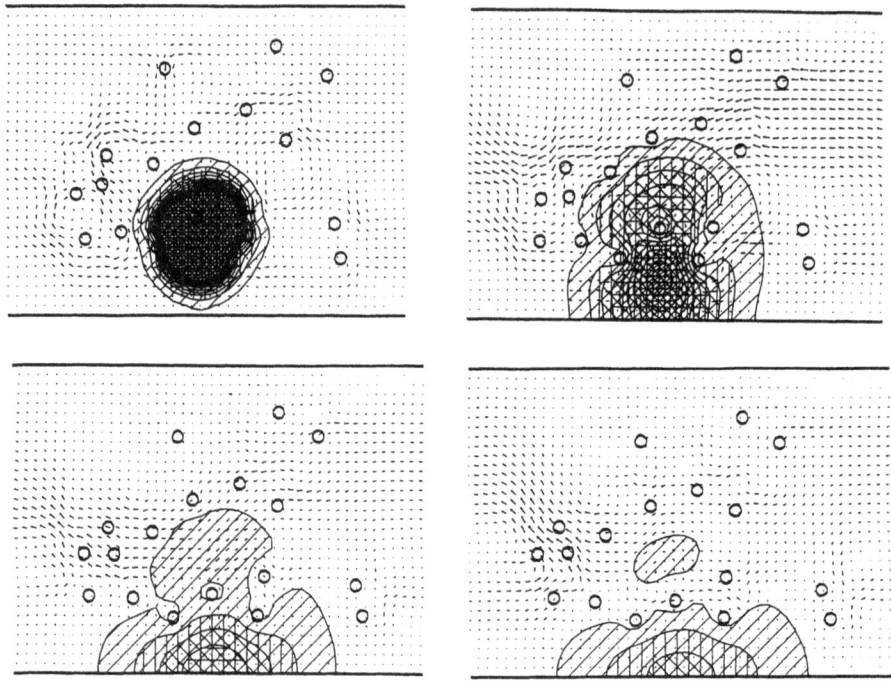

Figure 1: Solution to the coupled system Equations 1-7 with eighteen microbes at several time steps.

force, concentration of microbes, pore size, contaminant distribution, diffusion rate and uptake kinetics. We are currently extending this model to include cell-cell and cell-pore attachment. Agglomeration will be achieved by exerting appropriate binding stresses, or spring forces, between the organisms that may hold them together or, if fluid stresses are large, may yield and release the organisms. This binding stress will be a function of the ionic strength, since agglomeration and binding forces relate to this parameter. These forces will be applied to the fluid in a manner analogous to the way the elastic forces are transmitted to the fluid by each microbe through $\mathbf{F}_{\mathbf{microbe}(i)}$. This method has been used to model cell-to-cell adhesion within a fluid-filled channel [6]. The ability of this method to model complex fluid-structure interactions is shown in Figure 2 which shows a snapshot of the channel with flow streamlines, cell positions and binding structures. The numerical models described above can be generalized to three-dimensions.

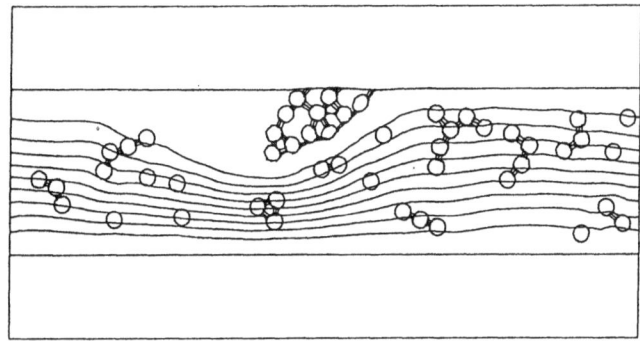

Figure 2: Snapshot of cell aggregate along a channel wall. Cells, binding links and flow stream-lines are shown. Adapted from Fauci and Fogelson [6].

ACKNOWLEDGEMENTS

The authors are grateful to Aaron Fogelson, Charlie Peskin, Richard Ewing, Kyriakos Papadopoulos and Zewen Liu for helpful discussions. The work of R. Dillon, L. Fauci and D. Gaver was supported in part by DOE grant FG-01-93EW53023. The work of L. Fauci was supported in part by NSF grant DMS9208340 and an Alfred P. Sloan Foundation Fellowship. The work of D. Gaver was supported in part by NSF grant BCS-9358207.

References

[1] M. Alexander, R. J. Wagenet, P. C. Baveye, J. T. Gannon, U. Mingelgrin, and Y. Tan. *Movement of bacteria through soil and aquifer sand.* U. S. Environmental Protection Agency, 1991.

[2] P. Baveye and A. Valocchi. An evaluation of mathematical models of the transport of biologically reacting solutes in saturated soils and aquifers. *Water Resour. Res.*, 25:1413–1421, 1989.

[3] H. C. Berg. Chemotaxis in bacteria. *Ann. Rev. Biophys. Bioeng.*, 4:119–136, 1975.

[4] H. C. Berg and L. Turner. Chemotaxis of bacteria in glass capillary arrays. *Biophys. J.*, 58:919–930, 1990.

[5] M. Y. Corapcioglu and A. Haridas. Microbial transport in soils and groundwater: a numerical model. *Adv. Water Resources*, 8:188–200, 1985.

[6] L. J. Fauci and A. L. Fogelson. Truncated Newton methods and the modeling of complex immersed elastic structures. *Comm. on Pure and Appl. Math.*, 46:787–818, 1993.

[7] L. J. Fauci and C. S. Peskin. A computational model of aquatic animal locomotion. *J. Comp. Phys.*, 77:85–108, 1988.

[8] A. L. Fogelson. A mathematical model and numerical method for studying platelet adhesion and aggregation during blood clotting. *J. Comp. Phys.*, 56:111–134, 1984.

[9] R. M. Ford and D. A. Lauffenburger. Measurement of bacterial random motility and chemotaxis coefficients: II. *Application of single-cell-based mathematical model. Biotechnology and Bioengineering*, 37:661–672, 1991.

[10] G. Harkes, J. Dankert, and J. Feijen. Bacterial migration along solid surfaces. *Applied and Environmental Microbiology*, 58:1500–1505, 1992.

[11] C. S. Harwood, R. E. Parales, and M. Dispensa. Chemotaxis of *pseudomonas putida* toward chlorinated benzoates. *Appl. Environ. Microbiol.*, 56:1501–1503, 1990.

[12] C. S. Harwood, M. Rivelli, and L. N. Ornston. Aromatic acids are chemoattractants for *pseudomonas putida. J. Bacteriol.*, 160:622–628, 1984.

[13] E. F. Keller and L. A. Segel. Traveling bands of chemotactic bacteria: a theoretical analysis. *J. Theor. Biol.*, 30:235–248, 1971.

[14] H. G. Othmer, S. R. Dunbar, and W. Alt. Models of dispersal in biological systems. *J. Math. Biol.*, 26:263–298, 1988.

[15] T. J. Pedley and J. O. Kessler. Hydrodynamic phenomena in suspensions of swimming microorganisms. *Annu. Rev. Fluid Mech.*, 24:313–58, 1992.

[16] C. S. Peskin. Numerical analysis of blood flow in the heart. *J. Comp. Phys.*, 25:220–252, 1977.

[17] M. A. Rivero, R. T. Tranquillo, H. M. Buettner, and D. A. Lauffenburger. Transport models for chemotactic cell populations based on individual cell behavior. *Chemical Engineering Science*, 44:2881–2897, 1989.

[18] S. Soby and K. Bergman. Motility and chemotaxis of *rhizobium meliloti* in soil. *Appl. Environ. Microbiol.*, 46:995–998, 1983.

SOLUTION METHODS FOR MULTISCALE POROUS MEDIA FLOW

R.E. EWING and M. ESPEDAL
Institute for Scientific Computation
Texas A&M University
College Station, Texas 77801–3404

M. CELIA
Department of Civil Engineering
Princeton University
Princeton, New Jersey 08544–0001

The ability to numerically simulate multiphase flow of fluids in porous media is extremely important in developing an understanding of the complex phenomena governing the flow. The flow is complicated by the presence of heterogeneities in the reservoir and by phenomena such as diffusion, dispersion, and viscous fingering. These effects must be modeled by terms in coupled systems of nonlinear partial differential equations which form the basis of the simulator. The simulators should be able to handle both single and multiphase flows and the transition regimes between the two. A discussion of some of the aspects of modeling dispersion is presented along with directions for future work.

The partial differential equation models are convection-dominated and contain important local effects. An operator-splitting technique is used to address these different effects accurately. Convection is treated by time stepping along the characteristics of the associated pure convection problem, and diffusion is modeled via a Galerkin method for single phase flow and a Petrov-Galerkin technique for multiphase regimes. Eulerian-Lagrangian techniques are discussed to effectively treat the advection-dominated processes. Accurate approximations of the fluid velocities needed in the Eulerian-Lagrangian time-stepping procedure are obtained by mixed finite element methods. Adaptive local grid refinement techniques are then indicated to resolve important local phenomena around wells and large heterogeneities or to resolve the moving internal boundary layers which often govern the mass transfer between phases.

MODEL EQUATIONS

Let Ω in $\mathbf{R}^i$, $i = 1, 2, 3$, represent a porous medium. The global pressure p and total velocity $\mathbf{u}$ formulation of a two-phase water (w) and air (a) flow model in Ω is given

A. Peters et al. (eds.), Computational Methods in Water Resources X, 449–456.
© 1994 Kluwer Academic Publishers. Printed in the Netherlands.

by the following equations:

$$S_a c_a \frac{dp}{dt} + \nabla \cdot \mathbf{v} = -\frac{\partial \phi(p)}{\partial t} + q(x, S_w), \qquad x \in \Omega, \quad t > 0, \tag{1}$$

$$\mathbf{v} = -\mathbf{K}\lambda(\nabla p - \mathbf{G}_\lambda), \qquad x \in \Omega, \quad t > 0, \tag{2}$$

$$\phi \frac{\partial S_w}{\partial t} + \nabla \cdot (f_w \mathbf{v} - \mathbf{K}\lambda_a q_w \delta \rho \mathbf{g} - \mathbf{D}(S_w) \cdot \nabla S_w) = -S_w \frac{\partial \phi(p)}{\partial t} + q_w, \quad x \in \Omega, \quad t > 0. \tag{3}$$

The global pressure and total velocity are defined by

$$p = \frac{1}{2}(p_w + p_a) + \frac{1}{2} \int_{S_c}^{S} \frac{\lambda_a - \lambda_w}{\lambda} \frac{dp_c}{d\xi} d\xi \qquad \text{and} \qquad \mathbf{v} = \mathbf{v}_w + \mathbf{v}_a. \tag{4}$$

Further, $\frac{d}{dt} \equiv \phi \frac{\partial}{\partial t} + \frac{\mathbf{v}_a}{S_a} \cdot \nabla$, S_w and S_a are the saturation for the water and air phases, $\phi(p)$ is the porosity, $\lambda = \lambda_w + \lambda_a$ is the total mobility, $\mathbf{K}$ is the absolute permeability tensor, and $\lambda_i = \frac{k_{ri}}{\mu_i}$, $i = w, a$, is the mobility for water and air, where k_{ri} is the relative permeability. The capillary pressure p_c is given by $p_c = p_a - p_w$.

The gravity forces G_λ and capillary diffusion term $D(S)$ are expressed as

$$\mathbf{G}_\lambda = \frac{\lambda_w \rho_w + \lambda_a \rho_a}{\lambda} \mathbf{g} \qquad \text{and} \qquad \mathbf{D}(S) = -\mathbf{K}\lambda_a f_w \frac{dp_c}{dS} \tag{5}$$

and the compressibility c_a is defined by

$$c_a = \frac{1}{\rho_a} \frac{d\rho_a}{dp_a}. \tag{6}$$

The phase velocities for water and air, which are needed in transport calculations, are given by:

$$\begin{aligned}
\mathbf{v}_w &= f_w \mathbf{v} + \mathbf{K}\lambda_a f_w \nabla p_c - \mathbf{K}\lambda_a f_w \delta \rho \mathbf{g}, \\
\mathbf{v}_a &= f_a \mathbf{v} - \mathbf{K}\lambda_w f_a \nabla p_c + \mathbf{K}\lambda_w f_a \delta \rho \mathbf{g},
\end{aligned} \tag{7}$$

where $f_\alpha = \lambda_\alpha / \lambda$, $\alpha = w, a$, and $\delta \rho = \rho_a - \rho_w$. Within the groundwater literature, the pressure normally is scaled by the gravity potential function. Equation (1) would then be given in terms of the pressure head. We should also note that if the Richards approximation, $p_a = 0$, is valid, Equation (3) can be replaced by: $p_c(S_w) = -p_w$. We may note that the phase velocity for air is given by Equation (7) even if the Richards approximation is used.

BOUNDARY CONDITIONS

If Γ is the boundary of Ω, general boundary conditions for Equations (1)–(3) can be given by a combination of the following expressions:

$$p = p_\Gamma(x, t), \qquad x \in \Gamma_1, \quad t > 0, \tag{8}$$

$$\mathbf{v} \cdot \boldsymbol{\nu} + b(x, t, S_w)p = G(x, t, S_w), \qquad x \in \Gamma_2, \quad t > 0, \tag{9}$$

$$\int_{\Gamma_3} \mathbf{v} \cdot \boldsymbol{\nu} = g(t), \quad \text{and} \quad p = p_\Gamma(x, t) + d(t), \qquad x \in \Gamma_3, \quad t > 0, \tag{10}$$

$$S_w = S_\Gamma(x, t), \qquad x \in \Gamma_4, \tag{11}$$

$$(f_w \mathbf{v} + \mathbf{K}\lambda_a f_w(\nabla p_c - \delta\rho\mathbf{g})) \cdot \boldsymbol{\nu} + b_w(x, t, S_w)p = G_w(x, t, S_w), \quad x \in \Gamma_5, \quad t > 0, \tag{12}$$

where Γ_i, $i = 1, 2 \ldots 5$ are given partitions of Γ.

Normally the boundary conditions will be nonlinear functions of the physical boundary conditions for the original two-pressure formulation [5]. This means that we have to iterate on the boundary conditions as a part of the solution process. Our experience is that this does not cause problems.

MIXED METHODS FOR ACCURATE VELOCITY APPROXIMATIONS

The system in Equations (1)–(3) is solved sequentially. An approximation for $\mathbf{v}$ is first obtained at time level $t = t^n$ from a solution of Equation (2) with the mobility λ evaluated from the value of S_w at time level t^{n-1}. Equations (1) and (2) can be solved via a mixed finite element methods for the fluid velocity.

There are two major sources of error in the methods currently being utilized for finite difference discretizations of Equations (1)–(3). The first occurs in the approximation of the fluid pressure and velocity. The second comes from the techniques for upstream weighting to stabilize Equation (3). We first describe mixed finite element methods for the accurate approximation of the total velocity $\mathbf{v}$. We then discuss some alternate upstream-weighting techniques in a finite element context in the next section for use in Equation (3).

Since the transport term in Equation (3) is governed by the fluid velocity, accurate simulation requires an accurate approximation of the velocity $\mathbf{v}$. Because the lithology in the reservoir can change abruptly, causing rapid changes in the flow capabilities of the rock, the coefficient $\mathbf{K}$ in Equations (2) and (3) can be discontinuous. In this case, in order for the flow to remain relatively smooth, the pressure changes extremely rapidly. Thus, standard procedures of solving Equations (1) and (2) are to eliminate the velocity and solve the remaining second-order equation as an elliptic partial differential equation for pressure; the differentiation of $\mathbf{K}\lambda$ can produce very poor approximations to the velocity $\mathbf{v}$. In this section, a mixed finite element method for approximating $\mathbf{v}$ and p simultaneously, via the coupled system of first order partial differential equations (1) and (2), will be discussed.

We define certain function spaces and notation. Let $W = L^2(\Omega)$ be the set of all functions on Ω whose square is finitely integrable. Let $H(\text{div}; \Omega)$ be the set of vector functions $\mathbf{v} \in [L^2(\Omega)]^2$ such that $\nabla \cdot \mathbf{v} \in L^2(\Omega)$ and let

$$V = H(\text{div}; \Omega) \cap \{\mathbf{v} \cdot \mathbf{n} = 0 \text{ on } \Gamma\} \,.$$

Let $(v, w) = \int_\Omega vw\, dx$, $\langle v, w \rangle = \int_\Gamma wv\, ds$, and $||v||^2 = (v, v)$ be the standard L^2 inner products and norm on Ω and Γ. We obtain the weak solution form of Equations (1) and (2) by dividing each side of Equation (2) by $\mathbf{K}\lambda$, multiplying by a test function $\mathbf{u} \in V$, and integrating the result to obtain

$$\left((\mathbf{K}\lambda)^{-1}\mathbf{v}, u\right) = (p, \nabla u) + (G\lambda, u), \qquad u \in V. \tag{13}$$

The right-hand side of Equation (13) was obtained by further integration by parts and use of the boundary condition.

Next, multiplying Equation (1) by $w \in W$ and integrating the result, we complete our weak formulation, obtaining

$$(\nabla \cdot \mathbf{v}, w) = (q, w) - \left(SaCa\frac{dp}{dt}, w\right) - \left(\frac{\partial \phi(p)}{\partial t}, w\right). \tag{14}$$

For a sequence of mesh parameters $h > 0$, we choose finite dimensional subspaces V_h and W_h with $V_h \subset V$ and $W_h \subset W$ and seek a solution pair $(\mathbf{U}_h; P_h) \in V_h \times W_h$ satisfying

$$(\mathbf{K}\lambda)^{-1}\mathbf{V}_h, \mathbf{u}_h) - (P_h, \operatorname{div} \mathbf{u}_h) = (G\lambda, \mathbf{u}_h), \qquad \mathbf{u}_h \in V_h, \tag{15}$$

$$(\operatorname{div} \mathbf{V}_h, w_h) = (q, w_h) - \left(SaCa\frac{dp}{dt} + \frac{\partial \phi(p)}{\partial t}, w_h\right), \qquad w_h \in W_h. \tag{16}$$

Equations (15) and (16) lead to a saddle-point problem requiring care in solution. Preconditioning or efficient iterative methods are essential. Effective block preconditioners are presented in [12], and efficient multigrid techniques are being developed.

OPERATOR-SPLITTING TECHNIQUES

In finite difference simulators, the convection is stabilized via upstream-weighting techniques. In a finite element setting, we use a possible combination of a modified method of characteristics and Petrov-Galerkin techniques to treat the transport separately in an operator-splitting mode.

In miscible or multicomponent flow models, the convective part is a linear function of the velocity. An operator-splitting technique has been developed to solve the purely hyperbolic part by time stepping along the associated characteristics [9,12,13,21].

In immiscible or multiphase flow, the convective part is nonlinear. A similar operator-splitting technique to solve this equation needs reduced time steps because the pure hyperbolic part may develop shocks. Recently, an operator-splitting technique has been developed for immiscible flows [7,10] which retains the long time steps in the characteristic solution without introducing serious discretization errors.

The splitting of the convective part of Equation(3) into two parts: $\mathbf{f}^m(S) + \mathbf{b}(S)S$, is constructed [10] such that $\mathbf{f}^m(S)$ is linear in the shock region, $0 \leq S \leq S_1 \leq 1$, and $\mathbf{b}(S) \equiv 0$ for $S_1 \leq S \leq 1$.

The operator splitting is defined by the following set of equations:

$$\phi \frac{\partial \bar{S}_w}{\partial t} + \frac{d}{dS} \mathbf{f}^m(\bar{S}_w) \cdot \nabla \bar{S}_w \equiv \phi \frac{d}{d\tau} \bar{S}_w = 0, \tag{17}$$

$$\phi \frac{\partial S_w}{\partial \tau} + \nabla \cdot (\mathbf{b}^m(S_w) S_w) - \nabla \cdot (D(S_w) \nabla S_w) = -S_w \frac{\partial \phi(p)}{\partial t} + q_w \tag{18}$$

$t_m \leq t \leq t_{m+1}$, together with proper initial and boundary conditions. As noted earlier, the saturation S_w is coupled to the pressure/velocity equations, which will be solved by mixed finite element methods [8,11,12,13].

For a fully developed shock, the characteristic solution of Equation (17) will always produce a unique solution and, as in the miscible case, we may use long time steps Δt without loss of accuracy.

Equation (18) is solved by Petrov-Galerkin variational methods, where the time derivative and the nonlinear constants are approximated by the solution from Equation (17) [10].

An iterative solution procedure based on domain decmoposition methods [4] is used in the solution of the variational form of Equation (17).

It seems natural to relate the size of the coarse domains to the solution of the pressure-velocity equation [10], since the velocity varies slowly and defines a natural long space scale compared to the variation of the saturation S at a front. A local error estimate which determines if a coarse-grid block must be refined, is given in [10]. Normally, local refinement must be performed if a fluid interface is located within the coarse-grid block in order to resolve the solution there. A slightly different strategy is to make the region of local refinement big enough such that we can use the same refinements for several of the large time steps allowed by the method. The local grid-refinement strategy combined with the operator splitting is defined in the literature [7,10]. The solution at each of the coarse-grid vertices and the local refinement calculation may be sent to separate processors to achieve a high level of parallelism in the solution process.

The difficult problem with these techniques is the communication of the solution between the fine and coarse grids. The domain decomposition technique described in [4] gives accurate and efficient treatment of the communication problem.

UPSCALING

The solution procedure described above represents an excellent tool for handling multiscale phenomena. Often, data such as permeabilty, porosity, and capillary forces will have a multiscale dependence. Our multilevel solution procedure fits into this very well. However, the local refinement capabilities also mean that we must be able to give the appropriate model equations for different computational scales. Given a local computational grid, subgrid information has to be incorporated properly into the data representation.

Large scale groundwater or oil reservoirs may have a very complex structure, and the geological description is normally a subject of great uncertainty. A two-stage geostatistical model is often proposed [17]:

- Large-scale heterogeneities associated with facies are modeled from the information achieved from seismic data, well data, and analogous outcrops.
- Rock properties of the facies are modeled by a continuous multivariate Gaussian field or other statistical models. Seismic and well data can be used by a conditioning technique, and core data and other available data should be used to determine the statistical properties of the random field (mean, variance, correlation, etc.).

The coarsest computational domains should coincide with the facies of the model. The level of refinement of these domains and the grid within a given domain have to be decided from the knowledge of geometry, permeability variation, pressure gradients, etc.

Pressure-Velocity Equation

In both groundwater and petroleum modeling, a substantial amount of research has been done on the upscaling of the permeability field to give a grid-block permeability [6,20], which could be used within our computational framework. The homogenization type of upscaling [1,2], which leads to a symmetric block-tensor for the permeability, seems to be especially well suited. The additive Schwarz type of domain decomposition methods leads to zero-Dirichlet boundary conditions for the local computations, consistent with the periodic boundary condition needed for the homogenization technique. One should note that within our computational framework, we need only the assumption of a periodic media locally on a given domain. We want to extend this single-scale homogenization technique to a multiscale model. Based on a wavelet representation of permeability, we have started research within this area, and so far the results look promising.

Saturation Equation

The upscaling of the saturation equation gives a new macrodispersion term in Equation (3), originating from the subgrid permeability variation. For a single-phase model this has been successfully studied [3,6,14,19,20].
Within two-phase models little work has been done. Using a multi-fractal hypothesis, scaling laws for macrodispersion terms have recently been presented in the literature [15,16]. Also, macrodispersion models have recently been derived for a model where the permeability has a lognormal distribution [18]. The derivation is based on the solution technique given above. It gives a saturation dependent block-tensor dispersion coefficient. From the numerical experiments that are performed,

we can conclude that the weakly correlated saturation fluctuations, on average, can be adequately described by this dispersion term.

Upscaling, leading to block-tensor dispersion terms, falls naturally into our computational setup, and we will continue our work based on this kind of modeling.

REFERENCES

1. Amaziane, B. and Bourgeat, A. (1988) "Effective behavior of two-phase flow in heterogeneous reservoir", Numerical Simulation in Oil Recovery (M.F. Wheeler, ed.), IMA Volumes in Mathematics and Its Application, 11, Springer Verlag, 1–22.

2. Amaziane, B., Bourgeat, A., and Koebbe, J. (to appear) "Numerical simulation and homogenization of two-phase flow in heterogeneous porous media", Transport in Porous Media.

3. Binning, P.J. (1994) Modeling Unsaturated Zone Flow and Contaminant Transport in the Air and Water Phases, Ph.D. Dissertation, Department of Civil Engineering and Operations Research, Princeton University.

4. Bramble, J.H., Ewing, R.E., Pasciak, J.E., and Schatz, A.H. (1988), "A preconditioning technique for the efficient solution of problems with local grid refinement", Computer Methods in Applied Mechanics and Engineering, 67, 149–159.

5. Chen, Z., Ewing, R.E., and Espedal, M. (to appear) "Multiphase flow simulation with various boundary conditions", Proceedings X International Conference on Computational Methods in Water Resources, Heidelberg, Germany, July 19–22, 1994.

6. Dagan, G. (1989) Flow and Transport in Porous Formations, Springer-Verlag, Berlin-Heidelberg.

7. Dahle, H.K., Espedal, M.S., Ewing, R.E., and Sævareid, O. (1990) "Characteristic adaptive sub-domain methods for reservoir flow problems", Numerical Methods for Partial Differential Equations, 6, 279–309.

8. Douglas, Jr., J., Ewing, R.E., and Wheeler, M.F. (1983) "A time-discretization procedure for a mixed finite element approximation of miscible displacement in porous media", R.A.I.R.O. Analyse Numerique, 17, 249–265.

9. Douglas, Jr., J. and Russell, T.F. (1982) "Numerical methods for convection dominated diffusion problems based on combining the method of characteristics with finite element or finite difference procedures", SIAM J. Numer. Anal., 19, 871–885.

10. Espedal, M.S. and Ewing, R.E. (1987) "Characteristic Petrov-Galerkin subdomain methods for two-phase immiscible flow", Comp. Meth. Appl. Mech. and Eng., 64, 113–135.

11. Ewing, R.E. and Heinemann, R.F. (1984) "Mixed finite element approximation of phase velocities in compositional reservoir simulation", Computer Meth. Appl. Mech. Eng., 47, 161–176.

12. Ewing, R.E., Koebbe, J.V., Gonzalez, R., and Wheeler, M.F. (1985) "Mixed finite element methods for accurate fluid velocities", Finite Elements in Fluids, 4, John Wiley, 233–249.

13. Ewing, R.E., Russell, T.F., and Wheeler, M.F. (1984) "Convergence analysis of an approximation of miscible displacement in porous media by mixed finite elements and a modified method of characteristics", Computer Meth. Appl. Mech. Eng., 47, 73–92.

14. Ewing, R.E., Russell, T.F., and Young, L.C. (1989) "An anistropic coarse-grid dispersion model of heterogeneity and viscous fingering in five-spot miscible displacement that matches experiments and fine-grid simulations", SPE 18441, Proceedings Tenth SPE Symposium on Reservoir Simulation, Houston, Texas, February 6–8, 1989, 447–446; and SPE Res. Eng., (to appear).

15. Glimm, J. and Lindquist, W.B. (1992) "Scaling laws for macrodispersion", Proceedings of the Ninth Int. Conf. on Comp. Meth. in Water Resources, Vol. 2, 35–49.

16. Glimm, J., Lindquist, W.B., Pereira, F., and Zhang, Q. (to appear) "A theory of macrodispersion for the scale up problem", Transport in Porous Media.

17. Haldorsen, H. and Damseth, E. (1990) "Stochastic modeling", J. Pet. Tech., 42, 404–413.

18. Langlo, P. and Espedal, M.S. (submitted) "Macrodispersion for two-phase, immiscible flow in porous media", Advances in Water Resources.

19. Lasseter, T.J., Waggoner, J.R., and Lake, L.W. (1986) "Reservoir hetero-geneities and their influence on ultimate recovery", Reservoir Characterization (L.W. Lake and H.B. Carroll, eds.), Academic Press.

20. Rubin, Y. (1990) "Stochastic modeling of macrodispersion in heterogeneous porous media", Water Resources Res., 26, 133–141.

21. Russell, T.F. (1985) "The time-stepping along characteristics with incomplete iteration for Galerkin approximation of miscible displacement in porous media", SIAM J. Numer. Anal., 22, 970–1013.

A PORE-SCALE ALGORITHM FOR SIMULATION OF DISSOLUTION IN POROUS MEDIA

L. A. FERRAND
Center for Water Resources and Environmental Engineering Resesarch, Dept. of
Civil Engineering, City University of New York (Y-120), New York, NY 10031, USA
M.A. CELIA, H. RAJARAM[1] and P.C. REEVES
Water Resources Program, Dept. of Civil Engineering and Operations Research
Princeton University, Princeton, NJ 08544, USA
[1]currently at: Dept. of Civil, Environmental and Architectural Engineering, University
of Colorado, Boulder, CO 80309, USA

ABSTRACT
    Remediation of soils and aquifers contaminated by slightly soluble non-aqueous phase liquids (NAPLs) often relies on dissolution of residual NAPLs into a flowing aqueous phase. A pore-scale computational model for emplacement of a residual NAPL, and its subsequent dissolution and transport through a porous medium is described. Phenomena simulated include the existence of a high concentration of dissolved NAPL immediately adjacent to NAPL-water interfaces, slow diffusion of dissolved NAPL through stagnant pores and advective/diffusive transport in flowing pores. Model results can be analyzed to explore correlations between pore geometry and size distributions, soil hydraulic properties, NAPL mass transfer to the water phase and subsequent transport away from the source.

1. INTRODUCTION

    An understanding of the phenomena of mass transfer from a non-aqueous phase liquid (NAPL) to another fluid phase (the aqueous phase or a gas phase) in a porous medium is important in assessing and remediating subsurface contamination from organic contaminants. NAPLs released into the unsaturated zone are transported primarily downward, although significant horizontal spreading may be observed in heterogeneous soils. The fate of a body of NAPL once it encounters the water table is determined by the density of the NAPL and by capillary forces. A NAPL lighter than water spreads horizontally over that water table and moves down the regional hydraulic gradient, leaving behind a region of residual NAPL in the unsaturated zone and capillary fringe. A dense NAPL may sink to the bottom of an aquifer leaving a region of residual separate phase contaminant which continues through the saturated zone. Slightly soluble components of the NAPL are transferred to water flowing through a zone of residual by slow dissolution or to the soil gas phase by volatilization.

A. Peters et al. (eds.), Computational Methods in Water Resources X, 457–463.
© 1994 Kluwer Academic Publishers. Printed in the Netherlands.

Attempts to quantify these phenomena have related mass transfer to the volume of NAPL relative to total pore volume (NAPL saturation), the flow velocity and the kinetics of the dissolution process (e.g., Miller et al., 1990; Powers et al., 1991; Imhoff, 1992). In general, the highly discontinuous and disordered distribution of residual NAPL has been ignored. The kinetics of mass transfer is often modeled using a local equilibrium assumption (LEA) which implies that the concentration of a contaminant in one phase is related to that in a neighboring phase by an equilibrium partition coefficient. While it is widely accepted that the LEA is valid at the pore scale in the immediate vicinity of an interphase interface, it is not necessarily true at the porous medium scale. For instance, a region of high NAPL saturation is characterized by a low relative permeability to water and hence would be bypassed by flowing water. The mass transfer rate in this case would be controlled by diffusion through relatively stagnant water in this region to the surrounding flow.

In the research reported here, the porous medium dissolution and transport processes are modeled in a network of pores. This lattice model was originally developed to study the displacement of one fluid by another in heterogeneous media (Ferrand and Celia, 1992), therefore, it simulates the emplacement of residual NAPL in a physically meaningful way. Following this step, the pore-scale flow field for the water phase is simulated for a given hydraulic gradient across the medium. Thus, the NAPL distribution and its influence on hydrodynamic conditions are consistently specified.

The transport of dissolved NAPL is simulated using a particle-tracking scheme. That is, contaminant mass is represented by a finite number of particles, each of which has a constant mass and time-varying position. These particles are generated and moved through the lattice according to a set of computational rules which represent various physical phenomena. Network-based particle-tracking simulations in which all particles originate at a single inflow boundary have been reported by Sahimi and Imdakm (1988) and Sorbie and Clifford (1991). In the model described below, particles are introduced into the flow field at locations adjacent to NAPL-occupied pores. The rate at which they are introduced is determined by the estimated rate of diffusion through intervening stagnant pores.

2. THE LATTICE MODEL

In the computational model, pore space is represented by a cubic lattice of spherical pore bodies or sites connected by pore throats or bonds. Figure 1 (right) shows a single pore body with its six associated half-throats. Figure 2 (below) shows a pair of adjacent sites with radii $R_{s(1)}$ and $R_{s(2)}$ connected by a single bond.

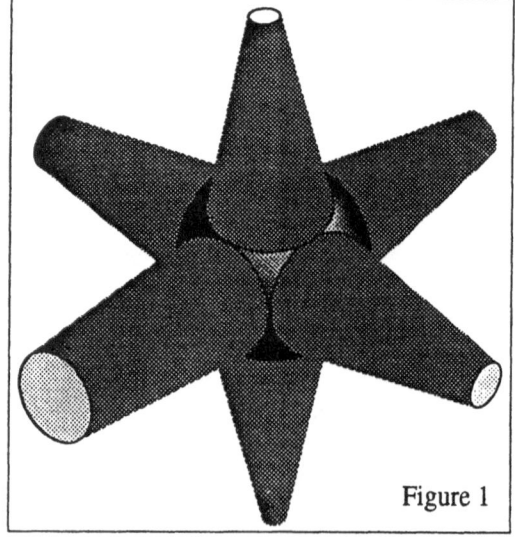

Figure 1

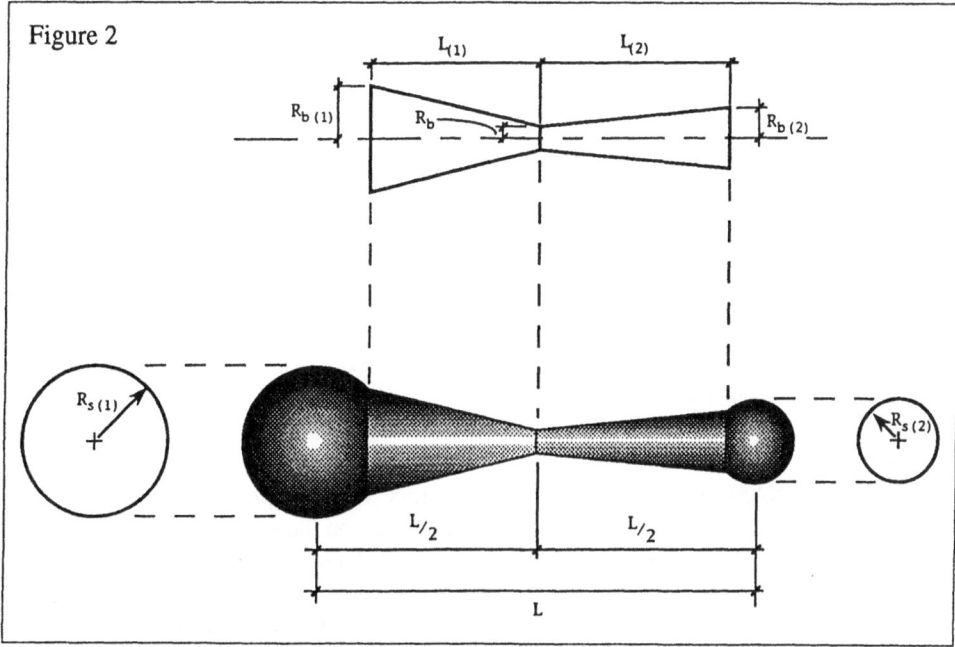

Figure 2

Spherical pore bodies are centered a constant distance L apart. Pore throats have circular cross-sections with a specified minimum radius (R_b) midway between adjacent site centers and radii equal to a specified fraction of adjoining pore body diameters ($R_{b(i)}=R_{s(i)}/\sqrt{2}$ i=1,2) on either end. The length of each half throat ($L_{(1)}$, $L_{(2)}$) is,therefore, a function of the radius of the associated pore body. The random nature of the pore space is captured by assigning pore body diameters and minimum pore throat diameters from separate probability distributions.

A preliminary step in the dissolution algorithm is computation of a carrier-fluid flow solution for the lattice. Given the pore-by-pore fluid distribution and a fluid pressure gradient across the lattice, the fluid flux in pore throats is calculated by solving a fluid mass balance equation at each pore body

$$\left[\sum_{j=1}^{6} q_j\right]_k = \left[\sum_{j=1}^{6} K_j \nabla p_j\right]_k = 0 \qquad \text{(EQ 1)}$$

where q_j is the volumetric flow rate in one of six pore throats intersecting site k, K_j is the conductance of pore throat j, and Δp_j is the pressure difference across throat j. If we assume one-dimensional flow in pore throats, the conductance of a pore throat for the geometry described above is given by

$$K_j = \frac{3\pi (L_{(1)} + L_{(2)})_j}{8\mu} \left((\mathfrak{R}_{b,1})_j + (\mathfrak{R}_{b,2})_j \right)^{-1}$$

(EQ 2)

where

$$(\mathfrak{R}_{b,i})_j = \left(\frac{L_{(i)}}{R_{b(i)} - R_b} \left(R_{b(i)}^{-3} - R_b^{-3} \right) \right)_j ; i = 1, 2$$

(EQ 3)

and μ is the fluid viscosity.

3. DISSOLUTION ALGORITHM

Residual NAPL blobs are represented by lattice sites completely filled with a non-aqueous phase. The mass of each blob is calculated based on the spherical site volume and the density of the NAPL of interest. This mass is translated into a specified number of computational particles.

The flow calculation described in Section 2 above gives a flux value for each pore throat in the network. For a lattice where one or more intersections is closed to flow by the presence of a NAPL blob, this solution results in six stagnant pore throats surrounding each NAPL-filled pore body. In order for the NAPL to enter the flow domain, if must diffuse through one of these throats. The time-dependent rate of flux of NAPL into a pore body adjacent to a blob is estimated by analytically solving a diffusion equation in a cylindrical tube with a diameter equal to the mean throat diameter and a length equal to $L_{(1)}+L_{(2)}$ for the given pore throat. Mass flux from a NAPL blob into the flowing domain may vary with time as a result of varying NAPL concentration in a blob-adjoining pore body. The transient flux is represented as a (potentially different) steady-state flux over each time step. The concentration boundary condition at the inner (blob) end of the hypothetical tube is set to the solubility limit of the NAPL of interest for all time steps. The concentration at the outer end is calculated based on the mass of particles in the adjoining pore unit at the end of the last time step.

Solution of the diffusion equation gives a constant mass flux into a pore body for a given time step. This must be translated into a particle flux into the flow domain. In the present algorithm, a particle originates, i.e., comes into existence, at the center of a pore body in response to the diffusive flux of a mass equivalent of a single particle into that pore. At the end of a time step, a mass equal to one-half a particle mass (or more) 'becomes' a particle; a mass of less that one-half a particle mass 'disappears'.

At its time of origination each particle is given an identification number. At this and all subsequent times each particle has a location, specified by five indices: j,k,l,m and d. Integers j,k and l indicate intersection location within the lattice and m indicates whether the particle is in a site (m=0) or in a bond oriented in the j (m=1), k (m=2) or l (m=3) direction. Real number d indicates the distance that the particle has traveled in its current pore throat.

Upon origination in a pore body, and on each subsequent occasion when the particle enters an intersection, it must 'choose' which pore throat it will enter next. In the current algorithm, several assumptions are made. First, a pore body is perfectly mixed, that is, the direction from which a particle enters an intersection does not influence the direction by which it leaves. Some experimental evidence seems to contradict this assumption (e.g., photomicrographic studies of colloid tranport in etched plate micromodels (Wilson, 1992), although these experiments have generally been carried out at flow rates significantly higher than those being simulated here. Second, particles will not enter a pore in which the fluid flow direction is toward the intersection, i.e., an inflow pore. This ignores the possibility of molecular diffusion of mass against a mean flow direction. Finally, the probability that a particle will choose a given outflow pore ($P(b_i)$) is proportional to its relative flow rate

$$P(b_i) = \frac{q_{i(o)}}{\sum_j q_{j(o)}} \tag{EQ 4}$$

where $q_{i(o)}$ is the flow rate in outflow bond i and $\Sigma q_{j(o)}$ is the total outflow rate from the current intersection location of the particle.

The transport of a dissolved component in a pore is influenced by the local fluid flow field (advection, diffusion due to variability in the velocity field within a pore) as well as by molecular-scale effects (diffusion due to Brownian motion). The interaction of these mechanisms results in dispersion as well as movement in the bulk flow direction. This spreading can be represented as a specified distribution of travel times for particles once they have entered a given pore throat.

The mean particle travel time for a throat is given by the pore length L divided by the mean velocity of the fluid in the bulk flow direction (v_m). Assuming one-dimensional flow in a tube of varying radius, this is given by

$$v_{m(j)} = \frac{q_j}{2\pi (L_{(1)} + L_{(2)})_j} (\Im_{1,2})_j \tag{EQ 5}$$

where

$$(\Im_{1,2})_j = \left(\frac{L_{(1)}}{R_{b(1)} - R_b} \left(R_b^{-1} - R_{b(1)}^{-1} \right) + \frac{L_{(2)} - L_{(1)}}{R_{b(2)} - R_b} \left(R_b^{-1} - R_{b(2)}^{-1} \right) \right)_j \qquad \text{(EQ 6)}$$

Longitudinal dispersion within a pore results from the existence of a laminar velocity profile within a pore throat. That is, part of the spreading within the pore is due to the fact that the magnitude of velocity varies with distance from the centerline of the pore throat. Thus, in the context of the model, one component of the variability in the distribution of travel times is equivalent to the 'choice' of a streamline at the entrance to the pore throat. Note that once within a throat the velocity along this streamline varies due to the variability of cross-sectional area of the pore. An additional component of longitudinal dispersion arises from molecular diffusion (Sorbie and Clifford, 1991). While this may be relatively insignificant for most pores, it may be very important for pores in which fluid velocities are very small. In this case, the part of the variability in travel times accountable to this mechanism prevents particles from being trapped for excessively long periods of time in slow flowing pores.

Since we are not concerned with concentration profiles within pores, the role of transverse disperion in the model is to 'move' particles from one streamline to another as they are advected through throats. For the slow-flow case, assumed here, there is no macroscopic turbulent mixing contibution to transverse dispersion. However, molecular diffusion may spread dissolved components across the mean flow direction. A component of travel time variability must account for this phenomena.

The concentration field, i.e., the distribution of dissolved NAPL concentration within the simulated porous medium, must be recalculated at the end of each time step. A concentration value is assigned to each lattice intersection. As in the case of the outer boundary condition for diffusion in a stagnant pore, this value is given by the equivalent mass of particles within a pore unit (one pore body plus six half throats) divided by the total unit volume.

Concentration profiles at a given time are found by summing particle masses and pore unit volumes over slices of the lattice perpendicular to the mean flow direction. A breakthrough curve is generated by monitoring time-varying particle mass fluxes across a given slice during each time step. Division by the steady-state fluid flux across the same face gives a curve of concentration versus time. These results can be analyzed using techniques generally applied to laboratory column data to estimate macroscopic mass transfer rates and diffusion coefficients. Since the pore-scale geometry of the computational porous medium is known exactly and since the same lattice can be used to generate capillary pressure - saturation - relative permeability data, correlations between these soil properties and mass transfer and transport properties can be established.

4. ACKNOWLEDGEMENTS

This work was supported, in part, by the National Science Foundation under Grant EAR-9218803.

5. REFERENCES

Ferrand, L.A. and M.A. Celia, The effect of heterogeneity on the capillary pressure - saturation relation, *Water Resour. Res.* (28), 859-870, 1992.

Imhoff, P.T., *Dissolution of a nonaqueous phase liquid in saturated porous media,* Ph.D. dissertation, Princeton University, 1992.

Miller, C.T., M.M. Poirier-McNeill and A.S. Mayer, Dissolution of trapped non-aqueous phase liquids: Mass transfer characteristics, *Water Resour. Res.* (26), 2783-2796, 1990.

Powers, A.E., C.L. Loueiro, L.M. Abriola and W.J. Weber, Theoretical study of the significance of nonequilibrium dissolution of nonaqueous phase liquids in subsurface systems, *Water Resour. Res.* (27), 463-471, 1991.

Sahimi, M. and A.O. Imdakm, The effect of morphological disorder on hydromdynamic dispersion in flow through porous media, *J. Phys. A: Math. Gen.* (21), 3833-3870, 1988.

Sorbie, K.S. and P.J. Clifford, The inclusion of molecular diffusion effects in the network modeling of hydrodynamic dispersion in porous media, *Chem. Eng. Sci.* (46), 2525-2542, 1991.

Wilson, J., *Visualization of groundwater flow and transport through a microscope,* 1992 Henry Darcy Distinguished Lecture, given at Princeton University, Sept. 1992.

MODELING SMALL-SCALE PHYSICAL NON-EQUILIBRIUM AND LARGE-SCALE PREFERENTIAL FLUID AND SOLUTE TRANSPORT IN A STRUCTURED SOIL

J. P. GWO and P. M. JARDINE
Environmental Sciences Division
Oak Ridge National Laboratory
P.O. Box 2008, Oak Ridge, TN 37830
USA

G. V. WILSON
Department of Plant and Soil Science
University of Tennessee
Knoxville, TN 37901
USA

G.-T. YEH
Department of Civil and Environmental Engineering
Pennsylvania State University
University Park, PA 16802
USA

The deviation of non-reactive solute transport from that predicted by classical convection-dispersion equations is usually attributed to physical non-equilibrium caused by small- and large-scale pore structures in porous media. Diffusion of fluid and solute into micropores or rock matrix may occur locally, while fluid and solutes can also be channeled preferentially through interconnected macropores or fractures. A multiple-pore-region (MPR) approach with local advective-diffusive mass exchange is adopted to simulate soil column tracer breakthrough and field-scale tracer releases in the Melton Branch Subsurface Transport Facility within the Oak Ridge Reservation, Tennessee. The soil column simulation indicates that both inter-region mass exchange and intra-region convection-dispersion contribute to small-scale solute transport in approximately the same order of magnitude. The field-scale study suggests that advective mass exchange has minor effect on subsurface hydrographs, and that large diffusive mass exchange may retain tracers near the source area. Comparison of modeling results and field data suggests that subsurface bedding planes on the field site may be the cause of large-scale heterogeneity and preferential mass transport.

A. Peters et al. (eds.), Computational Methods in Water Resources X, 465–472.
© *1994 Kluwer Academic Publishers. Printed in the Netherlands.*

INTRODUCTION

A MPR conceptual model is composed of flow regions with different hydraulic parameters that result in discernible velocity fields exchanging fluid and solute with other flow regions. Intra-region and inter-region mass transfer are conceptualized as two separate but intermingled processes that govern the transport of mass within a flow region and across flow regions, respectively. As illustrated in Figure 1 for a three-pore-region case, each flow tube consists of a flow region, while mass exchange of various rates exists between every two flow tubes. Fluid and solute may stay in the same flow tube or detour through other flow tubes while they move to a downstream area.

macropores   mesopores   micropores

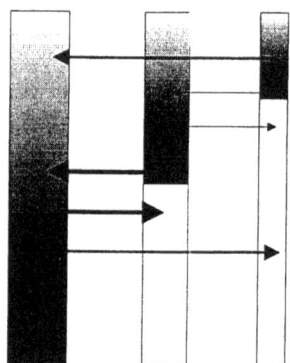

Figure 1. A three-pore-region mass transfer mechanism. Increased shading indicates an increase in solute concentration and position of the concentration front. Arrows indicate mass exchange among pore regions.

Sensitivity analysis on advective and diffusive exchange is conducted to investigate the effect of mass exchange on fluid and solute transport in a macropore-mesopore-micropore soil. The Melton Branch site, which has been a study site for subsurface storm flow dynamics since the 1980's, is used to demonstrate the application of a three-dimensional multi-region flow and solute transport model. Through the aforementioned two studies, we intend to gain insight on the effect of small-scale and large-scale heterogeneities on the movement of fluid and solute under transient flow and solute transport conditions in variably-saturated soils.

MATHEMATICAL MODELS

Under the concept of overlapped continua, one can apply volume-average procedures to each of the continua or pore regions in a porous medium. The following equation is obtained after the empirical Darcy's law is adopted to represent the intra-region isothermal fluid flow processes (Gwo, 1992).

$$F_\alpha \frac{\partial h_\alpha}{\partial t} = \nabla \cdot K_\alpha \nabla(h_\alpha + z) + \theta_\alpha q_\alpha - \sum_{\substack{j=1 \\ j \neq \alpha}}^{n} \varepsilon_{\alpha j}{}^f (h_\alpha - h_j), \qquad \alpha = 1, 2, .., n \qquad (1)$$

where, for pore region α, F_α is storage coefficient, h_α is fluid pressure head, z is elevation head, K_α is hydraulic conductivity tensor, t is time, θ_α is water content, q_α is external source or sink flow rate, and $\varepsilon_{\alpha j}{}^f$ is an empirical advective exchange coefficient.

The solute transport equations for the pore regions can be similarly derived, and they are (Gwo, 1992):

$$\theta_\alpha \frac{\partial c_\alpha}{\partial t} + \rho_{b\alpha} \frac{\partial s_\alpha}{\partial t} + \theta_\alpha v_\alpha \cdot \nabla c_\alpha = \nabla \cdot (\theta_\alpha D_\alpha \cdot \nabla \cdot c_\alpha) + \theta_\alpha q_\alpha c_{q\alpha}{}^* - \theta_\alpha q_\alpha c_\alpha$$

$$-\sum_{\substack{j=1 \\ j \neq \alpha}}^{n} \varepsilon_{\alpha j}{}^f (h_\alpha - h_j) c_\varepsilon{}^* - \sum_{\substack{j=1 \\ j \neq \alpha}}^{n} \varepsilon_{\alpha j}{}^l (c_\alpha - c_j) + \sum_{\substack{j=1 \\ j \neq \alpha}}^{n} \varepsilon_{\alpha j}{}^f (h_\alpha - h_j) c_\varepsilon, \quad \alpha = 1, 2, ..., n \qquad (2)$$

where, for pore region α, D_α is dispersion coefficient tensor, c_α is solute concentration, v_α is velocity, $c_{q\alpha}{}^*$ is the solute concentration of the incoming fluid if injecting and $c_{q\alpha}{}^* = c_\alpha$ if withdrawing, $\rho_{b\alpha}$ is bulk density of the soil, $\varepsilon_{\alpha j}{}^l$ is an empirical diffusive exchange coefficient, $c_\varepsilon{}^* = c_\alpha$ if $h_\alpha > h_j$, $c_\varepsilon{}^* = c_j$ if $h_\alpha < h_j$, and s_α is concentration in the sorbed phase.

Other formulations of mass exchange (e.g., Dykhuizen, 1990; and Zimmerman et al., 1993) can also be adopted to the fluid and solute exchange terms in equations (1) and (2). Equation (1) is solved using a Galerkin finite element method and has been implemented as computer codes MURF and 3DMURF. A hybrid Lagrangian-Eulerian finite element method is used to solve the solute transport equation (2) and the corresponding computer codes are MURT and 3DMURT. These computer codes have been verified (Gwo, 1992) using a fractured porous media model and a mobile-immobile model, respectively.

PARAMETER DETERMINATION

To obtain three sets of hydraulic conductivity and water retention relations for a three-pore-region soil, we divide pore regions, at -10 cm and -250 cm pressure heads, according to the classification scheme proposed by Luxmoore (1981). Both Fermi function and the van Genuchten equation (van Genuchten, 1980) are used for the θ (h)

relation (Wislon et al., 1992). The following Fermi function is used to represent the water retention relation in macropores:

$$\theta_a = \frac{\theta_{sa} - \theta_{sr}}{1 + \exp[-\gamma_a(h_a - h_{\theta a})]} \tag{3}$$

where θ_a is the water content in macropores, θ_{sa} and θ_{sr} are the saturated and residual water contents of macropores, respectively, γ_a and $h_{\theta a}$ are function parameters, and h_a is the pressure head in macropores. For mesopores and micropores, the following van Genuchten equation is used:

$$\theta_\alpha = \frac{\theta_{s\alpha} - \theta_{r\alpha}}{[1 + (-\beta_\alpha h_\alpha)^{n_\alpha}]^{m_\alpha}} \tag{4}$$

where, for pore region α, θ_α is water content, $\theta_{s\alpha}$ and $\theta_{r\alpha}$ are saturated and residual water contents, h_α is pressure head, β_α, n_α, and $m_\alpha = 1 - 1/n_\alpha$ are equation parameters. The hydraulic conductivity relation used in this study, for all of the pore regions, is also a Fermi function:

$$Log_{10}(K_{a\alpha} / K_{s\alpha}) = \frac{\xi_\alpha}{1 + \exp[-\kappa_\alpha(h_\alpha - h_{k\alpha})]} - \xi_\alpha \tag{5}$$

where for region α, $K_{a\alpha}$ and $K_{s\alpha}$ are absolute and saturated hydraulic conductivities, respectively, and ξ_α, κ_α, and $h_{k\alpha}$ are function parameters.

SENSITIVITY ANALYSES ON ADVECTIVE AND DIFFUSIVE EXCHANGE

In order to understand the effect of microscopic mass exchange processes on subsurface mass transport, we conduct a sensitivity analysis on advective and diffusive exchange coefficients. The fluid flow model MURF is used to study the transient variably-saturated flow fields of a vertical soil column. The flow fields, then, are imported into the solute transport model MURT for the computation of pore-region concentration fields. A total recharge rate of 9 cm³/hr/m² is applied to the soil column at the beginning of simulations, and is terminated 5 hours later. A non-reactive tracer of relative concentration 1 is also injected at the beginning of the simulations and is terminated 2 hours later. The simulations are stopped at time = 10 hr. The advective exchange coefficients are varied by 5 orders of magnitude, and the diffusive exchange coefficients are varied by 7 orders of magnitude. The other model parameters are obtained from soil samples and a soil column collected at the Melton Branch site (Wilson et al., 1992a; Jardine et al., 1993).

The times required to balance pore-region pressure and concentration fields are inversely related to advective and diffusive exchange coefficients, respectively. However, no apparent relation between advective exchange coefficients and the equilibration of concentration fields has been observed. The contributions to the increase/decrease of fluid and solute mass in macropores and mesopores by intra-region advection-dispersion and inter-region advective and diffusive exchange are in the same orders of magnitude (Figure 2), while in micropores the major contribution to mass change rates is the advective and diffusive mass exchange. Because of the high flow rates in macropores and mesopores and the large porosity in micropores, these results indicate that bypassing flow is very likely to occur in the soil during moderate to high infiltration rates and that not only diffusive but also advective exchange needs to be incorporated into a fluid and solute transport model if the micropores of a soil are to be considered as an immobile pore region without an associated velocity field.

APPLICATION TO A FORESTED WATERSHED

The Melton Branch subwatershed is underlain by Conasauga shale interbedded with Maryville Limestone formation. The strata of the Conasauga group weathers to a shallow (< 1m) soil, which overlies saprolite that retains the bedding and structure of the parent materials. A 2 m deep by 16 m long trench was excavated across the outflow region of the subwatershed to collect subsurface drainage. Two H-flumes equipped with Manning ultrasonic level recorders were positioned in the trench to give integrated flow rates of the soil horizons and a surface runoff collector (Wilson et al., 1992b). A buried line source of length 6 m at the depth of 0.5 m is located at the top of the eastern hillslope. In the numerical model, the subwatershed is divided into 10,500 elements and 12,090 nodes. Since each node consists of three pore regions, this requires the solving of 36,270 coupled equations simultaneously. Two series of rain storms in 1991, from February 13 to 20, are used as input to the system. The rain storms during February 13 to 14 (25 mm) lasted for 20 hours and those during February 17 to 20 (145 mm) lasted for 66 hours. The subsurface hydrograph of the former storm events is used to calibrate the model. Parameters thus obtained are used to predict the subsurface hydrograph for the latter storm events. A tracer release of 8000 mg Br/L (as $MgBr_2$) at a rate of 0.194 m^3/h was initiated on February 18 at 0950 (relative time 119 hours) over a 16.5 hour period through the buried line source near the ridge top (Wilson et al., 1992b).

For the calibration study using the rain storm events during Feb. 13-14, the peak arrival times of the simulated and measured hydrographs are in good agreement (not shown). The total outflow obtained from the simulated and measured hydrographs are also approximately the same. For the storm events during Feb. 17-20, the predicted peak arrival times agree well with those measured (Figure 3), but the predicted total flow is about 60% of the measured. Scenario studies with different advective exchange coefficients suggest that pore-scale mass exchange has little effect on subsurface hydrographs in field-scale problems. These results and the low fluid storage of the

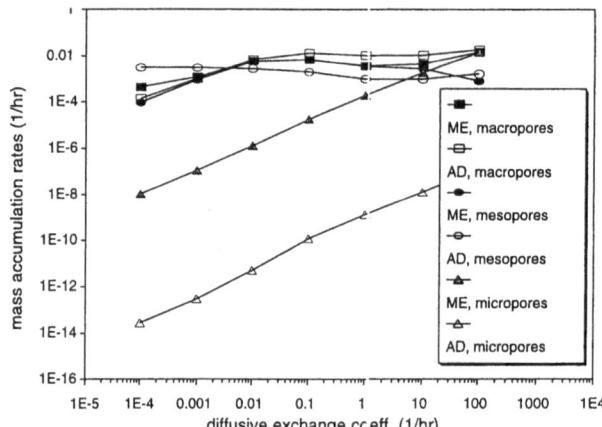

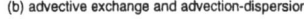

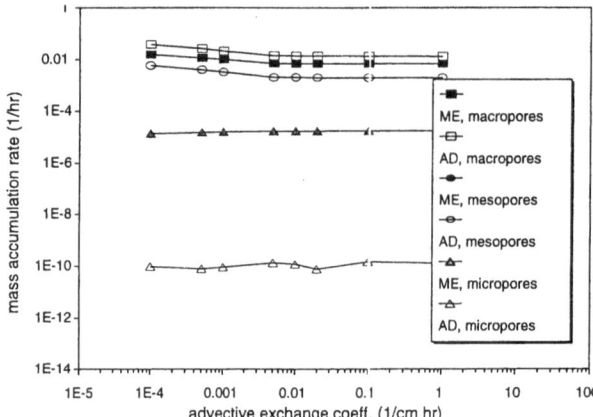

Figure 2. Effect of mass exchange on pore-region solute mass accumulations:
(a) diffusive exchange, and (b) advective exchange. The contributions of
inter-region and intra-region mass transfer (ME = mass exchange, AD =
advection-dispersion) in macropores and mesopores are approximately in the
same order of magnitude, while in micropores inter-region mass transfer
dominates.

storm flow zone suggest that large-scale heterogeneities, i.e., subsurface bedding planes
that dip perpendicular to the topography, may be responsible for the differences
between the predicted and measured subsurface hydrographs.

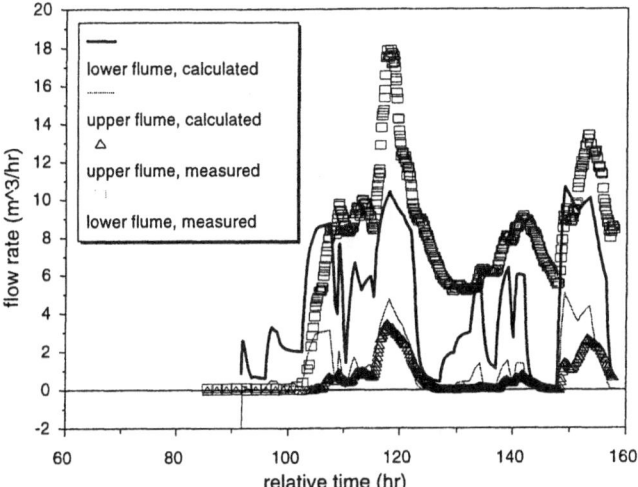

Figure 3. The calculated and measured subsurface hydrographs for the storm events during Feb. 17-20.

The peak concentrations of the predicted BTC for the C horizon (Figure 4) are about two times of those measured. The flowpath obtained during the flow simulations appears to follow the topography of the site (not shown), and the length of the flowpath is about half of the true flowpath that dips southwards along a subsurface bedding plane until it reaches the convergent zone. This large-scale feature has not yet been implemented into the numerical model.

Because hydrodynamic dispersion reaches beyond the locale of interest for storage of solutes and diffusive exchange utilizes local storage available in the micropores, high dispersivities may result in field-scale distribution of solutes, but high diffusive exchange tends to restrict solutes from moving far away from the source area. Scenario studies with various diffusive exchange coefficients suggest that micropore storage is utilized to restrain Br from moving further downstream in high diffusive exchange cases. This observation is in contrast to macroscopic dispersion that extends the area covered by the solute plume in low diffusive exchange cases.

ACKNOWLEDGMENT

This work is supported by the Subsurface Sciences Program, Office of Health and Environmental Research, Department of Energy, under contract DE-AC05-84OR21400 with Martin Marietta Energy System, Inc. and under subcontract No. 86X-SE414C with Pennsylvania State University. This research is conducted in the Oak Ridge National Environmental Research Park.

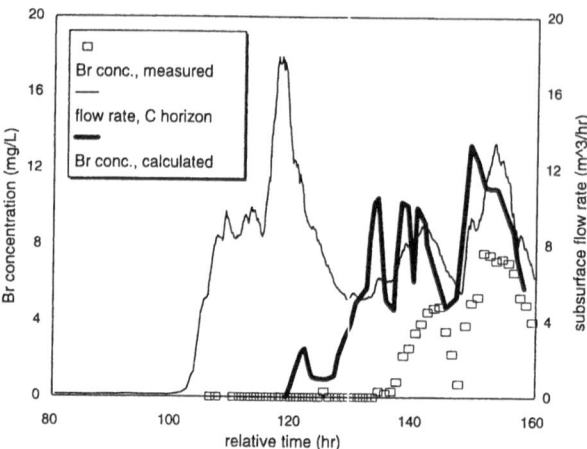

Figure 4. The calculated and measured Br BTCs through C horizon. The subsurface flow rate in the soil horizon is also shown.

REFERENCES

Dykhuizen, R.C. 1990. A new coupling term for dual-porosity models. Water Resour. Res., 26, p351-356.

Gwo, J. P., 1992. Multi-region Flow and Transport Modeling in Subsurface Media. Ph.D. Thesis. The Department of Civil and Environmental Engineering, The Pennsylvania State University, University Park, PA 16802, 283pp.

Jardine, P. M., G. K. Jacobs, and G. V. Wilson, 1993. Unsaturated transport processes in undisturbed heterogeneous porous media. I. Inorganic contaminants. Soil Sci. Soc. Am. J. 57:945-953.

Luxmoore, R.J., 1981. Micro-, meso-, and macroporosity of soil. Soil Sci. Soc. Am. J., 45:671-672.

van Genuchten, M. Th., 1980. A closed-form equation for predicting the hydraulic conductivity of unsaturated soils. Soil Sci. Soc. Am. J., 44:892-898.

Wilson, G. V., P. M. Jardine, and J. P. Gwo, 1992a. Modeling the hydraulic properties of a multiregion soil. Soil Sci. Soc. Am. J. 56:1731-1737.

Wilson, G. V., P. M. Jardine, J. D. O'Dell, and M. Collineau, 1992b. Field-scale transport from a buried line source in variable saturated soil. J. Hydrol., 145:83-109.

Zimmerman, R. W., G. Chen, T. Hadgu, and G. S. Bodvarsson, 1993. A numerical dual-porosity model with semi-analytical treatment of fracture/matrix flow. Water Resour. Res., 29, p2127-2137.

EFFECTS OF FAULTS ON NATURAL CONVECTION IN GEOLOGICAL STRUCTURES : COMPUTATION OF THE STABILITY CRITERIA AND OF THE CRITICAL FLOW PATTERNS USING FINITE ELEMENTS

N. JOLY and D. BERNARD
L.E.P.T.-ENSAM (URA 873)
Esplanade des Arts et Métiers
33405 TALENCE Cédex
FRANCE

INTRODUCTION

Real geological structures usually have complex geometry and heterogeneous physical properties. This complexity combined with the existence of the geothermal flux induces thermo-convection, whose amplitude is very sensitive to heterogeneities. In this paper we consider a peculiar case of heterogeneities: the fault zones.

The set of equations describing free convection in a saturated porous structure may have several solutions. In this case stability study permits the determination of the most stable one. A numerical code has been developed to treat the case of two-dimensional complex geological structures. The values of the different parameters for which free convection may develop (critical conditions) can be computed as well as the associated temperature, pressure, and velocity fields.

To illustrate the effects of faults and the capabilities our programs, we consider the case of two horizontal porous layers bounded by impervious conducting boundaries and crossed by a set of periodic vertical faults. The variation of the critical Rayleigh number is presented as a function of different geometrical parameters (fault period, porous layer thickness, distance between layers) and of the physical parameters of the media (permeabilities and thermal conductivities ratio).

MATHEMATICAL MODELLING

Water flows encountered in sedimentary basins are slow, and the local thermal equilibrium between fluid and solid matrix is observed. In this work, we assume a rigid porous matrix, the absence of chemical reaction or phase change, and a constant fluid density, except in the motion equation (Boussinesq assumption). Free convection in porous media is then described by the following dimensionless equations [1]:

continuity $$\nabla . \mathbf{V} = 0 \tag{1.1}$$

motion (Darcy's law) $$\mathbf{V} = -\mathbf{K}.(\nabla p + \mathrm{Ra} * \mathrm{T}\, \mathbf{e}) \tag{1.2}$$

heat transfer $$\mathrm{R} * \frac{\partial \mathrm{T}}{\partial \mathrm{t}} - \nabla . (\Lambda . \nabla \mathrm{T}) + \mathbf{V} . \nabla \mathrm{T} = \mathrm{F}_\Omega \tag{1.3}$$

A. Peters et al. (eds.), Computational Methods in Water Resources X, 473–480.
© 1994 *Kluwer Academic Publishers. Printed in the Netherlands.*

where $\mathbf{v}$ is the fluid velocity, T the temperature, p the fluid pressure, Λ the thermal conductivity tensor of the equivalent continuous medium, $\mathbf{K}$ the mobility tensor of the fluid in the porous medium, $\mathbf{e}$ the unit vector directing gravity, and F_Ω the volumic heat source. For free convection the filtration Rayleigh number Ra* quantifies the relative influence of the driving force to the stabilizing effects due to the fluid viscosity and the porous media thermal conductivity. The first two equations are only defined into the fluid saturated porous region Ω_s, whereas the third equation is verified into the whole studied domain Ω. The external boundaries of Ω are assumed to support Newton- (Γ_N) or Dirichlet (Γ_D) -type boundary conditions (2.1). In order to balance the fluid flow on the boundary Γ_π where the permeability tensor $\mathbf{K}$ presents a discontinuity (including impervious condition on the external boundary of Ω_s), the condition (2.2) is applied.

on Γ_N $\qquad\qquad -(\Lambda.\nabla T).\mathbf{n} + hT = q$

$$(2.1)$$

on Γ_D $\qquad\qquad\qquad T = \overline{T}$

on Γ_π $\qquad\qquad -(\mathbf{K}.\nabla p).\mathbf{n} = Ra * T\left[(\mathbf{K}.\mathbf{e}).\mathbf{n}\right]$ $\qquad\qquad (2.2)$

Stability problem

Depending on geometry, physical properties and boundary conditions, the equation set (1) to (2) may have several mathematical solutions. For instance, pure conduction (corresponding to a zero filtration velocity) may be a solution if ∇T and $\mathbf{e}$ are colinear [1], nevertheless a convective solution can be more *stable* than conduction. Stability study permits to determine whether a solution is stable or not for some kind of 2D or 3D perturbations. It can be noted that in geological structures, conduction is seldom stable because the necessary stability criteria (∇T and $\mathbf{e}$ colinear) is generally not verified. The numerical method presented therein [2] is concerned with stability problems within geological structures whose geometry and properties do not vary along one direction (denoted y). This direction is assumed to be a principal axis for the permeability and the conductivity tensors. Permeability and thermal conductivity in this direction are noted, respectively, K_y and Λ_y.

Let $\mathbf{E}_b(T_b, \mathbf{V}_b, p_b)$ be a steady state solution of the equation set (1) completed by the boundary conditions (2). The stability of $\mathbf{E}_b$ is tested superimposing to it an arbitrary infinitesimal disturbance $\varepsilon(\theta,\phi,\pi)$ and studying the evolution of its amplitude with time. As both $\mathbf{E}_b$ and the disturbed state ($\mathbf{E}_b + \varepsilon$) are solutions of (1) and (2), the local equations and boundary conditions satisfied by the perturbation ε can be built by difference. ε being infinitesimal, this equation set can be linearized. Searching some solutions of the form [3]:

$$\pi_0(x,y,z,t) = \pi(x,z)\ \cos(\alpha y).e^{\sigma t} \qquad\qquad (3.1)$$

$$\theta_0(x,y,z,t) = \theta(x,z)\ \cos(\alpha y).e^{\sigma t} \qquad\qquad (3.2)$$

$$\phi_0(x,y,z,t) = (\phi(x,z)\ \cos(\alpha y) - \pi(x,z)\sin(\alpha y)\mathbf{y}).e^{\sigma t} \qquad\qquad (3.3)$$

i.e. $2\pi/\alpha$ periodic along y, π and θ are solutions of the following equation set:

$$\nabla.(\mathbf{K}.(\nabla\pi + Ra * \theta\ \mathbf{e})) - \alpha^2 K_y\pi = 0 \qquad\qquad (4.1)$$

$$R * \sigma\theta - \nabla.(\Lambda.\nabla\theta) + \mathbf{V}_b.\nabla\theta - \mathbf{K}.(\nabla\pi + Ra * \theta\ \mathbf{e}).\nabla T_b + \alpha^2\Lambda_y\theta = 0 \qquad\qquad (4.2)$$

completed by the boundary conditions:

on Γ_N $\qquad\qquad -(\Lambda.\nabla\theta).\mathbf{n} + h\theta = 0$

(4.3)

on Γ_D $\qquad\qquad\qquad \theta = 0$

on Γ_π $\qquad\qquad -(\mathbf{K}.\nabla\pi).\mathbf{n} = Ra^* \; \theta\left[(\mathbf{K}.\mathbf{e}).\mathbf{n}\right]$ $\qquad$ (4.4)

Depending on the different parameters, there exist or not perturbations $\varepsilon(\theta,\phi,\pi)$ associated to a positive time evolution coefficient (σ). The stability criterion will be obtained writing explicitly the marginal stability condition: $\sigma = 0$.

NUMERICAL MODEL

The set of equations (1) (2) is discretized using finite element modules from the general purpose finite elements program package MODULEF [4]. Since the studied domain is invariant along y, a 2D grid is sufficient for this problem. Temperature is interpolated in the whole domain using piecewise linear functions (P1), while the pressure field is approximated on the saturated porous sub domain Ω_s by piecewise second order polynomials (P2).

The basic steady state solution $\mathbf{E_b}$ being computed, a weak integral form of the linear set (4.1) (4.2) can be written on the same 2D grid. Its expression under matricial form is given below [3]. Applying Green's theorem, the boundary conditions (4.3) and (4.4) are taken into account when computing the matrices $[A_{22}]$ and $[A_{11}]$.

$$\left[A_{11}\right]\{\pi\} + \left[A_{12}\right]\{\theta\} = 0 \qquad (5.1)$$

$$\left[A_{21}\right]\{\pi\} + \left[A_{22}\right]\{\theta\} = -\sigma[B]\{\theta\} \qquad (5.2)$$

where $\qquad$ $[A_{11}]$ is a symmetric square matrix depending on $\mathbf{K}$, K_y and α, the horizontal
$\qquad\qquad$ wave number. This matrix can be inverted.
$\qquad\qquad$ $[A_{12}]$ is a function of $\mathbf{K}$, Ra^* and $\mathbf{e}$.
$\qquad\qquad$ $[A_{21}]$ depends on the basic state $\mathbf{E_b}$ and $\mathbf{K}$
$\qquad\qquad$ $[A_{22}]$ is a square non-symmetric matrix depending on $\mathbf{K}$, Ra^*, $\mathbf{e}$, Λ, Λ_y, $\mathbf{E_b}$
$\qquad\qquad$ and the wave number α
$\qquad\qquad$ $[B]$ $\quad$ is a diagonal matrix function of R^* only and having an inverse.

Since the matrix $[A_{11}]$ has an inverse and the fluid saturated porous region Ω_s is a sub domain of Ω, equation (5.1) permits the expression of π as a function of θ :

$$\{\pi\} = -\left[A_{11}\right]^{-1}\left[A_{12}\right]\{\theta\} \qquad (6)$$

Substituting π into (5.2) by its expression (6) yields:

$$\left(\left[A_{22}\right]-\left[A_{21}\right]\left[A_{11}\right]^{-1}\left[A_{12}\right]\right)\{\theta\} = -\sigma[B]\{\theta\} \qquad (7)$$

Hence, denoting the matrix

$$[M] = -[B]^{-1}\left(\left[A_{22}\right]-\left[A_{21}\right]\left[A_{11}\right]^{-1}\left[A_{12}\right]\right)^{-1} \qquad (8)$$

the equation set (5) can be then written as an eigenvalue problem:

$$[M] \{\theta\} - \sigma \{\theta\} = 0 \tag{9}$$

If N_t is the number of temperature grid nodes not supporting Dirichlet boundary condition ($\notin \Gamma_D$), this eigenvalue problem has exactly N_t solutions.

$$[M] \{\theta_i\} - \sigma_i \{\theta_i\} = 0 \qquad i = 1, N_t \tag{10}$$

where σ_i is an eigenvalue of the matrix $[M]$ and $\{\theta_i\}$ the corresponding eigenvector.

Stability criterion, construction of the critical perturbation

Denoting $\mathrm{Re}(\sigma_i)$ the real part of the eigenvalue σ_i, the formal expression (3) of the solutions implies that:
- if $\mathrm{Re}(\sigma_i) < 0$, the corresponding temperature disturbance θ_i is decreasing, and the disturbance ε_i (θ_i, ϕ_i, π_i) is going to vanish with increasing time: the steady state $\mathbf{E_b}$ is stable for the perturbation ε_i.
- if $\mathrm{Re}(\sigma_i) > 0$, the corresponding disturbance ε_i is growing, and it follows that the steady state $\mathbf{E_b}$ is unstable.

Considering now the whole set of solutions of (9),
- if, for all wave number α, all the solutions θ_i are such that $\mathrm{Re}(\sigma_i) < 0$ (the greatest eigenvalue real part is negative), then the basic state $\mathbf{E_b}$ is stable.
- in contrast, if it is possible to find a non negative wave number α admitting at least one eigenvector (temperature perturbation) whose associated eigenvalue has a positive real part (the greatest element of $[M]$ spectrum is positive), then $\mathbf{E_b}$ is unstable.

When stability study is meaningful, there exist a "critical value" for the Rayleigh number, denoted $\mathrm{Ra^*_c}$ such that:
for $\mathrm{Ra^*} < \mathrm{Ra^*_c}$, the spectrum of $[M]$ is negative, and the system is stable,
for $\mathrm{Ra^*} = \mathrm{Ra^*_c}$, the greatest eigenvalue real part is zero, and the system is on the stability margin,
for $\mathrm{Ra^*} < \mathrm{Ra^*_c}$, the greatest eigenvalue real part is positive, and the system is unstable.

To compute the kind of infinitesimal disturbance able to grow on the stability margin (critical disturbance), the equation (6) is used to express the pressure disturbance field $\{\pi_i\}$ from the eigenvector $\{\theta_i\}$. Then, using (3.3), the velocity disturbance field may also be obtained.

APPLICATION TO A FAULTED GEOLOGICAL STRUCTURE

In sedimentary basins, shearing, crustal extension or shortening often are accommodated by active parallel faults. The numerical method presented above is able to determine the favorable conditions for the onset of free convection in the geological structure schematized on fig. 1: two horizontal homogeneous saturated porous layers, having the same physical properties (equivalent thermal conductivity λ_2, permeability k_2) and the same thickness H, are included in an impervious medium (thermal conductivity λ_1). The distance in between those layers is B.H. A set of vertical parallel faults (period A.H along

x) crosses the horizontal sedimentary layers creating, for the saturating fluid, a possible flow path from one porous layer to the other. The region located in between two consecutive faults afterwards is called "block".

This faulted structure does not vary along y : the geometry is consistent with a 2D thermoconvective solution. It also is A.H periodic along x, so that the convective flow is assumed to be n.A.H periodic along x in the plane (x,z), i.e. structured on n blocks. Each fault is modelled as a drain whose thickness is 0.2 H , conductivity λ_1 and permeability k_1. The upper boundary is assumed to be isothermal, and the lower one submitted to a constant heat flux Φ. Lateral boundaries, corresponding to the symmetry axis of the extreme end faults, are assumed to be impervious and adiabatic.

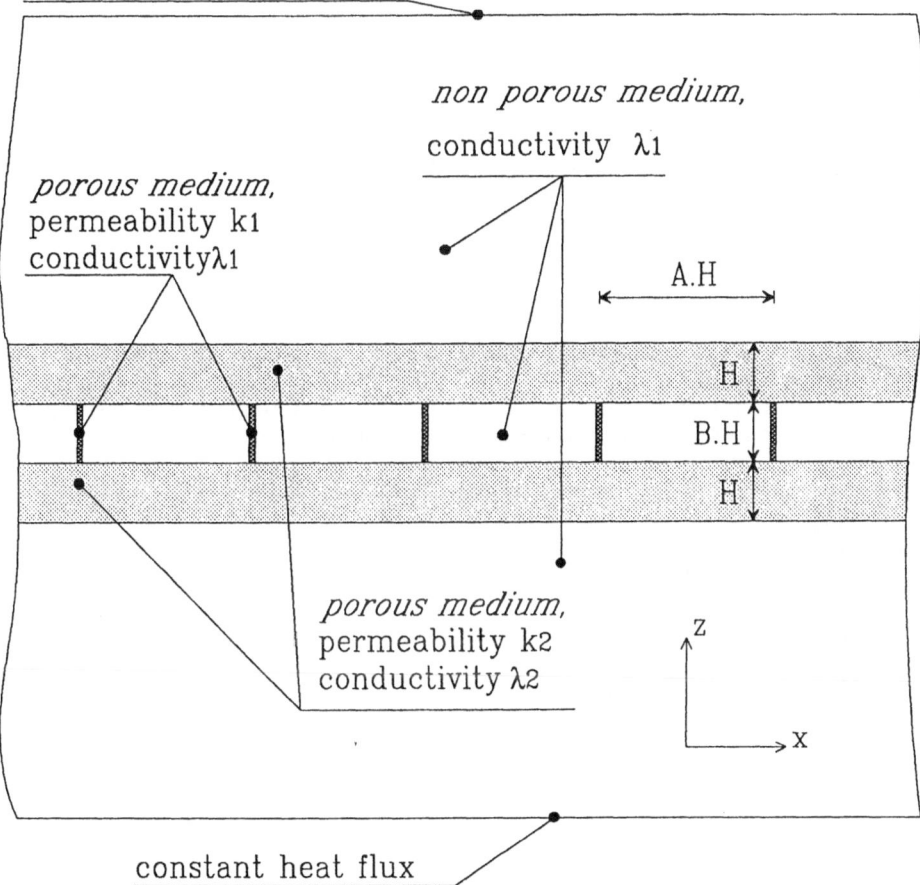

fig. 1 : geometry, physical properties and boundary conditions of the considered geological structure

The purely conductive isothermal surfaces are horizontal planes, and this study is concerned with the determination of the critical Rayleigh number corresponding to the onset of 2D free convection in the plane (x,z). The filtration Rayleigh number is defined as:

$$Ra^* = \frac{(\rho C_p)_f \; k_2 \, \rho_f \, \beta \, g \, \Phi \, H^2}{\mu_f \, \lambda_1^{\,2}} \tag{11}$$

where $(\rho C_p)_f$, ρ_f , β and μ_f are, respectively, the volumic fluid heat capacity, the fluid density, the fluid volumic expansion coefficient and the fluid dynamic viscosity. g is the gravity acceleration.

This configuration is characterized by five dimensionless parameters:
- the geometrical aspect ratios A and B,
- the thermal conductivity and permeability contrasts:

$$K_1 = \frac{k_1}{k_2} \quad ; \quad \Lambda_2 = \frac{\lambda_2}{\lambda_1}$$

- the filtration Rayleigh number Ra*.

This study is concerned with the determination of the critical Rayleigh number corresponding to the onset of 2D free convection ($\alpha=0$) in the plane (x,z). The basic configuration chosen to study the influence of the different parameters corresponds to $K_1=1$, $\Lambda_2=1$, A=3 , B=1 and a block number equal to one. Varying the parameters one by one, the following results have been obtained.

Permeability

For drains having a low permeability with respect to the porous layer ($K_1 \cong 0$), the critical filtration Rayleigh number tends to the value 26.3. As the two porous layers are intervening, this limit is lightly less than the critical Rayleigh number characterising the onset of natural convection in an infinite horizontal porous layer [5]. An increase of K_1 makes the onset of free convection easier, and reduces appreciably Ra*$_c$ (fig. 2).

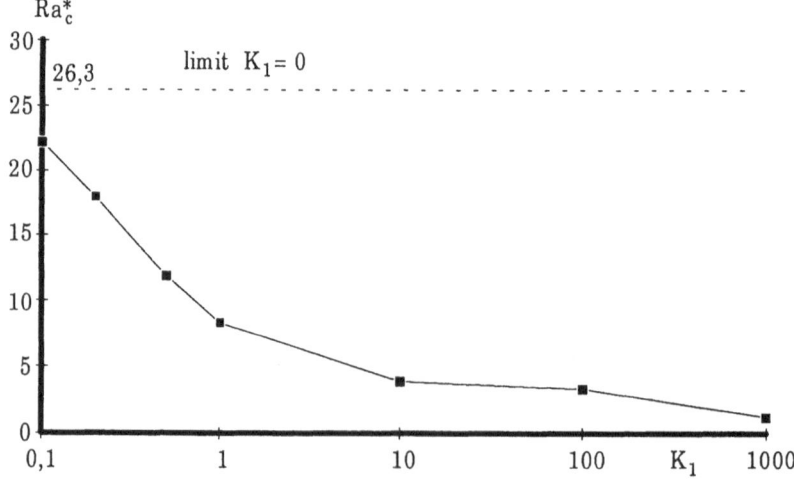

fig. 2 : Graph of Ra*$_c$ as a function of K_1 (B=1, A=3, Λ_2=1)

Thermal conductivity and geometrical aspect ratio B

An increase of the thermal conductivity contrast between the porous layers and the impervious zone (parameter Λ_2) increases the critical Rayleigh number and is then unfavourable for free convection.
The aspect ratio B has a low influence on the critical Rayleigh number ; Ra^*_c tends by excess to a constant when B increases.

Fault period A

The continuous line of the graph (fig. 3) presents the influence of A on Ra^*_c. The convective flow consists in one cell occupying the whole porous domain. One value (here about 9) is the more favourable one for the onset of free convection (minimum of Ra^*_c=6.6). This behaviour often is observed : the aspect ratio of the convective cells has a great influence on the corresponding critical Rayleigh number ; if geometrical constraints prevent the development of this "more favourable" flow pattern, Ra^*_c is increased.

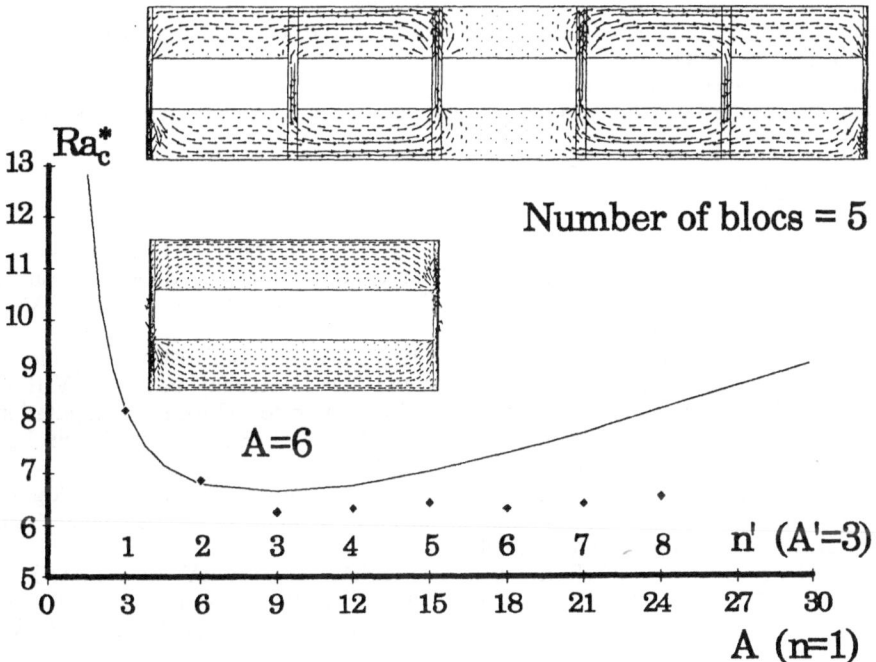

fig. 3 : Graph of Ra^*_c as a function of n' (A'=3) and A (n=1) ; Λ_2=1 , B=1

Effect of the number of blocs

On figure 3 we compare the critical Rayleigh numbers of domains having the same size, but different bloc numbers. The continuous line is Ra^*_c(A) for a one block domain , i.e. only containing the two extreme end faults, while the points correspond to a domain consisting on n' blocks (A'=3). The presence of more than two faults gives the convective flow more freedom to organise himself, what makes $Ra*_c$ smaller: the points are *on* (n'=1 and 2) or *below* (n'>2) the graph line. When the domain consists on more than 2 blocks, $Ra*_c$ remains slightly constant (about 6.3), and the flow pattern is mainly structured over 3 blocks. The existence of several faults allows the convective flow to take advantage of its "favourite" aspect ratio corresponding to (n=1, A=9) or (n'=3, A'=3); the critical Rayleigh number slightly is less than the curve's minimum, and the critical flow is as close as possible structured on the more favourable aspect ratio for the onset of free convection. This convective flow pattern is also observed in solving the nonlinear equations for low supercritical conditions.

CONCLUSION

The results presented in this paper clearly demonstrate that in large geological structures the effect of faults on the flow pattern can't be ignored. The existence of open faults favour the onset of thermo-convection and when a continuous network (putting into relation porous domains and faults) exists, large convective cells may be produced.
The numerical tools used for this study proved to be well adapted to this type of problem.
The study is continuing following three main directions;
1. estimation of the thermal effects for more realistic cases (simplified geometry of some North Sea reservoir),
2. coupling with dissolved salt transport,
3. estimation of the 3D stability of the 2D flow patterns put into evidence in this work.

REFERENCES

[1]   COMBARNOUS, M. and BERNARD, D. (1988) "Modeling of free convection in porous media : from academic cases to real configurations" 25 th ASME National Heat Transfer Conf. ASME, Int. Symp. Convection in porous media: Non DARCY effetcs, Houston, USA, 735-745

[2]   BERNARD, D. and MENEGAZZI, P. (1990) "Natural convection in geological porous structures: A numerical study of the 2D/3D stability problem" VIII Int. Conf. Comp. Methods in Water Res., June, Venice, ITALY, Computational methods in Subsurface Hydrology, GAMBOLATI et al. Eds., Comp. Mech. Pubs, Southampton, & Springer-Verlag, Berlin, 315-320

[3]   MENEGAZZI, P. (1989) "Convection naturelle dans les structures géologiques poreuses : étude numérique bidimensionnelle et stabilité des écoulements." PhD thesis University Bordeaux I, FRANCE

[4]   BERNADOU M., GEORGE P.L., HASSIM A., JOLY P., LAUG P., PERRONNET A., SALTEL E., DTEER D., VANDERBORCK G., VIDRASCU M., (1985), "MODULEF : une bibliothèque modulaire d'éléments finis.", INRIA, France

[5]   RIAHI, N. (1983) "Non linear convection in a porous layer with finite conducting boundaries.", J. Fluid. Mech. 129, 153,171

THE DYNAMICS OF DENSITY DRIVEN FINGER INSTABILITIES IN STOCHASTICALLY HETEROGENOUS POROUS MEDIA

MANFRED KOCH

Supercomputer Computations Research Institute and
Geophysical Fluid Dynamics Institute,
Florida State University, Tallahassee, FL 32306, U.S.

The dynamics of viscous fingers arising at a miscible interface in a density-stratified fluid in porous media is studied numerically. Both a deterministic or homogenous and a randomly heterogeneous (stochastic) porous medium are considered. For the latter case 2D Monte-Carlo realizations for the log-permeability distribution are generated with the turning bands method (TBM) and used as input in the numerical model. The results of the simulations show that, unlike in a homogenous porous medium, where the evolution and the morphology of the finger instabilities are mostly affected by the initial perturbation, for the random medium it is its stochastic realization that determines the fate of the fingers; i.e., the medium has an ordering effect on the finger pattern and fingers are essentially channeled through local sections of reduced hydraulic conductivity. To quantify the finger dynamics as a stochastic process, fractional dimensions of the finger morphology are computed. Whereas for a homogenous medium the fingers show a self-similar, or fractal behavior, with a fractal dimension of $D=1.56$, the ordering effect of a particular realization of the random medium destroys the fractality of the fingers.

1. INTRODUCTION

Hydrodynamic instability and fingering phenomena in porous media flow may arise when two fluids of different viscosities and densities are separated by an initially sharp interface that becomes unstable under unfavorable conditions. For a theoretical analysis of hydrodynamic instability and fingering phenomena in porous media flow, the distinction is important whether the two fluids are immiscible or miscible across the interface, respectively (cf. *Wooding and Morel-Seytoux, 1976*). As discussed by *Koch* (1993), the physical phenomena governing the onset and development of fingers in each case is quite different. For immiscible two-phase flow the interface instabilities (I-I) are also known as Rayleigh-Taylor (R-T) and Saffman-Taylor (S-T) instabilities (*Chandrasekhar*, 1961). Whereas in the R-T problem the (I-I) is triggered through an unfavorable density contrast, for the S-T problem the I-I is initiated when a less viscous fluid displaces a more viscous one. However, in both cases the interface is stabilized through the surface tension between the two immiscible phases. In most situations, such as two-phase flow through the unsaturated zone (wetting instabilities), the transport of immiscible contaminants in groundwater (*Parker, 1989*), and water-oil displacement flow in secondary petroleum recovery (*Perrine*, 1963) both R-T and S-T instabilities will act concurrently.

A. Peters et al. (eds.), Computational Methods in Water Resources X, 481–488.
© 1994 Kluwer Academic Publishers. Printed in the Netherlands.

Experimental studies of I-I in porous media show that pure immiscibility is rarely met (*Wooding, 1969*). Because of the pore structure and the ensuing tortuosity of the porous medium, interfacial hydrodynamic dispersion will lead to mixing of the two fluids and the surface tension is replaced by the dispersion, as the stabilizing parameter. Therefore, it has been common in petroleum reservoir modeling to treat viscous fingering as a 'partly-miscible' interfacial instability problem (*Wheeler*, 1988).

In groundwater hydrology density- or buoyancy-driven underground migration of highly concentrated leachates have been observed in salt brines (*Hassanizadeh and Leijnse*, 1988), and under unregulated landfills. As demonstrated by *Koch and Zhang* (1992a) and *Koch* (1992), migratory effects on a solute plume due to buoyancy are obtained for already moderate solute concentrations, corroborating experimental results of *Schincariol and Schwartz* (1990). The results of *Koch and Zhang* (1992a) and *Koch* (1992) corroborate qualitatively conclusions of earlier linear stability analyses (cf. *Wooding*, 1969), namely the stabilizing effect of the hydrodynamic dispersion and the destabilizing effect of the density contrast. The more detailed study of *Koch* (1993) illustrates that the onset, evolution and morphology of the instabilities are mostly affected by the dispersivity of the porous medium. Moreover, the instabilities show the typical behavior of a nonlinear dynamical system whose response depends essentially on the initial conditions imposed. However, it is found that the finger pattern shows a fractal behavior (see Section 3.1).

Here I will extend the simulations of *Koch* (1993) for a homogeneous medium to a randomly heterogeneous (stochastic) one, using a geostatistical description. Monte-Carlo realizations for the log-permeability distribution using the turning bands method will be used as input in the simulation model. Efforts will be made to investigate the fractality of the finger morphology in a random porous medium.

2. THEORY

2.1 Governing equations

Governing equations for the present problem of miscible flow and transport are the general *groundwater flow equation*

$$\nabla \cdot \left(\frac{k\rho}{\mu} (\nabla p + \rho g k_z) \right) = S \frac{\partial p}{\partial t} \tag{1}$$

for the the hydraulic pressure p $[M/L/T^2]$, and the *solute transport equation* for the solute concentration C [M/M] ([ppm])

$$\partial C / \partial t + v \cdot \nabla C = \nabla \cdot (D \nabla C) \tag{2}$$

(*Bear and Verruijt, 1987*), where t is the time [T]; k_z, the unit vector in the vertical direction; k $[L^2]$, the permeability tensor; g $[L/T^2]$, the gravity acceleration; S [dimensionless], the specific storativity of the aquifer; and D $[L^2/T]$ is the general hydrodynamic dispersion tensor $D = \alpha_T v \delta_{ij} + (\alpha_L - \alpha_T) v_i v_j / |v| + D^*$, where D^* is the coefficient of molecular diffusion, and α_L, α_T [L], denote the longitudinal and transversal dispersivities, respectively. Finally v $[L/T]$ denotes the flow velocity and is computed by Darcy's law, once the hydraulic head $h = p/\rho g + z$ (with z, the elevation head [L]) has been calculated from the pressure-solution of Eq. (1).

For hydrodynamic instabilities to occur, Eqs. (1) and (2) are to be coupled by equations of state for the density, ρ $[M/L^3]$, and the dynamic viscosity, μ $[M/L/T]$ as functions of the solute concentration C. As discussed by *Koch and Zhang* (1992a), for most practical applications, linear relationships of the form $\rho = \rho_0 + \eta C$ and $\mu = \mu_0 + \beta C$ appear to be appropriate and will be used here.

2.2 Geostatistical description and stochastic realization of random porous media

Eqs. (1) and (2) describe the solute transport problem within the continuum's or macroscopic characterization of a porous medium; i.e. they are valid for an appropriate REV. As such, they are also applicable for a heterogeneous medium, as long as the heterogeneity can be described by a macroscopic set of $REV's$. In such a situation the appropriate selection of an REV will depend on the geometrical scale of the porous medium (*Bear and Verruijt, 1987*). A geostatistical description of the porous medium (see below) allows to impose bounds on the REV and to quantify its underlying spatial stochastic process.

It is now universally accepted that the primordial factor governing advective-dispersive transport is the spatial variability of the permeability k ($\sim$ hydraulic conductivity K) of the porous medium (*Neuman, 1990*). Local variations of k will generate fluctuations of the local Darcian flow velocities; i.e., will eventually lead to an increase in the mechanical dispersion D_m of the solute. The positive correlation between the spatial randomness of k and D_m has been quantified theoretically in the last decade by stochastic theories on solute transport in random media (cf. *Gelhar and Axness, 1983. Welty and Gelhar* (1991) have extended this theory to include the effects of variable viscosity and density in one-dimensional miscible displacement flow. Efforts are presently being undertaken by the author to verify this 1D-parameterized stochastic theory through 2D numerical simulations.

In the geostatistical description of a porous medium the log-hydraulic conductivity K is defined as a spatial stochastic process (*Kitanitis*, 1992). With $Y = lnK$, the random field Y can be decomposed in a deterministic part, defined by the mean $E[Y] = \bar{Y}$, and a stochastic perturbation Y'; i.e. $Y = \bar{Y} + Y'$ with $E[Y'] = 0$. The spatial autocovariance Cov_Y of this random process is then given by $Cov_Y(\zeta) = E[Y'(\mathbf{x}+\zeta)Y'(\mathbf{x})]$. It it usually assumed that the covariance Cov_Y of lnK is stationary and anisotropic and given by $Cov_Y(\zeta) = \sigma_Y^2 exp[-(\zeta_1^2/\lambda_1^2 + \zeta_2^2/\lambda_2^2 + \zeta_3^2/\lambda_3^2)^{1/2}]$, where σ_Y^2 is the variance of lnK and λ_1, λ_2 and λ_3 are the correlation lengths of the random medium in the three space-directions. These four parameters have to be determined from a semivariogram analysis.

For the numerical solution of Eqs. (1) and (2) in heterogeneous media, 2D realizations of the stochastic process for $Y = lnK$ are required. Assuming that this process is stationary and ergodic (cf. *Dagan, 1986*), an unbiased representation of the physical model is then obtained by averaging over a large number of different realizations. Numerous procedures for the generation of a spatial stochastic process have been developed in the past (cf. *Journel and Huijbregt* (1978). Among these, the *turning bands method* (TBM), as coded by *Thompson et al.* (1987), has found widespread applications in stochastic hydrology and will be used here.

3. NUMERICAL SIMULATIONS

3.1 Method, parameters, geometry, boundary and initial conditions

The governing equations (1) and (2) are solved in a vertical 2D $x - z$ aquifer cross-section using an adaptation of the method of characteristics/finite difference 2D-model MOCDENSE (Sanford and Konikow, 1985) (cf. *Koch and Zhang*, 1992a and *Koch*, 1992, for a description). This method has the advantage of overcoming many of the numerical problems other classical FD or FE techniques for advection-dominated transport problems such as numerical dispersion and/or dissipation (cf. *Koch and Zhang*, 1992b) The base media and fluid parameters used are:

Aquifer dimensions: 200×100 m; mesh size $\Delta x = \Delta z = 0.2$; Constitutive constants $\eta = 7.52 \cdot 10^{-4}$ and $\beta = 1.62 \cdot 10^{-9}$; Permeability $k_x = k_z = 10^{-11}m^2$ (isotropic); Hydraulic conductivity $K_x = K_z = 10^{-4}m/sec$; Porosity $n=0.2$; Specific storativity $S = 10^{-5}$; Initial solute concentration $C_0 = 16000ppm$; Density of fresh water $\rho_0 = 1000kg/m^3$; Molecular dispersion $D^* = 10^{-9}m^2/sec$; Longitudinal dispersivity: $0.01 < \alpha_L < 100m$; Transversal dispersivity: $0.001 < \alpha_T < 10m$;

where the noted ranges for α_L and α_T reflect the 'scale effect' in real aquifers from local pore-scale values of $O(mm)$ to eventually the regional aquifer-scale of $O(1-100m)$ (cf. *Dagan, 1986; Neuman, 1990*).

Boundary and initial conditions for the pressure p and for the concentration C (1) are chosen to investigate instabilities at an interface at $0.5 z_{max}$ between an initially immobile layer of solute of concentration C_0 and density ρ_{sol0}, superimposed over a layer of pure water ($\rho = \rho_0$). Thus the initial conditions for the solute are $C(t_0) = C_0$ in the upper layer and $C(t_0) = 0$ elsewhere. Hydrostatic conditions for the pressure are assumed initially. All boundaries are taken as impervious and Dirichlet conditions $C = 0$ are imposed there. To initialize the instabilities a small random periodic perturbation δC of variable amplitude and wavelength λ_p is superposed on the concentration field at the interface. As discussed in *Koch* (1993), in a homogenous porous medium these parameters have a crucial impact on the finger development and their morphology.

Variogram analyses of the hydraulic conductivity field and moment analyses of solute tracer plumes have been performed for real aquifers in recent years (cf. *Rehfeldt and Gelhar*, 1992). Variances σ_Y^2 of $O(0.25\text{-}2.5)$, correlation lengths $\lambda_1 \sim \lambda_2 = O(3 - 5m) >> \lambda_3 = O(0.2 - 0.7m)$; i.e., vertical anisotropy, have been obtained in most of the cases. Because of computational limitations only values for σ_Y^2, λ at the lower and upper borders, respectively, of these intervals could only be mimicked in the simulations.

3.2 Finger Morphology

As concluded by *Koch* (1992) and *Koch and Zhang* (1992), for a homogeneous porous medium, hydrodynamic dispersivities, as well as the initial conditions (I-C), have the largest impact on the finger dynamics. Here I will focus on the sensitivity of the finger morphology to the I-C, where it has been found *Koch* (1993) that, unlike in other hydrodynamics stability problems, no criteria for the selection of a 'most-unstable' mode could be established for the Rayleigh-Taylor problem in a homogeneous random medium. Thus, the finger instabilities show typical characteristics of a nonlinear dynamical system whose behavior is determined by the I-C

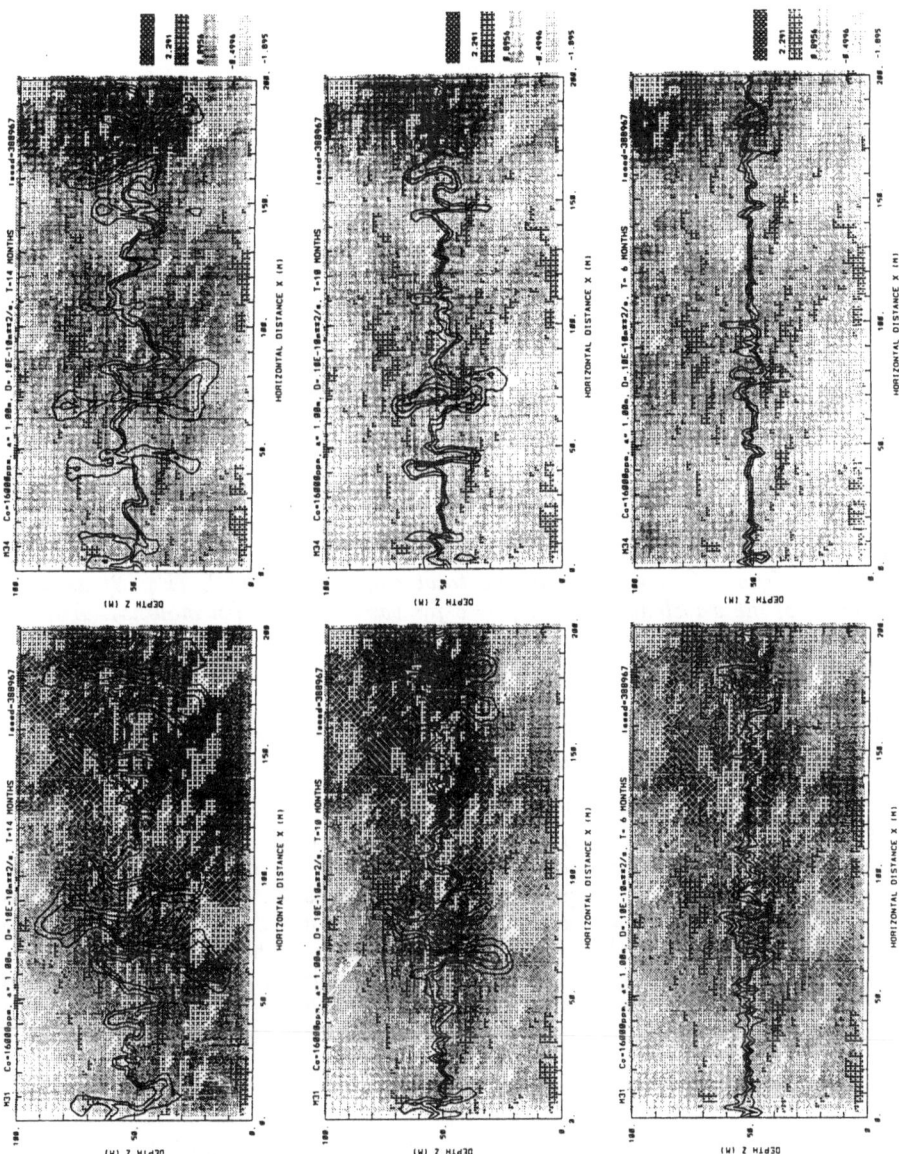

Figure 1. Evolution of fingers in two identical heterogeneous random media with $\sigma_Y^2=0.6$, $\lambda_1 = \lambda_3 =5m$, $\alpha_L=1$ m, $\alpha_T=0.1$ m, after three time spans (6, 10 and 14 months) for two models with initial perturbations of different λ_P. Isolines denote the normalized solute concentrations of $C_n = C/C_0=0.7$; 0.5 and 0.3 at the finger boundaries. Left panels: $\lambda_P/\Delta x=2$; Right panels: $\lambda_P/\Delta x=8$. Legend on the right sides denote the hafton-values of $Y = lnK$

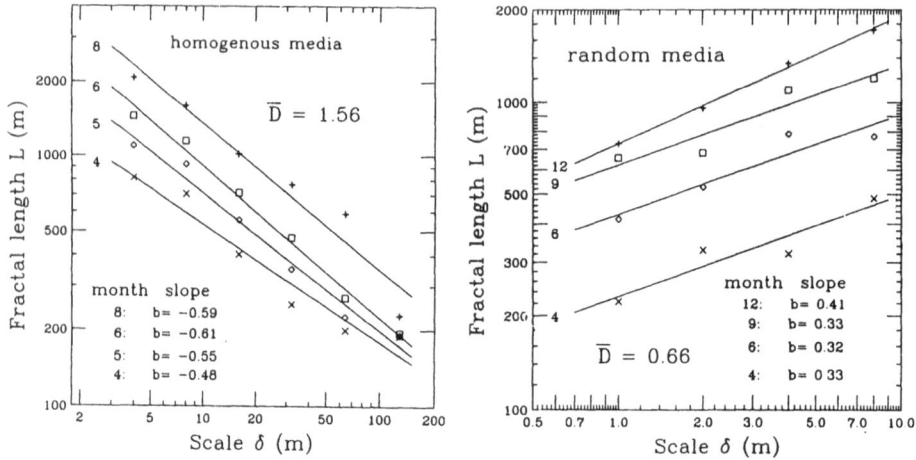

Figure 2. Fractal and self-similar behavior of the finger morphology for homogenous (left panel, adapted from Koch (1993)) and heterogenous (right panel) medium. Symbols denote the total contour lengths L (Eq. 3) as a function of the wavelength-scale $\lambda_p = \delta$ for various times (in months). Also shown are the regression lines with slopes $b = 1 - D$ (Eq. 3), as indicated. Dispersivities are $\alpha_L = 1$ m, $\alpha_T = 0.1$ m.

imposed. Since the last are random in nature an 'order'-parameter, other than a fractal dimension (see below) cannot be established.

The situation is somewhat different for the random heterogenous medium. One of the major conclusions from Fig. 1 and many other, similar models is that, while initial conditions may define the morphology and the dynamics of the fingers at the onset and in the early stages of the instabilities, soon after their fate is more or less imposed by the heterogeneous structure of the porous medium, as realized by the particular simulation of the turning bands generator. Therefore is appears that the porous medium itself has an ordering effect on the finger pattern and fingers are much channeled through the local sections of reduced hydraulic conductivity, with the initial conditions loosing their 'grip' on the finger morphology.

3.2 Fractal analysis

To further quantify the finger dynamics in Fig. 1 and the vanishing effect of the initial conditions, a fractal analysis of the finger morphology, similar to *Koch* (1993), is carried out. As discussed there, fractality is the property of a process, defined by a typical pattern or structure, to replicate itself on many scales. The appealing aspect of complex spatial processes to be fractal—as quantified by the fractal dimension D (see below)—is that, once a pattern has been identified on one length-scale, it can be transformed into another one by a self-similar transformation. There is now sufficient evidence that many spatial geological processes exhibit some kind of fractal, or self-similar behavior (cf. *Feder*, 1988 and *Koch*, 1993, for a review).

Using the classical 'coastline–divider rule' definition of a fractal dimension D which relates the measured length L of a fractal curve to the measurement divider rule δ:

$$L(\delta) = L_N \delta^{1-D} \tag{3}$$

where L_N is the classical length of the curve which would be obtained if $D=1$; i.e., when the measured length $L := L_N$ is independent of the divider rule used. For the analysis of possible fractality of the finger pattern, the total length $L(\delta)$ in Eq. 3 is evaluated through numerical line-integration along the finger boundaries for various wavelengths $\lambda_P(:= \delta$ in Eq. 3) of the initial interfacial perturbation (see *Koch* (1993) for details).

Fig. 2 shows the results of such a fractal analysis for both homogenous (left panel, adapted from Koch (1993)) and the heterogenous medium depicted in Fig. 1. Because the finger morphology is time-dependent, the linear regression of the log-fractal length L versus the log-scale length δ (Eq. 2) has been executed for different times. From the slopes b the fractal dimensions $D = 1 - b$ are computed. For the homogenous medium the negative slopes hint of the increase of the length of the finger boundaries with smaller scale length δ of the initial perturbation, showing that all possible instability modes are excited. An average fractal dimension of $\bar{D} = 1.56$ is obtained. This value for D results in a Hurst exponent H (see *Koch*, 1983) close to 0.5 which may hint of a random process with small spatial correlation for the fingers. For miscible velocity-driven displacement flow, *Maloy et al.* (1985), using a similar approach, have obtained a fractal dimension of $D=1.62$ for the finger pattern. For the heterogenous medium (Fig. 2., right), on the other hand, the situation is opposite: The slopes b are positive; i.e., the length of the finger boundaries increases slightly with increasing scale length δ of the initial perturbation. A theoretical value of $\bar{D} = 0.66$ is obtained for the fractal dimension which, because it it less than one, does not make sense with the notation of the fractal dimension of a 2D planar curve. Therefore fractality of the finger morphology in heterogenous medium with respect to the scale lengths of the initial conditions cannot be established, i.e. supporting the above argument of the 'ordering' influence of the porous media itself.

REFERENCES

Bear, J., and A. Verruijt, (1987) *Modeling Groundwater Flow and Pollution*, D. Reidel Publishing Company, Boston.

Chandrasekhar, S., (1961)*Hydrodynamic and Hydromagnetic Stability*, Oxford University Press, Oxford.

Dagan, G., (1986) Statistical theory of groundwater flow and transport: Pore to laboratory, laboratory to formation, formation to regional scale, *Water Resour. Res.*, 22, 120S–135S.

Feder, J., (1988) *Fractals*, Plenum Press, New York.

Gelhar, L.W. and C.L. Axness, Three-dimensional stochastic analysis of microdispersion in aquifers, *Water. Resour. Res.*, 19, 161–180, 1983.

Hassanizadeh, S.M., and T. Leijnse, 1988. On the modeling of brine transport in porous media, *Water. Resour. Res.*, 24, 321–330.

Hewett, T.A., (1986) Fractal distribution of reservoir heterogeneity and their influence on fluid transport, *SPE 15386, 61st Annual Technical Conference Soc. Petr. Eng., New Orleans, Oct. 5-8.*

Journel, A.B. and Ch.J. Huijbregts, *Mining Geostatistics*, Academic Press, New York, 1978.

Kitanidis, P.K. (1993) Geostatistics, in *Handbook of Hydrology*, Maidment, D.R. (ed.), Chapter 20, McGraw Hill, New York.

Koch, M., (1992) Numerical simulation of finger instabilities in density and viscosity dependent miscible solute transport, in *Proceedings of the 'IX International Conference on Computational Methods in Water Resources', Denver, CO, June 9-12, 1992, Vol. 2, Mathematical Modeling in Water Resources*, Russel, T.F., R.E. Ewing, C.A. Brebia, W.A. Gray and G.F. Pinder (eds.), pp. 155-162, Computational Mechanics Publications, Southampton, UK.

Koch, M. (1993) Modeling the dynamics of finger instabilities in porous media: Evidence for fractal and nonlinear system behavior, in *Advances in Hydroscience and -Engineering, Volume I*, Wang, Sam S.Y. (ed.), pp. 1763-1774, Center for Computational Hydrosciene and Engineering, The University of Mississippi.

Koch, M. and G. Zhang (1992a) Numerical simulation of the migration of density dependent contaminant plumes, *Ground Water*, 5, 731-742.

Koch, M. and G. Zhang (1992b) Forward and inverse modeling of the advection diffusion equation in the presence of sharp fronts, in: *Computational Issues in Geoscience*, Fitzgibbon, W.E., and M.F. Wheeler (eds.), pp. 154-184, SIAM, Philadelphia, PY.

Maloy, K., J. Feder and T. Jossang (1985) Viscous fingering fractals in porous media, *Phys. Rev. Lett.*, 55, 2688-2691.

Neuman, P.S. (1990) Universal scaling of hydraulic conductivities and dispersivities in geological media, *Water Resour. Res.*, 26, 1749-1758.

Parker, J.C. (1989) Multiphase flow and transport in porous media, *Rev. Geophys.*, 27, 311-328.

Perrine, R.L. (1963) A unified theory of stable and unstable miscible displacement, *Soc. Pet. Eng. J.*, 3, 205-213.

Rehfeldt, K.R. and L. W. Gelhar (1992), Stochastic analysis of dispersion in unsteady flow in heterogeneous aquifers, *Water Resour. Res.*, 28, 2085-2099.

Sanford, W.E., and L.F. Konikow (1985) A two-constituent solute transport model for groundwater having variable density, *US. Geol. Surv. Water Resour. Invest. Rep.*, 85-4279.

Schincariol, R.A., and F.W. Schwartz (1990) An experimental investigation of variable density flow and mixing in homogeneous and heterogeneous media, *Water Resour. Res.*, 26, 2317-2329.

Thompson, A.F.B, Ababou, R. and L.W. Gelhar, Implementation of the three-dimensional turning bands random field generator, *Water Resour. Res.*, 2227-2243, 1989.

Welty, C. and L.W. Gelhar (1991) Stochastic analysis of the effects of fluids density and viscosity variability on macrodispersion in heterogenous porous media, *Water Resour. Res.*, 27, 2061-2075.

Wheeler, M.F. (ed.) (1988) *Numerical Simulation in Oil Recovery*, Springer, New York.

Wooding, R.A. (1969) Growth of fingers at an unstable diffusive interface in a porous media or Hele-Shaw cell, *J. Fluid Mech.*, 39, 477-495.

Wooding, R.A., and H.J. Morel-Seytoux (1976) Multiphase flow through porous media, *Ann. Rev. Fluid Mech.*, 8, 233-274.

THE ONSET OF INSTABILITIES IN THE NUMERICAL SIMULATION OF DENSITY-DRIVEN FLOW IN POROUS MEDIA

A.Leijnse[1] and M.Oostrom[2]
[1] RIVM, National Institute of Public Health and Environmental Protection, PO Box 1, 3720 BA Bilthoven, The Netherlands
[2] Pacific Northwest Laboratory, PO Box 999, MS K6-77, Richland, WA 99352, U.S.A.

ABSTRACT

If density gradients play a role in the flow of groundwater and the transport of solutes, instabilities in the flow field may arise due to these density gradients. Although the final flow field in such cases will be completely controlled by the macroscopic physical system, the onset of the instabilities is usually due to local perturbations which may exist on a microscopic scale. In numerical simulations of such systems, instabilities will in general be triggered by numerical roundoff in the solution of the non-linear, discretized equations. The question arises to what extent the onset of instabilities and the final solution is influenced by the numerical solution scheme employed. The results of a series of numerical simulations show that both the element size and the timestep size influence the onset of instabilities in a physically instable system.

INTRODUCTION

In many cases, density gradients play an important role in the determination of the flow field and the transport of contaminants in porous media. Typical examples are the infiltration of contaminated water from landfills in shallow aquifers and the intrusion of sea water in coastal aquifers. Especially in cases where the density of the liquid phase decreases with depth, instable situations may occur, and the transport of the contaminants may be enhanced by the occurrence of free convection. Whether such an unstable situation exists depends on a number of physical parameters typical for the system considered (Oostrom et al., 1992). If such instabilities do occur, they will usually be triggered by perturbations in the density distribution as a result of very local (or even microscopic) phenomena. Typically,

A. Peters et al. (eds.), Computational Methods in Water Resources X, 489–496.
© 1994 Kluwer Academic Publishers. Printed in the Netherlands.

once the instabilities in the density distribution are triggered, the growth and final distribution of these instabilities is fully governed by the macroscopic physical parameters. In numerical simulations of such systems, instabilities can either be generated by perturbing the solution (e.g. by introducing blocks with deviating permeability) or by using the numerical roundoff in the solution of the (non-linear) equations as perturbation to the concentration (density) distribution. Since the equations are discretized equations, their behaviour regarding bifurcation is not necessarily the same as the behaviour of the partial differential equations describing the system.

The objective of this paper is to investigate the differences in behaviour of the PDE's describing the flow and transport in a porous medium and the discretized equations that result from a spatial discretization using a Galerkin weighted residual method and a temporal discretization using an Euler backward (implicit) finite difference approximation. To make the comparison possible, a simple 2-D vertical cross-section of a homogeneous, isotropic porous medium is considered. For such a system, the onset of instabilities can analytically be determined. A number of numerical simulations are carried out to investigate the onset of instabilities in the numerical system.

GOVERNING EQUATIONS

The equations describing the flow and transport of an inert solute in a porous medium follow from the mass conservation principles for the liquid and the solute (Hassanizadeh and Leijnse, 1988):

$$\frac{\partial}{\partial t}(n\rho) + \nabla \cdot (\rho \boldsymbol{q}) = 0 \tag{1}$$

$$\frac{\partial}{\partial t}(n\rho\omega) + \nabla \cdot (\rho\omega\boldsymbol{q}) + \nabla \cdot \boldsymbol{J} = 0 \tag{2}$$

where n=porosity, ρ=liquid density, q=specific discharge, ω=solute mass fraction and J=dispersive mass flux of solute. q and J are given by Darcy's and Fick's law respectively:

$$\boldsymbol{q} = -\frac{\boldsymbol{k}}{\mu} \cdot (\nabla p - \rho \boldsymbol{g}) \tag{3}$$

$$\boldsymbol{J} = -n\rho\boldsymbol{D} \cdot \nabla \omega \tag{4}$$

where k=permeability, μ=liquid viscosity, p=pressure, g=acceleration of gravity and D=dispersion tensor.

In general, both the liquid density ρ and the liquid viscosity μ are functions of the solute mass fraction ω. The dispersion tensor $\mathbf{D}$ is a function of the velocity q. For the present study, a homogeneous isotropic porous medium is considered, i.e. both the porosity n and the permeability k are constants. The dispersion tensor $\mathbf{D}$ and the liquid viscosity μ are assumed to be constants. Furthermore, the liquid density ρ is assumed to be a linear function of the solute mass fraction ω. Under these assumptions and Boussinesq's approximation, i.e. assuming that the variation in the liquid density can be neglected except in the gravity term in Darcy's law (3), the equations can be written in dimensionless form as:

$$\nabla \cdot \mathbf{q} = 0 \tag{5}$$

$$\frac{\partial \omega}{\partial t} + \nabla \cdot (\omega \mathbf{q}) - \nabla^2 \omega = 0 \tag{6}$$

$$\mathbf{q} = -\nabla p - A\omega \mathbf{e}_z \tag{7}$$

where $\mathbf{e}_z$ is the unit vector in the positive z-direction (positive upward) and A is the Rayleigh number defined by:

$$A = \frac{\Delta\rho g k H}{\mu D} \tag{8}$$

$\Delta\rho$ is the maximum density difference in the system and H is a characteristic length of the system. Note that in equations (5)-(7) all variables are dimensionless, where the solute mass fraction ω has a value between 0 and 1. For a given geometry and a given set of initial and boundary conditions, the solution of equations (5) through (7) is fully governed by the value of the Rayleigh number A.

PHYSICAL SYSTEM

The physical system considered here is a square box with dimensions H. Boundary conditions are no-flow and no-dispersive flux at the side boundaries, no-flow and prescribed solute mass fraction at top and bottom boundaries. The (scaled) solute mass fractions are $\omega=1$ at the top and $\omega=0$ at the bottom. The initial conditions are given by the basic solution to equations (5) through (7):

$$\omega = z \quad ; \quad p = p_0 - \tfrac{1}{2}A\,z^2 \quad ; \quad q_x = q_z = 0 \tag{9}$$

It is well known, that this basic solution is unstable for large enough values of the Rayleigh number A (Gebhart *et al.*, 1988). This means, that an initial perturbation in this basic solution will grow with time for large enough values of A. From a linearized perturbation analysis of the equations the initial growth rates of different possible forms of the instabilities can be given as a function of the Rayleigh number A. Two different instable modes will be considered here. The final flow fields and solute mass fraction distribution for these two modes are shown in fig. 1. Mode 1 has one convection cell and mode 2 has two convection cells.

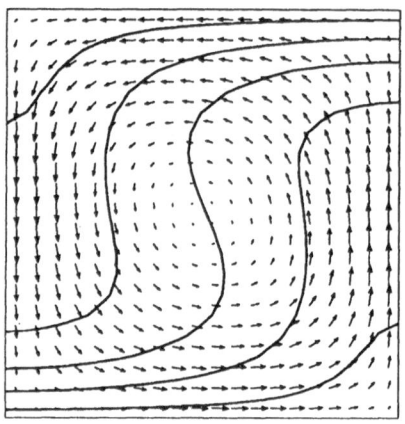 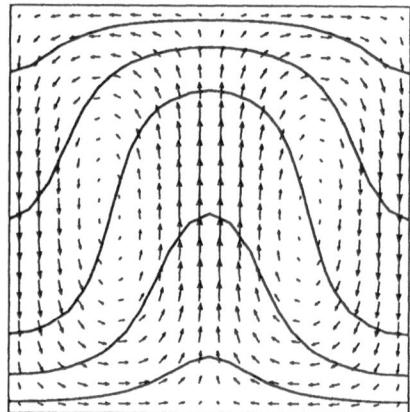

Figure 1   Mode 1 and mode 2 instabilities

The inital growth rates of the two modes are given by:

$$\lambda_1 = \frac{1}{2} A - 2\pi^2 \quad ; \quad \lambda_2 = \frac{4}{5} A - 5\pi^2 \tag{10}$$

where λ_1 and λ_2 are the initial growth rates for mode 1 and mode 2 respectively.

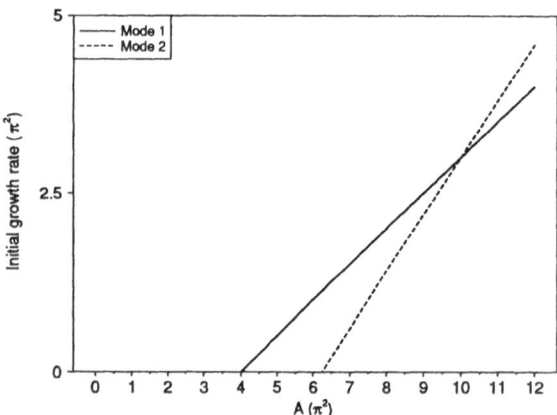

Figure 2   Initial growth rates of instabilities

Note, that $\lambda_1 > 0$ for $A > 4\pi^2$ and $\lambda_2 > 0$ for $A > 6.25\pi^2$. Hence, the mode 1 instability will occur for Rayleigh numbers larger than $4\pi^2$ only, and the mode 2 instability for Rayleigh numbers larger than $6.25\pi^2$. Fig. 2 shows the initial growth rates as a function of the Rayleigh number A. From fig. 2 it is evident that for Rayleigh numbers largen than $10\pi^2$ the growth rate of mode 2 is larger than the growth rate of mode 1.

NUMERICAL SIMULATIONS

A number of simulations have been carried out with the finite element code METROPOL-3 (Sauter *et al.*, 1993). The full set of equations (1) through (4) is solved with a Galerkin weighted residual method for the spatial derivatives and an Euler backward time stepping scheme (Leijnse, 1992). The maximum density difference in the simulations is small enough to assume that Boussinesq's approximation is valid. In all following simulations, changes in the Rayleigh number were obtained by changing the permeability of the porous medium.

The first series of simulations was carried out to check whether the numerical code was able to reproduce the switch from a stable linear solute mass fraction distribution to an unstable situation. Initial conditions were as given by (9) and the possible onset of an instability is controlled by numerical roundoff. Transient simulations were carried out for a large timespan until a (pseudo) steady state situation was reached. The element size in this simulations was 0.1 x 0.1 m. If a stable linear solute mass fraction occurs, the velocities in the system should all be zero. Hence, the switch from a stable to an unstable situation can be recognized by observing the value of the maximum velocity in the system. Fig. 3 shows a plot of this maximum velocity at the end of the simulation as a function of the Rayleigh number A. From this plot it is evident that the switch from a stable to an unstable solution occurs very close to the theoretical value of $A=4\pi^2$.

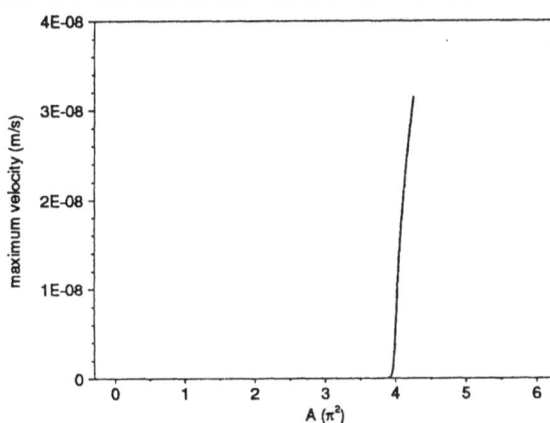

Figure 3 Maximum steady state velocity

A second series of simulations was carried out with the initial conditions as given by (9) and an initial perturbation added to the solute mass fraction distribution. The initial perturbation consisted of mode 1 and mode 2 instabilities in a known ratio. The total amplitude of the initial perturbation was less than 0.03 (scaled solute mass fraction). The simulations were carried out at a Rayleigh number of $10\pi^2$. Different element sizes were used: 10, 5 and 2.5 cm respectively. Theoretically, at a Rayleigh number of $10\pi^2$, the initial growth rates for the mode 1 and mode 2 instabilities are the same. Since the final solution is uniquely determined by the initial conditions (Clément *et al.*, 1992) it follows that the final solution is completely determined by the perturbation applied to the initial conditions. If mode 1 is dominant in this perturbation, the final solution should be a mode 1 instability. Simulations were

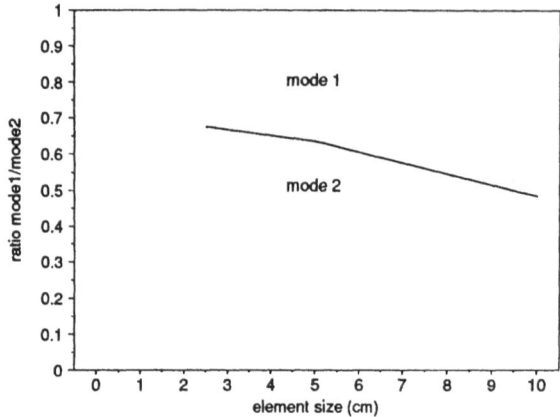

Figure 4 Switch from mode 2 to mode 1
vs. amplitude ratio

carried out for increasing ratio of the amplitudes of mode 1 and mode 2 in the perturbation in the initial conditions. Fig. 4 shows the ratio's for which a switch from a mode 2 instability to a mode 1 instability occurs as a function of the element size. Theoretically this switch should occur at a ratio 1. It is evident from fig. 4 that the switch occurs earlier, i.e. the numerical system has a "preference" for the mode 1 instability if compared with the behaviour of the partial differential equations. The difference is smaller for smaller size elements, as might have been expected.

A third series of simulations were carried out which were very similar to the second series of simulations. Again, the initial conditions are given by (9) with a perturbation added. In this case however, the ratio of the amplitudes of mode 1 and

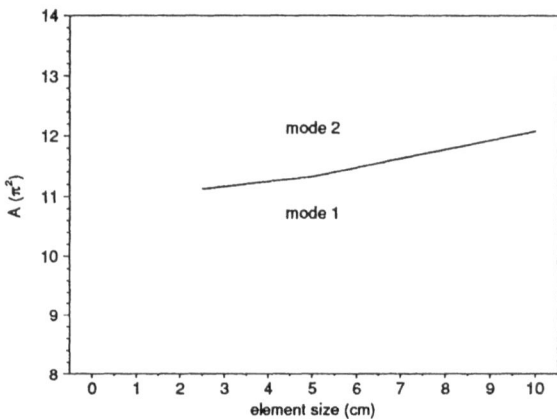

Figure 5 Switch from mode 1 to mode 2 vs. A

mode 2 perturbations is always 1, i.e. mode 1 and mode 2 are equally present in the initial condition. Simulations were now carried out for increasing Rayleigh number A. Theoretically, a switch from a mode 1 final solution to a mode 2 final solution should occur at a Rayleigh number of $10\pi^2$, where the initail growth rates for both modes are equal. Fig. 5 shows the values of the Rayleigh numbers at which the switch occurs for different

element sizes. Again, from this plot it is evident that the numerical system has a "preference" for the mode 1 instability.

A final series of numerical simulations were carried out with an element size of 10 cm and fixed timestep sizes. The Rayleigh number in all simulations was $A=10\pi^2$. The initial conditions are as given by (9) with a perturbation added. As in the second series of numerical simulations, the perturbation consisted of mode 1 and mode 2

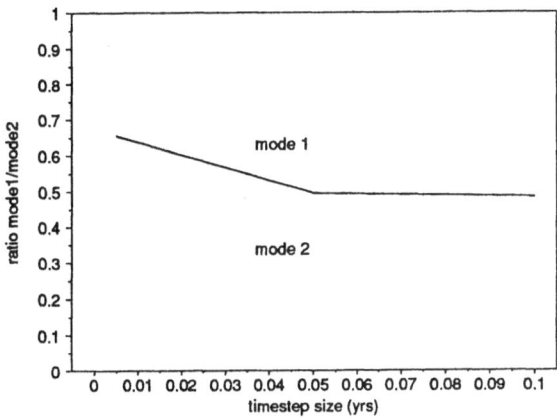

Figure 6   Effect of timestep size on switch from mode 2 to mode 1

instabilities in a known ratio. This ratio was increased and the switch from a mode 2 final solution to a mode 1 final solution was determined. Theoretical, for the given value of the Rayleigh number, this switch should occur at a ratio of 1. Fig. 6 shows the results obtained, and it is again evident that the switch occurs at a lower ratio, i.e. the numerical system has a "preference" for the mode 1 solution.

CONCLUSION

The results of the numerical simulations show that the behavior of the discretized equations describing density-driven flow in porous media differs slightly from the behavior of the partial differential equations. The effect decreases with decreasing element size and with decreasing timestep size.

REFERENCES

Clément, Ph., C.J. van Duijn and Shuanhu Li, On a nonlinear elliptic-parabolic partial differential equation system in a two-dimensional groundwater flow problem, *SIAM J. Math. Anal.*, 23(4), 836-851, 1992

Gebhart, B., Y. Jaluria, R.L. Mahajan and B.Sammakia, *Buoyancy-Induced Flows and Transport*, Harper and Row, New York, 1988

Hassanizadeh, S.M. and T.Leijnse, On the modeling of brine transport in porous media, *Water Resour. Res.*, 24, 321-330, 1988

Leijnse, A., Comparison of solution methods for coupled flow and transport in porous media, In: *Computational methods in surface and subsurface hydrology*, ed. Russell, Brebbia, Gray and Pinder, 1992

Oostrom, M., J.S.Hayworth, J.H.Dane and O.Guven, Behavior of dense aqueous phase leachate plumes in homogeneous porous media, *Water Resour. Res.*, 28(8), 2123-2134, 1992

Sauter, F.J., A. Leijnse and A.H.W. Beusen, METROPOL User's Guide, *RIVM Report 725205003*, Bilthoven, the Netherlands, 1993

MODIFIED EULERIAN LAGRANGIAN METHOD FOR FLOW AND TRANSPORT IN HETEROGENEOUS SATURATED AQUIFER

S. LUMELSKY and S. SOREK

Technology Institute of Israel, Civil Eng., Technion City, Haifa, 32000, Israel
J. Blaustein Inst. for Desert Research, and Mechanical Eng., Perlstone Center for
Aeronautical Studies, Ben-Gurion univ., Sede Boker Campus, 84993, Israel.

Eulerian Lagrangian (EL) and Modified Eulerian Lagrangian (MEL) numerical schemes are employed to the solution of decoupled 1-D flow and transport problems of a heterogeneous aquifer. Dispersion and conductivity coefficients are taken as continuous or discrete functions in space. Both schemes apply forward and backward particle tracking techniques. In the case of the transport problem, the EL scheme is associated with moving particles governed by fluid's velocity while for the MEL scheme, they are shifted by a combination of the fluid's velocity and the gradient of the hydrodynamic dispersion. In the case of the flow problem, we apply the MEL scheme with particle's velocity associated with the gradient of the hydraulic conductivity coefficient.

Comparisons between the MEL, EL and Eulerian Finite Elements (EFE) methods prove that the MEL scheme is superior in yielding almost no deviations from analytical solutions of 1-D flow and transport problems.

INTRODUCTION

All EL schemes are based on formal decomposition of the differential operator into "advection" along characteristic pathlines. In addition, Neuman (1981, 1984), Neuman and Sorek (1982), Sorek and Braester (1988) and Sorek (1985a,b; 1988) also decompose the dependent variable into advective and residual parts. The resulting advection problem is solved by methods, applicable to the Lagrangian formulation. The "dispersive" problem may be solved by conventional finite elements or finite difference methods at a fixed grid.

An EL least-square collocation method was proposed by Bentley, et al., (1990). The EL collocation method with the use of a characteristic method for the advective problem is documented by Allen and Khosravani (1990). The performance of the EL methods and sources of errors in such schemes are discussed by Bentley and Pinder (1992).

In the framework of this paper the following objectives will be cosidered:
- comparison of the performance of the EL, MEL, and EFE schemes for different spatial distributions of the dispersion coefficient concerning a 1-D transport problem;
- compare between the MEL and EFE schemes for the solution of a 1-D flow problem.

A. Peters et al. (eds.), Computational Methods in Water Resources X, 497–504.
© *1994 Kluwer Academic Publishers. Printed in the Netherlands.*

MATHEMATICAL STATEMENT

Consider the general advection-dispertion problem, without source terms, given by

$$\frac{\partial u}{\partial t} = -\nabla \cdot [(1 - f)\mathbf{V}u - \mathbf{K}\nabla u], \qquad (1)$$

where $u(\mathbf{x}, t)$ - denotes a dependent variable being a function of the spatial vector $\mathbf{x}$ and time t; $\mathbf{V}(\mathbf{x}, t)$ - denotes the velocity vector; $\mathbf{K}(\mathbf{x})$ - denotes a tensor associated with linear dependency between the dispersive flux and ∇u; f - denotes a control coefficient describing the transport problem $(f = 0)$ and the flow problem $(f = 1)$ The Eulerian form (1) may be conformed to different Eulerian-Lagrangian presentations, depending on φ.

$$\frac{Du}{Dt} = (1 - \varphi)\nabla \cdot [\mathbf{K}\nabla u] + \varphi \mathbf{K}\nabla^2 u - (1 - f)u\nabla \cdot \mathbf{V}, \qquad (2)$$

where the hydrodynamic derivative is defined by using the aparant velocity

$$\frac{D}{Dt} \equiv \frac{\partial}{\partial t} + \mathbf{V}^* \cdot \nabla; \qquad \mathbf{V}^* \equiv (1 - f)\mathbf{V} - \varphi\nabla \cdot \mathbf{K}. \qquad (3)$$

Note that in (2), u is interpreted as being carried, say by a particle, along a pathline defined by

$$D\mathbf{x} = \mathbf{V}^* Dt. \qquad (4)$$

In view of (2) to (3), we may construct various solution schemes. The MEL one, will utilize the modified hydrodynamic derivative incorporating $\nabla \cdot K$, the EL will be based on the standard hydrodynamic derivative and the EFE will conform to a fixed frame of reference. The various cases that may be obtained are described in table 1.

Problem	f	Control coefficients	
		$\varphi = 0$	$\varphi = 1$
Transport	0	EL scheme	MEL scheme
Flow	1	EFE scheme	MEL scheme

Table 1: Choice of control coefficients

Decomposition of the dependent variable

Neuman and Sorek, (1982) proposed to split the dependent variable into an *advection part* $\bar{u}$ and a residual $\mathring{u}$ regarded as the *dispersion part*

$$u = \bar{u} + \mathring{u}. \qquad (5)$$

The advection-dispersion, (2), can be formally decoupled into a purely hyperbolic "advection problem" defined in terms of $\bar{u}$

$$\frac{D\bar{u}}{Dt} = 0, \qquad (6)$$

and a predominantly parabolic "dispersion problem" defined in terms of $\hat{u}$ and u

$$\frac{D\hat{u}}{Dt} = (1 - \varphi)\nabla \cdot [\mathbf{K}\nabla u] + \varphi\mathbf{K}\nabla^2 u - (1 - f)u\nabla \cdot \mathbf{V}. \tag{7}$$

In what follows we will consider the case of a constant velocity and a 1-D domain. This, however, will not affect the development of the numerical scheme nor will it affect the implications resulting from the considered examples.

NUMERICAL IMPLEMENTATION

a. Advection by partical tracking techniques

Let a particle, p, located at point x_p at time t^k, be associated with $\bar{u}_p^k \equiv u(x_p, t^k)$. At the end of the time step t^{k+1}, each particle, p, reaches a new forward position, x_p^{k+1}, obtained from the solution of (4) by, say, the Runge-Kutta method.

This describes the *forward particles tracking* procedure, which projects $\bar{u}_p^{k+1}(= \bar{u}_p^k$, by virtue of (6)) onto nodal points. For nodes, that are not covered by these clouds, i.e., for any node located at x_n such that

$$x_{p\ min}^{k+1} > x_n > x_{p\ max}^{k+1}, \tag{8}$$

we use *backward particles tracking* procedure to project $u(\equiv {}^k u)$ from a backward location ${}^k x_p$ onto nodes obeying (8). The ${}^k x_p$ location is obtained by solving (4) with $x_p^{k+1} = x_n$ and ${}^k u$ is the interpolated value at the backward element.

b. Dispersion by Finite Elements

We approximate $u(x, t)$ and $K(x)$ by

$$u(x, t) \simeq \hat{u}(x, t) \equiv \sum_{j=1}^{N} u_j(t)\xi_j(x); \qquad K(x) \simeq \hat{K}(x) \equiv \sum_{l=1}^{N} K_l\xi_l(x). \tag{9}$$

where N - denotes the total number of the grid nodes; K_l - denotes the subscribed values of K at the nodal points and $\xi_j(x_i)(\equiv \delta_{ij}$, the Kroniker delta function) -denotes the shape functions.

We now substitute (9) into (7) together with Galerkin orthogonalization for the 1-D case spanning between $[x = 0, x = L]$, with $V = const$.

The hydrodynamic time derivative over $\hat{u}$ may be approximated by backward difference for each $\Delta t(= t^{k+1} - t^k)$

$$\frac{D\hat{u}}{Dt} \simeq \frac{\hat{u}^{k+1} - {}^k\hat{u}}{\Delta t}. \tag{10}$$

The resulting global algebraic set after integration by parts, in view of (10) and accounting for the forward particle projection, becomes

$$\sum_{j=1}^{N} u_j^{k+1} \int_{x=0}^{x=L} \left\{ \frac{1}{\Delta t}\xi_i\xi_j + \left[\sum_{l=1}^{N} K_l\xi_l\right]\frac{d\xi_i}{dx}\frac{d\xi_j}{dx} + \varphi\left[\sum_{l=1}^{N} K_l\frac{d\xi_l}{dx}\right]\xi_i\frac{d\xi_j}{dx} \right\}dx =$$

$$\frac{1}{\Delta t} \int_{x=0}^{x=L} \left(\sum_{j=1}^{N_1} u_j^k \xi(^k x_p) + \sum_{m=1+N_1}^{N} \bar{u}_m^{k+1} \delta_{im} \right) \xi_i dx + K \frac{\partial \hat{u}}{\partial x} (\delta_{iN} - \delta_{i1}), \tag{11}$$

where N_1, denotes the nodes that are not covered by cloudes of particles.

In view of (3) and (11), for $\varphi = 1$, we note that the MEL scheme may be viewed as the EL scheme to which we subtract and add the product $\nabla K \cdot \nabla u$. This is implanted, respectively, in the velocity of the particles and in the orthogonalization procedure.

NUMERICAL SIMULATIONS

We set ourselves at solving (2) for constant velocity in a 1-D domain (under general units system) spanning between $x \in [0, 2.5]$ and during a time interval of $t \in [0, \tau]$. Spatial and time steps were chosen to be $\Delta x = 0.05$ and $\Delta t = 0.5$, respectively. Adequate selection of the shape functions was found to be very important considering the discrete choice of K values (see, e.g., Fig.1a) and the need to approximate their gradient as described in (3). Use of linear shape functions essentially simplifies the matrix assembling procedure in a Galerkin FE scheme, but in our case it lead to numerical oscillations which was not experienced when the quadratic shape functions were applied as the interpolating functions (Fig.1b).

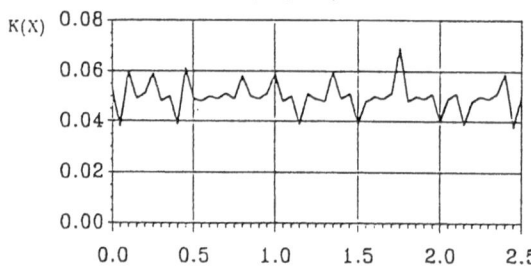

Fig.1a: Random spatial distribution of K.

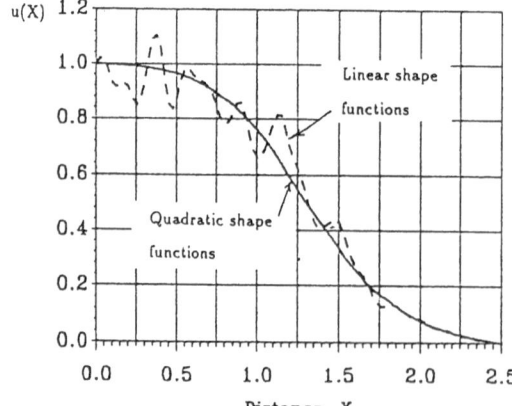

Fig.1b: Transport problem simulated by MEL for above spatial distribution of K.

For the transport problem delineated in Fig.1 we considered an initial concentration $u(x,0) = 0$ and boundary conditions $u(0,t) = 1$ and $u(2.5,t) = 0$.

Analytical solution of (2) for the 1-D transport $(f = 0)$ and flow $(f = 1)$ problems were developed concerning $K(x)$ as a cubic polynomial function of space. The general analytical solution for (2) may be given by

$$u = e^{at} \cdot [1 - \frac{x}{2L} - \frac{x^2}{2L^2}], \tag{12}$$

where $a(= -0.075)$ - denotes a constant time integration factor.

The dispersion coefficient $(K \equiv D)$ with $V(= 0.5)$ as the velocity is given by

$$D(x) = \frac{2ax^3 + (6V + 3aL)x^2 + (6VL - 12aL^2)x}{6(L + 2x)}, \tag{13}$$

Spatial distribution of $D(x)$ and the associated Peclet number $Pe(= \frac{\Delta x V^*}{D})$ are described in Fig.2.

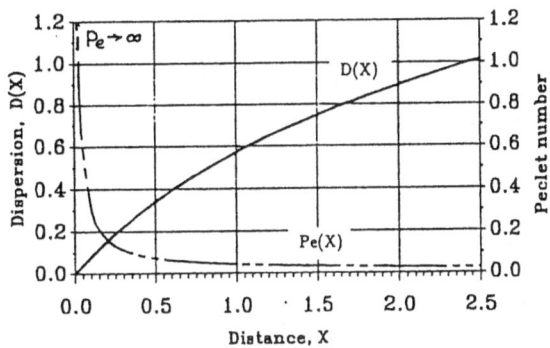

Fig.2: Spatial distribution of $D(X)$ and its associated $Pe(X)$.

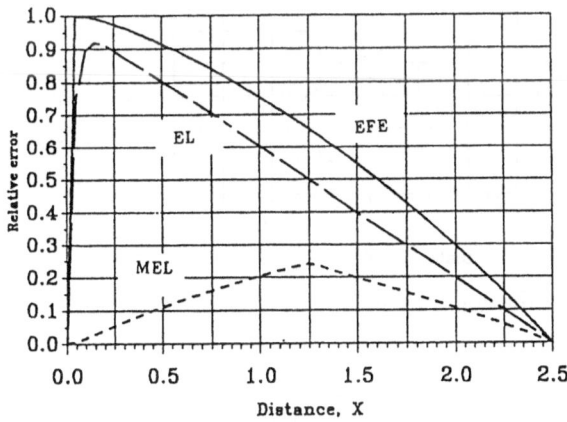

Fig.3: Relative error distribution for the transport problem.

The initial and boundary conditions are derived from (12). Comparisons between the solutions of MEL, EL and EFE schemes were performed and the relative deviation error

$$Relative \quad error = \frac{u_{numerical} - u_{analytical}}{|u_{numerical} - u_{analytical}|_{max}} \tag{14}$$

is delineated in Fig.3. The maximum absolute deviation for the MEL scheme was 0.0076, for the EL scheme was 0.0289, and 0.0314 for the EFE. Concerning Courant number $Cr(= \frac{V^{*}\Delta t}{\Delta x})$ based on particle's velocity V^{*}, we find that for the EL and EFE schemes this gives $Cr = 5$, while for MEL it decreases from $Cr = 72.25$ at $x = 0$ to $Cr = 3.77$ at $x = 2.5$.

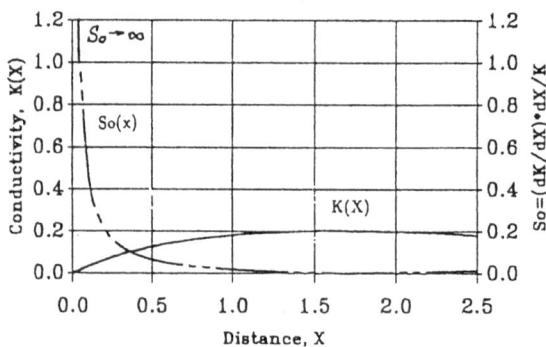

Fig.4: Spatial distribution of $K(X)$ and its associated $So(X)$.

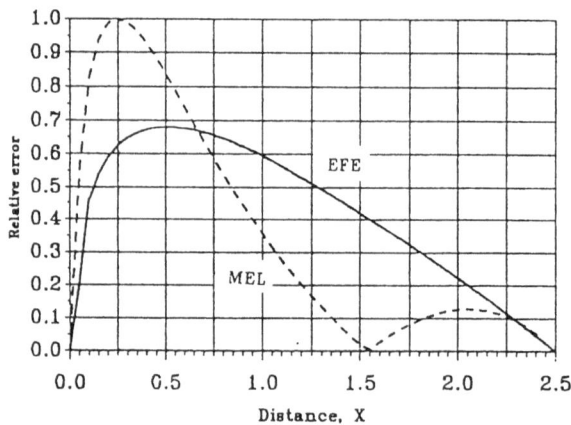

Fig.5: Relative error distribution for the flow problem.

For the flow problem the conductivity coefficient $K(x)$ was

$$K(x) = \frac{2ax^3 + 3aLx^2 - 12aL^2x}{6(L + 2x)}. \tag{15}$$

The analogous to Peclet number for the flow problem as proposed by Sorek and Braester (1988), reads

$$So = \frac{\Delta x \left[\frac{dK}{dx}\right]}{K}.$$ (16)

The spatial distribution of K and So is described in Fig.4. The initial and boundary conditions for ϕ, the hydraulic head, are obtained from (12). The relative error obtained by the MEL and EFE solutions is shown in Fig.5. Maximum absolute deviation for the EFE scheme was 0.0021 and 0.0030 for MEL. We note that in this case, the performance of both schemes is similar.

Let us now reconsider the choice of heterogeneous media which in our case is expressed by the random distribution of K in space (Fig.1a). We now employ also the EFE and EL schemes as numerical procedures to solve the 1-D transport problem (2) with respect to the same boundary and initial conditions. The comparison between the solutions obtained by the MEL, EL and EFE schemes is depicted in Fig.6. We note that the MEL scheme is superior in producing less numerical dispersion.

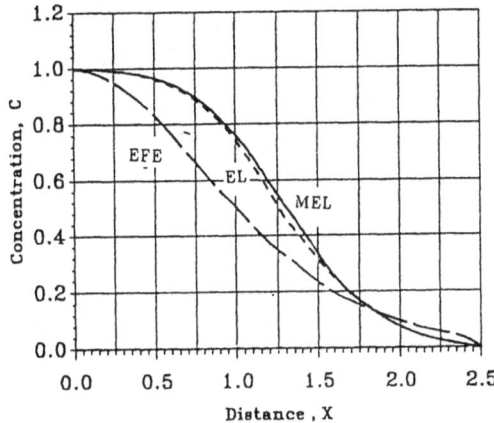

Fig.6: Transport problem simulated by the MEL, EL, and EFE for the arbitrary spatial distribution of K.

In the case of the EFE scheme we had also tried to apply point-wise values of K instead of continuous piecewise distribution. No significant difference was noticed.

CONCLUSION

The MEL scheme was developed specifically to address steep dispersion gradients in transport problems or steep conductivity gradients associated with flow problems. In such cases, the balance equation is hyperbolic dominated and simulations suffer from numerical dispersion problems. The MEL scheme is based on a modified hydrodynamic time derivative incorporating the dispersion (or conductivity) gradient as part of particle's velocity. It was employed to solve 1-D flow and transport problems involving continuous or discontinuous distribution in space of the conductivity and dispersion coefficients, respectively. Performance of the MEL scheme was compared against that of the EL and EFE schemes.

The MEL was proven to be superior in producing much less deviation from the analytical solution of the 1-D transport problem. In the case of random spatial distribution of the dispersion coefficient, we note that both the MEL and EL schemes are practically free of numerical dispersion, in compare to the EFE scheme. Yet, the MEL scheme produces a somewhat sharper front in better comply with the Peclet number. The MEL and EFE schemes were compared to an analytical solution of a 1-D parabolic flow problem. The MEL scheme yielded very small deviations, similar to the EFE scheme. We conclude that the MEL scheme produced better results in terms of numerical efficiency and accuracy.

REFERENCES

1. Allen, M.B., and Khosravani, A. (1990) "Eulerian-Lagrangian method for finite-element collocation using the modified method of characteristic", Computational Methods in Subsurface Hydrology Proc 8 Int Conf Comput Method Water Resour. Publ by Springer-Verlag Berlin, Dept ZSW, Berlin 33, GER. pp. 375-379.

2. Bentley, L.R., and Pinder, G.F. (1992) "Eulerian-Lagrangian solution of the vertically averaged groundwater transport equatios", Water Resourses Research, Vol. 28, No. 11, pp. 3011-3020.

3. Neuman, S.P. (1984) "Adaptive Eulerian - Lagrangian finite element method for advection - dispersion", Int. J. Num. Methods in Eng., 20, pp 321-337.

4. Neuman, S.P. (1981) "A Eulerian-Lagrangian numerical scheme for the dispersion-convection equation using conjugate space-time grids", Jour. Comp. Phys., No. 41(2), pp. 270-294.

5. Neuman, S.P., and Sorek, S. (1982) "Eulerian-Lagrangian methods for advection-dispersion", Pro. 4-th Inter. Conf. F. E. W. R., FGR, pp. 14.41-14.68.

6. Sorek, S. (1988) "Eulerian-Lagrangian method for solving transport in aquifers", Advances in Water Resources, V. 11, No. 2, pp. 67-73.

7. Sorek, S., and Braester, C. (1988) "Eulerian-Lagrangian formulation of the equations for groundwater denitrification using bacterial activity", Advances in Water Resources, V. 11, No. 4, pp. 162-169.

8. Sorek, S. (1985a) "Eulerian-Lagrangian formulation for flow in soil", Advances in Water Resources, V. 8, pp. 118-120.

9. Sorek, S. (1985b) "Adaptive Eulerian-Lagrangian method for transport problems in soils", in Scientific Basis for Water Water Resources Management, IASH Publ., 153, pp. 393-403.

A MODIFIED METHOD OF CHARACTERISTICS TECHNIQUE FOR SIMULATING CONTAMINANT TRANSPORT IN VARIABLY-SATURATED POROUS MEDIA

R.J. MITCHELL[1] and A.S. MAYER[2]
[1]Department of Civil and Environmental Engineering
[2]Department of Geological Engineering, Geology and Geophysics
1400 Townsend Drive
Michigan Technological University, Houghton, MI, 49931-1295, USA

A modified method of characteristics (MMOC) model was developed to simulate transport of nonreactive solutes in two-dimensional, unsaturated porous media. The advective concentration was determined using single-step reverse particle tracking and the dispersive flux was approximated using the Galerkin finite element method. The model accurately backtracks along characteristics in a variable flow field and employs a quadratic-linear interpolation scheme. Two-dimensional unsaturated flow fields were used to analyze the sensitivity of the MMOC model.

INTRODUCTION

Contaminant transport modeling in heterogeneous unsaturated porous media often is hindered by a wide variation in velocities. Conventional numerical methods for transport equations often produce errors when widely variable flow fields cause high enough Peclet numbers (Pe) to produce numerical dispersion and oscillations in solutions. Upwinding techniques have been shown to be effective in reducing numerical error. However, these techniques often are afflicted with artificial dispersion, or require small grid spacings and are often limited to Courant numbers (Cr) less than one.

Transport models based on the modified method of characteristics (MMOC) have been shown to be a suitable alternative for advective-dominated problems with variable flow fields in saturated (*e.g.*, Chiang *et al.*, 1989) and unsaturated subsurface systems (*e.g.*, Yeh *et al.*, 1993). The MMOC method, which employs a fixed grid system, is more computationally efficient than forward particle tracking or adaptive techniques (Zang *et al.*, 1993) and, unlike upwinding methods, the MMOC method is not limited to $Cr < 1$. The accuracy of the MMOC method in variably saturated media is a function of the technique used to backtrack a particle along a characteristic (Allen and Khosravani, 1992) and the interpolation method used in determining the advective concentration (Healy and Russel, 1989).

A. Peters et al. (eds.), Computational Methods in Water Resources X, 505–512.
© 1994 Kluwer Academic Publishers. Printed in the Netherlands.

When the velocity field is uniform, a simple one-step, Euler algorithm can be used to backtrack a particle accurately. However, in heterogenous conditions velocities can change rapidly and backtracking accurately under these conditions can be difficult. Various techniques have been suggested for tracking particles when the velocity fields are more complex, such as fourth-order Runge-Kutta methods (*e.g.*, Baptista, 1984; Yeh *et al.*, 1993), a multiple-step, Euler method (Allen and Khosravani, 1992) and semi-analytical techniques (*e.g.*, Pollock, 1988;).

A drawback of the MMOC method is that in the vicinity of sharp concentration fronts, linear interpolation will introduce numerical dispersion in the solution. Second-order accurate quadratic interpolation eliminates numerical dispersion but can produce oscillations (Cheng *et al.*, 1984). For one-dimensional transport, Healy and Russel (1989) have shown that a combination of quadratic and linear interpolation can produce oscillation-free solutions with reduced dispersion.

This paper will illustrate that when considering a variable, but piece-wise linear velocity field, a semi-analytical method is superior to other algorithms for backtracking along characteristics. It will be shown that the quadratic-linear interpolation technique outlined by Healy and Russel (1989) for reducing dispersion and eliminating oscillations can be extended to two dimensions. When these techniques are employed, a MMOC model can be used effectively in two-dimensional, unsaturated domains having a range of Pe and Cr numbers.

GOVERNING EQUATIONS AND MMOC ALGORITHM

The governing equations representing nonreactive solute transport in an unsaturated porous media can be written in Lagrangian form as

$$\theta \frac{dC}{dt} = \frac{\partial}{\partial x_i} \left(\theta D_{ij} \frac{\partial C}{\partial x_j} \right) \tag{1}$$

where t is time, x $(i, j = 1, 2)$ is the spacial distance, θ is the volumetric water content, C is the solute concentration, D_{ij} is the hydrodynamic dispersion tensor and

$$\frac{dC}{dt} = \frac{\partial C}{\partial t} + v_i \frac{\partial C}{\partial x_i} \tag{2}$$

represents the total derivative, which indicates the rate of change of C along velocity characteristics. Here, v_i $(= q_i/\theta)$ is the pore water velocity, where q_i is the Darcy flux.

Single-step reverse particle tracking (Neuman, 1984) is used to backtrack along velocity characteristics. With this approach, a fictitious particle from grid point x_i, is sent backward to the point

$$\bar{x}_i = x_i - \int_{t_k}^{t_{k+1}} v_i dt \tag{3}$$

which infers that a particle leaving $\bar{x}_i$ at t_k will arrive at the grid point x_i at t_{k+1}.

The solute concentration $\bar{C}_i$ at $\bar{x}_i$, is approximated by interpolation, using

$$\bar{C}_i = \sum_{j=1}^{N} C_j(x_i)\psi_j(x_i) \tag{4}$$

where ψ_j is a shape function and N represents the number of nodes used in the interpolation. The grid point x_i then assumes $\bar{C}_i$ as the advected concentration at time t_{k+1}. If $\bar{x}_i$ reaches across an inflow boundary it is assigned the concentration at the boundary. At no-flow boundaries the particle is reflected back into the domain.

Once the advected concentration $\bar{C}_i$ is determined at each node, the dispersive flux is approximated by solving (1) using a Galerkin finite element method. The total derivative in (2) is approximated by (Neuman, 1984)

$$\frac{dC_i}{dt} \approx \frac{C_i^{k+1} - \bar{C}_i}{\Delta t} \tag{5}$$

A backward-difference time-stepping scheme was used along with mass lumping to solve (5).

RESULTS AND DISCUSSION

Backtracking Analysis

The one-dimensional, steady-state velocity profile shown in Figure 1 was used to test three backtracking algorithms. The profile is typical of a rapid change in velocity that may be encountered in heterogeneous conditions. A time step size ($\Delta t = 200$ sec) was chosen such that a particle backtracking from node x_i ($x = 11$ cm) would reach $\bar{x}_i$ in Δt (Figure 1). The velocity field was assumed to be piece-wise linear for the analyses.

For the multiple-step, Euler method, N Euler steps of length Δt_s were taken, where

$$\Delta t = \sum_{s=1}^{N} \Delta t_s \tag{6}$$

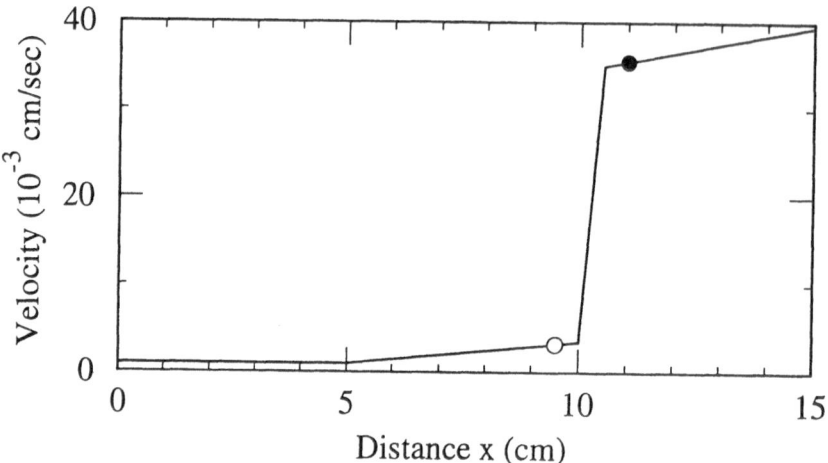

Figure 1. One-dimensional velocity profile showing x_i (•) and $\tilde{x}_i$ (o).

A fourth-order Runge-Kutta method was analyzed, also using N equal sub-intervals of the total Δt. For the Euler and fourth-order Runge-Kutta analyses, N was chosen such that the number of function evaluations were equal. The semi-analytical method used for the analysis is described in Pollock (1988). In brief, a particle travels from x_i to $\tilde{x}_i$ moving across entire or partial elements in varying sub-intervals of time until Δt is used up. The sub-intervals of time are calculated analytically by knowing the nodal velocities and assuming piece-wise linearity.

The results of the test are shown in Table 1. The error represented in the table is the difference between the semi-analytical method and the approximated value. For a given order of accuracy (*e.g.*, $\sim 10^{-2}$ cm) the semi-analytical method requires only a fraction of the number of function evaluations required by the other two methods. The semi-analytical method can be easily applied in two dimensions where each velocity component is assumed to independently vary linearly in an element. However, the semi-analytical method is limited to regions where the velocities are greater than zero.

The MMOC model used for the simulations described below utilized a simple one-step, Euler algorithm for solving equation (3) when the velocity field was uniform. When a varying velocity field was considered, the semi-analytical back-tracking approach was employed along with a multiple-step, Euler algorithm near boundaries. Unless stated otherwise, a quadratic-linear interpolation scheme was used to determine $\bar{C}_i$.

Table 1. Backtracking Algorithm Error Analyses

Method	Value of $\tilde{x}_i$ (cm)	Error (cm)
Multiple-Step Euler		
Δt_4	8.7696996	0.7268873
Δt_{16}	9.4348928	6.1694115×10^{-2}
Δt_{32}	9.4710032	2.5583425×10^{-2}
Δt_{48}	9.4801038	1.6483029×10^{-2}
Fourth-Order Runge-Kutta		
Δt_1	7.2831666	2.2134204
Δt_4	9.4148456	8.1741300×10^{-2}
Δt_8	9.4778071	1.8779867×10^{-2}
Δt_{12}	9.4968999	3.1292853×10^{-4}
Semi-analytical		
Δt_3	9.4965869	0.0000000

Model Verification

Analytic and numerical solutions were used to verify the MMOC model using uniform velocity fields. First, one-dimensional MMOC model results were compared to analytic solutions of Ogata and Banks (1961). When quadratic-linear interpolation was used along with non-integer $Cr \geq 1$, the solutions were oscillation free and had minimal dispersion at high Pe. Two-dimensional MMOC model results were in good agreement with the results produced by a Petrov-Galerkin finite element model when uniform velocities were considered in two dimensions, for $Pe = Cr = 1$.

2-D Steady-State Flow and Transient Transport

To test the MMOC model in a more complex flow field, the model was applied to simulate transport of a nonreactive solute in the domain shown in Figure 2. The domain is representative of a preferential flow path in dry, unsaturated porous media. Note (Figure 2) that the water contents and resultant pore water velocities change sharply at the flow path boundary which defines a severe test. Simulations were conducted with Dirchlet boundary conditions along the $z = 0$ cm boundary ($C_0 = 1$ for $0 \leq x \leq 5$ cm and $C_0 = 0$ for $5 < x \leq 10$ cm), and $z = 40$ cm boundary ($C_0 = 0$ for $0 \leq x \leq 10$ cm) and no-flux boundary conditions on other boundaries.

The domain contained 1701 nodes ($\Delta x = \Delta z = 0.5$ cm). Dispersivities were chosen such that a wide range of Peclet numbers ($0 < Pe_x < 75$ and $0 < Pe_z < 85$) and Courant numbers ($0 < Cr_x < 1.65$ and $0 < Cr_z < 3.3$) were attained. Simulations were run for a total time of 5000 seconds using a $\Delta t = 125$ seconds.

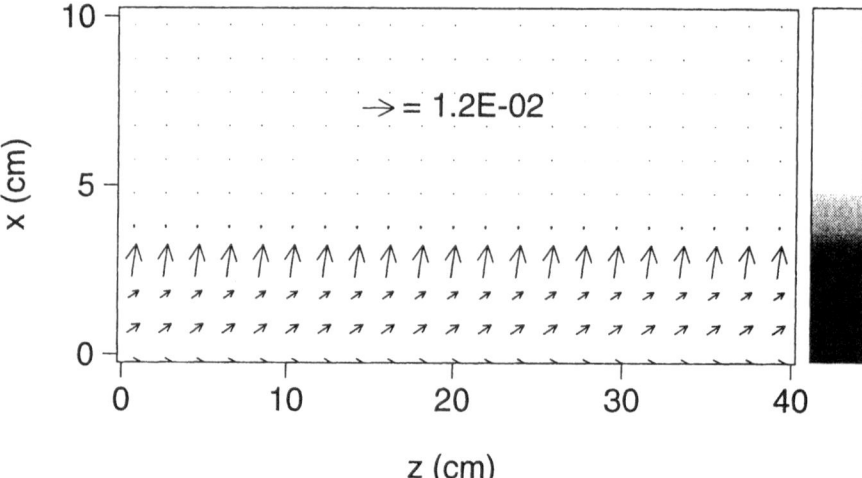

Figure 2. Two-dimensional simulation domain showing the resultant pore water velocity and water content distribution, where dark represents $\theta \approx 0.27$ and light indicates $\theta \approx 0.1$.

Simulations resulting from using linear interpolation in the model (not shown) displayed numerical dispersion. The dispersion was greatest along the z-axis. The surface plot shown in Figure 3 illustrates the oscillations that can occur at high Pe when only quadratic interpolation is used in the model. Figure 4 shows how oscillations can be eliminated (with reduced numerical dispersion) when the quadratic-linear interpolation method is used. An oscillation free solution could not be attained with a Petrov-Galerkin finite element model under these conditions.

CONCLUSIONS

A two-dimensional MMOC model was developed for simulating nonreactive solute transport in unsaturated porous media. A semi-analytical method was shown to be more efficient than other techniques for backtracking in a complex flow field, for a given order of accuracy. A semi-analytical backtracking algorithm and a quadratic-linear interpolation scheme was adopted for the MMOC model. Model simulations of transport in an unsaturated zone having a wide range of Pe and Cr produced oscillation free solutions with minimal dispersion.

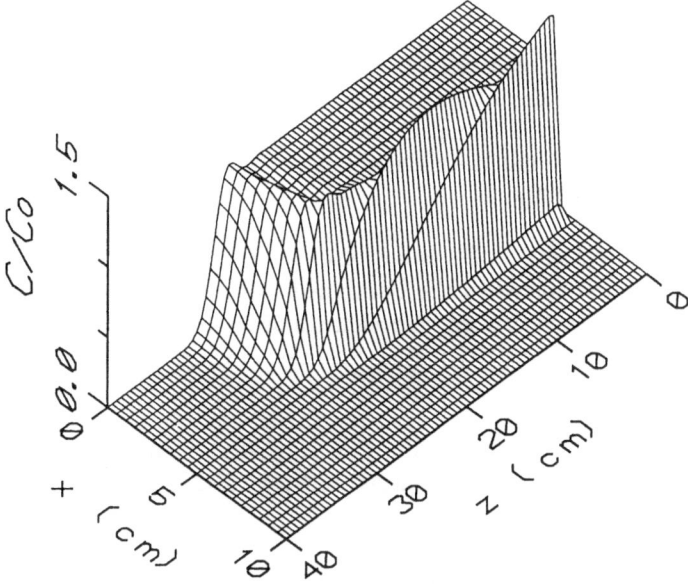

Figure 3. Surface plot indicating the solute concentration distribution when only quadratic interpolation is used.

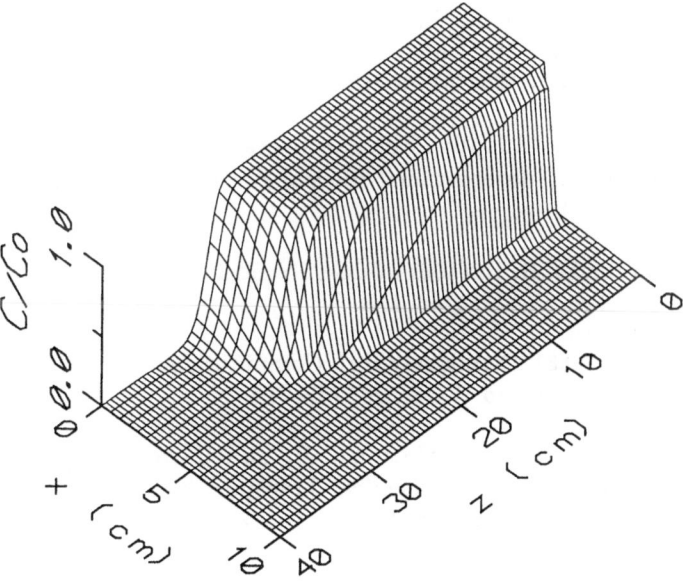

Figure 4. Surface plot indicating the solute concentration distribution when quadratic-linear interpolation is used.

REFERENCES

Allen, M. B., and Khosravani, A. (1990) "An Eulerian-Lagrangian Method for Finite-Element Collocation Using the Modified Method of Characteristics," Proceedings of the Eighth International Conference on Computational Methods in Water Resources, Computational Methods in Surface Hydrology, Springer-Verlag, Southampton, UK, pp. 375–380.

Baptista, A. M., Adams, E. E., and Stolzenbach, K. D. (1984) "The 2-D, Unsteady, Transport Equation Solved by the Combined Use of the Finite Element Method and the Method of Characteristics," Proceedings of the 5th International Conference on Finite Elements in Water Resources, Springer-Verlag, Berlin, Germany, pp. 353–362.

Cheng, R. T., Casulli, V., and Milford, S. N. (1984) "Eulerian-Lagrangian Solution of the Convective-Dispersion Equation in Natural Coordiantes," Water Resources Research, 20, 944–952.

Chiang, C. Y., Wheeler, M. F., and Bedient, P. B. (1989) "A Modified Method of Characteristics Technique and Mixed Finite Elements Method for Simulation of Groundwater Solute Transport," Water Resources Research, 25, 1541–1549.

Healy, R. W., and Russell, T. F. (1989) "Efficient Implementation of the Modified Method of Characteristics in Finite-Difference Models of Solute Transport," Proceedings from the NWWA Conference on Solving Ground Water Problems with Models, National Water Well Association, Columbus, OH, pp. 483–492.

Neuman, S. P. (1984) "Adaptive Eulerian-Lagrangian Finite Element Method For Advection-Dispersion," International Journal For Numerical Methods in Engineering, 20, 321–337.

Ogata, A., and Banks, R. B. (1961) USGS Professional Paper 411-A, U.S. Government Printing Office, Washington, DC.

Pollock, D. W. (1988) "Semianalytical Computation of Path Lines for Finite-Difference Models," Ground Water, 26, 743–750.

Yeh, J. T. C., Srivastava, R., Guzman, A., and Harter, T. (1993) "A Numerical Model for Water Flow and Chemical Transport in Variably Saturated Porous Media," Ground Water, 31, 634–644.

Zang, R., Huang, K., and van Genuchten, M. T. (1993) "An Efficient Eulerian-Lagrangian Method for Solving Solute Transport Problems in Steady and Transient Flow Fields," Water Resources Research, 29, 4131–4138.

EXACT CONDITIONAL MOMENT EXPRESSIONS FOR TRANSIENT UNSATURATED FLOW IN RANDOMLY HETEROGENEOUS SOILS WITH SCALABLE MOISTURE RETENTION CHARACTERISTICS

SHLOMO P. NEUMAN and COLLEEN LOEVEN
Department of Hydrology and Water Resources
The University of Arizona
Tucson, Arizona 85721
USA

ABSTRACT

We consider the effect of measuring randomly varying local soil hydraulic properties on one's ability to predict transient flow in a bounded unsaturated soil, driven by random source and boundary functions. More precisely, we consider the prediction of local pressure head $\psi(\mathbf{x},t)$, saturation $S(\mathbf{x},t)$ or volumetric moisture content $\theta(\mathbf{x},t)$, and flux $\mathbf{q}(\mathbf{x},t)$ by means of their unbiased ensemble moments $<\psi(\mathbf{x},t)>_c$, $<S(\mathbf{x},t)>_c$ or $<\theta(\mathbf{x},t)>c$, and $<\mathbf{q}(\mathbf{x},t)>_c$ conditioned on measurements of soil moisture characteristics at selected locations. It is often possible to scale the moisture retention curves of samples from a heterogeneous soil in a manner which greatly reduces their scatter. We show how this allows deriving a deterministic unsaturated (nonlinear) flow equation in terms of the above conditional predictors. The flux predictor $<\mathbf{q}(\mathbf{x},t)>_c$ is generally non-Darcian. As the conditional predictors are smooth relative to their random counterparts, they can be expressed in terms of finite-dimensional functions. Hence the flow equation should be amenable to deterministic solution by finite elements on a relatively coarse grid; there is no need for conditional Monte Carlo simulation or the use of a fine grid to account for random spatial variations. Explicit expressions are also given for the conditional covariances of pressure head and flux prediction errors.

INTRODUCTION

Fluid flow in unsaturated soils is strongly influenced by spatial variations in soil moisture characteristics. In practice, these characteristics are measured only at selected points in space. To estimate their values between the measurement points, it is sometimes possible to employ geostatistical methods such as kriging [1]. The latter yields smooth, unbiased estimates of the properties which honor, or are conditioned on, the measured values. Kriging also yields information about the variance and spatial covariance of the estimation errors. Though the latter have mean zero, they are generally nonstationary in that their variance and spatial correlation scales are smaller in the vicinity of measurement points than at distances from these points. Due to conditioning, the kriging estimation variance and correlation scales are generally smaller than their unconditional

513

A. Peters et al. (eds.), Computational Methods in Water Resources X, 513–520.
© 1994 Kluwer Academic Publishers. Printed in the Netherlands.

counterparts, *i.e.*, those that describe the actual spatial fluctuations of the medium properties. The usefulness of conditioning saturated flow and transport models on measured data is well-documented in the literature [2 - 6]. Recent work suggests that the same is true for unsaturated flow and transport [7].

Since conditioning involves nonstationary random functions, it has traditionally required the use of numerical Monte Carlo simulation. To obtain reliable results, this approach mandates the use of very fine space-time grids which are capable of resolving random fluctuations on scales much smaller than the (conditional) correlation scales of the corresponding soil variables. In the case of unsaturated flow in more than one space dimension, nonlinearity typically renders such high-resolution Monte Carlo simulation impractical unless one finds auxiliary ways to materially improve convergence [7 - 8]. An alternative to Monte Carlo simulation is to compute deterministically the leading conditional moments of key random variables such as head, flux and concentration. In their most general form, the equations which govern conditional first moments under saturated flow and transport were shown by Neuman and Orr [9] and by Neuman [10] to take on nonlocal, integro-differential forms. As conditional moments are smooth relative to their random counterparts, they can be expressed in terms of finite-dimensional functions. Hence the first moment equations are in principle amenable to deterministic solution by finite elements on a relatively coarse grid. In this paper we derive a local (differential) equation for the conditional first moments of pressure head and saturation (or water content) under transient unsaturated flow in a heterogeneous soil. For this, we assume that the moisture retention curves (but not necessarily other soil characteristics) of samples from such a soil scale in a particular manner. Our new theory also yields explicit expressions for the conditional covariances of pressure head and flux prediction errors.

THEORY

We start from the premise that, locally, the transient flow of water in unsaturated soils satisfies Darcy's law

$$q(x,t) = - K(x,\psi)\nabla h(x,t) \qquad\qquad h = \psi + x_3 \qquad\qquad \text{in } \Omega \qquad (1)$$

and the volume balance equation

$$-\nabla\cdot q(x,t) + f(x,t) = \phi\frac{\partial S(x,t)}{\partial t} \qquad\qquad \text{in } \Omega \qquad (2)$$

subject to initial and boundary conditions

$$\begin{aligned} h(x,0) &= h_0(x) & \text{in } \Omega & \qquad (3) \\ h(x,t) &= H(x,t) & \text{on } \Gamma_D & \qquad (4) \\ - q(x,t)\cdot n(x) &= Q(x,t) & \text{on } \Gamma_N & \qquad (5) \end{aligned}$$

where q is flux, K is hydraulic conductivity, S is saturation, h is hydraulic head, ψ is pressure head, x_3 is the vertical coordinate, ϕ is porosity, Ω is the flow domain, $\Gamma = \Gamma_D \cup \Gamma_N$ is the boundary (Dirichlet and Neumann) of Ω, and n is a unit outer normal to Γ. By saying that these equations hold locally we mean that all of the above quantities are measurable on some consistent local scale, or support, ω. This scale need not constitute an REV (Representative Elementary Volume) in the traditional sense [11].

We assume that $K(x,\psi)$ and $\psi(x,S)$ are random functions of the space coordinates x and consider $h_0(x)$, $f(x,t)$, $H(x,t)$ and $Q(x,t)$ to be independently prescribed random functions. We express the random hydraulic conductivity field as

$$K(x,\psi) = K_S(x)K_r[\alpha(x),\psi]$$ (6)

where K_S is hydraulic conductivity at saturation (S = 1), K_r is relative hydraulic conductivity (= K/K_S), and $\alpha(x)$ is a random vector field of parameters defining the (arbitrary, usually nonlinear) functional relationship between K and ψ at any x.

Assume that $K_S(x)$ and $\alpha(x)$ have been determined by measurement at selected points and that these data make possible obtaining (say via kriging) relatively smooth unbiased estimates, $<K_S(x)>_c$ and $<\alpha(x)>_c$ respectively, of these spatially varying parameters. Here the symbol $< >$ indicates ensemble mean (statistical expectation) and the subscript c implies that this mean is conditioned on measurements at specific points in space. Though these conditional estimates are deterministic functions, they are associated with zero-mean random estimation errors, $K_S'(x)$ and $\alpha'(x)$, respectively. If the estimates are obtained by kriging, the variance of the associated estimation errors generally increases with distance from the measurement points, rendering the estimation errors nonstationary (statistically nonhomogeneous). The errors may further exhibit spatial auto- and cross-correlations. They are related to the actual parameter values and their estimates via

$$K_S(x) = <K_S(x)>_c + K_S'(x) \qquad <K_S'(x)>_c \equiv 0 \qquad (7)$$
$$\alpha(x) = <\alpha(x)>_c + \alpha'(x) \qquad <\alpha'(x)>_c \equiv 0. \qquad (8)$$

The optimum predictors of h, ψ, and q are their conditional means. They relate to the former as

$$h(x,t) = <h(x,t)>_c + h'(x,t) \qquad <h'(x,t)> \equiv 0 \qquad (9)$$
$$\psi(x,t) = <\psi(x,t)>_c + \psi'(x,t) \qquad <\psi'(x,t)> \equiv 0 \qquad (10)$$
$$q(x,t) = <q(x,t)>_c + q'(x,t) \qquad <q'(x,t)> \equiv 0. \qquad (11)$$

By taking the conditional ensemble mean of (1) – (5) one finds that these predictors satisfy the deterministic equations

$$<q(x,t)>_c = - <K(x,\psi)>_c \nabla <h(x,t)>_c + r_c(x,t) \qquad \text{in } \Omega \qquad (12)$$

where

$$r_c(x,t) = - <K'(x,\psi)\nabla\psi'(x,t)>_c \tag{13}$$

and

$$- \nabla\cdot<q(x,t)>_c + <f(x,t)> = \phi\frac{\partial<S(x,t)>_c}{\partial t} \qquad \text{in } \Omega \tag{14}$$

subject to the initial and boundary conditions

$$<h(x,0)>_c = <h_0(x)> \qquad\qquad \text{in } \Omega \tag{15}$$
$$<h(x,t)> = <H(x,t)> \qquad\qquad \text{on } \Gamma_D \tag{16}$$
$$- <q(x,t)>_c\cdot n(x) = <Q(x,t)> \qquad \text{on } \Gamma_N. \tag{17}$$

Here we assumed that $\phi \equiv$ constant, as is common in the stochastic groundwater literature. However, this restriction is not difficult to relax.

A key assumption of our theory is that $\psi(x,S)$ scales according to

$$\psi(x,S) = \tilde{\beta}(x)<\psi(S)> + \epsilon(S) \qquad <\tilde{\beta}(x)> \equiv 1 \qquad <\epsilon(S)> \equiv 0 \tag{18}$$

where $\tilde{\beta}(x)$ is a random parameter field whose statistical properties can be inferred (by means of appropriate geostatistical methods) from the data, $<\psi(S)>$ is the space-independent unconditional ensemble mean of each local retention function (this implies that ψ is statistically homogeneous; if one further takes it to be mean ergodic, one can estimate $<\psi(S)>$ simply as the spatial average of all sample retention data), and $\epsilon(S)$ is white noise (pure, zero mean random error) whose statistical properties (distribution, variance, other moments) may depend on S. A similar method of scaling, without the formal consideration of $\epsilon(S)$, has been used successfully [12] to greatly reduce the scatter of pressure head (ψ) versus water content (θ) data from 36 undisturbed 100 cm³ cores of clayey loam soil taken along a 500 m transect in the Trebon region of southern Bohemia, the Czech Republic. We show in Fig.1 how such scaling reduces the scatter of saturation (S) versus pressure head (ψ) data from rock cores of Bandelier Tuff collected in Los Alamos County, New Mexico, USA [13]. The cores were extracted from depths of about 10 - 170 feet.

Define a normalized random parameter field

$$\beta(x) \equiv \frac{\tilde{\beta}(x)}{<\tilde{\beta}(x)>_c} \equiv 1 + \beta'(x) \qquad\qquad <\beta'(x)>_c \equiv 0. \tag{19}$$

Since $<\tilde{\beta}(x)>_c$ and $<\psi(S)>$ are inferred from the data, one can compute the conditional mean retention function

$$<\psi(x,S)>_c = <\tilde{\beta}(x)>_c<\psi(S)> \tag{20}$$

for any x in Ω. Likewise it follows that

$$\psi(\mathbf{x}, S) = \beta(\mathbf{x}) \langle \psi(\mathbf{x}, S) \rangle_c + \epsilon(S) \tag{21}$$
$$\psi'(\mathbf{x}, S) = \beta'(\mathbf{x}) \langle \psi(\mathbf{x}, S) \rangle_c + \epsilon(S) \tag{22}$$
$$\langle \psi'(\mathbf{x}, S)^2 \rangle_c = \langle \beta'(\mathbf{x})^2 \rangle_c \langle \psi(\mathbf{x}, S) \rangle_c^2 + \sigma_\epsilon^2(S) \tag{23}$$

where $\sigma_\epsilon^2(S) \equiv \langle \epsilon(S)^2 \rangle$ is the variance of $\epsilon(S)$. The left-hand side of (23) represents the variance of the prediction error associated with any $\mathbf{x}$ and S. As all terms on the right-hand side of (23) can be inferred from the data for any $\mathbf{x}$ and S, so can the corresponding prediction variance. However, to compute this variance at any $\mathbf{x}$ and t under prescribed mean initial and boundary conditions, one must first predict the corresponding value of S.

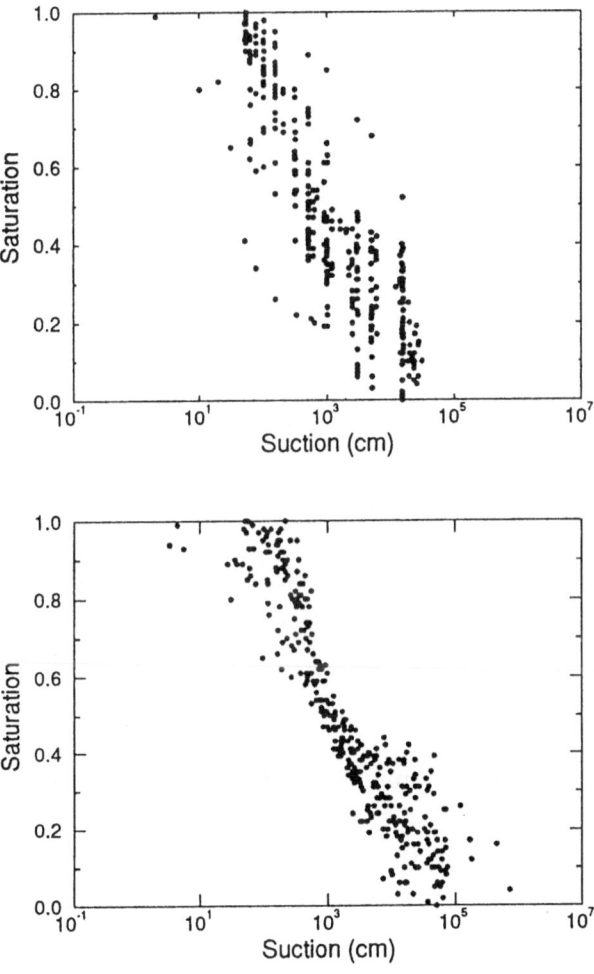

Figure 1: Original (top) and scaled (bottom) S versus ψ data from Bandelier Tuff cores.

Under our scaling assumption, $K(x,\psi)$ becomes a function of the deterministic quantities $<\psi(S)>$ or $<\psi(x,S)>_c$,

$$K(x,\psi) = K[x,<\psi(S)>] = K[x,<\psi(x,S)>_c] = K_S(x)K_r[\alpha(x),\tilde{\beta}(x)<\psi(S)> + \epsilon(S)]$$

$$= K_S(x)K_r[\alpha(x),\beta(x)<\psi(x,S)>_c + \epsilon(S)]. \tag{24}$$

If the (conditional) joint probability distribution of $K_S(x)$, $\alpha(x)$ and $\tilde{\beta}(x)$ is known at any finite number of points x, and if the probability distribution of $\epsilon(S)$ is also known, one can determine the (conditional) probability distribution of $K[x,<\psi(S)>]$ and/or $K[x,<\psi(x,S)>_c]$ at any x of interest by Monte Carlo simulation. This simulation can be done prior to solving the flow problem and is computationally much less demanding than solving the actual stochastic flow equation by Monte Carlo simulation. In particular, one can determine the (conditional) mean and variance

$$\kappa[x,<\psi(x,S)>_c] \equiv <K(x,\psi)>_c \tag{25}$$
$$\sigma_\kappa^2[x,<\psi(x,S)>_c] \equiv <K'(x,\psi)^2>_c \tag{26}$$

as functions of location x and $<\psi(x,S)>_c$ or $\psi<(S)>$. If only the first two moments of the requisite probability distributions can be inferred from the data (as is often the case in practice), we recommend that one adopt either normal or lognormal distributions as a working hypothesis (to be formally tested when sufficient data become available).

One can show that the conditional residual flux is given by

$$\begin{aligned} r_c(x,t) = & - <K[x,<\psi(x,t)>_c]\beta'(x)>_c \nabla<\psi(x,t)>_c \\ & - <K[x,<\psi(x,t)>_c]\nabla\beta'(x)>_c <\psi(x,t)>_c \\ & - <K[x,<\psi(x,t)>_c]\nabla\epsilon'(x)>_c \end{aligned} \tag{27}$$

and hence the flux predictor becomes

$$\begin{aligned} <q(x,t)>_c = & - \kappa[x,<\psi(x,t)>_c]\nabla<h(x,t)>_c \\ & - <K[x,<\psi(x,t)>_c]\beta'(x)>_c \nabla<\psi(x,t)>_c \\ & - <K[x,<\psi(x,t)>_c]\nabla\beta'(x)>_c <\psi(x,t)>_c \\ & - <K[x,<\psi(x,t)>_c]\nabla\epsilon'(x)>_c. \end{aligned} \tag{28}$$

Note that K is a function of β and ϵ so that the above conditional cross-covariances are generally nonzero. In a manner analogous to that described in the previous paragraph, one should be able to determine these conditional second moments by Monte Carlo simulation prior to solving the flow problem.

It is clear from (28) that $<q(x,t)>_c$ is generally non-Darcian and cannot be associated with an effective or equivalent unsaturated hydraulic conductivity function.

Allthough κ is a scalar, the above cross covariances need not be statistically isotropic. This may lead to ψ-dependent variation (anisotropy) in these coefficients.

Equations (12) - (17), coupled with (25) and (27) or (28), form a fully deterministic local system of nonlinear transient unsaturated flow equations that can be solved by standard numerical methods (such as finite elements) on a much coarser grid than is required for Monte Carlo simulation. The scale of grid discretization is controlled in a major way by the degree of smoothness of the deterministic dependent variable $<\psi(x,t)>_c$, which is much smoother than its random counterpart $\psi(x,t)$. The numerical implementation of this idea is now under testing and development.

Note from (28) that $\nabla<h(x,t)>_c = 0$ does not generally imply $<q(x,t)>_c = 0$. In other words, our model predicts (conditional mean) flow even in the absence of a predicted (conditional mean) head gradient. This is a direct consequence of the nonlinear dependence of q on a random ψ. It implies that even though locally there may not be any flow under a zero head gradient, this is not necessarily so when one considers (conditional) ensemble mean quantities.

The (conditional) covariance of the error in predicting ψ is simply

$$<\psi'(x,t)\psi'(y,s)>_c =$$
$$<\psi(x,t)>_c<\beta'(x)\beta'(y)>_c<\psi(y,s)>_c + \sigma_\epsilon^2[S(x,t)]\tilde{\delta}(x-y,t-s) \qquad (29)$$

where

$$\tilde{\delta}(x-y,t-s) \equiv 1 \qquad \text{if } x = y \text{ and } t = s$$
$$\delta(x-y,t-s) \equiv 0 \qquad \text{otherwise.} \qquad (30)$$

The random error in predicting flux is

$$q'(x,t) = - r_c(x,t) - K[x,<\psi(x,t)>_c]\beta'(x)\nabla<\psi(x,t)>_c$$
$$- K[x,<\psi(x,t)>_c]\nabla\beta'(x)<\psi(x,t)>_c - K[x,<\psi(x,t)>_c]\nabla\epsilon'(x)$$
$$- K'[x,<\psi(x,t)>_c]\nabla<h(x,t)>_c. \qquad (31)$$

It follows that once $<\psi(x,t)>_c$ has been determined deterministically, one can generate conditional values of $q'(x,t)$ explicitly by Monte Carlo simulation, then calculate the (conditional) covariance of the error in predicting q as $<q'(x,t)q'(y,s)^T>_c$. One can likewise generate explicitly (conditional) random velocities, and compute their moments, via

$$v(x,t) = \frac{q(x,t)}{\phi S(x,t)} \qquad (32)$$

ACKNOWLEDGMENTS

This work was supported jointly by the U.S. Nuclear Regulatory Commission under Contract NRC-04-90-51, the U.S. Geological Survey under Water Resources Research Grant 14-08-0001-G2092, and the U.S. Department of Agriculture under Grant 92-34214-7387.

REFERENCES

1. Warrick, A.W., D.E. Myers, and D.R. Nielsen, Geostatistical methods applied to soil science, in *Methods of Soil Analysis, Part 1, Physical and Mineralogical Methods, Second Edition*, edited by A. Klute, Am. Soc. Agronomy and Soil Sci. Soc. Am., Madison, Wisc. USA, 53-82, 1986.
2. Clifton, P.M. and S.P. Neuman, Effects of kriging and inverse modeling on conditional simulation of the Avra Valley aquifer in southern Arizona, *Water Resour. Res.*, 18(4), 1215-1234, 1982.
3. Graham, W. and D. McLaughlin, Stochastic analysis of nonstationary subsurface solute transport, 2, Conditional moments, *Water Resour. Res.*, 25(11), 2331-2355, 1989.
4. Rubin, Y., Prediction of tracer plume migration in disordered porous media by the method of conditional probabilities, *Water Resour. Res.*, 27(6), 1291-1308, 1991.
5. Zhang, D. and S.P. Neuman, Information-dependent prediction of solute transport in heterogeneous geologic media, in *Numerical Methods in Water Resources, Proc. CMWR'94*, edited by A. Peters, G. Wittum, B. Herling, and U. Meissner, Kluwer Academic Publ., this issue, 1994.
6. Neuman, S.P., O. Levin, S. Orr, E. Paleologos, D. Zhang, and Y.-K. Zhang, Nonlocal representations of subsurface flow and transport by conditional moments, in *Computational Stochastic Mechanics*, edited by A.H-D. Cheng and C.Y. Yang, Comp. Mech. Publ. and Elsevier Appl. Sci., New York, 451-473, 1993.
7. Harter, T., Unconditional and Conditional Simulation of Flow and Transport in Heterogeneous, Variably Saturated Soils, Ph.D. dissertation, The University of Arizona, Tucson, 1994.
8. Harter, T. and T.-C. J. Yeh, An efficient method for simulating steady unsaturated flow in random porous media: Using an analytical perturbation solution as initial guess to a numerical model, *Water Resour. Res.*, 29(12), 4139-4149, 1993.
9. Neuman, S.P. and S. Orr, Prediction of steady state flow in nonuniform geologic media by conditional moments: Exact nonlocal formalism, effective conductivities and weak approximation, *Water Resour. Res.*, 29(2), 341-364, 1993.
10. Neuman, S.P., Eulerian-Lagrangian theory of transport in space-time nonstationary velocity fields: Exact nonlocal formalism by conditional moments and weak approximation, *Water Resour. Res.*, 29(3), 633-645, 1993.
11. Baveye, P. and G. Sposito, Macroscopic balance equations in soils and aquifers: The case of space- and time-dependent instrumental response, *Water Resour. Res.*, 21(8), 1116-1120, 1985.
12. Vogel, L., M. Cislerova, and J.W. Hopmans, Porous media with linearly variable hydraulic properties, *Water Resour. Res.*, 27(10), 2735-2741, 1991.
13. Loeven, C.A. and E.P. Springer, Validation of Continuum Concepts for Flow and Transport in Unsaturated Fractured Bandelier Tuff, Internal Report, Los Alamos National Laboratory, New Mexico, 1993.

SOLUTE TRANSPORT IN 3D LABORATORY MODEL THROUGH AN HOMOGENEOUS POROUS MEDIUM : BEHAVIOUR OF DENSE PHASE AND SIMULATION

C. Oltean*, Ph. Ackerer* and M.A. Buès**
Institut de Mécanique des Fluides, Université Louis Pasteur, URA CNRS 854, 2, Rue Boussingault, F - 67000 Strasbourg
Laboratoire de Géomécanique, Ecole Nationale Supérieure de Géologie, Rue du Doyen Marcel Roubault, BP 40, F - 54501 Vandœuvre-lès-Nancy

INTRODUCTION

Nowadays, knowing a source of pollution, it is very difficult to forecast its propagation through an aquifer and to evaluate the risks incured by the various water catchments, because our knowledge of complex reservoirs is still very limited. A laboratory model permits for a reasonable cost to multiply experiments under various conditions, in order to improve our understanding on spreading of pollutions in porous formations and validate mathematical codes in complex configurations.

On physical model, we studied the behaviour of pollutant transport through an homogeneous and saturated porous medium, in 2D respectivelly 3D configurations. The contaminant plumes consisted of sodium chloride solutions introduced into the porous matrix from a source located on the top of the porous medium.

To analyze these experiments, a simulation code has been developed. The mathematical model takes into account the density and viscosity dependent transport problem and is based on the mixed hybrid finite element method. It treats non-steady problems and computes both the flow field and the pollutant concentration. We assume that, in the mixing zone, the viscosity and the density vary linearly with the salt concentration.

MATHEMATICAL MODEL

The displacement of two miscible fluids through an homogeneous and saturated porous medium is generaly described by the following equations system (Bear, 1972, Dorgarten and Tsang, 1991) :

$$\vec{V} = - \frac{k}{\mu} [\overrightarrow{grad}(P) + \rho g \overrightarrow{grad}(z)] \qquad \text{- generalized form of Darcy's law} \qquad (1)$$

$$div(\rho \vec{V}) + \frac{\partial(\rho \varepsilon)}{\partial t} + \bar{\rho}Q = 0 \qquad \text{- fluid continuity equation} \qquad (2)$$

$$\rho\varepsilon\frac{\partial c_m}{\partial t} + \rho\varepsilon\vec{V}\overrightarrow{grad}(c_m) - div[\rho\varepsilon D\overrightarrow{grad}(c_m)] = C^*Q \qquad (3)$$
$$\text{- convective-dispersive equation}$$

$$\rho = \rho_o + \frac{\partial\rho}{\partial c_m}(c_m - c_o) \qquad (4)$$

A. Peters et al. (eds.), Computational Methods in Water Resources X, 521–528.
© 1994 Kluwer Academic Publishers. Printed in the Netherlands.

$$\mu = f(\mu_o, c_m) \tag{5}$$

Here, ρ is the fluid density (M/L^3), g is the gravitational constant (L/T^2), k is the intrinsic permeability tensor of the porous medium (L^2), μ is the dynamic viscosity (ML^{-1}T^{-1}), z is the vertical spatial coordinate (L), ε is the kinetic porosity of the porous medium, P is the fluid pressure (ML^{-1}T^{-2}), c_m is the mass fraction of concentrated salt solution (M$_{salt}$/M$_{solution}$), $\vec{V}$ is the Darcy velocity (L/T), ρ_0 is the ambient density at c_0 concentration, Q is the external sinks and sources (L^3/T/L^3), $\widetilde{P}$, (M/L^3), and C* (M/L^3), denotes the density respectively the concentration of the sink/sources fluid.

The hydrodynamic dispersion tensor D is considered as the sum of a contribution from molecular diffusion and from hydrodynamic dispersion :

$$D = D_p I + (\alpha_L - \alpha_T) V V / |V| + \alpha_T |V| I \tag{6}$$

where D_p is the pore water diffusion coefficient (L^2/T), I is the unit tensor, α_L and α_T are the dispersivities parallel and perpendicular to the flow (L).

Inserting the generalized form of Darcy's law into the fluid continuity equation results in the density-dependent groundwater flow equation :

$$\rho S_{op}\frac{\partial P}{\partial t} + \varepsilon\frac{\partial \rho}{\partial c_m}\frac{\partial c_m}{\partial t} + \text{div}\{\frac{k\rho}{\mu} [\overrightarrow{\text{grad}}(P) + \rho g\overrightarrow{\text{grad}}(z)]\} = \widetilde{\rho}Q \tag{7}$$

where S_{op} is the specific pressure storativity ([M/LT^2]$^{-1}$). The groundwater flow equation and the contaminant transport equation are coupled by two constitutive relationships (4 and 5) that define the fluid density and dynamic viscosity.

In order to simulate the experiments in 2D configuration with density and viscosity contrasts, we have used the mixed hybrid finite element method. The physical model is discretized into K triangles, with the edges A_{Ki}, i=1,...3, having all of them the Raviart - Thomas space properties (Chavent and Roberts, 1991).

The mixte approximation consists in calculating simultaneously the pressure field and the velocity field. On each element K, P, and $\vec{V}$ are determinated by :

- an approximation of the mean of P on K ;
- an approximation of the mean of P on each edge A_{Ki} ;

- an approximation of $\vec{V}$ on the element K, determined by the knowledge of its flux through the edges A_{Ki}:

By using the same technique, we can also approximate the concentration field which is determinated as follows :

- an approximation of the mean of C on K ;
- an approximation of the mean of C on each edge A_{Ki} ;

Expressing these equations (1 to 3) in a variational form with the help of basis functions $\vec{w}$. defined by :

$$\int_{A_{Ki}} \vec{w}_i \cdot \vec{n}_{Kj} = \delta_{ij} \text{ for } j = 1,...3$$

where δ_{ij} is the Kronecker symbol, and using the continuity equations in pressure, concentration and flux between the adjacent elements, the equations system will be solved, in the first step, in pressure over edges. After, we use on each element K the local equation in order to determine the flux over each edge and the mean pressure over K. These values will be introduced into contaminant transport equation in order to estimate the concentration over edge, respectively the mean concentration over K.

PHYSICAL MODEL

The physical model was realized in order to study the behaviour of pollutant transport into an aquifer. The laboratory model includes three main parts (fig. 1) :
- a sand box which contains a porous medium sample;
- a feeding system;
- a measure system and a data acquisition system.

The flow channel is constituted by a sand box (with internal dimensions $1,63 \times 0,70 \times 0,40$ m) filled with a natural quartz sand (mean diameter d_{50} equal to $1,22$ mm and d_{60}/d_{10} equal to $1,64$). It constitutes the homogeneous porous medium. This channel is equipped with constant level tanks, regulated by weirs, located at each side of the set-up. The difference in level betwen the upstream and downstream tanks establishes the driving charge. The box is surrounded with a thermal isolation in order to reduce the temperature variation in the porous matrix. The porous medium is separated from tanks by a punched wall, covered with nylon screen.

The fresh water feeding of the model supplies constant flow with the help of tank connected to a weir.
Sodium chloride solutions are used as pollutant. They allow to realize the measurements of electrical conductivity which are then converted to salt concentrations. In the first step, the pollutant (high concentration about 250 g/l) is injected to a constant flow rate into the upper tank by using a peristaltic pump. Here, it is mixed with fresh water in order to obtain the same salt concentration over the whole cross-section of the medium. In this case, the initial and boundary conditions can be expressed as follows :

$$C(x, 0) = 0; \qquad C(0, t) = C_f \left[1 - \exp(-\beta t)\right]; \qquad C(\infty, t) = 0$$

where C is the salt solution concentration, C_f is the final concentration and β is the mixing cell coefficient. These parameters (C_0 and β) are determined in function of the injection flow rates of the fresh water and salt solution as well as their concentrations. The pollutant displacement can be considered as 1D configuration with constant velocity through an isotropic, homogeneous and saturated medium. These simplification permits us to estimate the hydrodynamic parameters (velocity and longitudinal dispersion) (Bues and Oltean, 1992), by using the analytical solution developed for a tracer case and fitted by an algorithm for least-squares estimation of nonlinear parameters (eg. Marquardt, 1963).
In the second step, we studied the behaviour of surface contaminant plumes in 2D, respectively 3D configuration. The sodium chloride solution is injected into flow channel from a source ($0.25 \times 0.4 \times 0.25$ m) located on the top of the porous medium, filled with glass balls ($\phi = 0.2 - 0.3$ mm) and equipped with a adjustable weir, in order to vary the injected flow rate. The little box was buried a few centimeters into the porous matrix, in order to have the injection surface underneath capillary fringe.

288 especially designed conductivity cells are distributed in 6 sections perpendicular to the mean flow displacement. These cells are connected to a computer by a data acquisition system whose main purposes are to : (i) read data from the experimental device, (ii) store

data in a file for subsequent analysis, (iii) control experimental parameters and (iv) display "on line" pollutant spreading.

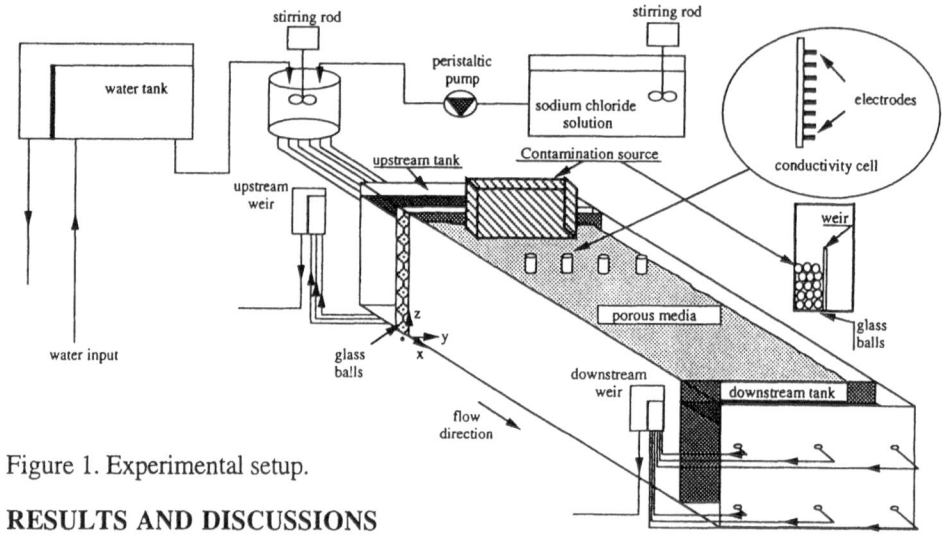

Figure 1. Experimental setup.

RESULTS AND DISCUSSIONS

The experiments carried out on the physical model were located in interference regim (regim 3) and in mechanical dispersion regim (regim 4) (Pfannkuch, 1963), *i.e.* for a molecular Peclet ($Pe_m = Vd_p/D_m$) range between 8.0 and 5.10^2 respectivelly between 5.10^2 and 2.10^5 - d_p is the mean diameter of grains and D_m is the molecular diffusion coefficient .

The hydrodynamic parameters (longitudinal dispersion coefficient and the horizontal component of velocity field), are presented in the table 1, in function of the travelling distance (L [cm]) and the depth of horizontal section (z). They are determined on hand of cross averaged concentration.

	z=1,3 cm		z=6,0 cm		z =10,7 cm		z=15,4 cm		z=20,1 cm		mean values	
L	V	K_L	V	K_L	V	K_L	V	K_L	V	K_L	V	K_L
40	2,0	2,5	2,1	1,5	2,0	2,1	2,1	2,4	1,9	2,6	2,0	2,3
65	2,0	2,1	2,0	2,4	2,1	1,4	2,1	2,5	1,9	2,1	2,0	2,1
90	2,0	3,4	2,0	2,2	2,0	1,9	2,0	2,9	2,0	2,5	2,0	2,6
115	2,0	3,0	2,1	2,0	2,0	2,1	2,1	2,9	2,0	3,0	2,0	2,6
140	2,0	4,1	2,1	2,6	2,0	2,1	2,1	3,2	2,0	4,0	2,0	3,3

Table 1. Longitudinal Dispersion Coefficient (cm^2/min) and the Horizontal Component of Velocity Field (cm/min).

The other parameters, as kinetic porosity, the boundary conditions for the groundwater flow (the piezometric head at the upper respectivelly the lower weir and at the little box for the source case) and the boundary conditions for the transport (concentrations into the upper weir and into the little box) were measured, so that no parameter fitting was required for simulation. The geometry of the system for the source case is shown in fig. 2, together with the finite element mesh that was used for the simulation.

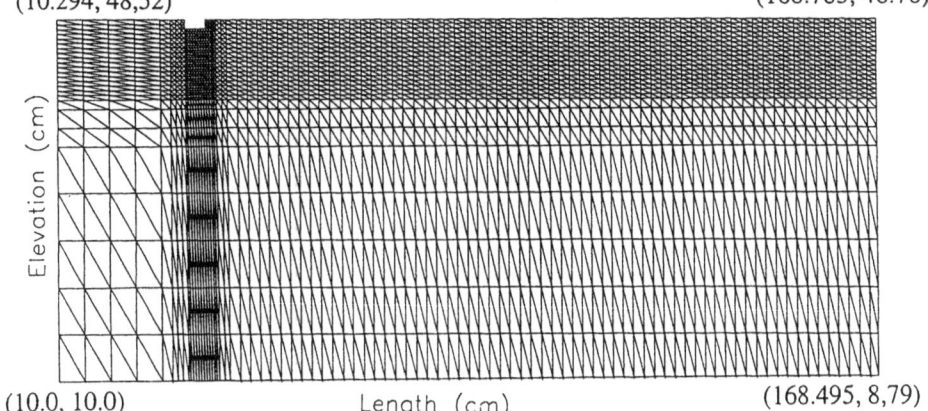

Figure 2. Geometry of the studied system with finite element discretisation.

All the parameters that were selected to specify the problem are listed in table 2 :

Parameter	Specified value
Soil parameters	
Horizontal permeability	$1.3 \cdot 10^{-9}$ m^2
Vertical permeanility	$1.3 \cdot 10^{-9}$ m^2
Kinematic porosity	0.42
Fluid parameters	
Ambient density (for T = 15°)	999.1 kg/m^3
Dynamic viscosity (Herbert et $al.$, 1988)	$1.002 \times (1. + 0.4819\ c_m) \cdot 10^{-3}$ Pa . s
Injection rate (for the source case)	$3.8 \cdot 10^{-6}$ m^3/s/m^3
Injection period	6 hours
Density factor $\partial\rho/\partial c_m$	700 kg/m^3
Transport parameters	
Longitudinal dispersivity	0.01 m
Transversal dispersivity	0.001 m
Molecular diffusivity in water	$1.5 \cdot 10^{-9}$ m^2/s
Hydraulic gradient	0.76%
Numerical model parameters	
Number of nodes	2222
Nomber of elements	4218
Simulation period	2 hours
Time step	Variable

Table 2. Data Specification for the Injection Simulation.

In the first case (injection over the whole cross-section of the medium), the mathematical model was tested for a injected concentration of 9.4 g/l. A good agreement was found between the simulated and experimental curves, as it shown in fig. 3 for two sections identified by their coordinates. The density effect can be seen only on the horizontal component of velocity field. In the first section (x = 40 cm) the concentration distribution can be compared with a tracer. More the travelling distance increases, more the horizontal component of velocity field increases (x = 140 cm), and the differences between concentration distribution and a tracer case becomes more significant.

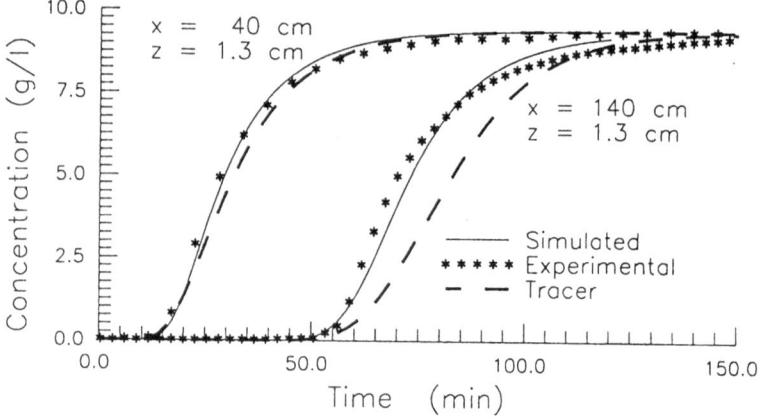

Figure 3. The simulated and experimental curves.

In the second case (injection from a source), the mathematical model was tested in two cases : stable and unstable displacement. The experimental and simulated concentrations evolution in several sections can be seen in fig 4. As long as the injected concentration is lower than 2.5 g/l, the pollutant displacement can be considered as stable. After a transition zone which varies in time from section to section, caracterized by a progressive decrease in concentration (fig. 4 a), the pollutant displacement keeps the same configuration. In the first section (x = 5 cm), the differences between the experimental and simulated concentrations can be considered practically insignificant, except the transition zone. The same comment for the last section (x = 105 cm), except the cell put in proximity of capillary fringe (z = 34.2 cm), where the differences are very important (about 20%). It is important to notice that the 2-D numerical model could be used in this case to simulate the experiments carried out in 3D configuration, because the recorded curves are symmetric.
If the injected concentration is greater then 7.5 g/l, the pollutant displacement becomes unstable. The experimental breakthrough curves are characterized by a "chaotic" behaviour (fig. 4 b, x = 105 cm). The pollutant displacement is also accompanied by a burying phenomenon. It appears, nevertheless, that these instabilities become blurred in the time. Taking into account the instability phenomenon, the use of the 2-D numerical model appears to be inadequate. Nevertheless, we shall presented in fig. 4, b, a few results for this case. The differences between the simulated and experimental curves are significant. It appears that the pollutant penetration depth is more important in the simulated case that in the experimental case.

CONCLUSIONS

A 2-D numerical model for density dependent transport problem based on the mixte hybrid element method has been developed. The code has been tested against the experiments carried out on physical 3D model, under various injection conditions. The

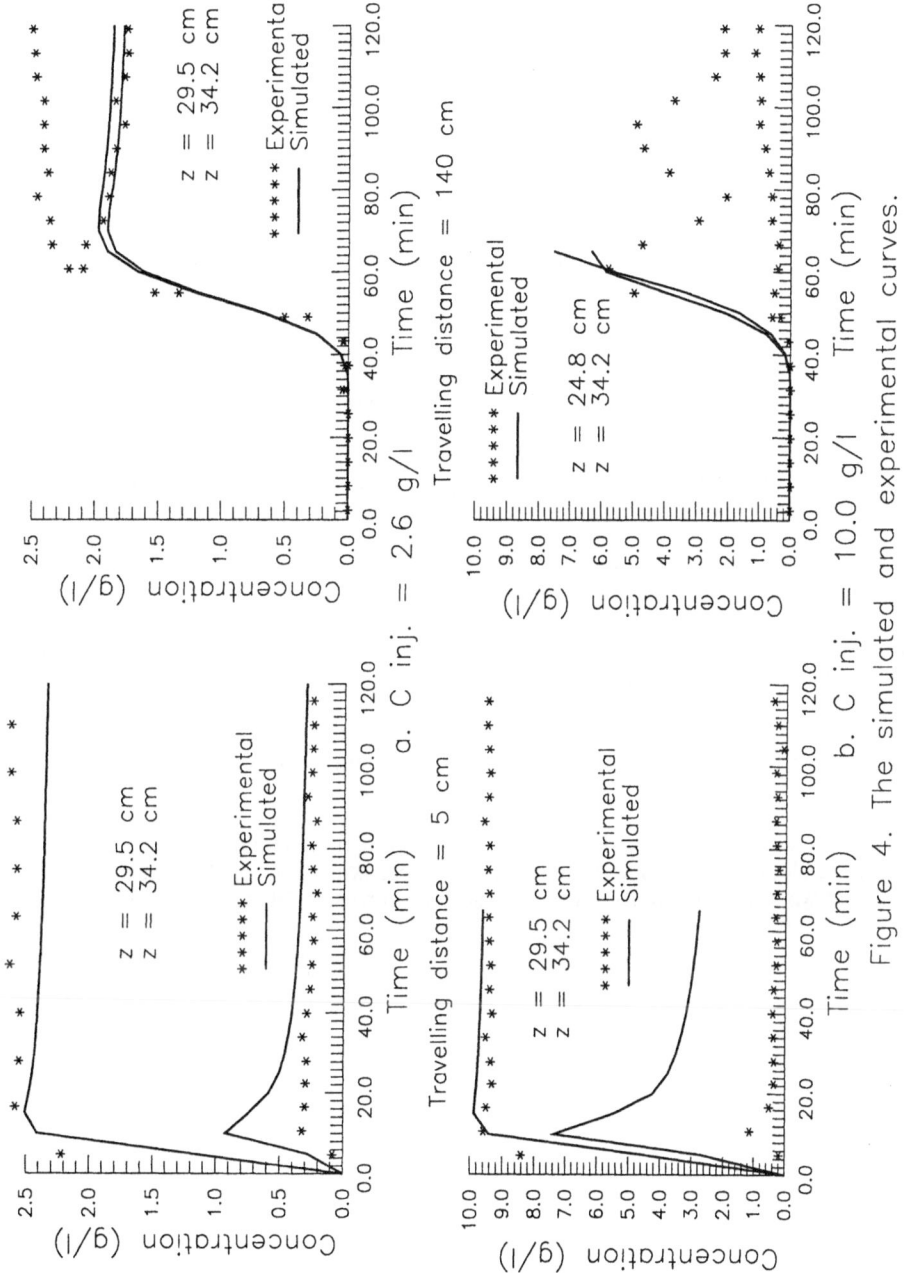

Figure 4. The simulated and experimental curves.

hydrodynamic parameters have been measured so that no parameter fitting has been required. In the first case (injection over the whole cross-section of the medium) a good agreement has been found between the simulated and experimental curves. The density effect can be seen only on the horizontal component of velocity field. In the second case (injection from a source), it appears that the mathematical code can be only used for the stable case. Excepting the transition zone as well as the cells put under the capillary fringe, a good agreement has been found betwen the simulated and experimental curves.

REFERENCES

Bear, J. (1972) "Dynamics of fluids in porous media". American Elsevier Publishing Compagny, Oxford, New-York.

Buès, M. A. and Oltean, C. (1992) "Solute transport in three dimensional model through an homogeneous porous medium : estimation of hydrodynamic parameters". Numerical Methods in Water Resources, ELSEVIER Applied Sciences, London, New-York, 1, pp. 395-402.

Marquardt, D. W. (1963) "An algorithm for least squares estimation of non-linear parameters". SIAM Journal of Applied Mathematics, 11, pp. 431-441.

Pfannkuch, H. O. (1963) "Contribution à l'étude des déplacements miscibles dans un milieu poreux". Revue de l'I. F. P., 2, pp. 1-54.

Dorgarten, H. W. and Tsang, C. F. (1991) "Modeling the density-driven movement of liquid wastes in deep sloping aquifers". Ground Water, v. 29, n° 5, pp. 655-662.

Herbert, A. W., Jackson, C. P. and Lever, D. A. (1988) "Coupled growndwater flow and solute transport with fluid density strongly dependent upon concentration". Water Resources Research, v. 24, n° 10, pp. 1781-1795.

Chavent, G. and Roberts, J. E. (1991) "A unified physical presentation of mixed, mixed-hybrid finite elements and usual finite difference approximation for the determination of velocities in waterflow problems". Adv. in Water Resources, v. 14, n° 6, pp. 329-348.

TRACER TRANSPORT MODELING OF THE DOUBLET WELL SYSTEM.

S. P. POZDNIAKOV and CHIN FU TSANG
Earth Sciences Division, Lawrence Berkeley Laboratory,
1 Cyclotron Road, Berkeley, CA 94720, USA

Steady state flow and tracer transport between an injection well and a pumping well in a heterogeneous confined aquifer is investigated using numerical modeling. Calculation of transport is based on advective model for heterogeneous aquifers. Dispersion is assumed to be controlled by micro scale velocity variation. An effective parameter of dispersion evaluated on the breakthrough curves is defined to account for the influences of heterogeneity. Breakthrough curves are calculated using numerical modeling of transport in strongly heterogeneous aquifer with spatial heterogeneous transmissivity fields. Results of modeling was processed by comparing with analytical solutions of doublet system to obtain the effective parameters. A special solution was developed for advective transport in aquifers with layered structure. Examples were given to show to its influence on breakthrough curves, with the resulting impact on of the effective macroscopic parameters.

INTRODUCTION

In the design of optimal groundwater remediation in heterogeneous aquifers, one needs to estimate in situ flow and transport parameters. One effective method for field scale determination of transport parameters is a tracer test during well injection. There are two common test schemes: radial flow test during injection in single well and horizontal doublet test with divergent-convergent flow pattern using two wells. The doublet test is considered to be more appropriate for heterogeneous aquifer as indicated by *Mironenko and Rumynin* (1986). The goal of present study is to investigate the influences of spatial heterogeneity on breakthrough curves in the pumping well of a doublet system, and to calculate the effective transport parameters for such a system.

Radial flow and transport in heterogeneous media has been investigated by several authors including *Shvidler* (1964), *Desbarats* (1992), *Moreno et. al.* (1988), *Vandenberg* (1977), *Mironenko and Rumynin* (1986). Flow in multiple-well system in heterogeneous media was studied by *Gomez-Hernandez and Gorelick* (1989), *Desbarats* (1993). Transport problems for such systems using advective and advective-dispersion approach studied by *Grove* (1971) and *Mironenko and Rumynin* (1986), who obtained analytical solutions for

A. Peters et al. (eds.), Computational Methods in Water Resources X, 529–536.
© *1994 Kluwer Academic Publishers. Printed in the Netherlands.*

advective-dispersive problem for the scale of heterogeneity much less than the doublet dimension (distance between injection and pumping wells). For a strongly heterogeneous aquifer it is much more common that the difference between the doublet dimension and the heterogeneity scale is not significant.

CURRENT STUDY

The advantages of the horizontal doublet test scheme for the entire thickness of aquifer had been discussed earlier by *Mironenko and Rumynin* (1986), *Gelhar* (1993). Given the same well yield, the flow lines of a doublet form a closed circuit between the wells, thus efficiently averaging the aquifer properties over the doublet's area of influence. It seems that this advantage may be only for the case that the doublet dimension is much longer than the scale of heterogeneity. However for many fractured and porous aquifers the horizontal scale of heterogeneity of hydraulic permeability or transmissivity can be tens of meter. From a practical point of view the doublet dimension usually does not exceed two hundreds meters because of the long time period required by the test.

The calculation assumes doublet dimension of 190m. and heterogeneity scale of 50m.. This is an extreme case as a practical field test but still permits a analyses of the influence of the realistic field scale heterogeneity. The aim is to study the effect of transmissivity heterogeneity on the results of tracer test. The influences of the porosity variation , natural flow gradient and vertical flow velocity are not considered in this paper..

Aquifer parameters used

The anisotropic exponential model of log transmissivity correlation was chosen for modeling of the random aquifer heterogeneity (*Tompson and Gelhar*, 1990). The variance of log transmissivity was 1,1. that exceeded the value for the estimation of the aquifer macrodispersion property within the framework of stochastic hydrology theory (*Gelhar*, 1993). Modeling the random field of transmissivity with such a large value of variance could lead to abnormally high (on order of $n*10^4 m^2/day$) and abnormally low (less than 10^{-3} m^2/day) transmissivity values. That is why a limited distribution was sought which would better describe the values near the small and large probability values but give values the lognormal distribution near median. The *Johnson's S-V* distribution was chosen which is normal for the transformed function F_T

$$F_T = \log \frac{T - T_{min}}{T_{max} - T} \qquad (1)$$

where T is transmissivity; T_{min} and T_{max} its minimum and maximum possible values.
The horizontal correlation scales of F_T were $L_x = 30$ and $L_y = 90$ meters. For the random field simulation "the source point method" (*Ghori et. al.*, 1993) was used. The transformation used to obtain the random transmissivity field is in two steps:
a) obtain F_T distribution for the F_T function:

$$F_T(x,y) = M_F + \sigma_F f(x,y) \qquad (2)$$

where $F_T(x,y)$ is the F_T function at the point with coordinate x, y ; M_F is the mean value of F_T ; σ_F is the variance; and $f(x,y)$ is a unit random process with defined autocorrelation function $R(\bar{r})$;

b) transform $F_T(x,y)$ into $T(x,y)$ according to

$$T = \frac{T_{max}\,\exp(F_T) + T_{min}}{1 + \exp(F_T)}$$
(3)

Parameters of the transmissivity distribution are:

M_F=-2.73, σ_F=2.0, T_{min}=0.4m^2/day, T_{max}=870m^2/day. The average transmissivity value is 53 m^2/day. Aquifer porosity n assumed to be equal to 0.004. Aquifer thickness m was also constant and equal to 40 meters. Microdispersivity parameters were defined as λ_L=0.05 m., λ_T=0.01m. The coefficient of molecular diffusion was assumed to be equal to 0. Sorption retardation and tracer decay were neglected.

Doublet test setup

The distance between well is R=190 m. The doublet rescharge-disharge flow rate Q was chosen to be 300 m^3/day. Such a value, given the actual well radius and the interwell spacing of 190 m., yields an average head difference value of 8 - 15 m. which is quite realistic for the field tests. The tracer injection period was assumed as 100 days. According to the analytical solution for homogeneous aquifer after this period the concentration in the pumping well certainly exceed 0.5 the injection concentration.

Simulation of the doublet system was conducted for 25 realizations of random transmissivity field. For each realization of field, two tests were simulated:

a) the doublet axis lies along the direction of the larger transmissivity correlation scale;

b) the doublet axis lies normal to the direction of the larger correlation scale.

NUMERICAL MODELING

To simulate the doublet tests, the numerical code **ASM** by *Kinzelbach and Rausch* (1991) is used. This code allows simulation of 2D steady/unsteady state flow by the finite difference method and the contaminant transport, by the random walk particle tracking technique (*Kinzelbach, 1988, Uffink, 1988*). Displacement of a particle is decomposed into the advective component calculated by particles tracking technique and dispersive component calculated by random walk part at each point.

A strongly divergent or convergent flow on a rectangular grid present a major problem in modeling. In such cases, as noted by *El-Kadi* (1988), the particles tracking can lead to serious errors. To avoid these errors interpolation nearest to source/sink nodes in original **ASM** code was improved by using logarithmic functional instead of the linear form. Another source of error in simulation of heterogeneous system could be connected with the bilinear velocity interpolation scheme. As was discussed by *Goode* (1990) the bilinear gradient interpolation scheme would be better for such modeling. After these improvement, numerical modeling using **ASM** of advective doublet flow for the homogeneous system gave results, which when compared with approximate analytical solution of *Mironenko and Rumynin* (1986) showed a good agreement in terms of breakthrough curves.

The size of model domain was 960*960 m. covered by a grid of 60*60. Two boundary blocks are defined at of the four sides with width 100m. All other internal 56 grid steps constant and equal to 10 m. The doublet axis was parallel to one of the grid directions. Halfway between wells is at the center of the grid domain. The ratio of the grid step and minimum correlation length is 0.3 which , according to a number of studies (*Tsang et. al.,* 1988, *Tompson and Gelhar*, 1990, *Desbarats* ,1993) is quite satisfactory. The external boundaries were defined as closed, and the wells flow rate is constant. Steady hydraulic head and flow velocity fields were obtained by solving the flow equation using the conjugate gradient method. The modeling time step for tracer injection during 100 days was 0.1 day (which is essentially lower then the Courant grid criterion), the number of particles used is 2000. The particles were initially distributed on a circle with a 10 m. radius around the injection well. To monitor the tracer arrival, two observation points were chosen on the doublet axis, halfway between the wells and the other at 1/4 the distance from the pumping well. Fields of head, velocity, concentration , breakthrough curves for pumping well and observation points were calculated and recorded in data base after each simulation.

RESULTS

As an example of the simulation results Fig. 1. shows the fraction of extracted over injected tracer as a function of time for a number of realizations . Averaged breakthrough curves in pumping well are found in Fig 2. The modeling results were processed to estimate the effective porosity value. This is done based on the approximate solution for advective transport (*Mironenko and Rumynin,* 1986). The choice of this particular solution as basis was determined by the fact that the solutions (*Grove,* 1971, *Mironenko and Rumynin,* 1986) showed that then is little influence of the longitudinal microdispersion within a broad range of the Peclet number values.

According to the above solution, the dimensionless concentration C in the pumping well is given by:

$$C(\tau) = \pi^{-1} \inf F(\psi), \quad \tau = \frac{tQ}{\pi R^2 nm}, \quad F(\psi) = \frac{\sin\psi - \psi\cos\psi}{\sin^3\psi} = \tau \quad (4)$$

where inf is the function inverse to F

Here, $C=0$ for time $t<t_0=1/3$ and $C>0$ for $t>t_0$ (Fig. 2) which enables one to use the onset time of tracer arrival (t_0) as an interpretation parameter:

$$n = \frac{3t_0 Q}{\pi R^2 m} \qquad (5)$$

However, more reliable results could be found using the information on all breakthrough curves in terms of the extracted dimensionless volume of tracer V:

$$V = (\int_{\tau_0}^{\tau} C(\tau)d\tau)/(Q\,\tau_0) = F_v(\tau/\tau_0) \qquad (6)$$

From the calculated extracted volume V_f as a function of t one can use the nonlinear curve fitting method find the best fit to the $V_f(t/t_0)$ and $V(\tau/\tau_0)$ curves. After the porosity n can be calculated trough the estimated t_0 value.

Both procedures was used for processing the modeling results. The results are showed in Table 1.

Table 1 *

	Doublet axis along the direction of largest correlation scale	Doublet axis normal the direction of largest correlation scale
Mean value n/n_0	0.70	1.04
$n_0 = 0.004$	0.74	0.95
Standard deviation	0.26	0.50
	0.23	0.17

* The first value is based on the tracer arrival time method and the second is based on extracted volume curves fitting method.

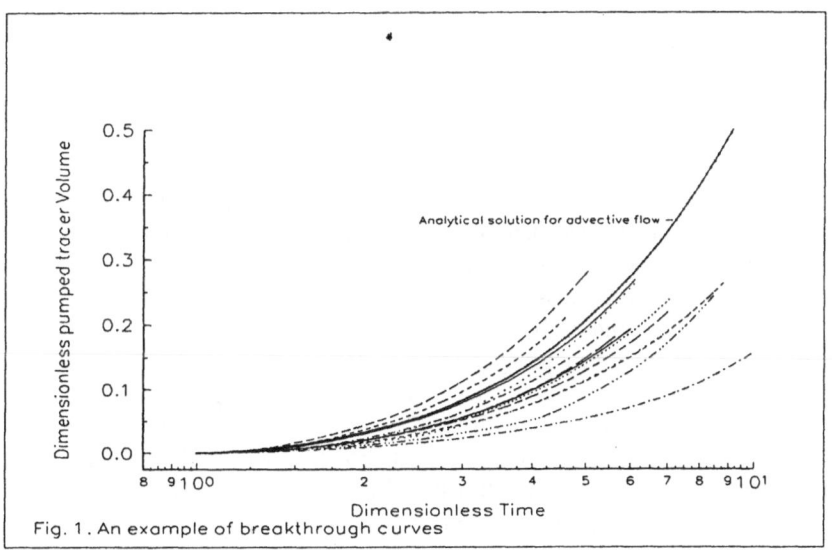

Fig. 1. An example of breakthrough curves

One can find from table 1 that the difference between results of arrival time methods and integral method is not significant but the second method demands a longer testing time . The modeling results confirmed the advantages of doublet tests. Even under such highly heterogeneous system the doublet tests results show an high stability of the calculated parameter for porosity.

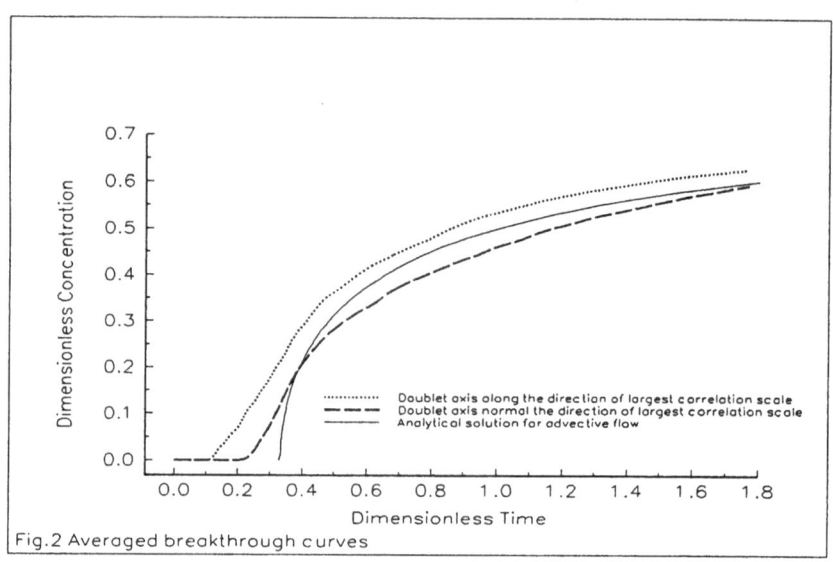

Fig.2 Averaged breakthrough curves

DISCUSSION

One can see from Fig. 3 a very asymmetrical shape of the tracer plume body. An analytical solution would give a symmetric picture according to flow lines. Taking into account the changing lengths of flow lines could improve the analytical solution in terms of macrodispersivity. The question is how can one correlate the change of flow paths with stochastic parameters of transmissivity field.

A serious question is the influence of the vertical structure of the aquifer on breakthrough curve. In most porous and fractured aquifers the permeability and porosity change considerably with depth. To estimate the influence of vertical heterogeneity a layers transport calculation scheme could be used. Let us to consider an aquifer with the thickness m and the total transmissivity T to be composed of N layers ($N \rightarrow \infty$) with porosity n_i and hydraulic conductivity k_i. According to this scheme, the Darsy velocity v_i is equal to Q_i / m_i in the i-th interlayer. This is determined as Q^*k_i /T. The advective velocity for this layer is equal to v_i / n_i . Change in advective velocities for various layers as determined by k_i / n_i changes with depth are the reason of pathline's in different layers have different arrival times to the pumping well. For this scheme let us introduce time dependent dimensionless concentration function $C(\tau)$ for a vertically homogeneous with depth aquifer. Then one can calculate the time dependent $C(\tau, \chi)$ function according to :

$$C(\tau, \chi) = (m^{-1}) \int_0^m C(\tau^* \chi(z)) dz, \quad \chi(z) = \frac{k(z)}{n(z)} \qquad (6)$$

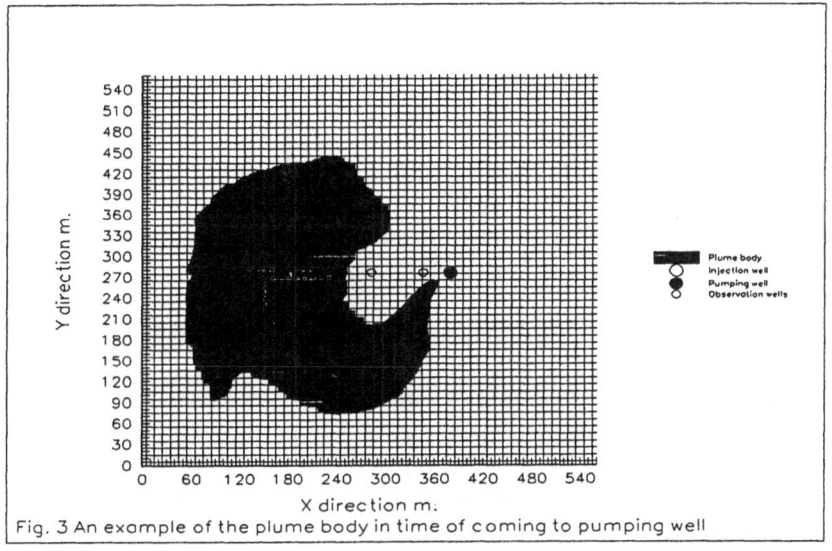

Fig. 3 An example of the plume body in time of coming to pumping well

Acknowledgments. *This work was developed for the LBL Russian - American Center for Contaminant Transport Studies under the auspices of the US Department of Energy, Office for Environmental Restoration and Waste Management, Office of Technology Development through Contract No DE-AC03-76SF00098 . The authors are grateful to V.A. Mironenko for discussion .*

REFERENCES

Desbarats A.J.(1992) " Spatial averaging of the transmissivity in heterogeneous fields with flow toward well", Water Resources Research,3, 757-767

Desbarats A.J.(1993) "Geostatistical Analysis of Interwell Transmissivity in Heterogeneous Aquifers", Water Resources Research, 4, 1239-1246

El-Kadi A.I (1988) "Applying the USGS mass-transport model (MOC) to remedial actions by recovery wells," Ground Water, 3, 281-288

Gelhar L.J. (1993) "Stochastic subsurface hydrology", Prentice - Hall Inc., New Jersey

Ghlori S.G., Heller J.P., Singh A.K.(1993) "An efficient Method of Generating random permeability Fields by the Source Point Method", Mathematical Geology,5, 559-572

Gomez-Hernandez J.J, Gorelick S.M. (1989) "Effective ground water model parameter values: Influence of spatial variability of hydraulic conductivity , leakance an resharge", Water Resources. Research, 3, 405-419

Goode D.J.(1990) "Particular velocity interpolation in blok-centred finite-difference ground water models", Water Resources. Research, 5, 925 -940

Grove D.V.(1971)" An analysis of the flow field of a disharging-recharging pair of wells", USGS Report 474-99 ,NTIS, 1971, 52pp.

Kinzelbach W.(1988) "The Random Walk Method in Pollutant Transport Simulation", in E. Custodio et al. (eds.), Ground Water Flow and Quality Modeling, D. Reidel Publ. Comp.,pp.227-245,

Kinzelbach W., Rausch R.(1989) "Aquifer Simulation Model ASM", Documentation, Universitat, Kassel, FRG

Mironenko V.A., Rumynin V.G.(1986) "Opitno Migracionnie raboty v vodonostnyx plastax", Nedra, Moscow

Moreno L., Tsang C.F., Tsang Y., Hale F.V., Neretnieks I. (1988) " Flow and tracer transport in signle fracture: a stochastic model and its relation to some field observations", Water Resources Research,12, 2033-2048

Shwidler M.I. (1964) "Filtration flow in heterogeneous media", Cons. Bureu, New York

Tsang Y.W., Tsang C.F., Neretnicks I.,.Moreno L. (1988) "Flow and tracer transport in fractured media: a variable aperture chanel model and its properties", Water Resources Research,12, 2049-2060

Tompson A.B., Gelhar L.W.(1990) "Numerical Simulation of Solute Transport in Three - Dimensional, Randomly Heterogeneous Porous Media", Water Resources Research, 10, 2541-2562

Uffink G.M.(1988) "Modeling of Solute Transport With the Random Walk Method", in E. Custodio et al. (eds.), Ground Water Flow and Quality Modeling, by D. Reidel Publ. Comp., pp.247-265

Vandenberg (1977) "Pump test in heterogeneous aquifers", Journal Hydrology (1/2),45-62

ON THE VALIDITY OF A FICKIAN DIFFUSION MODEL FOR THE SPREADING OF LIQUID INFILTRATION PLUMES IN PARTIALLY SATURATED HETEROGENEOUS MEDIA

K. PRUESS
Earth Sciences Division, Lawrence Berkeley Laboratory
University of California, Berkeley, CA 94720
U.S.A.

ABSTRACT

Localized infiltration of aqueous and non-aqueous phase liquids (NAPLs) occurs in many circumstances. Examples include leaky underground pipelines and storage tanks, landfill and disposal sites, and surface spills. Because of ever-present heterogeneities on different scales such infiltration plumes are expected to disperse transversally and longitudinally.

This paper examines recent suggestions that liquid plumes are being dispersed from medium heterogeneities in a manner that is analogous to Fickian diffusion. Numerical simulation experiments on liquid infiltration in heterogeneous media are performed to study the dispersive effects of small-scale heterogeneity. It is found that plume spreading indeed tends to be diffusive. Our results suggest that, as far as infiltration of liquids is concerned, broad classes of heterogeneous media behave as dispersive media with locally homogeneous (albeit anisotropic) permeability.

INTRODUCTION

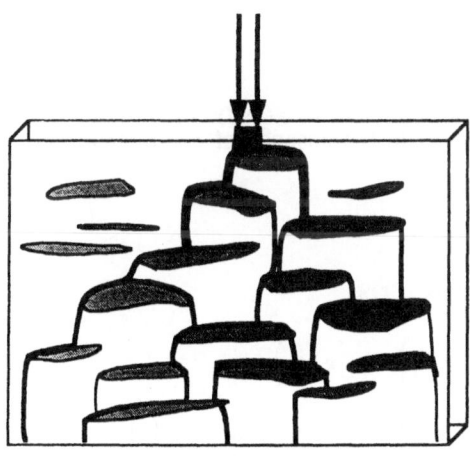

Figure 1. Schematic of liquid infiltration in an unsaturated heterogeneous medium. Regions of low permeability (shaded) divert flux sideways and cause a lateral spreading of the infiltration plume.

Liquids flowing downward through the unsaturated zone in coarse soils, or in large (sub-)vertical fractures, may encounter low-permeability obstacles, such as silt or clay lenses in soils, or asperity contacts between fracture walls. Water will pond atop the obstacles and be diverted sideways, until other predominantly vertical pathways are reached (Fig. 1). The conventional treatment of two-phase flow includes gravity, pressure, and capillary effects. Horizontal flow diversion from media heterogeneities can be represented only if such heterogeneity is modeled in full explicit detail. In practical applications, explicit numerical modeling of small-scale reservoir heterogeneities would require prohibitively large numbers of grid blocks, because heterogeneities occur on many different scales (small-scale laminations and layering in soils, low-permeability lenses, capillary barriers, fractures, fracture networks, lithologic units, etc.).

A. Peters et al. (eds.), Computational Methods in Water Resources X, 537–544.
© 1994 Kluwer Academic Publishers. Printed in the Netherlands.

It has recently been suggested that plume spreading due to media heterogeneities may be approximated as a diffusive process (Espedal et al., 1991; Langlo and Espedal, 1992; Pruess, 1993, 1994). In this paper we briefly summarize our convection-dispersion model for infiltration of liquids in the vadose zone. The validity of the Fickian dispersion hypothesis is then examined by means of numerical simulation experiments in media with fully-resolved small-scale heterogeneity.

FICKIAN DISPERSION MODEL

In two-phase immiscible flow of liquid and gas, mass fluxes F_β (β = liquid, gas) are customarily written as a multiphase version of Darcy's law,

$$F_\beta = -k\frac{k_{r\beta}}{\mu_\beta}\rho_\beta\left(\nabla P_\beta - \rho_\beta g\right) \qquad (1).$$

Here, k is the permeability tensor, $k_{r\beta}$ is relative permeability in phase β, μ is viscosity, ρ is fluid density, P_β is pressure in phase β, and g is acceleration of gravity. In the following we specialize to anisotropic media with principal axes of the permeability tensor in the horizontal and vertical directions.

$$k = k_h\left[e_x e_x + e_y e_y\right] + k_v e_z e_z \qquad (2)$$

In Eq. (2) we have introduced unit vectors e in the x, y, and z-directions. k_h and k_v are the horizontal and vertical permeabilities, respectively. Our proposed Fickian-type diffusion model for phase dispersion involves adding a dispersive flux term for liquid phase to Eq. (1) which, in analogy to solute dispersion in miscible flow (de Marsily, 1986), is written as

$$F_{l,dis} = -\rho_l \phi D_{dis} \nabla S_l \qquad (3).$$

Specializing to conditions where advective flow is dominated by gravity, we introduce the propagation velocity v of saturation disturbances in the absence of capillary effects (Pruess, 1991),

$$v = \frac{k_v}{\phi} \frac{\rho_l g}{\mu_l} \frac{d k_{rl}}{dS_l} \frac{g}{g} \qquad (4),$$

The dispersion tensor D_{dis} is then written as (Pruess, 1993)

$$D_{dis} = v\left(\alpha_T\left[e_x e_x + e_y e_y\right] + \alpha_L e_z e_z\right) \qquad (5).$$

Here we have introduced transverse (horizontal) and longitudinal (vertical) dispersivities α_T, α_L. g and v are the magnitude of the gravitational acceleration and velocity vectors, respectively. Note that the proposed phase-dispersive flux has the same structure as capillary flux. Indeed, the capillary-driven flux component of Eq. (1) can be written in a form like Eqs. (3-5), with transverse capillary dispersivity given by

$$\alpha_{cap,T} = \frac{k_h}{k_v} \frac{1}{\rho_l g} \frac{d\,P_{cap}}{d\ln k_{rl}} \tag{6}.$$

An analogous equation, without the anisotropy factor k_h/k_v, holds for the longitudinal capillary dispersivity. From the correspondence between phase-dispersive and capillary fluxes, we expect that phase dispersion effects from medium heterogeneities will be most important when capillary effects are weak, i.e., for non-wetting liquids, and for "coarse" heterogeneous media such as coarse-grained soils or large fractures. Longitudinal phase dispersion will modify the predominant downward advective flow. Transverse dispersion may lead to qualitatively new behavior, causing a lateral spreading of liquid plumes even when capillary pressures are weak. In the remainder of the paper we will focus mainly on transverse dispersion effects.

Numerical Simulation Experiments

To examine the validity of the proposed phase dispersion model we have performed numerical simulation experiments. The calculations were done with our multiphase flow code TOUGH2 (Pruess, 1991), enhanced with a set of preconditioned conjugate gradient routines for efficient solution of multidimensional flow problems with 10,000 or more grid blocks (Moridis, private communication, 1993). The flow experiments involve placing a localized plume of liquid into an unsaturated heterogeneous medium, such as shown in Fig. 2, and then allowing the plume to migrate under the combined action of pressure, capillary, and gravity forces (Eq. 1). Small-scale medium heterogeneity is resolved in detail, and no explicit allowance for phase dispersion as in Eq. (3) is made. Many numerical experiments were carried out for media with different parameters and style of heterogeneity, both regular and random.

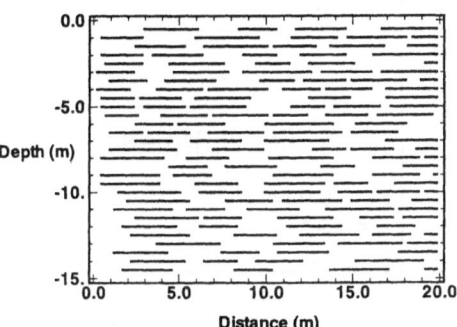

Figure 2. Two-dimensional vertical section of a heterogeneous medium with a random distribution of impermeable obstacles (black segments). Length of obstacles is uniformly distributed in the range of 2-4 m.

The spreading of the plumes is analyzed by evaluating spatial moments (Sahimi et al., 1986; Freyberg, 1986; Essaid et al., 1993). Specifically, an effective transverse diffusivity for a localized plume is calculated as

$$D_T = \frac{1}{2} \frac{d}{dt} \left(\sigma_T^2 \right) \tag{7},$$

where σ_T^2 is the mean square plume size in the transverse (horizontal) direction. Dividing by the downward velocity $d\langle z\rangle/dt$ of plume movement yields the transverse dispersivity

$$\alpha_T = \frac{D_T}{(d\langle z \rangle / dt)} \tag{8}.$$

Equations analogous to (7, 8) are used to calculate longitudinal dispersivities. Table 1 gives specifications for a test problem that involves infiltration of water at ambient

Table 1. Parameters for test problems with detailed explicit heterogeneity.

Permeability Porosity	$k = 10^{-11}$ m^2 $\phi = 0.35$
Relative Permeability	
van Genuchten function (1980) $k_{rl} = \sqrt{S^*}\left\{1-\left(1-\left[S^*\right]^{1/\lambda}\right)^\lambda\right\}^2$ irreducible water saturation exponent	$S^* = (S_l - S_{lr})/(1 - S_{lr})$ $S_{lr} = 0.15$ $\lambda = 0.457$
Capillary Pressure	
van Genuchten function (1980) $P_{cap} = -(\rho_w g/a)\left(\left[S^*\right]^{-1/\lambda} - 1\right)^{1-\lambda}$ irreducible water saturation exponent strength coefficient	$S^* = (S_l - S_{lr})/(1 - S_{lr})$ $S_{lr} = 0.0$ or 0.15 $\lambda = 0.457$ $a = 5$ m^{-1}
Geometry of Flow Domain	

2-D vertical (X-Z) section
          width (X)
          depth (Z)
          gridding
	
	20 m
	15 m
	80 x 120 = 9600 blocks
	$\Delta X = .25$ m
	$\Delta Z = .125$ m

heterogeneity: stochastic distribution of impermeable obstacles	
Initial Water Saturation	
for $6.5 \leq X \leq 13.5$m and $-3.5 \leq Z \leq 0$ m remainder of domain	$S_l = 0.99$ $S_l = 0.15$

conditions (T = 15 °C) into a soil of 10^{-11} m^2 absolute permeability, with embedded impermeable obstacles. A simplified treatment with soil gas as passive bystander at constant pressure is used, corresponding to solving Richards' equation. Results are given in Figs. 3 and 4.

RESULTS

Water infiltration plumes for cases with and without capillary pressures show quite different structure (Fig. 3). When capillary pressures are neglected (or unimportant, as in coarse high-permeable media or large fractures), flow proceeds in the form of narrow fingers (Glass et al., 1989). Capillary suction pressures will tend to dampen out the narrower fingers. Overall, transverse spreading of plumes from medium heterogeneities increases when capillary pressures are included, while longitudinal spreading is seen to diminish (Fig. 4). Transverse dispersivities go through transient changes at early times, then stabilize at nearly constant values. Longitudinal dispersivities on the other hand are

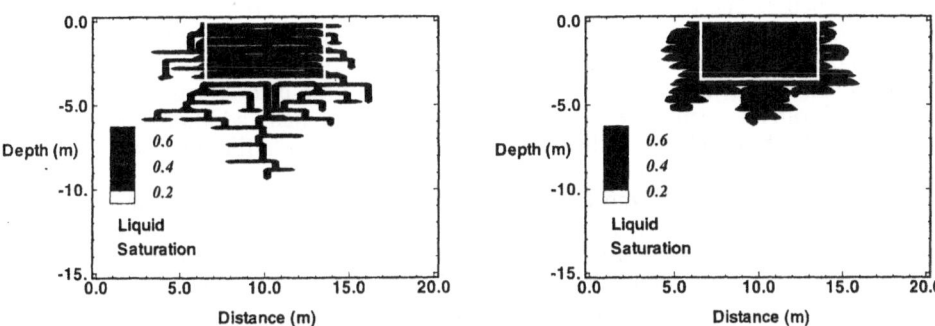

Figure 3.   Simulated infiltration plumes in the medium of Fig. 2 after 2×10^5 seconds. The plume on the left is for the case without capillary pressure, the one on the right has a capillary pressure as given in Table 1, with $S_{lr} = 0$. Initially, the plume has a uniform water saturation of $S_l = .99$ and occupies the region indicated by the white rectangle at the top of the figures.

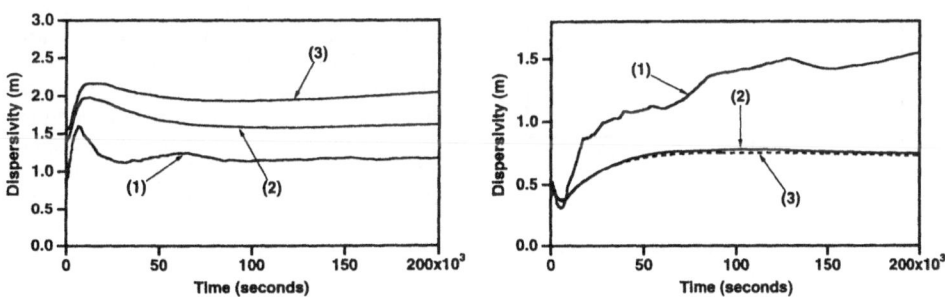

Figure 4.   Effective dispersivities for the liquid infiltration plumes of Fig. 3. Transverse (left) and longitudinal (right) dispersivities are shown for cases with increasingly strong capillary pressures: (1) no P_{cap}, (2) P_{cap} with $S_{lr} = 0$, (3) P_{cap} with $S_{lr} = .15$.

seen to stabilize only when capillary effects are included; otherwise they continue to increase with time. Fig. 4 shows that transverse dispersivities stabilize at approximately 1.2, 1.7, and 2 m, respectively, for the cases of (1) no capillary pressure, (2) weaker and (3) stronger capillary pressure.

DISCUSSION

The most important result is that transverse dispersivities stabilize, after a period of transient changes at early times, at nearly constant values. This stabilization occurs regardless of the strength of capillary pressure, and indicates that transverse plume spreading from the intrinsic heterogeneities of the medium indeed gives rise to a Fickian diffusion process. The changes at early times are numerical artefacts, caused by the extreme discontinuity of the initial saturation distribution. For the large initial water saturation of $S_l = .99$ in the plume water flow rates are large, leading to rapid saturation changes which are poorly resolved with the space and time discretization used in our simulation. From the results presented here as well as from many additional simulations we conclude that broad classes of heterogeneous media disperse infiltrating liquid plumes transversally in a Fickian manner.

It should be emphasized that Fickian-type dispersive behavior from medium heterogeneities is by no means inevitable or universal. In fact, for certain heterogeneity conditions and spatial scales infiltration plumes may show "anti-dispersive" behavior, becoming more narrowly focussed with depth (Kung, 1990). Flow behavior depends on the nature of the heterogeneities and the strength of capillary forces. When capillary effects are neglected, our simulations show "anomalous diffusion" behavior for longitudinal dispersion, with dispersivities growing with time.

To further elucidate the interplay between heterogeneity and capillarity, we have in Fig. 5 plotted transverse capillary dispersivities calculated from Eq. (6) for the medium of Fig. 2. The applicable anisotropy ratio was determined by running single-phase horizontal and vertical flow simulations to steady state. Effective horizontal and vertical permeabilities were found to be $k_h = 7.50 \times 10^{-12}$ m^2 and $k_v = 1.17 \times 10^{-12}$ m^2. Accordingly, the data in Fig. 5 are plotted for an anisotropy ratio of $k_h/k_v = 6.41$. For the function with $S_{lr} = 0$, transverse capillary dispersivity is seen to be approximately 0.5 m over a wide range of saturations. This value agrees closely with the difference between transverse dispersivities with (for $S_{lr} = 0$) and without capillary effects (Fig. 4), indicating that heterogeneity- and capillary-derived transverse dispersivities are additive.

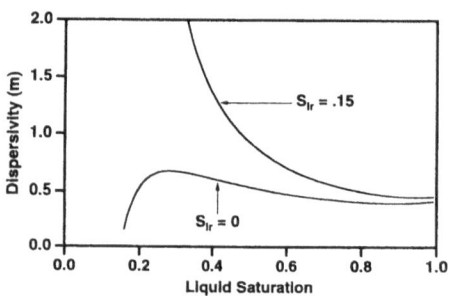

Figure 5. Transverse capillary dispersivities for a typical soil with permeability $k = 10^{-11}$ m^2. Two curves are shown, for weaker ($S_{lr} = 0$) and stronger ($S_{lr} = .15$) capillary pressure. Dispersivity is proportional to the anisotropy ratio of horizontal to vertical permeability, which is here assumed as 6.41, appropriate for the medium in Fig. 2.

Capillary dispersivities are expected to correlate inversely with permeability. For aqueous liquids, a rough estimate for the dependence of transverse capillary dispersivity on horizontal and vertical soil permeabilities k_h, k_v may be given from the data shown in Fig. 5, using Leverett's scaling relationship $P_{cap} \propto k^{-1/2}$ in Eq. (6) (Scheidegger, 1974). We obtain

$$\alpha_{cap,T} \approx \frac{k_h}{k_v} \sqrt{\frac{10^{-11} \, m^2}{k_h}} \times .08 \, m \tag{9}.$$

For nonaqueous phase liquids (NAPLs) capillary pressures are usually weaker. Dispersive effects of permeability heterogeneity are expected to be similar for aqueous and nonaqueous phase liquids. With capillary dispersivities being smaller for NAPLs, we expect that heterogeneity-derived dispersion would become relatively more important than it is for aqueous liquids.

CONCLUSIONS

Liquid phases infiltrating in the unsaturated zone disperse from capillary effects and from medium heterogeneities. Our high-resolution numerical simulation experiments suggest that for broad classes of medium heterogeneities transverse fluid phase dispersion tends to be Fickian. Further numerical and physical experiments, and field observations, are needed to determine the range of heterogeneity conditions under which the Fickian dispersion model is applicable.

ACKNOWLEDGEMENT

This work was supported through U.S. Department of Energy Contract No. DE-AC03-76SF00098 by the Director, Office of Civilian Radioactive Waste Management, Office of External Relations, administered by the Nevada Operations Office in cooperation with the Swiss National Cooperative for the Disposal of Radioactive Waste (NAGRA). The author thanks his colleague George Moridis for providing a package of preconditioned conjugate gradient routines that made possible the simulation of problems with of the order of 10,000 grid blocks. For a review of the manuscript and the suggestion of improvements thanks are due to Drs. Stefan Finsterle, Lynn Gelhar, and Yvonne Tsang.

REFERENCES

de Marsily, G. *Quantitative Hydrogeology*, Academic Press, Orlando, FL, 1986.

Espedal, M. S., P. Langlo, O. Saevareid, E. Gislefoss and R. Hansen. Heterogeneous Reservoir Models: Local Refinement and Effective Parameters, paper SPE-21231, presented at Society of Petroleum Engineers 11th Symposium on Reservoir Simulation, Anaheim, CA, February 1991.

Essaid, H.I., W.N. Herkelrath and K.M. Hess. Simulation of Fluid Distributions Observed at a Crude Oil Spill Site Incorporating Hysteresis, Oil Entrapment, and Spatial Variability of Hydraulic Properties,*Water Resources Res.*, Vol. 29, No. 6, pp. 1753-1770, June 1993.

Freyberg, D.L. A Natural Gradient Experiment on Solute Transport in a Sand Aquifer. 2. Spatial Moments and the Advection and Dispersion of Nonreactive Tracers, *Water Resources Res.,* Vol. 22, No. 13, pp. 2031-2046, 1986.

Glass, R.J., T.S. Steenhuis and J.Y. Parlange. Wetting Front Instability, 2. Experimental Determination of Relationships between System Parameters and Two-Dimensional Unstable Flow Field Behavior in Initially Dry Porous Media, *Water Resources Res.,* Vol. 25, No. 6, pp. 1195-1207, 1989.

Kung, K. J. S. Preferential Flow in a Sandy Vadose Zone: 2. Mechanism and Implications, *Geoderma,* Vol. 46, pp. 59-71, 1990.

Langlo, P. and M.S. Espedal. Heterogeneous Reservoir Models, Two-Phase Immiscible Flow in 2-D, in T.F. Russell et al. (eds.), Computational Methods in Water Resources IX, Vol. 2, pp. 71-79, 1992.

Pruess, K. TOUGH2 - A General Purpose Numerical Simulator for Multiphase Fluid and Heat Flow, Report No. LBL-29400, Lawrence Berkeley Laboratory, Berkeley, CA, May 1991.

Pruess, K. Grid Orientation and Capillary Pressure Effects in the Simulation of Water Injection into Depleted Vapor Zones, *Geothermics,* 20 (5/6), 257-277, 1991.

Pruess, K. Dispersion of Immiscible Fluid Phases in Gravity-Driven Flow: A Fickian Diffusion Model, *Water Resources Res. (submitted),* Report No. LBL-33914, June 1993.

Pruess, K. Liquid-Phase Dispersion During Injection into Vapor-Dominated Reservoirs, presented at 19th Annual Workshop on Geothermal Reservoir Engineering, Stanford University, Stanford, CA, January 1994.

Sahimi, M., B. D. Hughes, L. E. Scriven and H. T. Davis. Dispersion in Flow Through Porous Media - I. One-Phase Flow, *Chem. Eng. Sci.,* Vol. 41, No. 8, pp. 2103-2122, 1986.

Scheidegger, A. E. *The Physics of Flow Through Porous Media,* University of Toronto Press, Toronto and Buffalo, Third Edition, 1974.

van Genuchten, M. Th. A Closed-Form Equation for Predicting the Hydraulic Conductivity of Unsaturated Soils, Soil Sci. Soc. Am. J., Vol. 44, pp. 892-898, 1980.

INFORMATION-DEPENDENT PREDICTION OF SOLUTE TRANSPORT IN HETEROGENEOUS GEOLOGIC MEDIA

DONGXIAO ZHANG and SHLOMO P. NEUMAN

Department of Hydrology and Water Resources
The University of Arizona, Tucson, AZ 85721
U.S.A.

Recently, a unified Eulerian-Lagrangian theory has been developed [1] for solute transport in random, space-time nonstationary velocity fields. We describe a computational method based on this theory for the special case of steady state flow in a mildly fluctuating, statistically homogeneous, lognormal hydraulic conductivity field. We take the unconditional mean velocity to be uniform but allow it become nonuniform by conditioning on measurements of log hydraulic conductivity (or transmissivity) and/or hydraulic head. This renders the transport equation information-dependent. Since we consider only advection, Peclet numbers are initially infinite but diminish gradually with time. We avoid numerical difficulties by solving the problem analytically at early time. At later time, we adopt a pseudo-Fickian representation which involves a conditional dispersion tensor. The latter is evaluated numerically along mean trajectories, and the conditional transport equation is solved by finite elements on a relatively coarse grid. The final step is an explicit numerical computation of lower bounds on the conditional concentration prediction variance and coefficient of variation, solute mass flow rate across a "compliance surface," cumulative mass release and travel time distribution across this surface, the corresponding error variance, and plume spatial moments. The method can account for uncertainty in initial mass and/or concentration when predicting the future evolution of a plume. Conversely, the method can quantify uncertainty in locating the source of solute "particles" that have been detected in the subsurface. We illustrate some of these capabilities with the aid of two-dimensional examples that involve instantaneous point and nonpoint sources.

INTRODUCTION

Stochastic models of solute transport in randomly heterogeneous media are becoming applicable to realistic field situations. This is in large measure due to the development in recent years of methods to condition such models on actual field data [2,3]. Previous methods of conditioning have been either Eulerian [2] or Lagrangian [3]. Recently, Neuman [1] developed a unified Eulerian-Lagrangian

A. Peters et al. (eds.), Computational Methods in Water Resources X, 545–552.
© 1994 Kluwer Academic Publishers. Printed in the Netherlands.

theory of transport conditioned on hydraulic data in space-time nonstationary velocity fields. Neuman's theory yields an exact, closed form transport equation in terms of the first two conditional Lagrangian velocity moments and the conditional Lagrangian cross-covariance between velocity and its forcing terms. In the hypothetical case where these three Lagrangian moments are known, the equation can in principle be solved exactly for the conditional ensemble mean solute concentration and flux. This is a formal advantage over Eulerian theories which cannot be expressed in closed form without approximation. The Eulerian-Lagrangian theory has two formal advantages over the pure Lagrangian approach: It explicitly conserves fluid mass, and does not require any distributional assumption about displacements to predict solute concentration or flux. However, Lagrangian velocities in the subsurface are not possible to measure. To compute them requires Monte Carlo simulation of particle trajectories on a fine grid. We avoid entirely the use of such high-resolution Monte Carlo simulation by solving the problem deterministically. At early time we employ an analytical solution which is rigorously valid in the limit as time goes to zero and is easy to compute. At later time, we use a pseudo-Fickian approximation which yields an advective-dispersive transport equation with an information and space-time dependent dispersion tensor. The latter is amenable to solution by finite elements on a relatively coarse grid. Our method also allows computing relevant ensemble, spatial and temporal moments explicitly. We describe and illustrate some of these capabilities below.

ANALYTICAL-NUMERICAL APPROACH

At early time $(t - t_o)$ relative to the conditional Lagrangian velocity correlation time, the conditional mean concentration $\langle c(\mathbf{x}, t | \mathbf{x}_o, t_o) \rangle_v$ due to an instantaneous point source of mass M_o introduced at $(\mathbf{x}_o, t_o)$ is given by [1,4]

$$\langle c(\mathbf{x}, t | \mathbf{x}_o, t_o) \rangle_v = \frac{M_o}{\phi(t - t_o)^k} \, p \left(\frac{\mathbf{x} - \mathbf{x}_o}{t - t_o} \right)_v \tag{1}$$

where $\langle \, \rangle$ represents ensemble mean, the subscript v indicates conditioning on hydraulic data, ϕ is porosity (assumed constant), k is the number of space dimensions, and $p(\mathbf{v})_v$ is the conditional *pdf* of the velocity $\mathbf{v}$. Numerical studies [5] found that (1) is consistent with Monte Carlo results up to dimensionless time (defined later) of order 0.5. For an instantaneous nonpoint (line, area, volume) source, one simply integrates (1) over the source domain.

For an instantaneous point and nonpoint source under a pseudo-Fickian regime, the conditional mean concentration $\langle c(\mathbf{x}, t | t_o) \rangle_v$ satisfies the pseudo-Fickian advection-dispersion equation

$$\frac{\partial \langle c(\mathbf{x}, t | t_o) \rangle_v}{\partial t} + v(\mathbf{x}, t) \cdot \nabla \langle c(\mathbf{x}, t | t_o) \rangle_v - \nabla \cdot \mathbf{D}_v(\mathbf{x}, t; t_o) \nabla \langle c(\mathbf{x}, t | t_o) \rangle_v = \langle g(\mathbf{x}) \rangle \delta(t - t_o) \tag{2}$$

where v is the conditional mean velocity, g is a solute source term given by $(M_o/\phi)\delta(\mathbf{x} - \mathbf{x}_o)$ for a point source and $s_o(\mathbf{x})/\phi$ for a nonpoint source with a specific strength s_o, and $\mathbf{D}$ is a conditional dispersion tensor given in linearized form by

$$\mathbf{D}_v \approx \int_{t_o}^t \langle \mathbf{v}'(\mathbf{x},t)\mathbf{v}'^T(\langle\tilde{\chi}(\tau)\rangle_v,\tau)\rangle_v d\tau. \tag{3}$$

Here $\langle\mathbf{v}'\mathbf{v}'^T\rangle_v$ is the conditional velocity covariance and $\langle\tilde{\chi}(\tau)\rangle_v$ the conditional mean position at time τ, $t_o \leq \tau \leq t$, of a particle which, at a later time t, is found at the downstream location $\mathbf{x}$. The linearization generally improves with conditioning. Similar to (3), one can define a particle origin covariance to measure uncertainty about the location at time t_o (the origin) of a particle known to have reached the downstream location $\mathbf{x}$ at later time t [5]. This covariance quantifies uncertainty about the sources of detected groundwater contamination.

The conditional covariance of predicted concentrations due to an uncertain instantaneous source introduced at t_o is given by [1,5]

$$\langle c'(\mathbf{y},s|t_o)c'(\mathbf{x},t|t_o)\rangle_v = F(\mathbf{y},s;\mathbf{x},t|t_o) + G(\mathbf{y},s;\mathbf{x},t|t_o). \tag{4}$$

Here F is written in linearized form (which yields a lower bound) as

$$F(\mathbf{y},s;\mathbf{x},t|t_o) = \nabla^T \langle c(\mathbf{y},s|t_o)\rangle_v \int_{t_o}^s \int_{t_o}^t \langle \mathbf{v}'(\langle\tilde{\eta}(\theta)\rangle_v)\mathbf{v}'^T(\langle\tilde{\chi}(\tau)\rangle_v)\rangle_v d\theta d\tau \nabla \langle c(\mathbf{x},t|t_o)\rangle_v \tag{5}$$

while G, the term accounting for source uncertainty, is given by (5.3) in [5]. When $\mathbf{y} = \mathbf{x}$ and $s = t$, $G \geq 0$. Hence uncertainty in the source or initial condition always causes an increase in the concentration prediction variance.

Another way to characterize the movement and spread of a plume is by its spatial moments. The zeroth moment corresponds to the total solute mass, the first moment to the plume center of mass, and the second moment measures spread about the plume center of mass. As the velocity field and initial solute source are uncertain, so are all spatial moments. The plume center can be estimated by its conditional mean $\langle\chi(t)\rangle_v = \phi/M_o \int \mathbf{x}\langle c(\mathbf{x},t)\rangle_v d\mathbf{x}$. The second spatial moment of the conditional mean plume about its center of mass can be estimated as $\mathbf{M}_v = \phi/M_o \int [x - \langle\chi(t)\rangle_v][x - \langle\chi(t)\rangle_v]^T \langle c(\mathbf{x},t)\rangle_v d\mathbf{x}$.

One may also estimate cumulative mass release across a "compliance surface" S during the period $t_o < t \leq \tau_S$, due to an instantaneous source of total mass M_o located upstream of S at time t_o, by its conditional mean $\langle M_S(\tau_S|t_o)\rangle_v = M_o - \phi \int_R \langle c(\mathbf{x},\tau_S|t_o)\rangle_v d\mathbf{x}$. The associated variance is $\sigma_M^2 = \phi^2 \int_R \int_R \langle c'(\mathbf{x},\tau_S|t_o)c'(\mathbf{y},\tau_S|t_o)\rangle_v d\mathbf{x}d\mathbf{y}$, R being the domain of the plume upstream of S. Total mass flux across the compliance surface can be computed according to $\langle Q_S(\tau_S|t_o)\rangle_v = [d\langle M_S(t|t_o)\rangle_v/dt]_{t=\tau_S}$. The conditional travel time distribution across the surface is related to the cumulative mass release via $G(\tau_S; S, t_o)_v = \langle M_S(\tau_S|t_o)\rangle_v/M_o$ [1,6].

NUMERICAL IMPLEMENTATION

In this paper we restrict consideration to steady state uniform mean flow through mildly fluctuating, statistically homogeneous, lognormal permeability fields having an unconditional variance not much larger than one. This allows us to linearize the pseudo-Fickian dispersion coefficient about its conditional mean trajectories, to use available analytical expressions for the unconditional Eulerian velocity covariance obtained by linearizing the stochastic flow equation, and to condition this covariance on permeability and/or hydraulic head data via cokriging. When the variance σ_Y^2 of log permeability is small, the velocity field is Gaussian or nearly so [5,6]. Then the pseudo-Fickian equation (2) is valid, to first order, at all times.

As the time-dependent dispersion coefficient is initially zero, the grid Peclet number P_e is initially infinite. For a finite element scheme to yield nonoscillatory results without excessive numerical dispersion, it is necessary to satisfy $P_e = \langle v_1 \rangle \Delta x_1 / D_{11} < 2$ and $C_o = \langle v_1 \rangle \Delta t / \Delta x_1 < 1/3$ simultaneously, where C_o is courant number, $\langle v_1 \rangle$ is mean velocity parallel to x_1, D_{11} is longitudinal dispersion coefficient, Δx_1 is longitudinal space discretization interval, and Δt is time discretization interval. For example, in the unconditional 2-D case, at early time $t = \Delta t$, $2D_{11}t \approx \sigma_{v_1}^2 t^2$ and $\sigma_{v_1}^2 = 3/8 \langle v_1 \rangle^2 \sigma_Y^2$ [6], hence $D_{11} \approx 3/16 \langle v_1 \rangle^2 \sigma_Y^2 \Delta t$. Here the above two criteria cannot be satisfied simultaneously unless $\sigma_Y^2 > 8$. This numerical difficulty is eliminated by using the analytical solution (1) at early time. The result of (1) is then introduced as initial condition for the later time pseudo-Fickian Galerkin finite element solution. When the velocity is non-Gaussian, the early time analytical solution becomes not only a numerical, but also a theoretical, necessity.

Once the conditional velocity covariance and the the conditional mean concentration become available, lower bounds on the concentration (co)variance and coefficient of variation, solute mass flux across a compliance surface, cumulative mass release and travel time distribution across this surface, the corresponding error variance, and spatial moments can be computed explicitly by numerical integration.

COMPUTATIONAL EXAMPLES

We illustrate the proposed analytical-numerical approach on some two-dimensional unconditional and conditional examples. Our examples concern transport from deterministic and uncertain sources of unit mass under steady state prior (unconditional) uniform mean flow. The posterior flow is made non-uniform by conditioning on hydraulic data. We take the prior log transmissivity $Y = \ln T$ to be a statistically homogeneous field with zero mean $\langle Y \rangle$, unit variance σ_Y^2 and unit integral scale λ. The effective porosity ϕ is arbitrarily set equal to one. Solute sources are introduced at time $t_o = 0$ into a domain of size 20λ by 10λ at a prior uniform mean velocity $\langle v_1 \rangle = 0.1$ parallel to x_1. The background concentration is zero, the velocity is Gaussian, and pore-scale dispersion is neglected.

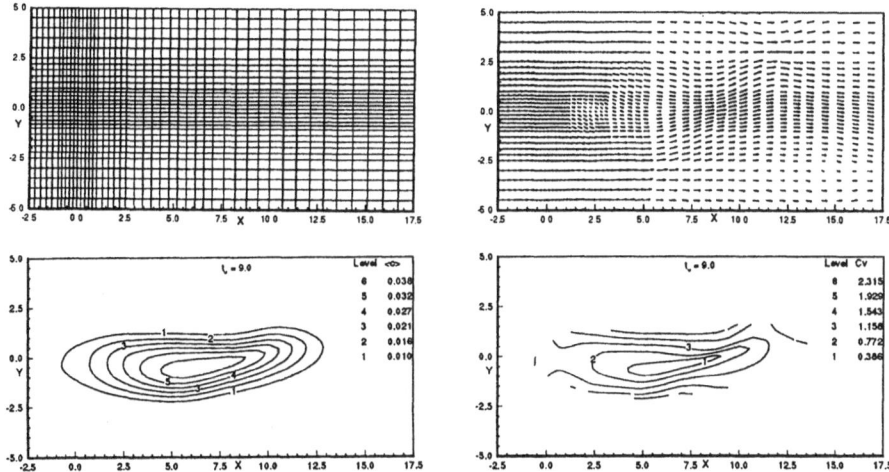

Figure 1: Finite element mesh (top left), conditional mean velocity (top right), conditional mean concentration $\langle c \rangle$ at dimensionless time 9.0 (bottom left), and lower bound coefficient of variation Cv (bottom right).

Fig. 1 shows the finite element mesh used to compute mean concentration, the mean velocity conditioned on 3 x 6 log transmissivity data shown in Fig. 2a, the conditional mean concentration at dimensionless time $t_v = \langle v_1 \rangle t / \lambda = 9.0$ due to a deterministic instantaneous source at the origin of the normalized coordinates $X = x_1/\lambda$ and $Y = x_2/\lambda$, and a lower bound on the corresponding coefficient of variation (Cv). The log transmissivity data shown in Fig. 2a are residuals, $Y' = \ln T - \langle \ln T \rangle$, arbitrarily selected from a random log transmissivity field generated by a turning bands method. As a result of conditioning, the mean velocity is no longer uniform. The conditional mean streamlines are attracted toward confirmed high permeability zones and away from confirmed low permeability zones. Hence the plume no longer travels along the prior uniform mean trajectory but along a curved conditional mean trajectory. The conditional dispersion coefficient differs from its unconditional counterpart in that it depends on location, not only on time. For this reason, the plume is no longer a symmetric ellipse but a skewed, asymmetric one. As will soon become apparent, conditioning materially reduces the predicted spread, or dispersion, of the plume as is implied by theory [1]. The contours of Cv are more distorted than those of $\langle c \rangle_v$. Conditioning on hydraulic head data or a combination of log transmissivity and head data has similar effects. This effect generally increases with the number and density of conditioning points [5].

Figs. 2b-c show the longitudinal and transverse components of conditional (solid) and unconditional (dashed) mean plume centers of mass, respectively, as functions of t_v. The conditional components deviate significantly from their unconditional counterparts. The conditional mean plume center is attracted first toward the identified high permeability zone at the lower middle portion of the do-

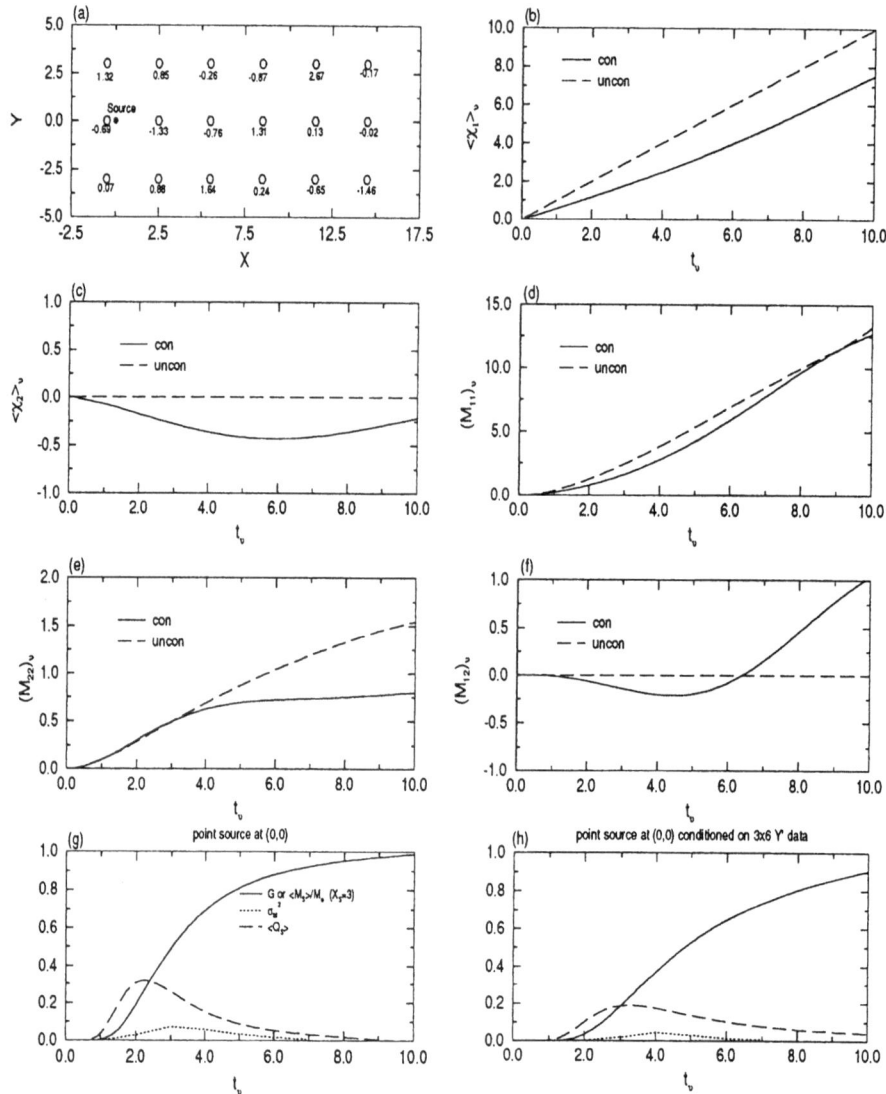

Figure 2: Spatial moments, normalized cumulative release across $X_S = 3$ (and travel time distribution), its variance, and total solute flux.

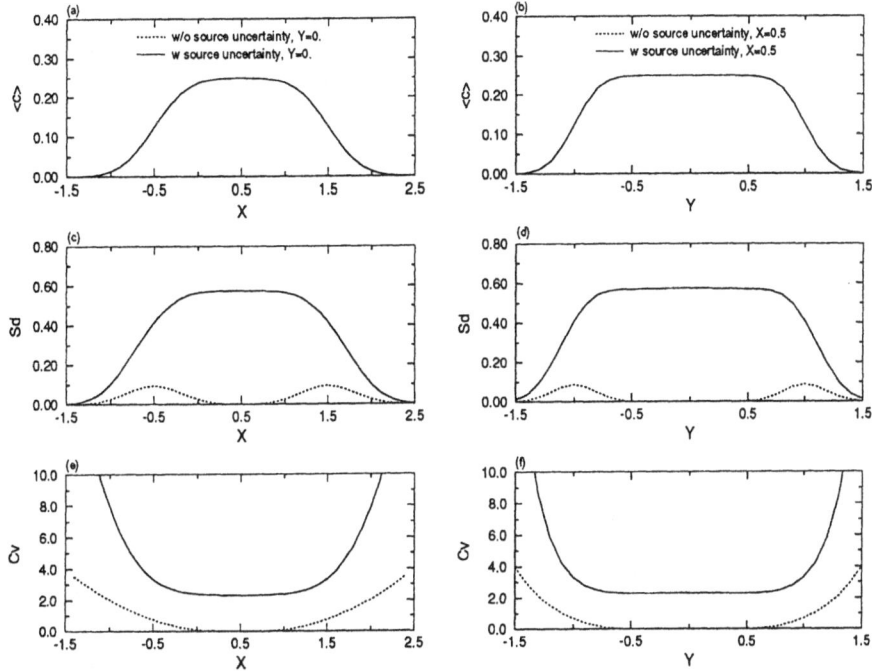

Figure 3: Spatial profiles of unconditional $\langle c \rangle$, and lower bound uncertainty measures Sd and Cv, at $t_v = 0.5$ due to uncertain and deterministic instantaneous square sources.

main and then toward the upper right corner, and its conditional rate of movement is slower than the unconditional rate. Figs. 2d-f depict components of the second spatial moment $\mathbf{M}$ of the conditional (solid) and unconditional (dashed) mean plumes about their centers of mass. Both the conditional longitudinal (M_{11}) and transverse (M_{22}) components are generally smaller than their unconditional counterparts while the conditional off-diagonal component M_{12} deviates significantly from zero. This is a reflection of our earlier finding that conditioning generally reduces spread (or dispersion) and renders the predicted plume skewed and asymmetric. Figs. 2g-h show conditional and unconditional mean solute fluxes $\langle Q_S \rangle$ across a compliance surface placed normal to the mean direction of flow at a downstream location from the source ($X_S = 3$), normalized cumulative mass releases $\langle M_S \rangle / M_o$ or equivalently travel time distributions G across this surface, and associated error variances σ_M^2. Conditioning on 3 x 6 Y' data brings about a decrease in peak mean flux and an increase in arrival time. This observation is data specific and should not be interpreted as a general pattern. However, conditioning generally reduces the maximum prediction variance σ_M^2. Though this is not shown here, its influence generally increases with the number of measurement points between the source and the compliance surface, and the prediction variance decreases consistently with source dimension [5].

Fig. 3 compares longitudinal and transverse plume profiles of unconditional mean concentration $\langle c \rangle$, lower bounds on standard deviation (Sd) and coefficient of variation (Cv) due to an uncertain instantaneous square source (solid), of size 2λ by 2λ centered about the origin, with those due to a deterministic source (dashed) at $t_v = 0.5$. The uncertain source has a uniform mean initial concentration $\langle c(\mathbf{x}, t_o) \rangle = \langle g(\mathbf{x}) \rangle = 0.25$ and a constant variance $\sigma_g^2(\mathbf{x}, t_o) = 0.125$ inside the square whose location is known with certainty. The mean concentration is clearly not affected by the initial uncertainty. The latter however causes an increase in the lower bound on the concentration prediction error standard deviation and coefficient of variation. The same holds true for point sources and under conditioning [5].

Space does not permit us to demonstrate here more than just these elementary aspects of our approach. To summarize, conditioning renders the velocity field statistically nonhomogeneous with reduced variances, renders the predicted plume irregular and non-Gaussian, and generally reduces dispersion due to a reduction in uncertainty.

ACKNOWLEDGMENTS

This work was supported jointly by the U.S. Nuclear Regulatory Commission under Contract NRC-04-90-51, and the U.S. Geological Survey under Water Resources Research Grant 14-08-0001-G2092.

REFERENCES

1. Neuman, S. P., Eulerian-Lagrangian theory of transport in space-time nonstationary velocity field: Exact nonlocal formalism by conditioning moments and weak approximations, *Water Resour. Res.*, 29(3), 633-645, 1993.

2. Graham, W., and D. McLaughlin, Stochastic analysis of nonstationary subsurface solute transport: 2. Conditional moments, *Water Resour. Res.*, 25(11), 2331-2355, 1989.

3. Rubin, Y., Prediction of tracer plume migration in disordered porous media by the method of conditional probabilities, *Water Resour. Res.*, 27(6), 1291-1308, 1991.

4. Batchelor, G. K., Diffusion in a field of homogeneous turbulence: II. The relative motion of particles, *Proc. Cambridge Philos. Soc.*, 48, 345-363, 1952.

5. Zhang, D., Conditional stochastic analysis of solute transport in heterogeneous media, *Ph.D. dissertation*, 231pp., Univ. of Ariz., Tucson, 1993.

6. Dagan, G., *Flow and Transport in Porous Formations*, Springer-Verlag, New York, 1989.

4. GEOSTATISTICS

Stochastic Finite Element Analysis for the Transport of Trichloroethylene Vapors

Roger Ghanem
Civil Engineering Department,
State University of New York,
Buffalo, NY, USA.

Hitoshi Seya, Yoshimasa Shigeno, Tadahiko Shiomi
Technical Research Laboratory,
Takenaka Corporation,
Tokyo, Japan.

Contamination of groundwater from TCE vapor transport in the unsaturated zone constitutes a serious environmental hazard, and has recently been the subject of a number of investigations. This transport, however, is highly sensitive to random variations in the characteristics of the surrounding porous medium, as well as, to the surrounding environment. In this paper, these variations are modeled as stochastic processes, and a stochastic finite element method is developed to provide an efficient probabilistic characterization of the solution to the problem. The flow and mass equations governing the transport of TCE in the unsaturated zone are coupled through the dependence of viscosity and density of the gas mixture on the TCE concentration, as well as, through advection. This results in a set of nonlinear partial differential equations with coefficients that feature stochastic fluctuations. The approach followed in solving these equations is an extension of one of the authors' work on linear stochastic finite elements. It consists of expanding the process representing the medium's randomness using the Karhunen-Loeve expansion. Further, the solution process is expanded using the Polynomial Chaos expansion which features a complete basis in the space of random variables. The coefficients in this latter expansion are then obtained through a Galerkin scheme. Once they are computed, these coefficients can be utilized to efficiently simulate realizations of the solution process. They can also be used to readily provide accurate estimates of the second order statistics of the solution process.

INTRODUCTION

This paper presents a new approach for solving coupled nonlinear equations with stochastic coefficients. The specific physical problem addressed in this paper is that of the transport of Trichloroethylene in the unsaturated zone with the physical parameters of the porous medium, such as dispersion and permeability, modeled as stochastic processes with spatial variability. The approach followed consists of using the Karhunen-Loeve expansion to represent the processes modeling these

A. Peters et al. (eds.), Computational Methods in Water Resources X, 555–562.
© 1994 Kluwer Academic Publishers. Printed in the Netherlands.

material properties. This has the desired effect of casting the problem in terms of a discrete set of random variables, which can be truncated in a mean-square optimal sense. Following that, the processes representing the concentration and effective head throughout the domain are expanded by relying on the Polynomial Chaos representation of stochastic processes. This representation is a generalized Fourier expansion for random functions with respect to a basis consisting of orthogonal random polynomials. The coefficients in the expansion of the solution along this basis are calculated by relying on a Galerkin scheme to minimize the error resulting from truncating the expansion after a finite number of terms. Application of the resulting form of the solution to risk assessment and decision making issues is also explored.

TCE TRANSPORT IN A VARIABLY SATURATED MEDIUM

The transport of TCE in the unsaturated zone can be modeled by the following two non-dimensional equations representing continuity and conservation of mass, respectively,

$$\nabla . \boldsymbol{\kappa} \nabla \chi - \frac{\partial}{\partial \eta} (\kappa_{\eta \eta} \rho_r) = S_s L \dot{\chi} \tag{1}$$

$$\nabla . \boldsymbol{\Delta} \nabla c - \boldsymbol{\nu} . \nabla (\theta_g c) = R_v \dot{c} , \tag{2}$$

where

$$\xi = x/L, \qquad \eta = z/L, \qquad \tau = t/T, \qquad \chi = h/L, \tag{3}$$

with h denoting the effective head, and L denoting a representative length scale of the problem. In the above equations, the following symbols are used

$$\nu_\xi = -\kappa_{\xi\xi} \frac{\partial \chi}{\partial \xi}, \qquad\qquad \nu_\eta = -\kappa_{\eta\eta} \left(\frac{\partial \chi}{\partial \eta} - 1 \right) \tag{4}$$

$$\boldsymbol{\kappa} = \frac{T}{L} \frac{\rho_a g}{\mu} \begin{bmatrix} k_\xi & 0 \\ 0 & k_\eta \end{bmatrix} \tag{5}$$

$$\Delta_{ij} = \alpha_T^* |\boldsymbol{\nu}| \delta_{ij} + (\alpha_L^* - \alpha_T^*) \frac{\nu_i \nu_j}{|\boldsymbol{\nu}|} + \Delta_a^* \delta_{ij} . \tag{6}$$

Finally, the statement of the problem is completed by invoking the following constitutive relationships, (Mendoza, 1990),

$$R_v = 1 + \frac{\theta_w}{\theta_g} K_h + \frac{\rho_b}{\theta_g} K_h K_d , \tag{7}$$

$$S_s = \theta_g \rho_a g \gamma_c \tag{8}$$

$$\rho_r = \frac{\rho}{\rho_a} - 1 = \frac{X_c M_c + (1 - X_c) M_a}{X_a M_a} \tag{9}$$

$$\mu = \frac{X_c \mu_c}{X_c + (1 - X_c)\Phi_{ca}}(1 - X_c)\mu_a X_c \phi_{ac} + (1 - X_c) \tag{10}$$

$$\phi_{ij} = \frac{1}{\sqrt{8}}\left(1 + \frac{M_i}{M_j}\right)^{-1/2}\left[1 + \left(\frac{\mu_i}{\mu_j}\right)^{1/2}\left(\frac{M_j}{M_i}\right)^{1/4}\right]^2, \quad i,j = c,a \tag{11}$$

$$X_c = \frac{cM_a}{\rho_a M_c + cM_a}. \tag{12}$$

In these equations, the following symbols are used. The inverse dimensionless Henry's constant is denoted by K_h, ρ_b denotes the bulk density of the medium, K_d denotes the solid-liquid partitioning coefficient, θ_w denotes the water-filled porosity, γ_c denotes the macroscopic compressibility of the gas phase. Furthermore, X_c and X_a denote the mole fraction of TCE and air, respectively, while M_c and M_a denote the molecular weights, respectively, and μ_c and μ_a are their respective viscosities.

FINITE ELEMENT FORMULATION

The governing equations (1) and (2) are now projected onto a deterministic finite-dimensional space, thus transforming them into algebraic equations. The finite element formalism will be used for that purpose. Accordingly, these equations are recast in the following finite-dimensional form,

$$\mathbf{C}\dot{\chi} + \mathbf{K}(\mathbf{c})\chi = \mathbf{F}(\chi, \mathbf{c}) \tag{13}$$

$$\mathbf{D}\dot{\mathbf{c}} + \mathbf{H}(\chi)\mathbf{c} = \mathbf{G}(\chi, \mathbf{c}). \tag{14}$$

The elemental matrices corresponding to the global matrices appearing in the last two equations are given by,

$$\mathbf{K}^e = \int_{\Omega^e} \nabla \mathbf{N}^T \boldsymbol{\kappa} \nabla \mathbf{N} d\Omega^e, \tag{15}$$

$$\mathbf{F}^e = \mathbf{F}_1^e(\mathbf{c}) + \mathbf{F}_2^e(\mathbf{c}) + \mathbf{F}_3^e(\mathbf{c}). \tag{16}$$

$$\mathbf{F}_1^e = \int_{\Gamma^e} \boldsymbol{\kappa}(\nabla \chi \cdot \mathbf{n})\mathbf{N}^T d\Gamma^e, \qquad \mathbf{F}_2^e = \int_{\Omega^e} \kappa_{\eta\eta}\rho_r \frac{\partial \mathbf{N}^T}{\partial \eta} d\Omega^e \tag{17}$$

and,

$$\mathbf{F}_3^e = \int_{\Omega^e} \kappa_{\eta\eta}\rho_r \mathbf{N}^T d\xi = -\int_{\Gamma_{top}} \kappa_{\eta\eta}\rho_r \mathbf{N}^T d\xi + \int_{\Gamma_{bottom}} \kappa_{\eta\eta}\rho_r \mathbf{N}^T d\xi . \qquad (18)$$

Furthermore,

$$\mathbf{H}^e = \mathbf{H}_1^e + \mathbf{H}_2^e , \qquad (19)$$

where,

$$\mathbf{H}_1^e = \int_{\Omega^e} \boldsymbol{\nabla}\mathbf{N}^T \boldsymbol{\Delta}\boldsymbol{\nabla}\mathbf{N}d\Omega^e , \qquad \mathbf{H}_2^e = \int_{\Omega^e} \mathbf{N}^T \boldsymbol{\nu}^T \boldsymbol{\nabla}\mathbf{N}d\Omega^e , \qquad (20)$$

and,

$$\mathbf{G}^e = \int_{\Gamma^e} (\boldsymbol{\Delta}\boldsymbol{\nabla}c).\mathbf{n}\mathbf{N}^T d\Omega^e . \qquad (21)$$

Finally, the following symbols are also used,

$$\mathbf{C}^e = \int_{\Omega^e} S_s L \mathbf{N}^T \mathbf{N}d\Omega^e , \qquad \mathbf{D}^e = \int_{\Omega^e} R_v \mathbf{N}^T \mathbf{N}d\Omega^e . \qquad (22)$$

The above treatment has not taken into account the stochastic nature of any of the parameters. In the next section, the permeabilities, and the effective diffusion coefficient will be assumed to be realizations of a stochastic field; they will be represented using their respective Karhunen-Loeve expansions, and equations (1) and (2) will be modified accordingly.

EXPANSION OF THE RANDOM FIELDS

The permeabilities k_x and k_y will be assumed to be second-order stochastic processes (i.e. with bounded covariance function) with the following means and covariance functions,

$$\bar{k}_\alpha(\mathbf{x}) = <k_\alpha(\mathbf{x})> , \qquad R_{k_\alpha k_\alpha}(\mathbf{x}_1\mathbf{x}_2 = <k_\alpha(\mathbf{x}_1)k_\alpha(\mathbf{x}_2)>, \qquad \alpha = \xi,\eta . \ (23)$$

The Karhunen-Loeve expansion of either of these random processes is obtained upon solving the following integral eigenvalue equation,

$$\int_\Omega R_{k_\alpha k_\alpha}(\mathbf{x}_1,\mathbf{x}_2)\phi_i(\mathbf{x}_2)d\mathbf{x}_2 = \lambda_i\phi_i(\mathbf{x}_1) . \qquad (24)$$

Using the following notation,

$$k_{\alpha_i}(\mathbf{x}) = \lambda_i\phi_i(\mathbf{x}) , \qquad \alpha = \xi,\eta , \qquad (25)$$

the two permeabilities and the effective diffusion coefficient can then be represented, in an optimal mean-square convergent expansion, as follows,

$$k_\xi(\mathbf{x}) = \sum_{i=0}^{\infty} \xi_i k_{\xi i}(\mathbf{x}) \,, \qquad k_\eta(\mathbf{x}) = \sum_{i=0}^{\infty} \xi_i k_{\eta i}(\mathbf{x}) \,, \qquad \Delta_a^*(\mathbf{x}) = \sum_{i=0}^{\infty} \xi_i \Delta_{ai}^*(\mathbf{x}) \,. \quad (26)$$

The covariance functions of the solution processes are, in general, not known a-priori. Therefore, the associated Karhunen-Loeve expansions cannot be implemented, and a more general expansion is relied upon to represent these processes, namely the Polynomial Chaos expansion (Ghanem and Spanos, 1991). Similarly to the square-integrable functions in L_2, second-order random variables define a Hilbert space Θ. Any function $c(\theta)$ in Θ can be expanded in a uniformly convergent series

$$c(\theta) = \sum_{i=0}^{\infty} c_i \, \Psi_i(\theta), \qquad (27)$$

where c_i is some constant independent of θ, and $\{\Psi_i(\theta)\}_{i=1}^{\infty}$ is a basis in Θ. The Polynomial Chaos has been introduced (Wiener, 1938; Kallianpur, 1980) as a means for expanding random variables in terms of polynomials that are orthogonal with respect to the Gaussian measure.

The conductivity tensor depends in a nonlinear fashion on the concentration, and this latter is itself a nonlinear function of dispersion and conductivity. Furthermore, the dispersion tensor is a nonlinear function of velocity. Therefore, the conductivity tensor, the dispersion tensor, along with the unknown head and concentration values can be expanded in their respective Polynomial Chaos series. Due to the orthogonality of the Polynomial Chaoses, expressions for the coefficients in these expansions can be readily obtained upon multiplying both sides of the equations by Ψ_i and averaging. This leads to the following expansions for the various quantities,

$$\boldsymbol{\kappa} = \sum_{i=0}^{\infty} \Psi_i \boldsymbol{\kappa}_i \,, \qquad \boldsymbol{\kappa}_i = \frac{<\boldsymbol{\kappa}\Psi_i>}{<\Psi_i^2>} \qquad (28)$$

and

$$\boldsymbol{\Delta} = \sum_{i=0}^{\infty} \Psi_i \boldsymbol{\Delta}_i \,, \qquad \boldsymbol{\Delta}_i = \frac{<\boldsymbol{\Delta}\Psi_i>}{<\Psi_i^2>} \,, \qquad (29)$$

$$\chi(\mathbf{x}, t) = \sum_{i=0}^{\infty} \Psi_i \chi_i(\mathbf{x}, t) \,, \qquad \chi_i(\mathbf{x}, t) = \frac{<\chi(\mathbf{x}, t)\Psi_i>}{<\Psi_i^2>} \,, \qquad (30)$$

and

$$c(\mathbf{x}, t) = \sum_{i=0}^{\infty} \Psi_i c_i(\mathbf{x}, t) \,, \qquad c_i(\mathbf{x}, t) = \frac{<c(\mathbf{x}, t)\Psi_i>}{<\Psi_i^2>} \,. \qquad (31)$$

The denominators on the right hand sides of the second equations in (28)-(31) can be readily obtained and have been tabulated by Ghanem and Spanos (1991). To evaluate the numerator, it is first noted that the conductivity, being a function viscosity, depends in a nonlinear fashion on the concentration, and is therefore a nonlinear function of $\boldsymbol{\xi}$. The dependence of κ on c is explicitly known. Therefore, once the coefficients in the expansion for $c(\boldsymbol{\xi})$ have been obtained, this dependence can be calculated, at least in an algorithmic form. This allows for the calculation of the following expected value,

$$<\kappa\Psi_i> \equiv <\kappa(\boldsymbol{\xi})\Psi_i(\boldsymbol{\xi})> = \int_{-\infty}^{\infty} \kappa(\boldsymbol{\xi})\Psi_i(\boldsymbol{\xi})f_{\boldsymbol{\xi}}(\boldsymbol{\xi})d\boldsymbol{\xi} . \tag{32}$$

Depending on the assumed functional dependence of κ on c, and hence on $\boldsymbol{\xi}$, the above integral may be explicitly evaluated. In general, however, it will have to be numerically evaluated. A similar treatment to the one detailed above can be applied to the dispersion tensor resulting in the following equations,

$$<\boldsymbol{\Delta}\Psi_i> \equiv <\boldsymbol{\Delta}(\boldsymbol{\xi})\Psi_i(\boldsymbol{\xi})> = \int_{-\infty}^{\infty} \boldsymbol{\Delta}(\boldsymbol{\xi})\Psi_i(\boldsymbol{\xi})f_{\boldsymbol{\xi}}(\boldsymbol{\xi})d\xi n . \tag{33}$$

Substituting the above expansions in the finite element integrals over each element results in the following expansions for those integrals,

$$\mathbf{K}^e = \sum_{i=0}^{\infty} \Psi_i \int_{\Omega^e} \boldsymbol{\nabla}\mathbf{N}^T\kappa_i\boldsymbol{\nabla}\mathbf{N}d\Omega^e = \sum_{i=0}^{\infty} \Psi_i K_i^e \tag{34}$$

$$\mathbf{F}^e = \sum_{i=0}^{\infty}\sum_{j=0}^{\infty} \Psi_i\Psi_j\mathbf{F}_{ij}^e \tag{35}$$

$$\mathbf{H}^e = \sum_{i=0}^{\infty} \Psi_i\mathbf{H}_i^e \tag{36}$$

and,

$$\mathbf{G}^e \equiv \sum_{i=0}^{\infty}\sum_{j=0}^{\infty} \mathbf{G}_{ij}^e . \tag{37}$$

These elemental integrals can be assembled, following standard finite element procedures, into global matrices resulting into the following two ordinary differential equations,

$$\sum_{i=0}^{\infty} \Psi_i\mathbf{C}\dot{\chi}_i + \sum_{i=0}^{\infty}\sum_{j=0}^{\infty} \Psi_i\Psi_j\mathbf{K}_i\chi_j = \sum_{i=0}^{\infty}\sum_{j=0}^{\infty} \Psi_i\Psi_j\mathbf{F}_{ij} , \tag{38}$$

and

$$\sum_{i=0}^{\infty} \Psi_i\mathbf{D}\dot{c}_i + \sum_{i=0}^{\infty}\sum_{j=0}^{\infty} \Psi_i\Psi_j\mathbf{H}_ic_j = \sum_{i=0}^{\infty}\sum_{j=0}^{\infty} \Psi_i\Psi_j\mathbf{G}_{ij} . \tag{39}$$

Truncating each of the summations in the above equations at the Pth term results in an error which is minimized by requiring it to be orthogonal to each of the Polynomial Chaoses used in the expansions. Thus, multiplying each of the above equations by Ψ_k, and averaging, yields,

$$\mathbf{C}\dot{\boldsymbol{\chi}}_k + \sum_{i=0}^{P}\sum_{j=0}^{P} d_{ijk}\mathbf{K}_i\boldsymbol{\chi}_j = \sum_{i=0}^{P}\sum_{j=0}^{P} d_{ijk}\mathbf{F}_{ij}, \quad k = 1, 2, \ldots, \tag{40}$$

and

$$\mathbf{D}\dot{\mathbf{c}}_k + \sum_{i=0}^{P}\sum_{j=0}^{P} d_{ijk}\mathbf{H}_i\mathbf{c}_j = \sum_{i=0}^{P}\sum_{j=0}^{P} d_{ijk}\mathbf{G}_{ij}, \quad k = 1, 2, \ldots . \tag{41}$$

These last two equations represent two sets of coupled nonlinear algebraic equations to be solved for vectors $\boldsymbol{\chi}_k$ and $\mathbf{c}_k$. Once these vectors are obtained, simulations of the heads and concentrations throughout the domain of the problem can be readily obtained by relying on equations (30) and (31). Equations (40) and (41) can be integrated in time using standard time-stepping techniques (Russell and Wheeler, 1983.)

EXPANSION FOR RELIABILITY CALCULATIONS

It is now shown how the new method can also be used to obtain a probabilistic description of the solution process through a characterization of the corresponding response surface. First, note that second order properties of the solution process can be readily obtained once the coefficients in the Polynomial Chaos expansion have been computed. This covariance matrix has the concise representation given by the equation

$$\mathbf{R_{cc}} = \sum_{i=0}^{M} \mathbf{c}_i \, \mathbf{c}_i^H <\gamma_i^2(\theta)> , \tag{42}$$

where the superscript H indicates complex conjugation, and $<\gamma_i^2(\theta)>$ represents the variance of the Polynomial Chaos and has been tabulated (Ghanem and Spanos, 1991). Further probabilistic information about the solution process can be obtained by noting that a symbolic representation of this solution at any given point within the medium is given by

$$c(\mathbf{x}, t, \theta) = h_t[\alpha(\mathbf{x}, \theta)] , \tag{43}$$

where $h_t[.]$ is some function of its argument. Equation (43) is an expression for the response surface as a function defined on the infinite dimensional space spanned by the random process $\alpha(\mathbf{x}, t, \theta)$. The Karhunen-Loeve expansion of $\alpha(\mathbf{x}, t, \theta)$ consists of an optimal expression of this response surface as a function defined on a finite dimensional sub-space, resulting in

$$c(\mathbf{x}, t, \theta) = s_t[\{\xi_i\}] . \tag{44}$$

The Polynomial Chaos spectral expansion then provides a convergent expansion of this continuous function in terms of a discrete number of coefficients. These coefficients are computed using the Galerkin scheme presented in the previous section. Once an explicit expression is obtained for the response surface $s[.]$, an expression for the failure surface $g[\{\xi\}]$ is immediately available by noting that

$$g_t[\{\xi\}] = s_t[\{\xi\}] - c_{failure}(\mathbf{x}, t) , \qquad (45)$$

where $c_{failure}(\mathbf{x}, t)$ denotes the capacity at failure at point $\mathbf{x}$ of the hydrological system. The probability of failure can then be obtained by direct integration of the failure surface as

$$P_{failure} = \int_{g[\{\xi_i\}] > 0} p_{\xi_{i_1}, \ldots, \xi_{i_n}} d\xi_{i_1} \ldots d\xi_{i_n} , \qquad (46)$$

where $p_{\xi_{i_1}, \ldots, \xi_{i_n}}$ is the joint probability distribution function of the orthogonal random variables $\{\xi_{i_1}, \ldots, \xi_{i_n}\}$. In general the shape of the failure surface will be such that closed form integration is not possible. A versatile numerical integration procedure is through simulation. This simulation, however, is readily implemented at a minimum computational cost. Specifically, realizations of the set $\{\xi_i\}_{i=1}^M$ are generated, and equation (45) is used to obtain the corresponding realizations of the failure surface. These realizations are then employed in a nonparametric estimation scheme (Becker et.al, 1988) to obtain an approximation to the probabilistic content of the failure surface.

REFERENCES

Becker, R.A., Chambers, J.M., and Wilks, A.R. (1988), *The New S Language: A Programming Environment for Data Analysis and Graphics,* Wadsworth and Brooks/Cole.

Cameron, R. and Martin, W. (1947), "The orthogonal development of nonlinear functionals in series of Fourier-Hermite functionals", *Ann. Math,* **48**, 385-392.

Ghanem, R., and Spanos, P. (1991), *Stochastic Finite Elements: A Spectral Approach,* Springer Verlag.

Kallianpur, G.(1980), *Stochastic Filtering Theory,* Springer-Verlag, Berlin.

Mendoza, C., and McAlary, T. (1990), "Modeling of ground-water contamination caused by organic solvent vapors," *Ground Water,* **28**, (2), 199-206.

Russell, T.F., and Wheeler, M.F. (1983), "Finite element and finite difference methods for continuous flows in porous media," in R. Ewing (ed) *The Mathematics of Reservoir Simulation,,* SIAM, Philadelphia, pp. 35-106.

Wiener, N., "The homogeneous chaos" (1938), *Amer. J. Math,* **60**, 897-936.

COMPARISON OF TWO INDICATOR SIMULATION METHODS

S. OPHEYS, A. F. B. TOMPSON[1] and G. ROUVÉ
Institute for Hydraulic Engineering and Water Resources Management
University of Technology, RWTH Aachen
Mies-van-der-Rohe-Straße 1, Aachen, 52056
Germany

In this paper two simulation algorithms based on indicator kriging are presented. In one of these non-parametric approaches an indicator is used to associate a spatial node with one of two different populations. The second algorithm is the so-called Multiple Indicator Simulation technique. Here a multiple indicator is used to estimate a conditional probability distribution function of the unknown variable. Any sequentially simulated point is obtained by straightforward Monte-Carlo drawing, and is conditioned upon nearby measurements. The Turning Bands Method is used for synthetically generating conditioning point values, called hard data. The resulting logconductivity random fields are compared in terms of their statistical pattern and computational efforts, some implementation aspects are discussed.

INTRODUCTION

In the past years stochastic simulations have become an important tool in many fields of hydrogeology. Geostastistical methods have been widely used to describe spatially variable hydrogeologic properties and phenomena (*Isaaks and Srivastava, 1989; Tompson and Gelhar, 1990; Tompson, 1993*). They are routinely applied in 'upscaling' studies, in which information gained from local (or point) measurement are used to examine processes over a much larger regional scale. One classical method used for this purpose is kriging. Although pure kriging tends to smooth interpolation (and reduce simulated variability) in areas where no data exist, conditional spectral simulation (*Tompson et. al., 1989*) can be used to reproduce measurements within a structure framework of Gaussian, random and correlated variability in all areas of a domain.

However, problems may arise when trying to infer spatial statistics of highly variable attributes, where the data feature longtailed distribution with large coefficients of variation. Extreme values may correspond to materials with specifically distinct integral scales. Using multinormal-related models with a unique integral scale of all classes leads to miscalculation of flow and transport phenomena, especially when calculating breakthrough curves. In order to reproduce flow patterns caused by different integral scales, indicator models have been developed. In these stochastic models the unknown variable $Z(\mathbf{x})$ is assumed to be a random function (RF). To any $Z(\mathbf{x})$ a binary indicator transform is defined. One indicator can be used to distinguish between different populations (a), or multiple indicators for different thresholds of the variable can be used

---

[1]Lawrence Livermore National Laboratory, California

A. Peters et al. (eds.), Computational Methods in Water Resources X, 563–570.
© *1994 Kluwer Academic Publishers. Printed in the Netherlands.*

to calculate a conditional probability distribution function at each simulated point (b). The aim of this paper is to shortly describe both algorithm and to illustrate their differences. In the following, method (a) is referred as the Population Indicator Simulation and method (b) as the Multiple Indicator Simulation. Both implemented algorithms can be used for generating three-dimensional conditional random fields, but in our work we used only two-dimensional fields.

MULTIPLE INDICATOR SIMULATION

The simulation algorithm used in our work is based on the paper of *Gómez-Hernandez and Srivastava (1990)*. Sequential simulations consist of calculating a value to a unsampled location using prior data. Once the value is simulated, it is added to the initial data set as an additional conditional sample, and the procedure is repeated. The resulting random field will honor the initial data and have the spatial structure desired. This is obtained by calculating a conditional probability function (cpdf) at each unsampled location. The cpdf provides the probability that the variable does not exceed a certain threshold. This probabiltiy is conditioned to the initial data set and to all previously simulated values. Then the value itself is estimated by drawing a random number from a uniform distribution between 0 and 1 and reading from the inverse of the cpdf.

Breaking the range of the variable $Z(\mathbf{x})$ into classes, the cpdf values are estimated at the class limits. With $\{Z(\mathbf{x}_\alpha), \alpha \in (n)\}$ as the conditional data values and z_0 as one class limit the value of the cpdf $P\{Z(\mathbf{x}) \le z_0 \mid Z(\mathbf{x}_\alpha), \alpha \in (n)\}$ at the unsampled location $\mathbf{x}$ is estimated by indicator kriging as in equation (3).

Assuming that $Z(\mathbf{x})$ is a random function, a binary indicator transform for $Z(\mathbf{x})$ at location $\mathbf{x}$ and for threshold z_0 is defined by:

$$I(\mathbf{x}; z_0) = \begin{cases} 0 & \text{if} & Z(\mathbf{x}) > z_0 \\ 1 & \text{if} & Z(\mathbf{x}) \le z_0 \end{cases} \tag{1}$$

The conditional expected value of $I(\mathbf{x}; z_0)$ is:

$$\begin{aligned} & E\{I(\mathbf{x}; z_0) | Z(\mathbf{x}_\alpha), \alpha \in (n)\} \\ &= 0 \times \quad P\{Z(\mathbf{x}) > z_0 | Z(x_\alpha), \alpha \in (n)\} \\ &+ 1 \times \quad P\{Z(\mathbf{x}) \le z_0 | Z(x_\alpha), \alpha \in (n)\} \\ &= \quad P\{Z(\mathbf{x}) \le z_0 | Z(x_\alpha), \alpha \in (n)\} \end{aligned} \tag{2}$$

For this reason the estimation of the value of the cpdf can be done by simply calculating the corresponding indicator conditional expectation. For this purpose the indicator kriging method is applied, and the estimate is obtained as a linear combination of the indicator data:

$$p^*\{Z(\mathbf{x}) \le z_0 | Z(\mathbf{x}_\alpha, \alpha \in (n)\} = \sum_{\alpha=1}^{n} \lambda_\alpha(\mathbf{x}; z_0) \times i(\mathbf{x}_\alpha; z_0) \tag{3}$$

where the superscript asterisk indicates an estimated value, $i(\mathbf{x}_\alpha; z_0)$ is the indicator transform of the sample value $Z(\mathbf{x}_\alpha)$ for the threshold z_0 and $\lambda_\alpha(\mathbf{x}; z_0)$ is the corresponding indicator kriging weight. The weights are obtained by solving a ordinary kriging system using the indicator covariance function $C_I(\mathbf{h}; z_0)$ specific to the binary random function $I(\mathbf{x}; z_0)$:

$$\sum_{\beta=1}^{n}\lambda_\beta(\mathbf{x};z_0) \times C_I(\mathbf{x}_\beta - \mathbf{x}_\alpha;z_0) + \mu(\mathbf{x};z_0) = C_I(\mathbf{x} - \mathbf{x}_\alpha;z_0) \tag{4a}$$

$$\sum_{\beta=1}^{n}\lambda_\beta(\mathbf{x};z_0) = 1 \tag{4b}$$

This formulation allows use of as many indicator covariance functions $C_I(\mathbf{h};z_0)$ as there are thresholds z_0 that discretize the range of $Z(\mathbf{x})$ values. The spatial statistical structure of the final simulated field is controlled by the choice of the discretization and the determination of the corresponding covariance functions.

POPULATION INDICATOR SIMULATION

In this algorithm (*Rubin and Journel, 1991*), which is also only briefly described, an indicator is used as the probability that one value $Z(\mathbf{x})$ refers to one of set of population values. Here, the RF $Z(\mathbf{x})$ is defined by:

$$Z(\mathbf{x}) = L(\mathbf{x}) \times Z_1(\mathbf{x}) + [1 - L(\mathbf{x})] \times Z_2(\mathbf{x}) \tag{5}$$

where $L(\mathbf{x})$ is a stationary binary RF defined by:

$$L(\mathbf{x})=1, \text{ if } \mathbf{x} \text{ refers to population } Z_1 \tag{6a}$$
$$L(\mathbf{x})=0, \text{ otherwise} \tag{6b}$$

The indicator L distinguishes the two populations Z_1 and Z_2, each having its own spatial structure defined by the covariance functions $C_1(\mathbf{h})$ and $C_2(\mathbf{h})$. The mean $p=E\{L(\mathbf{x})\}$ is the probability that $Z(\mathbf{x})=Z_1(\mathbf{x})$, and $(1-p)$ is the probability that $Z(\mathbf{x})=Z_2(\mathbf{x})$.

Therefore, the mean of $Z(\mathbf{x})$ takes the form:

$$E\{Z(\mathbf{x})\}=m_1 \times p + m_2 \times (1-p) \tag{7}$$

with $m_1=E\{Z_1(\mathbf{x})\}$ and $m_2=E\{Z_2(\mathbf{x})\}$.

The spatial structure of $L(\mathbf{x})$ is defined by the indicator covariance function $C_L(\mathbf{h})$. Assuming that the three random functions Z_1, Z_2 and L are globally independent, the covariance function of $Z(\mathbf{x})$ may be expressed as a function of the covariances of L, Z_1, and Z_2 by:

$$C_Z(\mathbf{h}) = C_1(\mathbf{h})\left[C_L(\mathbf{h}) + p^2\right] + C_2(\mathbf{h})\left[(1-p)^2 + C_L(\mathbf{h})\right] + C_L(\mathbf{h})\left[(m_1 - m_2)^2\right] \tag{8}$$

Estimation a value $Z(\mathbf{x})$ consists of two main steps:
1. Calculating the conditional probability $p^c(\mathbf{x})$ of location $\mathbf{x}$ to be in either population Z_1 or population Z_2. The value p^c is estimated by a first order approximation (*Journel and Alabert, 1989*).

$$p^c(\mathbf{x}) - p \cong \sum_{\alpha=1}^{n}\lambda_\alpha \times [l(\mathbf{x}) - p] \tag{9}$$

where $l(\mathbf{x})$ denotes the realization of L at $\mathbf{x}$, n is the number of conditioning data values. This includes the initial data and all previously generated data. Therefore this simulation method is a sequential simulation, too. The weights λ_α are obtained by solving the ordinary kriging system:

$$\sum_{\beta=1}^{n}\lambda_\alpha \times C_L(\mathbf{x}_\alpha - \mathbf{x}_\beta) + \mu(\mathbf{x}) = C_L(\mathbf{x} - \mathbf{x}_\beta) \quad \alpha = 1,..,n \tag{10a}$$

$$\sum_{\beta=1}^{n} \lambda_\beta(\mathbf{x}) = 1 \qquad\qquad (10b)$$

Once $p^c(\mathbf{x})$ is defined, a random number η from a uniform distribution between 0 and 1 is drawn. The location $\mathbf{x}$ is assigned to the Z_1 group if $\eta \le p^c(\mathbf{x})$, or to the Z_2 group otherwise. In the first case $l(\mathbf{x}) = 1$, and in the latter, $l(\mathbf{x}) = 0$.

2. Once the location is assigned to one population, the first two moments of its conditional distribution are estimated by ordinary kriging, but only the data are used which belong to the same population. The kriging estimate and variance then define a normal distribution from which the simulated value $Z(\mathbf{x})$ is drawn.

The simulated value is then added to the initial data set as an additional conditional sample.

TEST EXAMPLE

SETUP

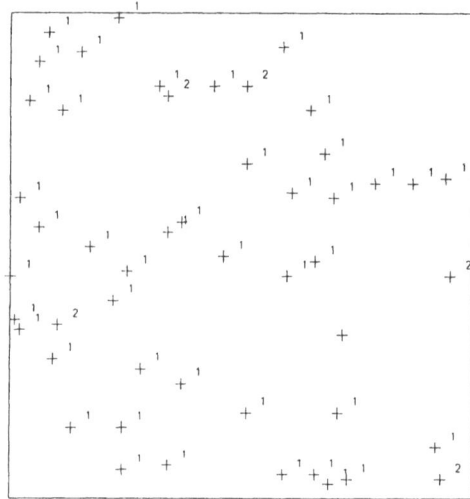

Fig.1: spatial distribution of hard data

Both methods were used for generating two-dimensional random fields on a rectangular, equidistant grid consisting of 100×100 nodes. The grid spaces dx and dy are both 1. Conditioning data, also called hard data, are generated using the Turning Bands Algorithm (*Tompson, 1989*). 90% of the data belong to population Z_1, the remaining 10% to population Z_2 with mean m_1=-2 and m_2=4 and variance $Var(Z_1)=Var(Z_2)=1$. All 50 conditioning points are randomly distributed over the domain. Figure 1 provides the spatial distribution of all hard data, the number indicates the associated population

Table 1 shows the parameters of all simulations carried out. For the Multiple Indicator Simulations the discretization of the range of $Z(x)$ is defined by the threshold values. The spatial structure is controlled by the correlation length λ_x und λ_y. Any random field generated by the Population Indicator Simulation is controlled by the indicator covariance function C_L and the covariance functions C_1 for population Z_1 and C_2 for population Z_2. and their corresponding correlation length λ_x und λ_y. An exponential covariance model is used for all covariance functions.

Table 1: parameter for simulation runs

Multiple Indicator Simulation

RUN ID	1	5	6	7
z_0	7;4;1;-2;-5	7;5;4;3	7;5;4;3; 1;-1;-2; -3;-5	7;5;4;3; 1;-1;-2; -3;-5
λ_x	20	50	10	20
λ_y	20	50	50	20
z_0		1;-1;-3;-5		
λ_x		20		
λ_y		20		

Population Indicator Simulation

RUN ID	12	13	14	15
C_L				
λ_x	50	10	20	10
λ_y	50	10	20	50
C_1				
λ_x	20	20	20	10
λ_y	20	20	20	50
C_2				
λ_x	20	20	20	10
λ_y	20	20	20	50

RESULTS

Figures 2a-d display the simulated random fields from runs 1,5,14 and 15. Comparing figure 2a and 2b one see the influence of the discretization of the range of the variable Z. The higher the resolution of the discretization the smoother the random field tends to be. This effect can be also seen in the histograms provided in figure 3. The coefficient of variation is reduced by an increasing number of classes. Comparing both methods it can be seen that he gap between the two subpopulations occurs at different values. Using Multiple Indicator Simulation this value is larger, and the subpopulation Z_1 has its own tail. Anisotropy is obtained by using different correlation lengths for x- and y-direction (RUN ID 7 and 15). The pattern shows a vertical stratified field, which is caused by choosing λ_y five times larger than λ_x. The values Z are correlated over larger distances in the y-direction than in the x-direction, a fact which is also proved by the calculated covariance functions in figure 4.

(a)RUN ID 1   (b) RUN ID 5

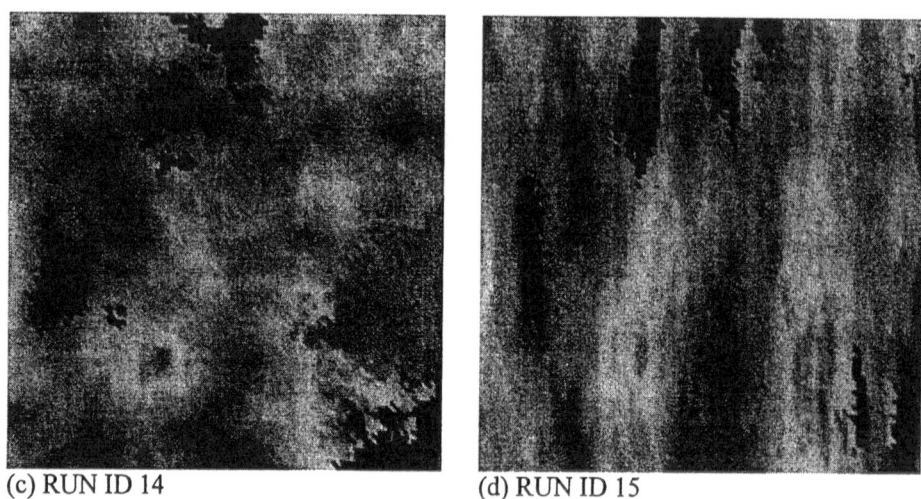

(c) RUN ID 14   (d) RUN ID 15

Fig. 2: greyscale map of 10.000 point values

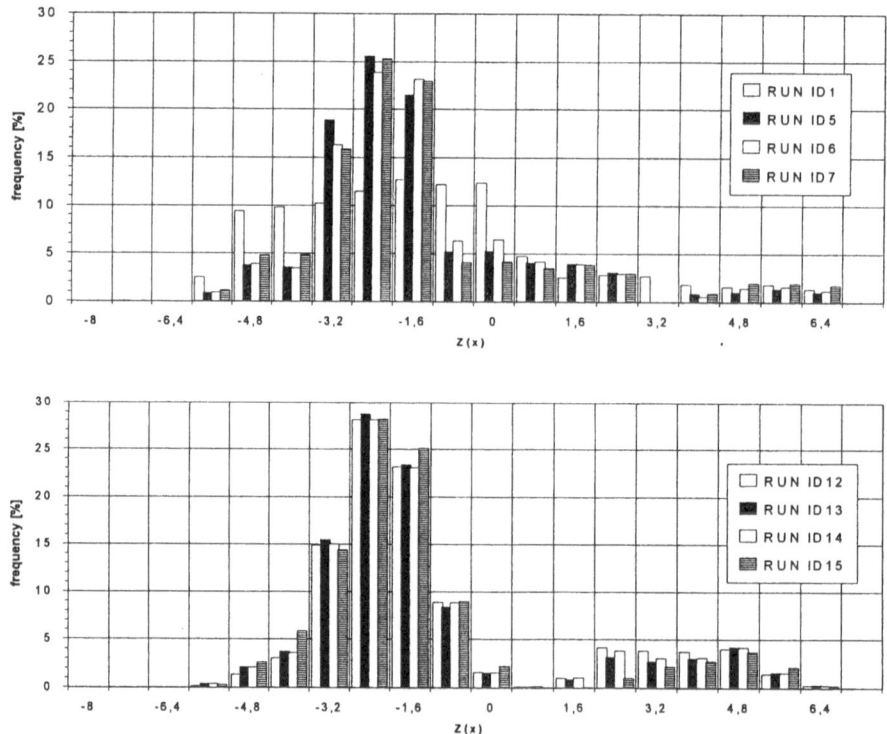

Fig. 3: frequency histogram

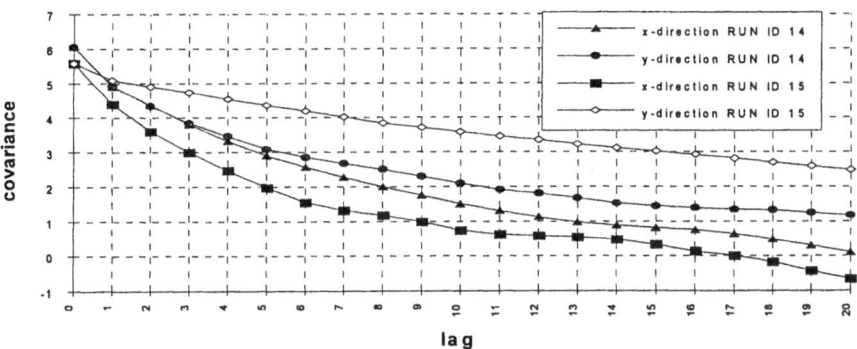

Fig. 4: autocovariance functions

The univariate statistics of all simulation runs are listed in table 2. The univariate statistics of simulation 1 show differences to any other simulation. Note that the coefficient of variation is globally larger for the Population Indicator Simulation, while the variance is smaller. All other values do not strongly vary. The minimum and the maximum of the Multiple Indicator Simulation are limited by the lower and upper class limits.

Table 2: univariate statistics

RUN ID	1	5	6	7	12	13	14	15
mean	-0.92	-1.18	-1.07	-1.10	-0.70	-0.89	-0.83	-1.07
variance	7.08	4.66	4.86	5.81	6.17	6.00	6.06	5.58
coeff. of variation	-2.88	-1.83	-2.06	-2.19	-3.53	-2.76	-2.97	-2.21
skewness	0.81	1.50	1.45	1.49	1.35	1.49	1.41	1.70
curtosis	3.40	5.67	5.55	5.20	3.75	4.33	4.06	5.28
minimum	-5.00	-5.00	-5.00	-5.00	-5.19	-5.32	-5.32	-5.37
1st quartile	-2.95	-2.46	-2.39	-2.46	-2.23	-2.30	-2.28	-2.37
median	-1.19	-1.68	-1.55	-1.66	-1.53	-1.62	-1.58	-1.64
3rd quartile	0.44	-0.68	-0.35	-0.81	-0.58	-0.75	-0.67	-0.88
maximum	6.99	6.99	6.99	6.99	7.63	7.40	7.37	7.15

IMPLEMENTATION ASPECTS

The implemented search strategy is very simple, because for each node only the closest points are used and no declustering method is applied. For the Population Indicator Simulation the maximum number of points used for the kriging estimate of the indicator has to be set much more larger than for the Multiple Indicator Simulation, because this set of points is reduced to the points of the actual population. Only this smaller set can be used for estimating the actual value. For this reason much more CPU-time is needed.

Both algorithms require generating a sequence of nodes to be simulated. For this purpose the random path generator of the form (*Bratley, Fox and Schrage, 1987*):

$$\text{index}_i = (5 * \text{index}_{i-1} + 1) \text{mod} 2^n \qquad (11)$$

is used. By selecting the smallest power of 2 that is larger than the total number of nodal points, it will generate one and only one each integer between 1 and 2^n (*see also Gómez-Hernandez and Srivastava, 1990*).

SUMMARY AND CONCLUSIONS

In this paper we described two implemented non-parametric indicator algortithms for generating three-dimensional conditional random fields. In the Population Indicator Simulation (a) an indicator is used to associate a spatial location with one of two different populations. The Multiple Indicator Simulation (b) uses a binary indicator transform of the unknown variable for estimating conditional probability distribution functions at any unsampled location. Both algorithms were compared in terms of the final spatial structure. The histograms of the random field generated by method (b) show a dependence on the number of chosen thresholds. Althoug the spatial structure can be fully controlled by the defined covariance functions at any class limit, this method requires a lot of input data gained from sparse, local or point measurements. Method (a) requires only the covariance function of any population and a covariance function defining the relative geometry of the two populations. The field pattern tends to be smoother than in the case of method (a). The search strategy has to be improved in both algorithms. The influence of differently generated random fields will be studied in the future.

Acknowledgement:
Part this work was funded by the Friedrich-Wilhem-Stiftung in Aachen by a scholarship given to the first author for a two-month stay at the Lawrence Livermore National Laboratory, California, in 1992.

REFERENCES

Bratley, P., B. Fox and L. E. Schrage (1987) A Guide to Simulation, Springer-Verlag, New York.

Gómez-Hernández, J. Jaime and R. Mohan Srivastava (1990) "ISIM3D: An ANSI-C Three-Dimensional Multiple Indicator Conditional Simulation Program", Computers & Geosciences, 16(4), 395-440.

Isaaks, Edward H. and R. Mohan Srivastava (1989) An Introduction to Applied Geostatistics, Oxford University Press.

Journel, A. and Ch. Huijbregts (1978) Mining Geostatistics, Academic Press, New York.

Journel, A. and F. G. Alabert (1989) "Non-Gaussian data expansion in the Earth Sciences", Terra Nova, 1(2),123-134.

Rubin, Yoram and André G. Journel (1991) "Simulation of Non-Gaussian Space Random Functions for Modeling Transport in Groundwater", Water Resources Research, 27(7), 1711-1721.

Tompson, Andrew F. B., Rachid Ababou and Lynn W. Gelhar (1989) "Implementation of the Three-Dimensional Turning Bands Random Generator", Water Resources Research, 25(10), 2227-2243.

Tompson, Andrew F. B. and Lynn W. Gelhar (1990) "Numerical Simulation of Solute Transport in Three-Dimensional Randomly Heterogeneous Porous Media", Water Resources Research, 26(10), 2541-2562.

Tompson, Andrew F. B. (1993) "Numerical Simulation of Chemical Migration in Physically and Chemically Heterogeneous Porous Media", Water Resources Research, 29(11), 3709-3726.

STOCHASTIC SOLUTE TRANSPORT IN NATURAL FORMATIONS: FINITE ELEMENT AND SPECTRAL METHOD SOLUTION

P. Salandin (*) and V. Fiorotto (**)
(*) Istituto di Idraulica "G.Poleni", Università di Padova - via Loredan 20, I-35131 PADOVA
(**) Dipartimento di Ingegneria Civile, Università di Parma - viale delle Scienze, I-43100 PARMA

Transport of non reactive solutes in natural porous formations is numerically investigated by a particle tracking approach using statistics of the flow field. Darcian velocity is conventionally calculated in the finite element modeling by taking the derivative of computed pressure field. Adopting three-node element the velocity is constant at each element and the related discontinuous discretization of the velocity field affects the particle tracking analysis. A continuous approximation of the velocity field can be obtained via spectral methods. However this method proves time-consuming and for a realistic approach with prescribed boundary conditions may be not applicable. In order to increase the accuracy in the Lagrangian analysis of solute transport and to limit the computer time, a bilinear finite element solution and the Cordes and Kinzelbach[3] postprocessor are here applied. By a Monte Carlo procedure the residual displacement tensor $X_{jl}(t)$ is computed in three different ways: i) by triangular finite element and velocities obtained taking the derivative of piezometric field; ii) by spectral analysis; iii) by bilinear finite element and the Cordes and Kinzelbach postprocessor. The results demonstrate the noteworthy capabilities of the latter approach in order to describe the Lagrangian flow field and the dispersion parameters.

INTRODUCTION

The dispersion analysis in natural formations, modeled as a random permeability field, has been extensively investigated by analytical and numerical approach. The permeability field $K(\mathbf{x})$ is assumed, as usually, lognormally distributed and characterized by the mean value $<Y>$ and the covariance structure $C_Y(r) = \sigma_Y^2 exp(-r/l_Y)$, where $Y=\ln K$ is the logpermeability, $r = |\mathbf{x}|$ is the separation vector, l_Y is the integral scale and σ_Y^2 is the logpermeability variance[5]. The dispersion process can be characterized[11] by the second moments of residual displacement tensor $X_{jl}(t)$, that is, by the time evolution of the moment of inertia of the plume around its centroid. The validity of the analytical solution[4] is limited to small value of σ_Y^2, i.e. for small fluctuations of the permeability around its mean value; as σ_Y^2 increases, the numerical approach is actually the only suitable procedure adopted. In order to compute the dispersion parameters, the particle's trajectories are computed by integration procedure over a known Eulerian velocity field, obtained by the solution of Laplace equation. The factors that influence the numerical solution are: i) the choice of procedure adopted to obtain the velocity field; ii) the discretization level of the input permeability field and of the solution grid. The computed trajectories depend on both factors.

A. Peters et al. (eds.), Computational Methods in Water Resources X, 571–578.
© 1994 Kluwer Academic Publishers. Printed in the Netherlands.

Assuming that the discretization levels are the same, differences in the results may appear when a continuous or discontinuous approximation of the velocity field is adopted. This fact becomes relevant following a finite element (FE) approach in order to solve the Laplace equation adopting linear or bilinear elements. In this case the velocity obtained by derivation of the shape function is discontinuous along the boundary of each element and this may cause inaccuracies in the trajectory determined by the integration process. Obviously, on reducing the size of the elements, this effect is mitigated, but the required computer memory and time grow rapidly.

In order to solve this problem the time evolution of the trajectories are here computed via the Cordes and Kinzelbach[3] postprocessor. The authors demonstrate that, also adopting FE solution of the piezometric field based on linear or bilinear elements, one can bypass the shortcomings of poor results in the path line reconstruction.

Preliminary investigations, performed in an heterogeneous permeability field, have shown relevant differences in the trajectories computed using linear triangular elements and velocities obtained by numerical differentiation. In Fig. 1 an homogeneous field is illustrated where some low permeability regions (hatched blocks) are inserted: the ratio between the two values of permeability is set equal to 100. The time evolution of a set of particles is computed in two different ways: *i*) via linear triangular elements and particle tracking procedure (triangle symbol); *ii*) via bilinear elements and Cordes and Kinzelbach postprocessor (square symbol). From Fig.1 the smoothing of the trajectories marked by triangles at each time step is manifest. The smoothing is mostly due to the discretization of the constant velocity at each element. The limited longitudinal and transversal displacement of the particle trajectories marked by triangles with to respect to those marked by squares, may be a relevant effect on the computed dispersion parameters as the conductivity field becomes more irregular.

In this paper the relevance of accurate computation of trajectories on the second moments of residual displacement tensor $X_{jl}(t)$ is investigated via the bilinear FE solution and the mentioned postprocessor. In order to check the convergence of computations different discretization levels are considered. We adopt different ratios l_Y / np, where np is the number of discretization point for a single integral scale. Comparison is carried out with the spectral method solution and the solution obtained by linear shape function elements and velocities obtained by numerical derivation[1].

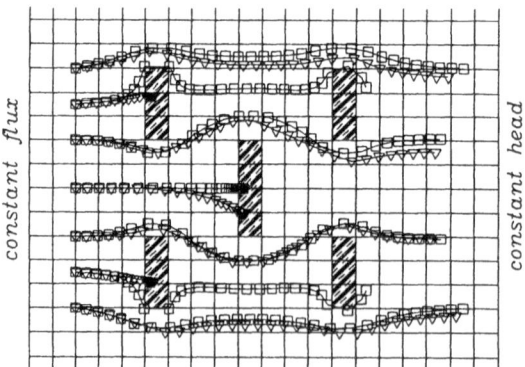

Fig. 1   Trajectories around some low permeability regions computed via *i*) linear triangular element and particle tracking procedure (triangle symbol); *ii*) bilinear element and velocity postprocessor (square symbol).

THE NUMERICAL METHOD

In order to numerically compute the steady state dispersion parameters in an heterogeneous flow field, two equations must be solved. The first is the Laplace equation

$$\nabla(K\nabla H) = 0 \tag{1}$$

defined over an Ω domain with prescribed condition on the boundary Γ. The second one, assuming the Lagrangian stationarity and neglecting the molecular dispersion[5], is the integral relationship between the particle displacement $X(x_0, t)$ and the Eulerian flow field $V(x)$

$$X(x_0, t) = \int_0^t V(X(x_0, \tau))d\tau \tag{2}$$

where x_0 is the coordinate of the starting point at $t=0$. In our simulations we analyze the two-dimensional case with $< V >= (U, 0)$, where U and 0 are respectively the constant expected values of longitudinal and transversal velocities.

The three methods here applied are characterized by the following different numerical approach in order to solve equations (1) and (2).

In the case 1, we solve the Laplace equation by linear triangular elements over a grid of $50 l_Y \times 50 l_Y$, using a ratio $l_Y / np = 4$. The boundary conditions are flux imposed at the upstream boundary, piezometric head set constant at the downstream and no flux at the lateral boundary. The total number of unknown nodes is 40,401 and we obtain by numerical differentiation 80,000 velocity values, constant in each element. The trajectories are computed via particle tracking procedure over a central region not affected by boundary condition, adopting a constant dimensionless integration time step $t U / l_Y = 0.05$, where t is the travel time.

In the case 2 the equation is solved via spectral method. The computation domain is taken as periodic within a finite "period box" of $64 l_Y \times 32 l_Y$, where the longer dimension is in the mean flow direction[6]. In this case the permeability, the piezometric and the velocity fields are described by continuous periodic functions. The velocities in the longitudinal and transversal directions x_1 and x_2 can be computed from the spectral components obtained by the solution of the spectral flux equation[7].

If $\hat{u}_{m,n}$ and $\hat{v}_{m,n}$ are the spectral components along x_1 and x_2 directions and m, n are the wave numbers along the same directions, the longitudinal and transversal velocities, u and v, can be computed by

$$u(x_1, x_2) = \sum_{m,n} \hat{u}_{m,n} \, exp[i(mx_1 + nx_2)] \quad \text{and} \quad v(x_1, x_2) = \sum_{m,n} \hat{v}_{m,n} \, exp[i(mx_1 + nx_2)] \tag{4}$$

where $i = \sqrt{-1}$.

In this manner the velocity components are available at each point $x = (x_1, x_2)$ of the flow field, but the procedure is hardly time consuming. The integration step in the particle tracking procedure (eq. (2)) was chosen equal to $t U / l_Y = 0.025$, thereby according the accuracy of the solution and the required computer time.

In case 3 the Laplace equation is solved in the same domain of case 1, using a bilinear finite element in order to compute the piezometric field. A continuous velocity field is obtained adopting the method proposed by Cordes and Kinzelbach[3]. Following this method the velocity field is given from a known distribution of piezometric head subdividing each quadrilateral element into four subelements. A continuous flux distribution across the boundary of subelements is computed by imposing the continuity and irrotationality conditions. The stream function is interpolated over each subelement by the same bilinear shape function used for the piezometric field. As results the Eulerian and the Lagrangian continuous flow fields are available on the entire domain. The procedure, applied in a first time by Pollok[9] to finite difference method, is utilized by Cordes and Kinzelbach[3] in order to reduce the error in the path lines computed from finite element solution of an homogeneous permeability field. One can show the generalization of the solution to heterogeneous porous formations by modification of the irrotationality condition, that is, by adopting as weighting coefficients the armonic mean of the permeabilities between adjacent elements encountered in a closed path[2]. Following this approach the usual particle tracking procedure is not necessary: on assuming statistical Lagrangian stationarity, the residual displacement tensor can be estimated from the Lagrangian velocity covariance $<V_j'[\mathbf{X}(0)]V_l'[\mathbf{X}(t)]>$ by the relationship[5]

$$\frac{d^2 X_{jl}(t)}{dt^2} = 2 <V_j'[\mathbf{X}(0)]V_l'[\mathbf{X}(t)]>,\tag{5}$$

where, $V_j'=V_j- <V_j >$.

In order to check the accuracy of the method applied here, a $50l_Y \times 50l_Y$ domain was solved with different ratios $l_Y / np=1, 2, 4$ and 8. Each square of dimension $l_Y \times l_Y$ is subdivided respectively into 1, 4, 16 and 64 elements with different permeability values. In the case with $l_Y / np=8$ the total number of elements is 160,000 and the number of unknowns grows to 160,801.

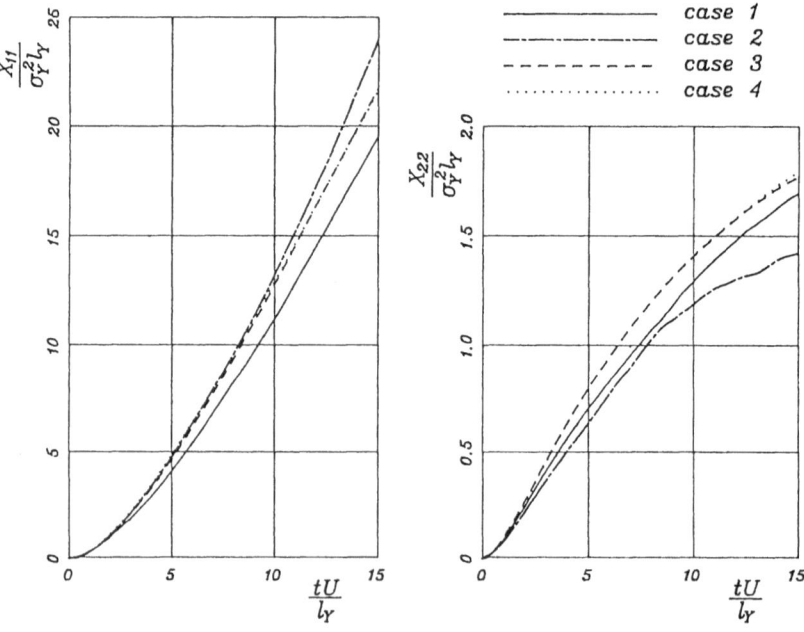

Fig.2   Dimensionless longitudinal (X_{11}) and transversal (X_{22}) displacement variance computed for $\sigma_Y^2=1.2$ and $l_Y / np=4$ as follows. Case 1: linear element and particle tracking; case 2: spectral method; case 3: bilinear element and velocity postprocessor; case 4: Dagan's linear solution.

The comparison among the three cases described, is posed on the residual displacement tensor $X_{jl}(t)$ computed via a Monte Carlo procedure. The logpermeability field is obtained at each iteration by a FFT generator[8] with a dense discretization grid, subdividing each integral scale in 8 parts in order to limit variance reduction effects. The total number of Monte Carlo runs is at the maximum set equal to 685. The statistics are computed from about 3,400 independent trajectories, releasing 5 particles starting at fixed point equally spaced of 5 l_Y along a transversal line to the mean flow direction at each Monte Carlo run.

RESULTS

Fig.2 shows the longitudinal (X_{11}) and transversal (X_{22}) displacement variances computed via the methods previously illustrated for $\sigma_Y^2 = 1.2$ and a ratio $l_Y / np = 4$ and Dagan's[4] linear solution. Although the behavior is similar, some differences (about the 10%) are evident between Dagan's solution and the cases 1 and 2. Moreover, case 3 shows a close agreement with case 4 (the theoretical solution). In case 1 it is not difficult to explain the differences with the piecewise constant discretization of the velocity field and the following inaccuracy implicit in the particle tracking procedure. This fact is manifest in Fig. 1. The smoothing effect related to the piecewise constant velocity approximation and the error in the particle tracking procedure become relevant as σ_Y^2 increases. Instead, the negligible cutoff error of the amplitude spectra in case 2 ensures the accuracy of the Eulerian velocity solution. Limited errors may be caused by the finite dimension of the time step in the particle tracking.

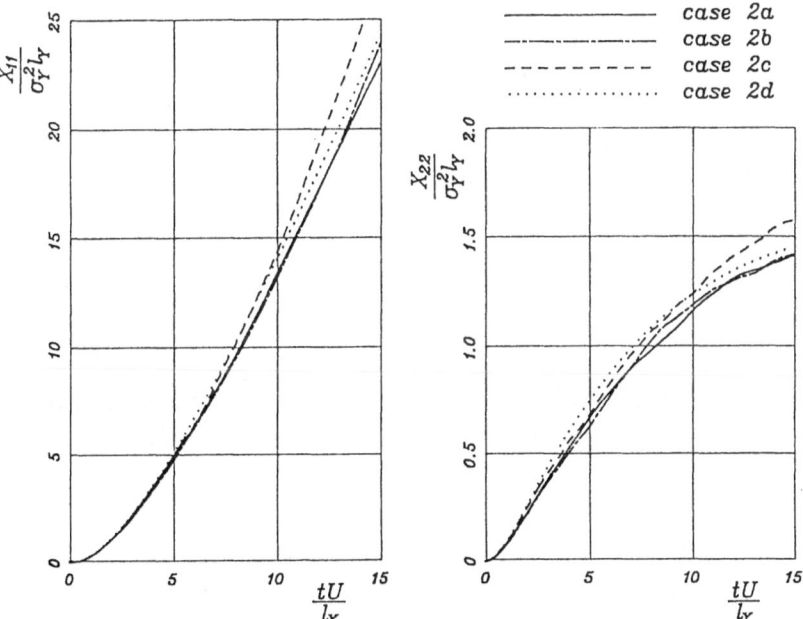

Fig.3   Dimensionless longitudinal (X_{11}) and transversal (X_{22}) displacement variance computed for $l_Y / np = 4$ via spectral method. Case 2a: $\sigma_Y^2 = 0.2$; case 2b: $\sigma_Y^2 = 1.2$; case 2c: $\sigma_Y^2 = 1.6$; case 2d: linear solution.

In order to explain the differences with X_{jl} described by the theoretical solution, the same equation integrated by Dagan[4] in a closed form was solved via the spectral method in the limited domain adopted to solve the nonlinear equations (1) and (2). The results of these computations are shown in Fig. 3. One can note that the linear solution is close to the non-linear values computed for $\sigma_Y^2=0.2$ and $\sigma_Y^2=1.2$. As σ_Y^2 increases, dimensionless X_{11} and X_{22} increases. This fact is demonstrated in the case 2c for $\sigma_Y^2=1.6$. The differences between the spectral solution of the linear equation case 2d and the Dagan's solution case 4 are due to the finite dimension of the computational domain and to the implicit periodicity of the variables. As a consequence the covariance structures become odd or even functions. Thus, only half of the domain is useful to obtain realistic dispersion parameters. On increasing the dimension of the computational domain, the results tend in the linear case to closed form solution computed in an infinite domain. In the non linear case this requires large computer memory and large computer time.

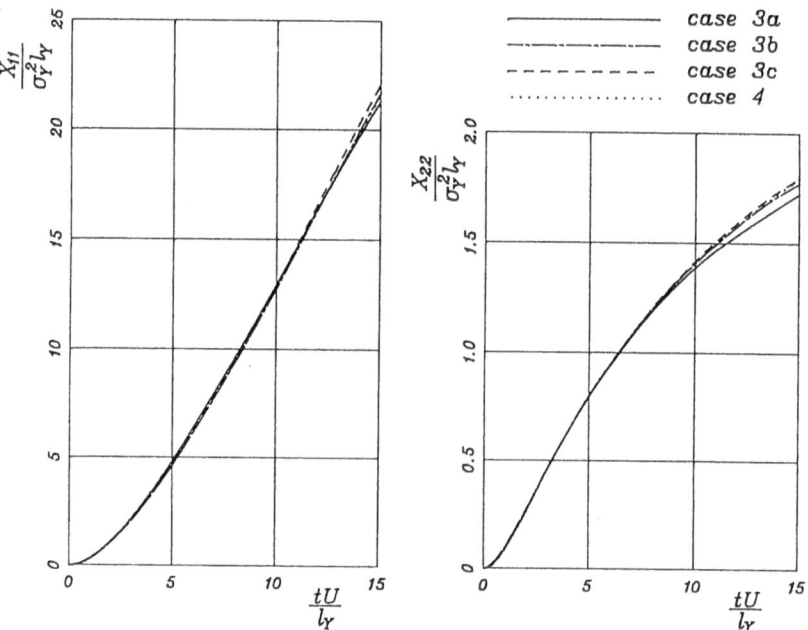

Fig.4   Dimensionless longitudinal (X_{11}) and transversal (X_{22}) displacement variance computed for $l_Y / np=4$ using bilinear element and velocity postprocessor. Case 3a: $\sigma_Y^2=0.2$; case 3b: $\sigma_Y^2=1.2$; case 3c: $\sigma_Y^2=1.6$; case 4: Dagan's linear solution.

The solutions obtained for the same σ_Y^2 range computed via FE and the Cordes and Kinzelbach method (case 3) are compared with Dagan's solution in Fig. 4. One can note the agreement with the linear solution for all the values of σ_Y^2. In analogous manner of the spectral analysis we note that dimensionless X_{11} increases as σ_Y^2 increases. X_{22} is substantially constant and equal to Dagan's solution for all the investigated range of σ_Y^2.

The accuracy of the solutions obtained for the case 3 was checked solving the same problem with different ratios l_Y / np, assumed equal to 1, 2, 4 and 8, that is adopting a different solution grid.

As previously mentioned the grid used in order to generate the logpermeability field remains the same in every case. The results of the computation are shown in Fig. 5 for $\sigma_Y^2=1.2$. From all the cases considered, the differences among the longitudinal residual dispersion variances X_{11} are negligible. The influence of the size of blocks of constant conductivity is more evident in the X_{22} transversal displacement variance, because $\sqrt{X_{22}}$ (a measure of deviation in lateral direction) becomes comparable with the dimension of the blocks when $l_Y/np \to 1$.

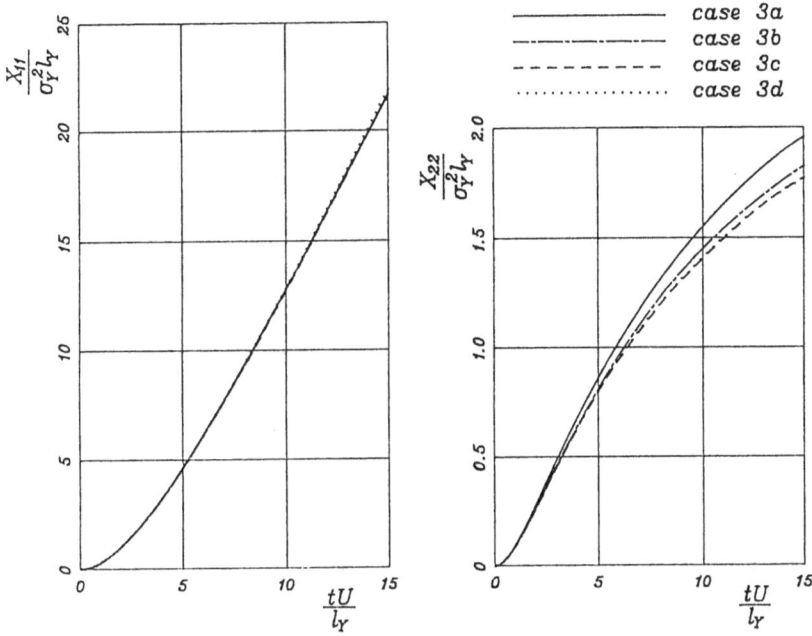

Fig.5   Dimensionless longitudinal (X_{11}) and transversal (X_{22}) displacement variance computed for $\sigma_Y^2=1.2$ using bilinear element and velocity postprocessor. Case 3a: $l_Y/np=1$; case 3b: $l_Y/np=2$; case 3c: $l_Y/np=4$; case 3d: $l_Y/np=8$.

For $\sigma_Y^2=1.2$, adopting an accurate method of solution for the velocity field, a domain subdivision of the random permeability field in a set of blocks of size up to the integral scale dimension is sufficient in order to obtain a suitable statistical characterization of the dispersion parameters. This result is in agreement with those obtained in the upscaling investigation by Salandin[10].

CONCLUSIONS

In this paper numerical approaches to analyze the transport of non reactive solute pollutants in heterogeneous porous formations are compared. Three different approaches are considered: the triangular FE solution and velocities obtained taking the numerical derivative of piezometric field (case 1); the spectral solution (case 2); the new bilinear FE solution coupled with the Cordes and Kinzelbach[3] postprocessor (case 3). The following conclusions can be drawn:

a) case 3 solution proves in close agreement with Dagan's[4] theoretical solution. The agreement is better than one find with case 1 and case 2 solutions. The case 2 solution tends to underestimate the residual displacement tensor X_{jl} as the logpermeability variance σ_Y^2 increases;

b) the spectral method seems to overestimate the theoretical X_{jl} in the longitudinal direction. A comparative analysis demonstrates that this fact is possibly due to the finite size periodic domain, but the general behavior of the solution is in agreement with case 3 results;

c) some numerical experiments performed to demonstrate the accuracy of computation in case 3, adopting different discretization level, ensure the reliability of the method. The better performance to respect the case 1 is probably due to: the bilinear element adopted to solve the piezometric field; the continuity of the velocity field; and the analytical computation of the particle's displacement;

d) up to $\sigma_Y^2 = 1.2$ the differences in logpermeability discretization level, ensuring that the blocks are of dimension smaller than one integral scale, have not relevant influence in the dispersion phenomena, if the accuracy of the computation is ensured. This is in agreement with the results on the upscaling effects previously obtained by Salandin[10].

The new proposed approach seems to give superior results among the tested numerical methods and requires acceptable computer resources and time. Further research is in progress to extend the case studied to higher logpermeability variance values.

Acknowledgments. This paper is a spin-off of the project MURST 40% "Moto dei fluidi negli ammassi filtranti". The writers are grateful to prof. Andrea Rinaldo for his advice and support.

REFERENCES

1.   Bellin A., P. Salandin and A. Rinaldo (1992). "Dispersion in heterogeneous porous formations: statistics, first order theories, convergence of computations", Water Resour. Res., 28(9), 2211-2227.

2.   Bear J. (1972). *Dynamics of fluids in porous media*, American Elsevier, New York.

3.   Cordes C. and W. Kinzelbach (1992). Continuous groundwater velocity fields and path lines in linear, bilinear and trilinear finite elements, Water Resour. Res., 28(11), 2903-2911.

4.   Dagan G. (1984). "Solute transport in heterogeneous porous formations", J.Fluid Mech., (145), 151-177.

5.   Dagan G. (1989). *Flow and transport in porous formations*, Springer-Verlag, Berlin.

6.   Fiorotto V. and A. Giorgini (1991). "Spectral analysis of flow through random aquifer: the method", Tech. Rep. CE-HSE-91-3, School of Civil Engrg., Purdue University, West Lafayette Indiana, USA.

7.   Fiorotto V. (1992). "Effetti non lineari dovuti all'eterogeneità in problemi di flusso in formazioni porose", Atti del XXIII Convegno di Idraulica e Costruzioni Idrauliche, Firenze, B37-B48 (in italian).

8.   Gutjahr A.L. (1989). "Fast Fourier Transform for Random Field Generation", Project Report for Los Alamos Grant, Department of Mathematics, New Mexico Tech., Socorro.

9.   Pollok D.W. (1988). "Semianalytical computation of path line for finite difference models", Ground Water, 26(6), 743-750.

10.   Salandin, P. (1992) "Upscaling conductivity on randomly heterogeneous porous media: results of 2-D numerical simulations", in Mathematical Modeling in Water Resources, proceedings of IX CMWR, 107-114, CMP, Southampton, U.K.

11.   Taylor, G.I. (1921) "Diffusion by continuous movements", Proc. London Math. Soc., 2(20), 196-212.

ANALYSIS AND PREDICTION OF EVAPOTRANSPIRATION RATES IN THE SOILS OF EASTERN CANADA

V. SILVESTRI, M. SOULIÉ, C. TABIB and M. H. BOUREZG
Civil Engineering Department
École Polytechnique
P.B. 6079, Station "Centre-Ville"
Montréal, Qc. Canada H3C 3A7

For the development of a soil water model which includes soil water uptake by plants, it was necessary to have reliable estimates of potential evapotranspiration (PET). On the basis of daily weather records for the period 1930-1988 in Montreal (Quebec, Canada), potential evapotranspiration was determined using the method of Penman. A time series analysis is carried out and models of the ARIMA type are developed for PET following examination of the various autocorrelation functions. Model's predictions are compared both with results of calculations performed using the Penman's approach and with field estimates from neutron probe measurements made at various locations in the region at study.

INTRODUCTION

The work presented in this paper addresses the question of evapotranspiration (ET) rates in the clay soils of Eastern Canada. The motivation for the work is one of development of an accurate soil water model for the prediction of seasonal drying of clay deposits in urban environments. For such development, it is necessary to have reliable estimates of evapotranspiration rates. The actual rates of evapotranspiration (AET) vary so much according to the variety of natural surfaces and the availability of soil moisture to different vegetation types that is impractical to attempt representative field sampling. Because of such problems, Thornthwaite(1948) introduced the concept of potential evapotranspiration (PET). According to Thornthwaite, AET increases and tends toward a maximum value termed PET when a reasonably uniform vegetation, which completely covers a drainage basin, has access to abundant soil moisture to satisfy its needs and to meet the demand of the atmospheric environment.

There are a number of ways to estimate PET but the most common method is through formulae that are based on weather or climate factors. While some of the formulae proposed are based on statistical adjustments and may be classified as empirical, other formulae evolved from a more rigorous approach. The only

A. Peters et al. (eds.), Computational Methods in Water Resources X, 579–586.
© 1994 Kluwer Academic Publishers. Printed in the Netherlands.

rigorous method treated herein is that of Penman (1948).

The original series containing 708 monthly PET calculations is divided into three sub-series, each covering a time span of 13 years, that is, 1930-1942, 1954-1968, 1976-1988, and comprising 156 observations. These sub-series are independently analyzed and univariate ARIMA models are obtained. Model predictions from the sub-series 1976-1988 are compared with field estimates obtained from neutron probe measurements made in the region at study.

POTENTIAL EVAPOTRANSPIRATION

Accurate and consistent estimates of evapotranspiration rates are important in irrigation planning and scheduling and in detailed water resources and hydrologic studies. Penman's combination of the energy balance and aerodynamic equations for estimating evapotranspiration (Penman 1948) alleviated the need for direct measurement of surface temperature and provided for theoretically based estimates of evapotranspiration using readily measured properties of air and solar radiation.

Potential evapotranspiration calculations have been carried out for the period 1930-1988 on the basis of Penman's approach. The various weather parameters needed for the analysis have been taken from the Monthly Meteorological Survey of the Atmospheric Environment Service of Environment Canada. Because of the enormous amount of daily observations, PET rates were averaged over monthly periods.

TIME SERIES ANALYSIS

As a first step, the monthly potential evapotranspiration rates for the period 1930-1988, calculated using Penman's formula for Montreal's region in Eastern Canada, were plotted against time. After unsuccessful attempts to fitting a model to the original observations, the series was divided into three sub-series for the periods 1930-1942, 1954-1966, and 1976-1988. These series were firstly independently analyzed and were secondly compared to each other in order to determine whether there was a change in behavior between the different periods considered. Due to space limitation, most of the discussion in this section will be concerned with the first time span, that is, 1930 - 1942.

Fig. 1 presents the monthly potential evapotranspiration rates for the period 1930 - 1942. There are $N = 156$ observations in this series. The sample mean and sample variance are, respectively, 42.735 mm and 1751.171 mm^2. Examination of the data shows that: a) the variance appears to be constant, b) the series is nonstationary due to a marked seasonal effect since PET is at its highest in summer months, and c) the series exhibits a periodic behavior with period $s = 12$ months. On the basis

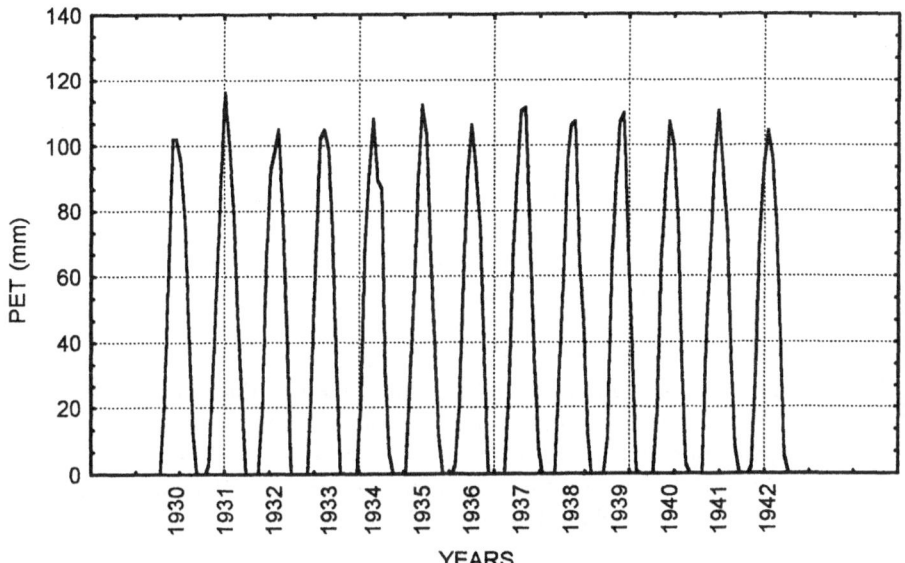

FIG.1: PET SERIES FOR 1930 - 1942

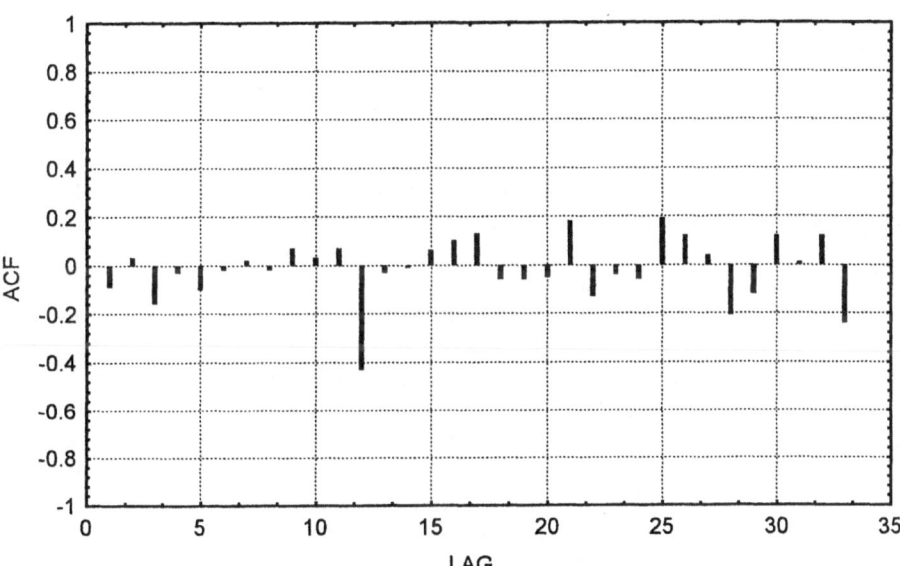

FIG. 2: AUTOCORRELATIONS FOR DIFFERENCED SERIES

of Fig. 1, it appears that a multiplicative model (Box and Jenkins, 1970) of the form

$$(1) \qquad \varphi \, (B) \, \Phi_p \, (B^s) \, \nabla^d \, \nabla_s^D \, Z_t = \Theta_q \, (B) \, \Theta_Q \, (B^s) \, a_t$$

might be adequate for the analysis of the data. The resulting multiplicative process is said to be of order (p, d, q) x (P, D, Q)^s. In this equation, φ (B) and Φ_p (B) = polynomials in B of degrees p and q, respectively; Θ_q (B^s) and Θ_Q (B^s) = polynomials in B^s of degrees P and Q, respectively; ∇ and ∇_s = 1 - B and 1 - B^s, respectively; d and D = degrees of nonseasonal and seasonal differencing, respectively; Z_t = observations; and a_t = white noise process.

The analysis of the correlograms of the original series for the period 1930 - 1942 indicated that the autocorrelations were large and failed to die out at higher lags. The series was then differenced: both a regular differencing of order d = 1 and a seasonal differencing of order D = 1 were applied. While this differencing reduced the correlations, a very heavy periodic component remained. This was evidenced particularly by very large correlations at lags 12, 24 and 36.

As the above procedure failed to produce a satisfactory model, the original series was reduced by 10%. After reduction, the PET series contained 144 observations. After several trials, the model that appeared the most appropriate was found to be a multiplicative ARIMA process of order (0, 0, 0) x (0, 1, 1)^{12}, which implied that the series contained n = N - d - sD values or n = 144 - 0 - 12 = 132. For this series, the autocorrelations are shown in Figs. 2 and 3. These two figures give the autocorrelations for lags k = 1, 2, ..., 33. Examination of the plot in Fig. 2 shows that the autocorrelations (ACF) seem to cut off after lag 12 with a significant correlation at lag 12. The partial autocorrelations (PACF) in Fig. 3 seem to die down at or near seasonal lags with significant partial autocorrelations at lags k = 12, 21, 24, 26, 28 and 30, the latter five being considerably smaller than the first.

The adequacy of the fit of the model was checked by examining the residuals of the fitted process. The Box - Pierce Chi-Square statistic (with 32 degrees of freedom) gives χ^2 = 36.9 when compared to a critical value of 43.8 at the 5% point (Bowerman and O'Connell, 1979). For the period 1976 - 1988, a similar model to the one just discussed appeared to be satisfactory. However, for the intermediate period spanning the years 1954 - 1966, a multiplicative ARIMA process of order (0, 1, 1) X (0, 1, 1)^{12} was found more appropriate to represent the observed PET rates. On the basis of the model found for the period 1976 - 1988, Fig. 4 shows the forecasts (denoted PETF), for lead times up to 12 months, all made at the arbitrarily selected origin, January 1989. It is seen by considering Table 1 that the model faithfully reproduces the seasonal pattern and supplies excellent forecasts.

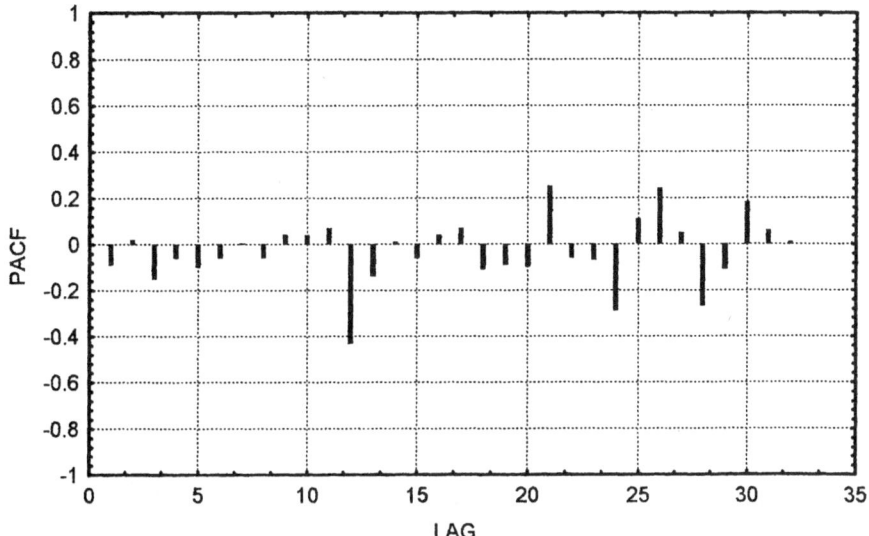

FIG.3: PARTIAL AUTOCORRELATIONS FOR DIFFERENCED SERIES

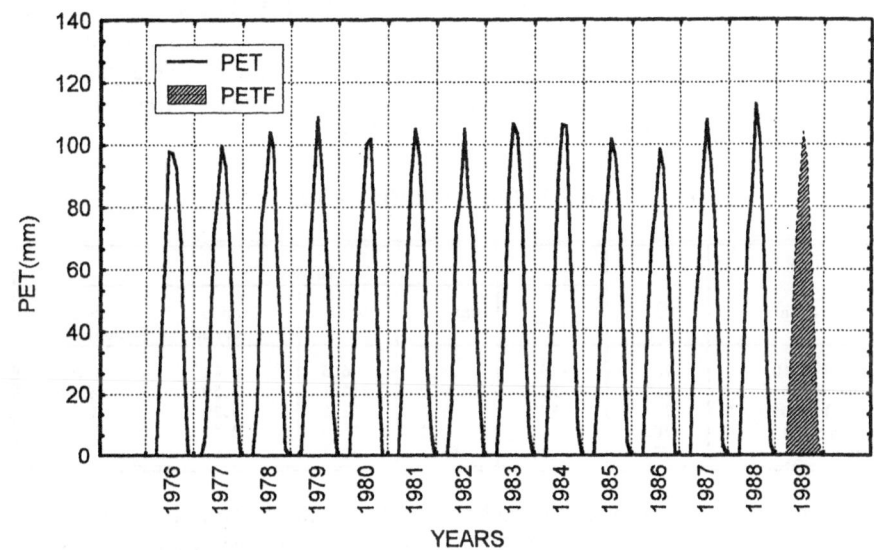

FIG. 4: 1976 - 1988 SERIES WITH FORECASTS FOR 1989

Table 1   PET forecasts for 1989

Months	PET (mm)			
	Penman	Forecast	95% confidence limits	
			Lower	Upper
January	0	0	-10.825	10.825
February	0	0	-10.825	10.825
March	0	0.197	-10.628	11.022
April	20.29	32.1932	21.368	43.018
May	69.25	63.815	52.99	74.64
June	92.94	86.609	75.784	97.434
July	108.35	103.999	93.174	114.824
August	95.68	95.681	84.856	106.506
September	75.11	70.428	59.603	81.254
October	43.31	37.364	26.539	48.189
November	0	5.41	-5.415	16.235
December	0	0	-10.825	10.825

REGRESSION ANALYSES

Since Penman's approach for the determination of PET rates is time consuming, it was deemed possible to obtain correlations between PET and other easily available meteorological data. The parameters used in this study are: temperature, number of hours of bright sunshine, relative humidity and wind speed. Table 2 presents the various statistical data. There are 708 entries, that is, 59 years (1930 -1988) x 12 months/year, for each parameter listed in this table.

Table 2 Basic monthly data

Variable	Cases	Value				
		Minimum	Maximum	Mean	Standard error	Standard deviation
Temperature, °C	708	-16.7	23.7	6.3	0.41	11.09
Sunshine, hrs	708	39.0	339.0	161.1	2.59	69.05
Relative humidity, %	708	53.0	85.0	72.6	0.29	5.70
Wind, m/s	708	1.8	7.0	16.0	0.11	3.17
PET, mm	708	0.0	155.3	50.3	1.93	51.40

Linear regression analyses were carried between PET and the remaining parameters. The results which are shown in Table 3 indicate that the best correlation is found between PET and the temperature ($r = 0.93$) while a slightly lower r-value is obtained between PET and sunshine hours.

Table 3 Coefficients of correlation

Variable	Coefficient of correlation r				
	Temperature	Sunshine	Relative humidity	Wind	PET
Temperature	1.00	0.79	-0.29	-0.54	0.93
Sunshine	0.79	1.00	-0.60	-0.46	0.85
Relative humidity	-0.29	-0.60	1.00	---	-0.31
Wind	-0.54	-0.46	---	1.00	-0.53
PET	0.93	0.85	-0.31	-0.53	1.00

SOIL MOISTURE CHANGES

Soil moisture changes at various locations distant from trees were monitored in a clay deposit using a series of neutron probe boreholes. The probe was lowered down 4.5 m deep aluminium access tubes which were permanently inserted into the ground. Neutron count readings were taken in each access tube at 10 cm intervals.

Modelling of the volumetric water content changes was carried out using the Versatile Soil Moisture Budget - Version Four (Boisvert et al., 1991). This model handles water extraction by plant roots in a similar way to evaporation from bare soil, except that beside PET and a soil factor, a plant factor is also included to limit PET. The model was initialized on the day of the first observations, that is, June 6, 1989 with started soil moisture contents equal to the first set of observations and was run until the end of December, 1990.

Typical results are shown in Fig. 5. Linear regression analyses, carried out between the estimated water contents and those observed in the access tubes, yielded the following relationship:

$$(2) \quad \Theta_{obs} \, (\%) = 0.96 \, \Theta_{est} \, (\%) + 1.84, \quad r = 0.85$$

where Θ_{obs} and Θ_{est} = observed and estimated water contents, respectively.

CONCLUSIONS

On the basis of the content of this paper, the following conclusions are drawn:

1. Time series analyses of monthly PET data permitted to obtain appropriate multiplicative ARIMA models, even though the original series had to be decomposed in different partial series.

2. Forecasts based on the time series for the period 1976-1988 were found to be quite adequate.

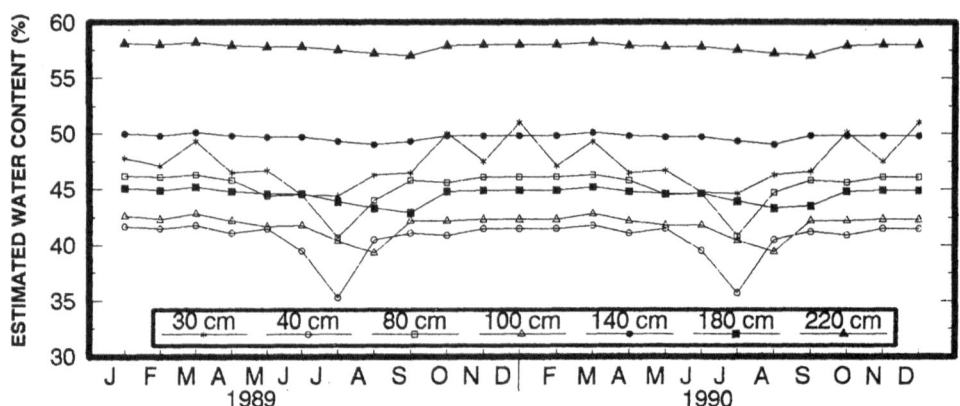

FIG.5: TIME VARIATION OF VOLUMETRIC WATER CONTENTS IN CLAY DEPOSIT

3. Regression analyses between PET and various weather parameters indicated that the best correlation existed between PET and temperature.

4. Comparison between measured and estimated volumetric water contents in a typical clay deposit of Eastern Canada showed a very good degree of correlation.

ACKNOWLEDGEMENTS

The authors wish to express their gratitude to the National Research Council of Canada and the Fonds pour la formation de chercheurs et l'aide à la recherche du Québec for the financial support received in the course of this study.

REFERENCES

Boisvert, J.B., Dyer, J.A., and Brewin, D. (1991) "The versatile soil moisture budget reference manual-version 4". Contribution 91, Research Branch, Agriculture Canada, Ottawa, 55 p.

Box, G.E., and Jenkins, G.M. (1970) "Time series analysis: forecasting and control", Holden-Day, San Francisco.

Bowerman, B.L., and O'Connell, R.T. (1979) "Time series and forecasting", Wadsworth, Inc., Belmont.

Penman, H.L. (1948) "Natural evaporation from open water, bare soil and grass", Proceedings of the Royal Society of London, Series A, Vol. 193, pp. 120-145.

Thornthwaite, C.W. (1948) "An approach toward a rational classification of climate", Geographical Review, Vol. 38, pp. 55-94.

5. REACTIVE FLOW

NUMERICAL ANALYSIS OF THE 1D DIFFUSION EQUATION: EXACT SOLUTION OF SEMI-DISCRETIZED EQUATION

BROC Daniel
COMMISSARIAT A L'ENERGIE ATOMIQUE
DRN/DMT/SEMT CEN SACLAY 91191 Gif-sur-Yvette Cedex FRANCE

ABSTRACT

Numerical solutions of the diffusion equation, obtained either by finite element or finite difference methods, exhibit numerical errors which may be very large at short time. It is particularly the case when the diffusion equation contains a radioactive decay term. An exact analysis of the numerical errors due to time discretization is presented for the semi-discretized 1-D diffusion equation (continuous in space, with an implicit scheme for the time discretization). The solution of the semi-discretized equation can be interpreted as a diffusion process characterized by a density of diffusion coefficients. This density has the property that, at time N dt, it can be expressed as a N-convolution product of the density at time dt. The integration of the convolution product is easily performed and leads to an analytical expression of the density. At the present time, this analysis has been applied in modelling studies for high level waste geologic repositories. In the case of a repository in a clay formation, this analysis has proved useful for radionuclides whith decay period smaller than their transfer time.

DIFFUSION EQUATION

The diffusion equation considered is:

$$(1) \qquad \frac{\partial C}{\partial t} = \operatorname{div}(D \,\overline{\operatorname{grad} C}) - \lambda C$$

with
C = concentration
t = time
D = diffusion coefficient
λ = radioactive decay constant

Numerical solution, by finite element or finite difference method, exhibit errors.

A. Peters et al. (eds.), Computational Methods in Water Resources X, 589–596.
© 1994 Kluwer Academic Publishers. Printed in the Netherlands.

ANALYTICAL SOLUTION

The 1-D equation admits analytical solutions. On an infinite domain $(-\infty < x < +\infty)$, with the initial condition $C_1(x, 0) = \delta(0)$, the solution is:

$$(2) \quad C_1(x,t) = \frac{1}{2\sqrt{\pi D t}} e^{-\frac{x^2}{4 D t}} e^{-\lambda t}$$

The flux at $x = L$ is: $\quad (3) \quad \varphi_1(L,t) = \frac{L}{4t\sqrt{\pi D t}} e^{-\frac{x^2}{4 D t}} e^{-\lambda t}$

On a semi-infinite domain $(-\infty < x \leq +L)$, with the initial condition $C_2(x, 0) = \delta(0)$, and, for $x = L$, the boundary condition $C_2(L, t) = 0$, the solution is:

$$C_2(x,t) = C_1(x,t) - C_1(x - 2L, t)$$

The released flux at $x = L$ is: $\varphi_2(L,t) = 2\,\varphi_1(L,t)$

NUMERICAL SOLUTION

Equation (1) is solved numerically, with a finite element method in space, and an implicit difference method in time. The discretization steps are $\Delta x = L/10$ and $\Delta t = 1/2\,\Delta x^2/D$. Two methods are used for the radioactive decay term.

a) Direct solution:
Equation (1) is solved by:

$$(4) \quad \frac{C_{t+dt} - C_t}{dt} = \mathrm{div}(D\,\overline{\mathrm{grad}}\,C_{t+dt}) - \lambda C_{t+dt}$$

Equation (4) can be solved using:

$$(5) \quad \frac{C'_{t+dt} - C'_t}{dt} = \mathrm{div}(\frac{D}{1+\lambda dt}\,\overline{\mathrm{grad}}\,C'_{t+dt})$$

with $C'_{n\,dt} = C_{n\,dt}(1+\lambda dt)^n$.

b) Solution in two steps (operator splitting):
Equation (1) give

$$(6) \quad \frac{\partial C'}{\partial t} = \mathrm{div}\,D\,\overline{\mathrm{grad}}\,C'$$

with $\quad C' = C\,e^{\lambda dt}$

In the first step, equation (6) is solved, and in the second step C is obtained from C'.

The two solution methods (a) and (b) use the same basic equation, i.e. diffusion without radioactive decay:

$$(7) \quad \frac{C_{t+dt} - C_t}{dt} = \mathrm{div}\,D\,\overline{\mathrm{grad}}\,C_{t+dt}$$

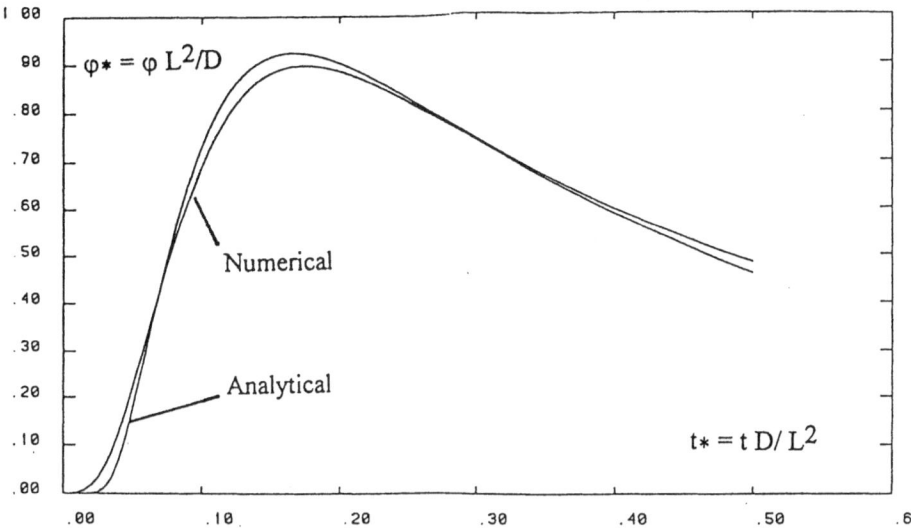

Figure 1a: 1-D diffusion, numerical and analytical solution

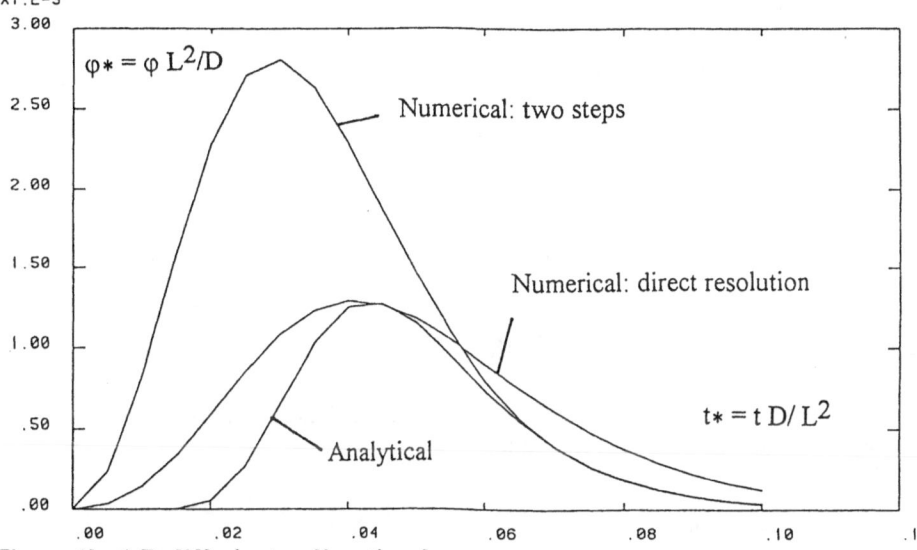

Figure 1b: 1-D diffusion, radioactive decay

The analytical and numerical solutions for the flux φ_2 are presented in figure 1. For $\lambda=0$ (figure 1a), the error is high for the low values of t. For $\lambda=D/(100\ L^2)$ (figure 1b), the numerical error is very large.

DEFINITION OF DENSITY OF DIFFUSION COEFFICIENTS

A semi-discretized solution of equation (7) gives: $\dfrac{C_{t+dt} - C_t}{dt} = D\dfrac{dC_{t+dt}}{dx^2}$.

In a 1-D infinite domain, with the initial condition $C_1(x, 0) = \delta(0)$, the solutions at $t=dt$ and $t= n\, dt$ are, respectively:

$$C_1(x, dt) = \frac{1}{2\sqrt{\pi\, D\, dt}}\; e^{-\frac{|x|}{\sqrt{D\, dt}}}$$

$$C_1(x, n\, dt) = P_n(x)\, e^{-\frac{|x|}{\sqrt{D\, dt}}}$$

where $P_n(x)$ is given by the recurrence formula:

$$D\, dt\, \frac{d}{dx^2}P_{n+1} - 2\sqrt{D\, dt}\,\frac{d}{dx}P_{n+1} + P_n = 0$$

However, this recurrence formula is not easy to use. The numerical solution will be interpreted below using a density of diffusion coefficients.

On an infinite 1-D domain, the analytical solution of the diffusion equation is, at time t, with the initial condition $C_1(x, 0) = \delta(0)$:

$$C(x,t) = \frac{1}{2\sqrt{\pi D t}}\; e^{-\frac{x^2}{4 D t}}\; e^{-\lambda t}$$

With the same initial condition, a diffusion process is governed by a density of diffusion coefficient $S(X)$, such that:

$$C(x,t) = \int_{X=0}^{\infty} S(X)\,\frac{1}{2\sqrt{\pi X t}}\; e^{-\frac{x^2}{4 X t}}\, dX$$

X has the dimension of a diffusion coefficient. For the analytical solution, $S(X)=\delta(0)$. For the numerical solution, at the first step time:

$$C_1(x, dt) = \frac{1}{2\sqrt{\pi D dt}}\; e^{-\frac{x^2}{4 D dt}} = \int_{X=0}^{\infty}\frac{1}{D}e^{-\frac{X}{D}}\,\frac{1}{2\sqrt{\pi X dt}}\; e^{-\frac{x^2}{4 X dt}}\, dX$$

and $S(X) = 1/D\, e^{-\frac{X}{D}}$

PROPERTIES OF THE DENSITY OF DIFFUSION COEFFICIENTS

If a diffusion process is governed by a density $S_1(X)$ from $t=0$ to $t=t_1$, and by a density $S_2(X)$ from $t=t_1$ to $t=t_2$, the concentration at the time t_1+t_2 is:

$$C(x) = \int_{y=-\infty}^{+\infty} C_1(y)C_2(x-y)dy$$

with: $C_i(x) = \int_{X=0}^{\infty} S_i(X)\frac{1}{2\sqrt{\pi X t_i}} e^{-\frac{x^2}{4X t_i}}dX$ (i=1, 2).

After some manipulation, we obtain:

$$C(x) = \int_{X=0}^{\infty} S_{t1+t2}(X)\frac{1}{2\sqrt{\pi X(t1+t2)}} e^{-\frac{x^2}{4X(t1+t2)}}dX$$

with

(9) $S_{t1+t2}(X) = \frac{t1+t2}{t1\,t2}\int_{z=0}^{(t1+t2)X} s1(\frac{z}{t1})s2(X(\frac{t1+t2}{t2})-\frac{y}{t2})dy$

Therefore, the density for two time steps is a convolution product of the elementary densities.

APPLICATION TO NUMERICAL SOLUTION OF DIFFUSION EQUATION

The concept of density of diffusion coefficients gives a physical meaning to the numerical error in the semi-discretized solution of equation (7). On a semi-infinite domain, at time step n, the density is:

(10) $S_n(X) = \frac{1}{(n-1)!}\frac{n}{D}(\frac{Xn}{D})^{n-1} e^{-\frac{Xn}{D}}$

The concentration is: (11) $C_2(x, n\,dt) = \int_{X=0}^{\infty} S_n(X)\frac{1}{2\sqrt{\pi Xndt}}(e^{-\frac{x^2}{4Xn\,dt}}e^{-\frac{(x-2L)^2}{4Xn\,dt}})dX$

The maximum value of $S_n(X)$ is obtained for: $\frac{dS_n(X)}{dX} = 0 \Rightarrow X = D(1-\frac{1}{n})$

For the high values of n, the numerical solution is close to the analytical one. We have:

$\lim_{n \to \infty} S_n(X) = \delta(D)$

Figure 2 shows the density $S_n(X)$ for different time steps .

The analytical expression of the numerical solution (equations (10) et (11)) can be used to estimate the error, with respect to the analytical solution.

APPROXIMATE ANALYTICAL SOLUTION

For the high values of n, the density Sn(X) can be approximated by a Gaussian fonction:

$$S_n(X) = \frac{n}{D\sqrt{2\pi(n-1)}} e^{-\frac{n-1}{2}\left(\frac{X-D(1-\frac{1}{n})}{D(1-\frac{1}{n})}\right)^2}$$

Using some other approximations, we obtain an expression for C that is to be compared with the analytical solution (equation (2)):

$$C(x, n\,dt) = \frac{1}{2\sqrt{\pi D'n\,dt}}\; e^{-\frac{x^2}{4D'n\,dt}\left(1-\frac{1}{2n}\frac{x^2}{4D'n\,dt}\right)} \quad \text{with } D' = D(1-\frac{1}{n}).$$

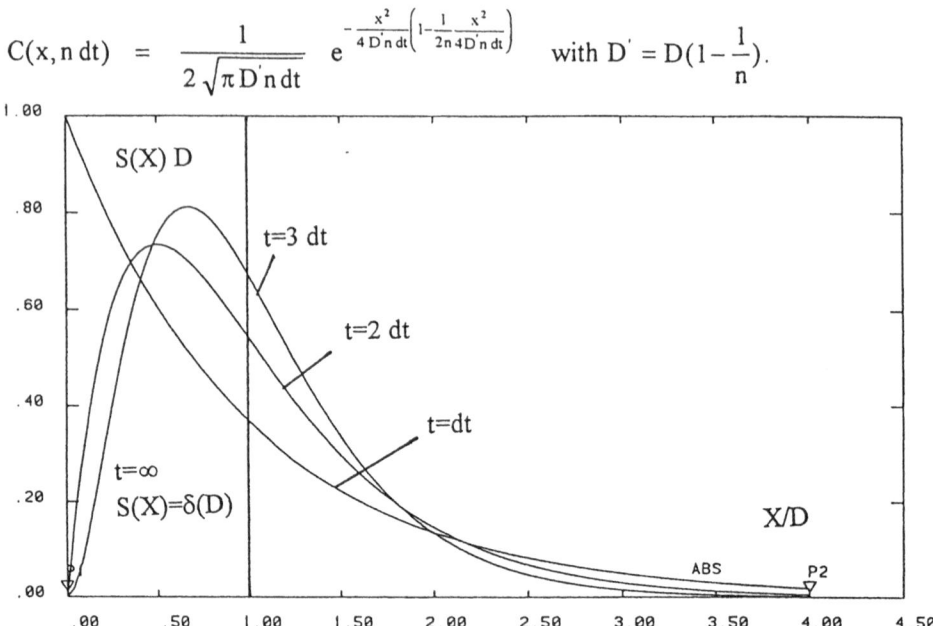

Figure 2: Density of diffusion coefficient at different discrete times

STUDY OF MAXIMUM VALUES OF THE NUMERICAL SOLUTION

From equations (10) and (11), we can deduce the recurrence formula:

$$n(n-1)C_{n+1}(x) = (n-\frac{1}{2})(n-1)C_n(x) + \frac{x^2}{4Ddt}C_{n-1}(x)$$

$$n(n-1)\varphi_{n+1}(L) = (n-\frac{3}{2})(n-1)\varphi_n(L) + \frac{L^2}{4Ddt}\varphi_{n-1}(L)$$

Those recurrence formula can be used to determine the time for which the maximum of C (concentration) or φ(flux) is reached.:

$$C_{n-1} = C_{n-1} = C_{n+1} \;\Rightarrow\; n(n-1) = (n-\frac{1}{2})(n-1) + \frac{x^2}{4Ddt}$$

$$x^2 = 2D(t-dt) \qquad \text{with} \quad t = ndt$$

For the analytical solution, the maximum of C is reached for:

$$x^2 = 2Dt$$

The maximum value of φ is obtained, for the numerical solution, for:

$$x^2 = 6D(t-dt),$$

and, for the analytical solution, for:

$$x^2 = 6Dt$$

Taking into account the radioactive decay, the recurrence formula gives:

for the direct solution:

$$n(n-1)C_{n+1}(x)(1+\lambda dt) = (n-\frac{1}{2})(n-1)C_n(x) + \frac{x^2}{4Ddt}C_{n-1}(x)$$

$$n(n-1)\varphi_{n+1}(L)(1+\lambda dt) = (n-\frac{3}{2})(n-1)\varphi_n(L) + \frac{x^2}{4Ddt}\varphi_{n-1}(L)$$

for the solution in two steps:

$$n(n-1)C_{n+1}(x)e^{2\lambda dt} = (n-\frac{1}{2})(n-1)C_n(x)e^{\lambda dt} + \frac{x^2}{4Ddt}C_{n-1}(x)$$

$$n(n-1)\varphi_{n+1}(L)e^{2\lambda dt} = (n-\frac{3}{2})(n-1)\varphi_n(L)e^{\lambda dt} + \frac{x^2}{4Ddt}\varphi_{n-1}(L).$$

The maximum values of C and φ are obtained:

for the direct solution:

$$\frac{x^2}{2D} = t(1+2\lambda t(1-\frac{dt}{t})) \text{ for C, and } \frac{L^2}{6D} = t(1+2\lambda t(1-\frac{dt}{t})) \text{ for } \varphi$$

for the solution in two steps:

$$\frac{x^2}{2D} = t(1+2\lambda t(1+\frac{3}{2}\lambda dt)) \text{ for C, and } \frac{L^2}{6D} = t(1+2\lambda t(1+\frac{3}{2}\lambda dt)) \text{ for } \varphi$$

for the analytical solution:

$$\frac{x^2}{2D} = t(1+2\lambda t) \text{ for the C, and } \frac{L^2}{6D} = t(1+2\lambda t) \text{ for } \varphi.$$

These results show that, for the high values of λ, the direct solution (relative error for time dt/t) is more precise than the solution in two steps (relative error for time λ dt/t).

APPLICATION TO SOLUTION WITH A SPACE TIME DISCRETIZATION

The diffusion equation is numerically solved, with a discretization in time $\Delta t=1/800$ L^2/D, a discretization in space: $\Delta x=L/20$ (case 1), and $\Delta x=L/200$ (case 2). Ten values of the radioactive decay constant are used. For the two cases, figure 3 gives the maximum value of φ, versus the time for which the maximum is reached. When the discretization in time is fine (case 1), the results correspond to the semi-discretized results. In case 2, the results are different, but, as in the previous case, the direct solution is more precise than that in two steps.

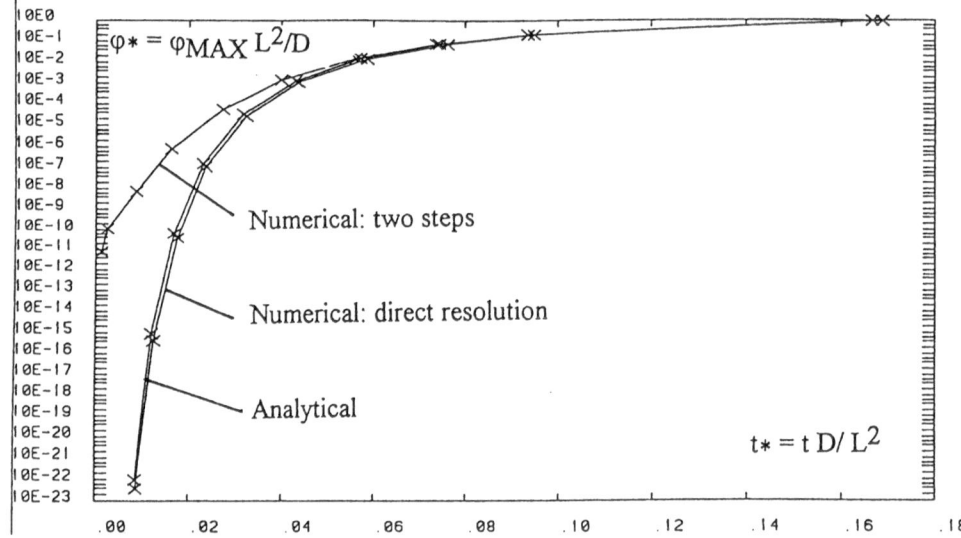

Figure 3a: Maximum values of φ, case 1

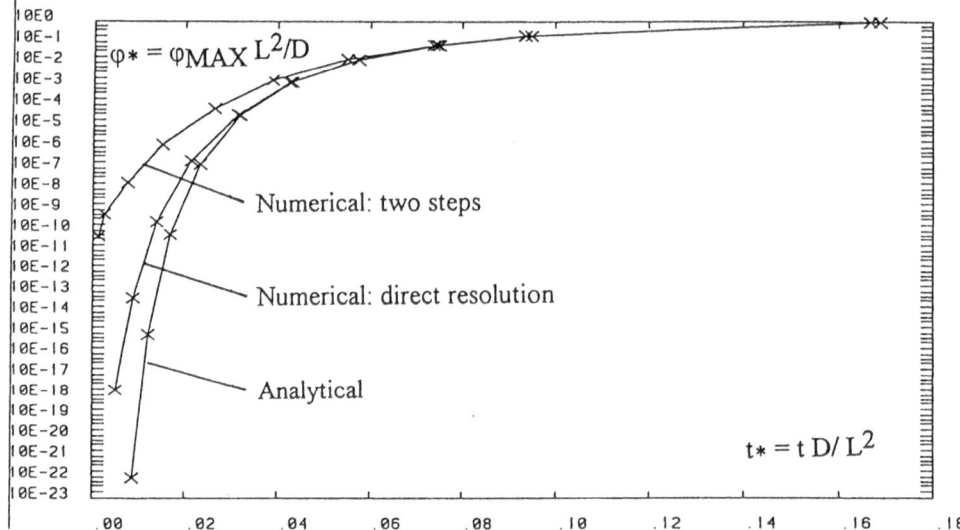

Figure 3b: Maximum values of φ, case 2

CONCLUSION

The concept of density of diffusion coefficients has been used for estimating the error in a numerical treatment of the diffusion equation in 1-D, and comparing the precision obtained with two numerical methods to take into account radioctive decay. This concept could be applied, in future work, to 2-D or 3-D problems.

AN APPROACH ON SIMULATING REACTIVE CHEMICAL TRANSPORT WITH PRECIPITATED SPECIES DOMINATING CHEMICAL EQUILIBRIUM

H. P. CHENG and G. T. YEH
Department of Civil and Environmental Engineering
The Pennsylvania State University
University Park, PA 16802
U.S.A.

When hydrological transport is coupled with chemical equilibrium, the basic requirement for computing the reactive chemical transport is the capability of computing chemical equilibrium at any point of interest. In other words, the model used for computing chemical equilibrium should be able to handle as many chemical systems as possible. Most models take the aqueous component concentrations as their independent variables to compute chemical equilibrium when they are coupled with hydrological transport. However, if the amount of an aqueous component dissolving in the aqueous phase is negligible when compared with that precipitating in the solid phase (i.e., the precipitated concentration dominates), then the model might fail to conserve the mass of that aqueous component. In order to deal with such a situation, which happens in real-world problems quite often, this study presents an essential approach. Based on the approach originated in the HYDROGEOCHEM model, we use the modified total analytical concentrations (with the concentrations of dominating precipitated species excluded) rather than the total analytical concentrations as the dependent variables of transport. In addition, we save the concentrations of all the dominating precipitated species at both the previous and current time steps for calculating the rate of change due to precipitation/dissolution. The rate, playing the role of a source/sink term, appears on the right hand side of the transport governing equation. This paper presents a brief derivation of this approach and includes two designed examples to verify the approach.

INTRODUCTION

During subsurface transport, reactive chemicals are subject to a variety of hydrophysical and chemical processes. The combined effects of all the processes on reactive chemical transport must satisfy the principle of conservation of mass. In the past decade, coupled models accounting for complex hydrophysical and chemical processes, with varying degrees of sophistication, have been developed [Yeh and Tripathi, 1991]. When chemical processes are coupled with solute transport in a mathematical/numerical model, the model needs to be constructed with some specific consideration (e.g., to generate symmetric positive definite global matrices) so that it is able to solve real-world problems for reasonably accurate solutions [Yeh and Tripathi, 1990]. No matter what consideration is taken into account, the basic requirement is that the coupled model should be able to handle as many chemical systems as possible. In computing chemical equilibrium, most models take the aqueous

A. Peters et al. (eds.), Computational Methods in Water Resources X, 597–604.
© 1994 Kluwer Academic Publishers. Printed in the Netherlands.

components as their independent variables, which are basic quantities to be checked for convergence when solving a set of nonlinear algebraic equations of chemical equilibrium. However, if dominating precipitated species exist in a chemical system, we might obtain a non-mass conservative but convergent solution which will introduce either nonconvergency or a wrong result in the transport process. This is so because there is always a nonzero error tolerance for checking the convergence of chemical equilibrium, which could make the order of solving nonlinear equations significant and result in initial guess-dependent solutions when dominating precipitated species exist.

In order to resolve this problem, an approach, which excludes the concentrations of dominating precipitated species from the related total analytical concentrations to eliminate the non-mass conservative problem caused by dominating precipitated species, is presented in this paper. A brief derivation, examples for verification, and conclusions for the approach are given in the next three sections.

DERIVATION

In the HYDROGEOCHEM model [Yeh and Tripathi, 1990], the chemical equilibrium is assumed to be achieved immediately, and it is computed by solving a set of nonlinear algebraic equations of (1) mass conservation of the total analytical concentrations of all components and (2) constitution describing the relationship between species and components. To overcome the non-mass conservative problem caused by dominating precipitated species, the mass conservation equations of the total analytical concentrations of aqueous components

$$T_j = c_j + \sum_{i=1}^{M_x} a_{ij}^x x_i + \sum_{i=1}^{M_y} a_{ij}^y y_i + \sum_{i=1}^{M_z} a_{ij}^z z_i + \sum_{i=1}^{M_p} a_{ij}^p p_i \quad i = 1, \ldots, N_a \quad (1)$$

are replaced by the mass conservation equations of the modified total analytical concentrations of aqueous components

$$\overline{T_j} = c_j + \sum_{i=1}^{M_x} a_{ij}^x x_i + \sum_{i=1}^{M_y} a_{ij}^y y_i + \sum_{i=1}^{M_z} a_{ij}^z z_i + \sum_{i=ndps+1}^{M_p} a_{ij}^p p_i \quad i = 1, \ldots, N_a \quad (2)$$

where T_j and $\overline{T_j}$, given as the input for computing chemical equilibrium, are the total and the modified total analytical concentrations of the j-th aqueous component, respectively; c_j is the concentration of the j-th aqueous component species; x_i, y_i, z_i, and p_i are the concentrations of the i-th complexed, adsorbed, ion-exchanged, and precipitated species, respectively; a_{ij}^x, a_{ij}^y, a_{ij}^z, and a_{ij}^p are the stoichiometric coefficient of the i-th complexed, adsorbed, ion-exchanged, and precipitated species, respectively, on the j-th component; and N_a, M_x, M_y, M_z, M_p, and ndps are the numbers of aqueous components, complexed species, adsorbed species, ion-exchanged species, precipitated species, and dominating precipitated species, respectively. Therefore, when checking the convergence of chemical equilibrium, the modified total analytical concentrations, rather than the total analytical concentrations, of aqueous components are examined for mass conservation. Although dominating precipitated species exist, they will not yield non-mass conservation because they are excluded from the equations of the modified total analytical concentrations of aqueous components. To couple chemical equilibrium to solute transport, we simply substitute Eq. (2) into the governing equations of the HYDROGEOCHEM model

$$\theta \frac{\partial T_j}{\partial t} + \mathbf{V} \cdot \nabla C_j - \nabla \cdot (\theta \mathbf{D} \cdot \nabla T_j) + q T_j + \frac{\partial \theta}{\partial t} T_j \qquad j = 1, \ldots, N_a$$
$$= -\nabla \cdot (\theta \mathbf{D} \cdot \nabla (S_j + P_j)) + q(S_j + P_j) + \frac{\partial \theta}{\partial t} C_j + q C_j^* \qquad (3)$$

to yield

$$\theta \frac{\partial T_j^-}{\partial t} + \mathbf{V} \cdot \nabla C_j - \nabla \cdot (\theta \mathbf{D} \cdot \nabla T_j^-) + q T_j^- + \frac{\partial \theta}{\partial t} T_j^- \qquad j = 1, \ldots, N_a$$
$$= -\nabla \cdot (\theta \mathbf{D} \cdot \nabla (S_j + \overline{P_j})) + q(S_j + \overline{P_j}) + \frac{\partial \theta}{\partial t} C_j + q C_j^* - \sum_{i=1}^{ndps} a_{ij}^p \frac{\partial (\theta p_i)}{\partial t} \qquad (4)$$

where C_j, S_j, P_j, and $\overline{P_j}$ are the total dissolved, total sorbed, total precipitated, and modified total precipitated concentrations of the j-th aqueous component; θ is moisture content; q is Darcy velocity; $\mathbf{V}$ is source/sink of water; and C_j^* is the total dissolved concentration of the j-th aqueous component of source/sink. Eq. (4) can be further written in the Lagrangian-Eulerian form as follows.

$$\theta \frac{DC_j}{Dt} + \theta \frac{\partial (S_j + \overline{P_j})}{\partial t} - \nabla \cdot (\theta \mathbf{D} \cdot \nabla T_j^-) + q T_j^- + \frac{\partial \theta}{\partial t} T_j^- \qquad j = 1, \ldots, N_a$$
$$= -\nabla \cdot (\theta \mathbf{D} \cdot \nabla (S_j + \overline{P_j})) + q(S_j + \overline{P_j}) + \frac{\partial \theta}{\partial t} C_j + q C_j^* - \sum_{i=1}^{ndps} a_{ij}^p \frac{\partial (\theta p_i)}{\partial t} \qquad (5)$$

EXAMPLES

In this section, two examples are used to verify the approach presented in this paper. The first example is to demonstrate how significantly the approach eliminates the problem of non-mass conservation due to the existence of dominating precipitated species. The second example is to demonstrate how properly the approach works for a clean-up problem.

Example 1: chemical equilibrium under complexation and precipitation

In this example, two components, namely Ca^{2+} and CO_3^{2-}, are considered. The given total analytical concentrations of Ca^{2+} and CO_3^{2-} are 10^{20} M and 10^{20} M, respectively. In addition to these two aqueous component species, one complexed species, $CaCO_3$, and one dominating precipitated species, $CaCO_{3(s)}$, are included for chemical equilibrium simulations. The associated equilibrium constants of $CaCO_3$ and $CaCO_{3(s)}$ are 10^3 M^{-1} and $10^{8.3}$ M^{-2}, respectively. If we directly use the total analytical concentrations of aqueous components to solve for chemical equilibrium and use 10^{-6} as the relative error tolerance, then we will obtain the following answer: $[Ca^{2+}] = 1.784 \times 10^{-12}$ M, $[CO_3^{2-}] = 2.809 \times 10^3$ M, $[CaCO_3] = 5.012 \times 10^{-6}$ M, $[CaCO_{3(s)}] = 10^{20}$ M. It is true that this convergent result is not mass-conservative for both components, and more obvious for CO_3^{2-}. This result will introduce either a wrong solution or nonconvergency if it is applied to solve for chemical transport. In the approach, however, the modified total analytical concentrations, which are set to be 0.0 M for both aqueous components before chemical equilibrium, are used to solve for equilibrium. The dominating precipitated species, $CaCO_{3(s)}$, has a concentration of 10^{20} M before equilibrium. In other words, we initially put all the concentrations into the dominating precipitated species rather than the component species. With the same relative error tolerance, 10^{-6}, the following answer can be obtained by using the approach: $[Ca^{2+}] = [CO_3^{2-}] = 7.079 \times 10^{-5}$ M, $[CaCO_3] = 5.012 \times 10^{-6}$ M, $[CaCO_{3(s)}] = 10^{20} - 7.580 \times 10^{-5}$ M. It is mass-conservative for both aqueous components. In the approach, the concentration of the dominating precipitated species, $CaCO_{3(s)}$, is recorded with two variables: one is to store the dominating amount, 10^{20} M, and the other one is to store the variable amount, -7.580×10^{-5} M, which is the amount of dissolution (if it is negative) or precipitation (if it is positive) relative to the situation before chemical equilibrium. With this recording, the difference of the concentration of a dominating precipitated species between two successive time steps can be correctly calculated in the transport process. In other words, the contribution from the change of a dominating precipitated species can be accurately computed, which cannot be appropriately calculated with the former approach.

Example 2: one-dimensional clean-up problem

This example considers the clean-up of a precipitated species in a one-dimensional horizontal column. The chemical system involves two aqueous components, A and B, and one precipitated species, D, which is composed of one component A and one component B and has $Ksp = 10^{10}$ M^{-2}. The simulation is conducted for 40 days with a time step size of 1.0 day. The region is made up of 100 elements of size 1 dm x 1 dm x 1 dm. In order to see how aptly the approach works, five different initial conditions are set up. They all have the same modified total analytical concentration, 1×10^{-5} M, for both components at every node. But they are given different amounts of precipitated species: 0.0 M, 1×10^{-5} M, 2×10^{-5} M, 5×10^{-5} M, and 1×10^{-4} M. As the simulation starts, clean water comes into the column through the left boundary plane. The flow velocity is 0.4 dm/day, the effective porosity is 0.4, and the dispersivity is assumed negligible. Fig. 1 shows the schematic description of this example.

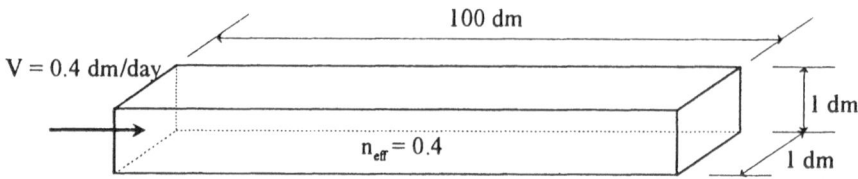

Figure 1   Schematic description of the clean-up example

Figures 2 through 6 illustrate the results of transient simulation for the five cases with initial precipitated species concentrations 0.0 M, 1×10^{-5} M, 2×10^{-5} M, 5×10^{-5} M, and 1×10^{-4} M, respectively. The concentrations of the precipitated species (i.e., the "Background" in Figures 2 through 6) for the five cases, varying from 0 up to 10 times of the designed modified total analytical concentration, 1×10^{-5} M, are not really dominant because this example is mainly designed for testing if the approach (i.e., solving Eq. (5)) has been correctly implemented. The computational results have been checked to be exactly the same as those from the former approach (i.e., solving Eq. (3) with the Lagrangian-Eulerian approach). Figures 7 and 8 display the difference among the five cases at time = 20 and 40 days, respectively. Since there is no precipitation during the simulation for case 1, both components can be thought as conservative components [Yeh and Tripathi, 1990]. Therefore, it is easy to determine the location of the clean water front.

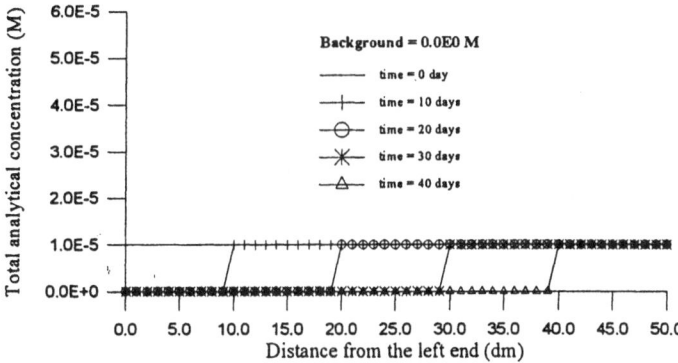

Figure 2   The transient simulation result of case 1

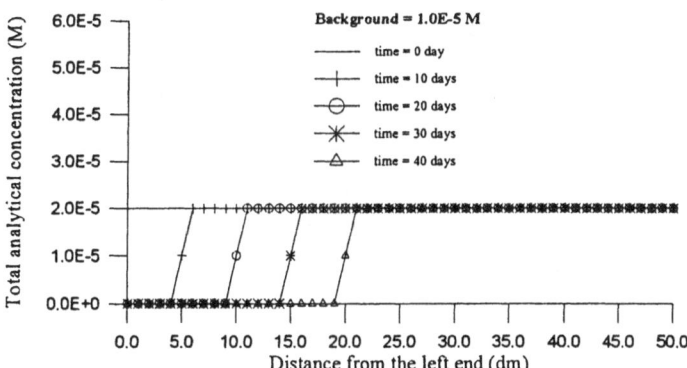

Figure 3   The transient simulation result of case 2

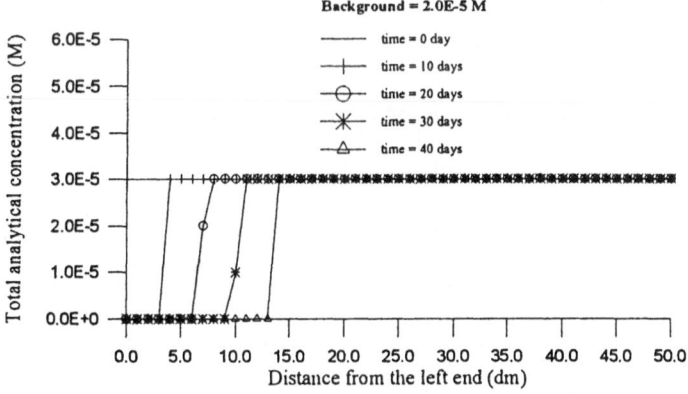

Figure 4   The transient simulation result of case 3

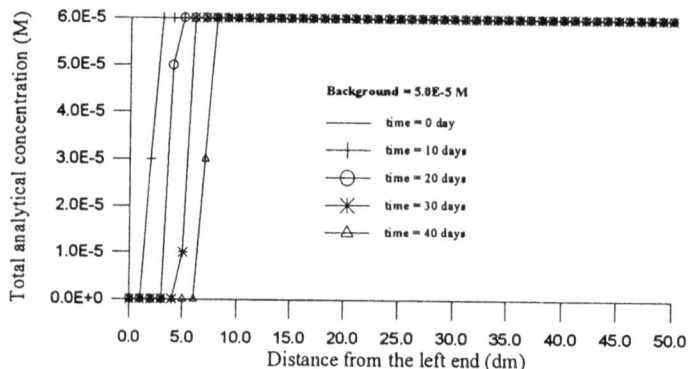

Figure 5 The transient simulation result of case 4

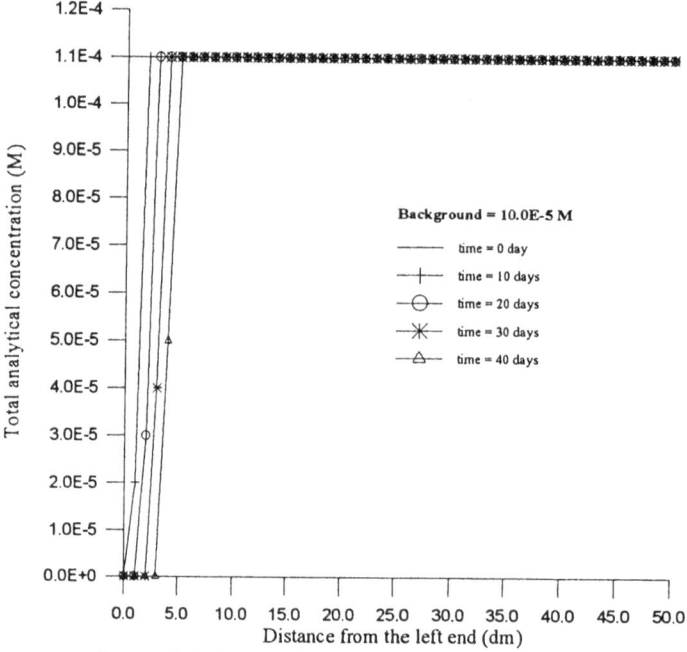

Figure 6 The transient simulation result of case 5

In this example, the mesh Courant number is set up to be 1.0 such that numerical dispersion can be avoided [Yeh, et al., 1992]. Figure 2 shows that the approach provides exactly the same result as expected for case 1. In Figures 7 and 8, the comparison between case 1 and the other four cases confirms that the approach offers reasonably accurate numerical solutions. That is so because the clean-up amounts of both components are 19.5×10^{-5}, 20×10^{-5}, 20.5×10^{-5}, 22×10^{-5}, and 23×10^{-5} moles at time = 20 days and 39.5×10^{-5}, 40.0×10^{-5}, 40.5×10^{-5}, 42×10^{-5}, and 44.5×10^{-5} moles at time = 40 days for cases 1 through 5, respectively. These data agree with the mass

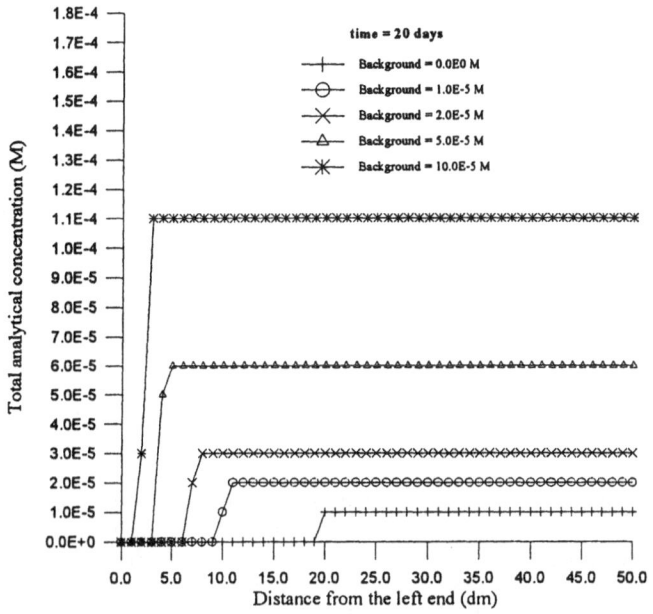

Figure 7 Comparison among the five cases at time = 20 days

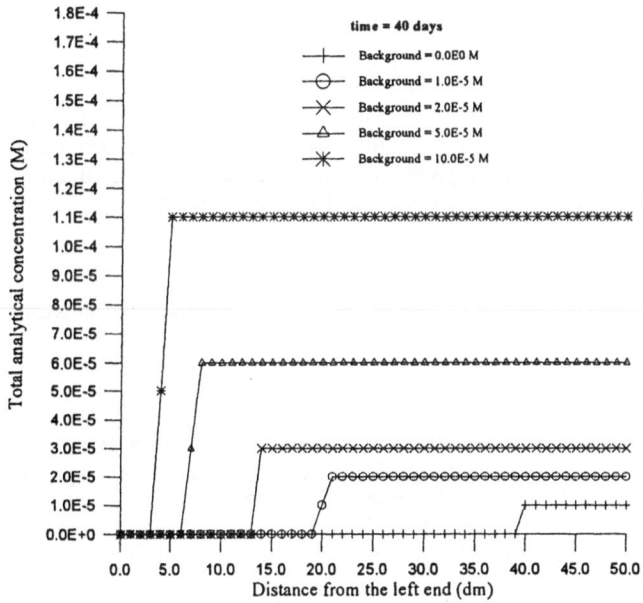

Figure 8 Comparison among the five cases at time = 40 days

conservation of the clean-up amount.

The difference of the clean-up amount between case 1 and any other case does not increase with time, instead, this difference increases with the increment of the concentration of the initial precipitated species (or the "Background" in the figures). In other words, this difference is generated due to the existence of the precipitated species which introduces the nonlinearity to the chemical transport equations. Only when finer grids and smaller mesh Courant numbers are used, more accurate solutions can be obtained if the "Background" is getting larger. In this case, for instance, if the adaptive local grid refinement technique is used to capture the clean water front through the simulation process, the difference mentioned above will be reduced [Yeh, et al., 1992]. The finer the local grid size, the less the difference.

CONCLUSIONS

This paper presents an approach to compute reactive chemical transport with dominating precipitated species taken into account. In the HYDROGEOCHEM model, the total analytical concentrations of aqueous components are used as the dependent variables in solving chemical transport equations. It helps to gain positive values of the total analytical concentrations of all components through the whole computational process, which is required to solve for chemical equilibrium. However, this approach is not suitable for dealing with a system containing any dominating precipitated species. Because a non-mass conservative solution might be obtained from computing chemical equilibrium, as illustrated in Example 1. Moreover, a non-mass conservative chemical equilibrium solution will introduce either nonconvergency or a wrong answer in solving transport equations. To overcome this problem, the approach presented in this paper uses the modified total analytical concentrations of aqueous components, which exclude the concentrations of dominating precipitated species, as the dependent variables in solving for chemical transport. Example 1 demonstrates that the approach provides a mass-conservative solution in computing chemical equilibrium under the existence of a dominating precipitated species. In addition, Example 2 shows that the approach offers reasonably accurate solutions for chemical transport if an appropriate spatial and temporal domain discretization is given, which is needed because of the nonlinearity precipitation reactions contribute to chemical transport equations.

ACKNOWLEDGEMENT

This research is supported by the Office of Health and Environmental Research, U. S. Department of Energy, Grant No. DE-FG02-91ER61197.

REFERENCES

Yeh, G. T., J. R. Chang, and T. E. Short (1992) "An Exact Peak Capturing and Oscillation-Free Scheme to Solve Advection-Dispersion Transport Equations", Water Resources Research, Vol. 28, No. 11, page 2937-2951.
Yeh, G. T. and V. S. Tripathi (1991) "A Model for simulating Transport of Reactive Multispecies Components: Model Development and Demonstration", Water Resources Research, Vol. 27, No. 12, page 3075-3094.
Yeh, G. T. and V. S. Tripathi (1990) "HYDROGEOCHEM: A Coupled Model of HYDROlogical Transport and GEOCHEMical Equilibria in Reactive Multicomponent Systems", ORNL-6371, Oak Ridge National Laboratory, Oak Ridge, Tenn. 37831.

NUMERICAL SIMULATION OF CONTAMINANT TRANSPORT AND BIODEGRADATION IN POROUS AND FRACTURED-POROUS MEDIA

O. CIRPKA and R. HELMIG
Institut für Wasserbau
University of Stuttgart
Pfaffenwaldring 61
D-70550 Stuttgart
Germany

A numerical method for the simulation of multicomponent advective-dispersive transport in groundwater coupled with chemical transformations is presented. The method is based on the following principles: Calculation of advective-dispersive transport by an improved finite element method (FEM); calculation of chemical transformations and biomass growth by multistep backward differentiation formulae and Newton's method; coupling of advective-dispersive transport and chemical transformations by the iterative two-step method; and approximation of geological structures by 1D, 2D and 3D finite elements in arbitrary combinations. Characteristics of the method are shown by an example for aerobic microbial degradation of a sorbing contaminant in a fractured-porous medium.

INTRODUCTION

In all industrialized countries pollution of the subsurface by organic compounds causes serious problems for the protection of groundwater. Many of these compounds like small aromatic and aliphatic hydrocarbons can be degradated in principle by indegenous microorganisms. Nevertheless it has been shown in many experiments that the degradation behaviour of microorganisms in natural aquifers differs considerably from that in small-scale laboratory investigations. Interaction between advective-dispersive transport and (bio-)chemical transformations as well as mass transfer between phases available and non-available for microbiota lead to a limitation of microbial activity. To understand these interactions a tool for the simulation of contaminant transport coupled with biomass growth and chemical transformations is necessary. This is of practical interest for the assessment of natural attenuation and for the set-up of *in-situ* bioremediation clean-up technologies.

GOVERNING EQUATIONS

Multicomponent transport in the saturated zone can be described by a set of partial differential equations PDE, one for each transported compound i (1). In equation (1) chemical reactions rates are included by the reaction term r_i, which acts as a source/sink term and is a function of all interacting compounds.

$$L(c_i) := n_e \frac{\partial c_i}{\partial t} + n_e \underline{v}_e^T grad(c_i) - div(n_e \underline{\underline{D}} grad(c_i)) - n_e r_i = 0 \qquad (1)$$

A. Peters et al. (eds.), Computational Methods in Water Resources X, 605–612.
© 1994 Kluwer Academic Publishers. Printed in the Netherlands.

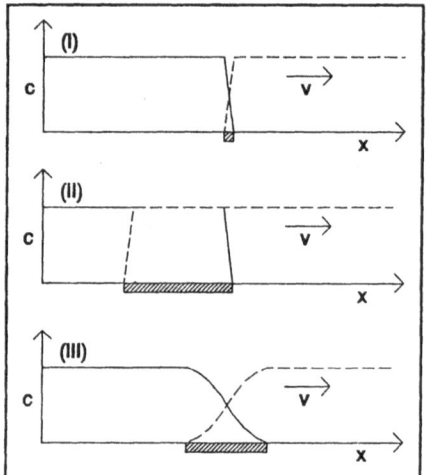

Fig. 1 Effects of sorption and dispersion on the extension of the mixing zone.

Determination of the reaction term in equation (1) leads to considerations about time scales of chemical transformations. Fast reactions like acid-bases reactions (e.g. inorganic Carbon balance) are often treated as processes in thermodynamic equilibrium, thus leading to algebraic equation systems. Microbial growth and related transformations are mostly treated as kinetically controlled processes, thus expressed by ordinary differential equation systems ODES. Depending on the time scale of transport and chemical reactions sorption and ion exchange can be treated as processes in equilibrium or as kinetically controlled processes. In general for chemical interactions a differential algebraic equation system DAES or at least a stiff ODES is to be expected:

$$\frac{\partial \underline{c}}{\partial t} = CHEM(\underline{c}) + \underline{T} \qquad (2)$$

In equation (2) $\underline{T}$ describes the contribution of advective-dispersive transport to the concentration change and is containing all spatial relations. $CHEM(\underline{c})$ describes the contribution of chemical transformations.

The following example (3) describes aerobic degradation of an instantaneous sorbing contaminant. The dependency of microbial growth on oxygen and on the primary substrate is expressed by a double Monod term. Microbial decay is expressed by a linear law. Oxygen and substrate consumptions are coupled to microbial growth by yield coefficients. Sorption is described by linear partitioning, and is assumed to be in equilibrium. For reasons of simplification the concentration of the sorbed substrate is related to the pore volume.

$$k_{gr} = \mu_{max}\left(\frac{c_{S_w}}{c_{S_w}+K_S}\right)\left(\frac{c_O}{c_O+K_O}\right) \qquad (3a); \qquad\qquad \frac{\partial X}{\partial t} = \left(k_{gr} - k_{dec}\right)X \quad (3b);$$

$$\frac{\partial c_O}{\partial t} = T_O - \frac{k_{gr}}{Y_O}X \quad (3c); \qquad \frac{\partial c_{S_w}}{\partial t} + \frac{\partial c_{S_s}}{\partial t} = T_{S_w} - \frac{k_{gr}}{Y_S}X \qquad (3d); \qquad c_{S_s} = K_d c_{S_w} \qquad (3e).$$

From equation (3a) it is obvious that microbial growth is taking place only where both oxygen and the substrate are available for microorganisms. In the case of bioremediation based on aerobic degradation oxygen is introduced into groundwater by injection of oxygen enriched water. The contaminant acts as primary substrate for the microbiota. Availability of both substrate and oxygen is reached only in a small mixing zone. The extension of the mixing zone depends on effective diffusion and for the direction of the streamlines on differences in sorption (chromatographic effect). The latter effect can be shown by combination of equation (1) and equation (3e) for the dissolved substrate leading to a transport equation (4) with a retardation factor. The corrected chemical transformation rate r_{Sw}^* includes no more sorption effects.

$$L(c_{S_w}) := n_e \frac{\partial c_{S_w}}{\partial t} + \frac{1}{1+K_d} n_e \underline{v}_e^T grad(c_{S_w}) - \frac{1}{1+K_d} div(n_e \underline{\underline{D}} grad(c_{S_w})) - \frac{1}{1+K_d} n_e r_{S_w}^* = 0 \qquad (4)$$

(Bio-)chemical reactions themselves reduce the extension of the mixing zone. The effect of chromatography and of dispersion is shown schematically in figure 1. Solid lines represent the concentration profile of the injected compound (oxygen) and dashed lines represent the concentration profile of the compound present in the domain in the initial state (substrate). (I) is a low dispersive case without retardation of any compound, (II) is a low dispersive case with retardation of the substrate, (III) is a high dispersive case without retardation. The extension of the mixing zone is marked on the x-axis.

To predict the extension of the mixing zone and hence of microbial growth a method is required that approximates front velocities and effective diffusion as exact as possible. Depending on the strength and direction of the chromatographic effect an overprediction of longitudinal dispersion might be tolerable. Since in the transversal direction effective diffusion is the only mixing process, transversal dispersion has to be approximated very accurately.

NUMERICAL FORMULATION

To our knowledge up to now no analytical solution is available for the equation system (1)+(3). Hence a numerical approximation is necessary. As has been shown by KINZELBACH ET AL. [1991] and others decoupling of advective-dispersive transport and chemical transformations by a two step method is an appropiate approach for solving the equation system. The nonlinear character of chemistry is not to be tackled by the transport modelling method, wheras the spatial dependency of transport is not influencing the chemistry modelling method.

Modelling of Advective-Dispersive Transport

For calculating advective-dispersive transport a semidiscrete finite element method (FEM) was chosen. The spatial approximation is based on the concept of a porous medium in conjunction with a discrete fracture system [ZIELKE & HELMIG, 1991]. By following the method of weighted residuals, applying the isoparametric concept for spatial coordinates, concentrations and reaction rates, carrying out partial integration to eliminate second order spatial derivatives, neglecting boundary integrals and integrating in time by finite differences equation (1) can be transformed to an equation for each element (5):

$$L(\underline{c}_i) := \left(\frac{1}{\Delta t} \underline{M} + \theta(\underline{\underline{A}} + \underline{\underline{B}}) \right) \hat{c}_i(t+\Delta t) - \left(\frac{1}{\Delta t} \underline{M} - (1-\theta)(\underline{\underline{A}} + \underline{\underline{B}}) \right) \hat{c}_i(t) - \underline{\underline{M}} \hat{R}_i = 0$$

with

$$\underline{M} = \frac{n_e}{\Delta t} \int_{V_{El}} \Phi^T \Omega \; dV ; \qquad\qquad \underline{\underline{A}} = \frac{n_e}{2} \int_{V_{El}} \Phi^T \underline{v}_e^T grad(\Omega) \; dV ; \qquad (5)$$

$$\underline{\underline{B}} = \frac{n_e}{2} \int_{V_{El}} grad(\phi)^T \underline{\underline{D}} \, grad(\Omega) \; dV \hat{c}_i(t+\Delta t) ; \qquad \hat{R}_i = \frac{1}{\Delta t} \int_t^{t+\Delta t} \hat{r}_i(\tau) d\tau .$$

Integrated reaction rates on the right side of equation (5) have to be calculated by the chemistry step.

Transport in groundwater is mostly advection dominated. Hence using a Standard Galerkin Method leads to strong oscillations in the solution especially if sharp fronts occur. Even if no step or pulse signals are introduced into the domain by boundary conditions, chemical reactions can sharpen concentration fronts. Oscillations can lead to negative concentrations in the solution of transport calculation, causing unstability in the chemistry calculation. Hence a monotone convergent method is required.

Different improved linear FE methods like Taylor-Galerkin formulations, smooth artificial logitudinal dispersion [PERROCHET & BÉROD, 1993], inconsistent upwinding by quadratic or cubic weighting functions, mass-lumping and consistent streamline upwind Petrov Galerkin (SUPG) methods [BROOKS & HUGHES, 1982] have been applied by the authors to the multicomponent reactive transport model CONTRACT (Contaminant Transport and Chemical Transformations), which is based on the FE transport model ROCKFLOW-TM developed by KRÖHN [1991]. Currently a combination of streamline weighted quadratic upwinding of the advection term and mass-lumping of the storage term is prefered. This combination has been successfully introduced by HELMIG [1993] for simulation of multiphase flow. As monotone convergent linear methods can be accurate at least of order one, the solution is overdiffusive. Nonlinear methods will be tested in the near future.

Modelling of Microbial Growth, Sorption and Chemical Transformations

For calculating chemical transformations the DAES solver DASSL developed by PETZOLD [1982] has been used. DASSL is described in detail by BRENAN ET AL. [1989]. It's based on multiple backward differentiation of order one to five and Newton's method for solving the resulting non-linear equation system. DASSL includes an error estimator for an automatic time step discretization and for an automatic order adaptation depending on the behaviour of the solution. It requires concentrations and their time derivatives at the beginning time level. If no information about consistent time derivatives for the initial values is available DASSL provides a numerical estimator based on a small implicit time integration step.

Coupling of Chemistry and Transport

Following an approach of HERZER [1989] chemical transformations are taken into account in the transport calculation by an additive sink/source term r which has to be evaluated in the chemistry step for each transported compound. Transport is taken into account in chemistry calculation by a constant transport rate which has to be evaluated in the transport step at every node. Iteration between transport step and chemistry step is repeated until a convergence criterion is reached. Because of stiffness in the system of equations, chemistry calculations require smaller time steps

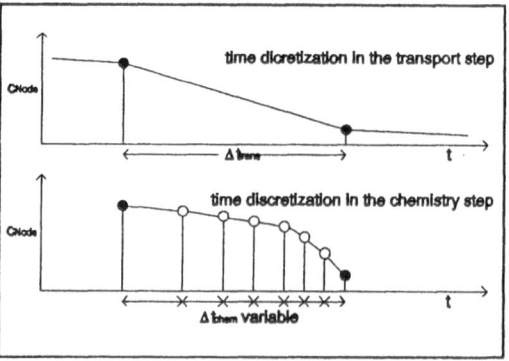

Fig. 2 Time discretization in the two half steps.

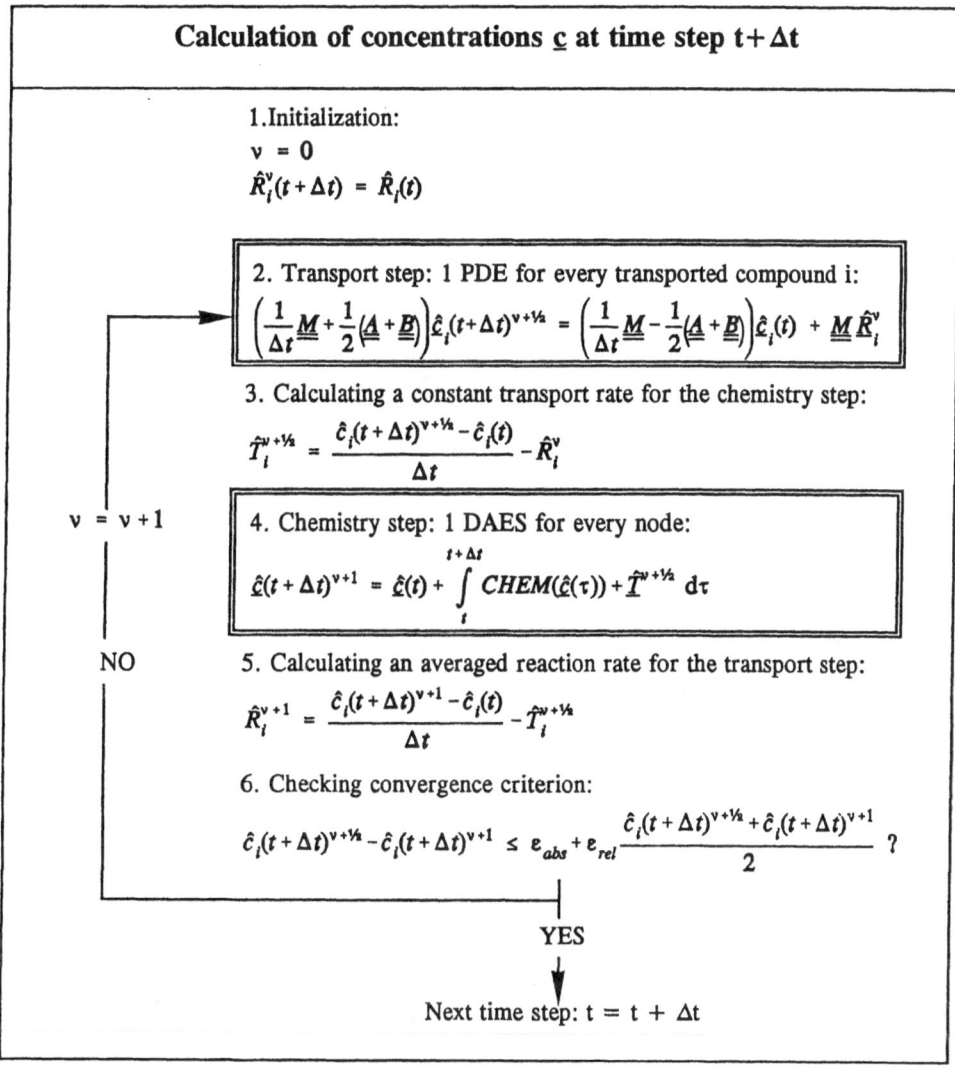

Fig. 3 Iterative two-step method for solution of the coupled system of multicomponent reactive transport as introduced by HERZER [1989].

than transport calculations. It's not necessary to apply these small time steps to the transport step. Time discretization is demonstrated by figure 2. It is obvious that the evaluation of the reaction rate in the transport step should be based on averaging over all sub-time steps of chemistry modelling for one transport time step. The resulting iteration scheme is shown in figure 3.

EXAMPLE

Figure 4 shows an artificial example which illustrates the effect of the spatial variability of permeability on the degradation behaviour of an organic contaminant in a fractured-porous medium with one fracture diagonal to the direction of streamlines in the rock matrix. The two-dimensional domain is 3m wide and 10m long. Hydraulic permeability in the porous matrix is 10^{-4}m/s. The fracture is 1cm in width, its permeability is 10^{-2}m/s. The difference in the piezometric head between the two short boundaries is 20cm, there are no-flow boundary conditions at the two long boundaries. Transport parameters are listed in table 1.

Transport Parameters			
$n_e = 0.3$	$K_d = 4.0$	$\alpha_l = 0.1$m (matrix) $= 0.01$m (fracture)	$\alpha_t = 0.01$m (matrix)
Microbiological Parameters			
$\mu_{max} = 2$/d	$Y_S = 0.09$	$Y_O = 0.032$	$K_S = 2.0$mg/L
$K_O = 0.5$mg/L	$k_{dec} = 0.1$/d		
Initial Conditions			
$c_{Sw} = 2$ mg/L	$c_O = 0$mg/L	$c_{Ss} = 8$mg/L	$X = 0.001$mg/L
Inflow Concentrations		Outflow boundary condition	
$c_{Sw} = 0$mg/L	$c_O = 5$mg/L	Pure advection	

Tab. 1   Parameters for transport calculation in the example.

The domain was discretized with 983 nodes in 926 bilinear 2D elements and 82 linear 1D elements. Figure 4 shows the model geometry and the concentrations of the substrate, oxygen and biomass after 20 days. It is obvious that the fracture acts as preferential flowpath. The substrate in the fracture itself is almost completely removed after 20 days. The penetration of oxygen takes place preferentially in the fracture, but it is limited by oxygen consumption of the biomass. Upstream of the fracture oxygen concentration in the matrix is hardly influenced by the concentration in the fracture, downstream of the fracture the influence is significant. As the fracture is diagonal to the flow direction, the concentration distribution in the matrix around the fracture is asymmetrical. The growth of biomass reaches its maximum in the fracture. Here both substrate and oxygen are available. The substrate is delivered from the matrix whereas oxygen is preferentially transported in the fracture. This is a typical situation for aerobic biodegradation in a heterogeneous aquifer: Low permeable zones (here: the matrix) act as source for the substrate, they are hardly penetrated by oxygen; high permeable zones (here: the fracture) act as transport pathways for oxygen. The preferential sites of biomass growth are the interfaces between low and high permeable zones, which coincide with the fracture in our example. The results of biomass concentrations show non-negative oscillations indicating discretization problems.

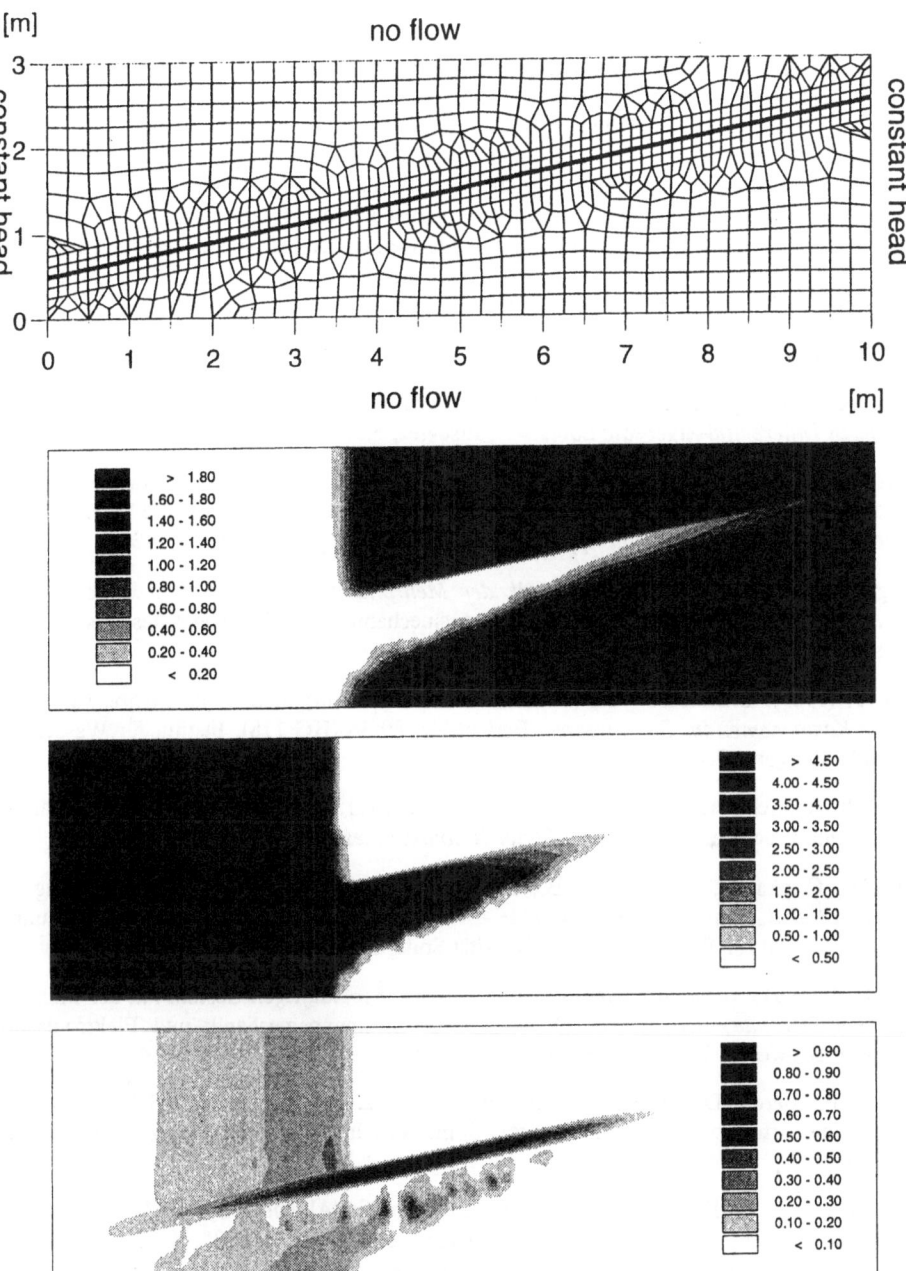

Fig.4 Model network and concentrations after 20 days for the calculated example. Top figure: model network and boundary conditions for flow calculation; 2nd figure: dissolved substrate; 3rd figure: dissolved oxygen; bottom figure: biomass. All concentrations in mg/L. Flow direction from left to right. Model parameters in the text.

CONCLUSIONS

A method has been developed which enables to simulate biodegradation of organic contaminants in highly heterogeneous aquifers like fractured-porous media. First model calculations show significant differences to calculations based on homogeneous lumped parameter fields. These differences are to be seen as a result of the nonlinear character of the governing equations describing biomass growth and microbial degradation. This might be an explanation for differences observed in the degradation behaviour of microorganisms between laboratory and field experiments. This topic will be investigated by large-scale well controlled physical experiments in the research facility for subsurface remediation *VEGAS* currently under construction at the University of Stuttgart [KOBUS ET AL., 1993].

REFERENCES

Brenan, K.E., Campbell, S.L., Petzold, L.R. (1989): *Numerical Solution of Initial-Value Problems in Differential-Algebraic Equations*, Elsevier, New York.

Brooks, A.N., Hughes, T.J.R. (1982): "Streamline Upwind/Petrov-Galerkin Formulations for Convection Dominated Flows with Particular Emphasis on the Incompressible Navier-Stokes Equations", *Computer Methods in Applied Mechanics and Engineering* 32, 199-259.

Helmig, R. (1993): *Theorie und Numerik der Mehrphasenströmungen in geklüftet-porösen Medien*, Bericht Nr. 34, Institut für Strömungsmechanik und Elektronisches Rechnen im Bauwesen, Universität Hannover, Germany.

Herzer, J. (1989): *CHEMFLO - Dokumentation eines Schadstoffmodells für mehrere wechselwirkende Komponenten im Grundwasser*, Bericht Nr. 89/34 (HG 118), Institut für Wasserbau, Universität Stuttgart, Germany.

Kinzelbach, W., Schäfer, W., Herzer, J. (1991): "Numerical Modeling of Natural and Enhanced Denitrification Processes in Aquifers", *Water Resources Research* 27(6), 1123-1135.

Kobus, H., Cirpka, O., Barczewski, B., Koschitzky, H.-P. (1993): *Versuchseinrichtung zur Grundwasser- und Altlastensanierung VEGAS - Konzeption und Programmrahmen*, Mitteilungsheft Nr.82, Institut für Wasserbau der Universität Stuttgart, Germany.

Kröhn, K.-P.(1991): *Simulation von Transportvorgängen im klüftigen Gestein mit der Methode der Finiten Elemente*, Bericht Nr. 29, Institut für Strömungsmechanik und Elektronisches Rechnen im Bauwesen, Universität Hannover, Germany.

Perrochet, P., Bérod, D. (1993): "Stability of the Standard Crank-Nicolson-Galerkin Scheme Applied to the Diffusion-Convection Equation: Some New Insights", *Water Resources Research* 29(9), 3291-3297.

Petzold, R. (1982): "A Description of DASSL: A Differential/Algebraic System Solver - presented at IMACS World Congress, Montreal, Canada, August 8-13, 1982", *SANDIA REPORT* SAND82-8637, Sandia National Laboratories, Livermore, California.

Zielke, W., Helmig, R. (1991): "Grundwasserströmung und Schadstofftransport - FE-Methoden für klüftiges Gestein", in: Universität Karlsruhe (Hrsg.) *Wissenschaftliche Tagung "Finite Elemente - Anwendung in der Baupraxis"*, Verlag Ernst & Sohn, Berlin.

PASS THE TRASH

DAVID E. DOUGHERTY
Department of Civil & Environmental Engineering, University of Vermont,
Burlington, VT 05405 USA

ANDREW F. B. TOMPSON
Earth Sciences Department, Lawrence Livermore National Laboratory,
Livermore, CA 94550 USA

A numerical method for solving advection-diffusion-reaction transport models in one, two, and three dimensions is described and applied to several test problems. Operator splitting is employed to separate the advection and diffusion operators from the reaction term. The approach is to discretize the domain using a mesh of points. A first-order splitting moves from one time level to another in two fractional steps. In the first, advection-diffusion is evaluated using a combination of backward tracking along characteristics and numerical quadrature of the Green's function for diffusive effects. Variable coefficients can be handled by local approximations. The second fractional step evaluates changes in mass due to reaction processes on the same grid. Higher-order splittings can be defined. Numerical simulations of the one-dimensional real Ginzburg-Landau equations compare very well with analytical solutions in the form of a steady traveling wave and a perturbed system undergoing a transition between two different steady waves.

Introduction

Consider a general advection-diffusion-reaction equation of the form

$$\frac{\partial c}{\partial t} + \nabla \cdot (\mathbf{v}c) - \nabla \cdot (\mathbf{D} \cdot \nabla c) - r = 0, \tag{1}$$

where $c(\mathbf{x}, t)$ is the concentration of an arbitrary dissolved constituent in a flowing fluid, $\mathbf{v}(\mathbf{x})$ is the steady fluid velocity, $\mathbf{D}(\mathbf{x})$ is a diffusion tensor dependent on location and fluid velocity, and $r(\mathbf{x}, t)$ is a general rate of production.

An explicit time integration procedure comprising n fractional steps may be used to advance a chemical transport system such as (1) from one time level to the next. In one method, for example, concentration changes due to advection and diffusion

613

A. Peters et al. (eds.), Computational Methods in Water Resources X, 613–620.
© 1994 *Kluwer Academic Publishers. Printed in the Netherlands.*

are first computed over the interval Δt (Step I). These results are then used to calculate changes caused by reactions over the same interval (Step II). We focus on the advection-diffusion step in this paper.

It is important to identify several important issues regarding this two-step approach. First, it is inherently explicit, leading to concerns regarding timestep size and numerical stability. Secondly, because concentration estimates made after *Step I* are used to predict reactions, estimation errors e will be propagated through the reaction calculation, and an error feedback loop will be established. The magnitude of this error will depend on the type of reaction being considered, the function ζ used to evaluate concentrations, and the precision of the calculations. Next, the overall accuracy of the operator split method presented above is Δt. A higher-order accurate splitting can be obtained by symmetrizing the operator splitting [12]. For example, integration over the interval Δt can be computed to second-order accuracy with the sequence of four fractional steps—Step I, followed by II, followed by II, followed by I—each using a timestep of $\Delta t/2$. Another issue involves the specific way in which the reaction term is integrated in time. The smaller the time scale of the reaction process relative to Δt, the stiffer the integral becomes, and more accurate ODE integration schemes are needed [7]. Alternatively, when reaction rates are sufficiently fast, the kinetic terms may be replaced with simpler equilibrium models [4]. Moreover, if equilibria among several constituents can be considered reversible, partitioning relationships may be introduced. Under these circumstances, concentrations of individual components may be extracted from total concentrations using the partitioning relationship. More general circumstances may require both kinetic and equilibrium treatments, leading to systems of differential-algebraic equations [11].

In this paper we present a new method for numerically solving the advection-diffusion step. To keep the exposition simple and to stay within space limitations, we focus on a one-dimensional setting, in which (1) reduces to

$$\frac{\partial c}{\partial t} + \frac{\partial}{\partial x}(vc) - \frac{\partial}{\partial x}\left(D\frac{\partial c}{\partial x}\right) - r = 0, \tag{2}$$

where v and D may still be functions of space. For numerical purposes, $c(x,t)$ may be represented on a discrete grid of uniform spacing Δx and updated over small time increments Δt. The quantity c_i^n will be used to denote the value of c at $i\Delta x$ and $n\Delta t$. Irregular meshes may be used in general.

Advection Diffusion Operators

The advection-diffusion problem has often been subdivided for the development of robust numerical methods. Following this precedent, we address advection first, followed by diffusion.

Advection Only

The method of characteristics can be used to obtain the solution of the pure advection case, described by (2) with $D = 0$. Here, the value of c at the point $(\mathbf{x}, t^{n+1})$

is found from the value of c at the point $(\mathbf{x}_f, t^n)$ that lies on the characteristic passing through $(\mathbf{x}, t^{n+1})$. The point $(\mathbf{x}_f, t^n)$, or an approximation to it $(\mathbf{x}_f^*, t^n)$, is called the *trace-back point* or *foot of the characteristic*.

In an infinite spatial domain, this solution can be expressed as a convolution

$$c(x_i, t^{n+1}) = \int_{-\infty}^{\infty} c(x', t^n) G(x', x_i, t^n, t^{n+1}) dx'. \tag{3}$$

involving the concentration at time t^n and a kernel, or Green's function, G. The latter is defined by

$$G(x', x_i, t^n, t^{n+1}) = \delta(x_i - x_f). \tag{4}$$

Therefore, the solution for c at a set of grid points $\mathbf{x}_g$ at time t^{n+1} can be obtained from the concentration field at time t^n by looking back along the characteristics from every point $\mathbf{x}_g$, i.e., integrating approximately

$$\frac{d\mathbf{x}}{dt} = \mathbf{v} \text{ for } \mathbf{x}(t^{n+1}) = \mathbf{x}_g, \tag{5}$$

from t^{n+1} to t^n. Hence, the method of characteristics may be viewed as a change of coordinates

$$\frac{\partial c}{\partial t} + \mathbf{v} \cdot \nabla c = \frac{Dc}{Dt} = \frac{\partial c}{\partial \tau}, \tag{6}$$

where τ is the characteristic-oriented time. Note that each integration for the foot of the characteristic is independent and can be performed in parallel (except possibly for data accesses).

In practice, the solution $c(\mathbf{x}, t^n)$ is not available. Rather, an approximation to it based on the grid and some interpolation method is available.

The accuracy of the solution at $c(\mathbf{x}, t^{n+1})$ depends upon the quality of the trace-back and on the quality of the approximation $c(\mathbf{x}, t^n)$. A great deal of effort can be expended in evaluating the trace-back integral numerically, particularly in heterogeneous flow fields.

Diffusion Only

Consider the solution of (2) on an infinite domain in the case of no velocity, no reaction, constant diffusion D, infinite spatial domain, and a nonnegative initial concentration distribution, $c(x, 0)$. At any arbitrary time $t > 0$, c may be expressed exactly as the convolution

$$c(x, t) = \int_{-\infty}^{\infty} c(x', 0) G(x' - x, t) dx'. \tag{7}$$

This convolution involves the initial concentration condition and a kernel, or Green's function, G, defined by

$$G(x' - x, t) = \frac{1}{\sqrt{4\pi Dt}} e^{-(x'-x)^2/4Dt}. \tag{8}$$

The Green's function G represents the solution at time t at location x' due to a unit impulse δ initially located at x.

In a time-discrete sense, the result (7) may be used to explicitly update nodal values of c over a time interval Δt via

$$c_i^{n+1} = \int_{-\infty}^{\infty} c^n(x')G(x'-x, \Delta t)dx', \tag{9}$$

where $c^n(x')$ in an integrable approximation of $c(x', n\Delta t)$, provided a suitable approximation to the integral can be developed. Several explicit quadrature schemes of the form

$$c_i^{n+1} \approx \sum_{p=1}^{P} W_p \, c^n(x_p) \tag{10}$$

may be used for the evaluation of (9).

- *Finite Difference.* Standard finite difference methods may be viewed as the quadrature (9) with quadrature points x_p restricted to grid points $x_p = p\Delta x$, where p is an integer.

- *Monte Carlo Quadrature.* A statistical method may be used in which P hypothetical particles are released from point i at time $(n+1)\Delta t$ and "randomly-walked" (with a step variance of $2D\Delta t$) to determine the sampling locations x_p. Because the distribution of locations follows the form of G, the weight factors W_p all equal $1/P$ and ensure that the updated values remain non-negative. This is simply a reverse implementation of a conventional forward-in-time random walk particle approach. Since the accuracy of the result is controlled by the number of samples used, a relatively large number will be needed to get a resonably accurate solution, and that this would reduce the efficiency of computation. Variance reduction methods could also be employed, of course.

- *Riemann Quadrature.* A more obvious approach for approximating the integral in (9) would be to directly construct a Riemann sum involving the values of $c^n(x)$ at a series of P regularly spaced points, weighted by the the interval length and the value of G at each location. Again, the updated solutions would remain nonnegative. Accuracy in the integration will be controlled by the degree to which variations in G and $c(x, t^n)$ are simultaneously resolved by the spacing of the sampling and grid points. Generally, the larger the support of G with respect to Δx, the more sampling points required; for smaller supports, accuracy will be also be affected by the way in which $c(x, t^n)$ is interpolated between neighboring grid points.

- *Gauss-Hermite Quadrature.* A alternative to the Riemann approach is to use Gauss-Hermite quadrature to approximate (9). If the known solution $c^n(x)$ is suitably well behaved [1], then P properly chosen sampling points centered around each node i may be used with a set of weights to approximate (9) to high (polynomial) order. For example, a three-point formula will take the form

$$c_i^{n+1} \approx \frac{1}{6}(c_{i-A}^n + 4c_i^n + c_{i+A}^n), \tag{11}$$

where "A" signifies a spatial shift of magnitude $\sqrt{6D\Delta t}$ on either side of point i.

As the Gauss-Hermite method states, it is not necessary that quadrature points be located at nodes. Traditional finite difference methods impose such a restriction. If off-node quadrature is used however, some interpolation of c^n should be available. If c^n is nonnegative, this implies the interpolant must be nonnegative. Historically, the utilization of off-node values c_p^n has usually been avoided. However, with increasing use of the modified method of characteristics (MMOC) [3], local adjoint methods [2], and particle-grid methods [10, 9], this issue is no longer being skirted.

Pass The Trash

"Pass The Trash" (PTT) is a method that uses local approximations to characteristics together with the kernel method above for advection diffusion equations. By using characteristics to, in effect, symmetrize the advection diffusion operator, we can focus on the treatment of the diffusion term.

Stability. The stability of the general three point formula (11) may be examined with a von Neuman analysis by setting $c_i^{n+1} = \xi\, c_i^n$ and $c_{i+A}^n = c_i^n\, e^{i\sigma A}$. Substitution yields a purely real, nonnegative amplification factor of the form

$$\xi = \frac{2}{3} + \frac{1}{3}\cos(\sqrt{6D\Delta t}\,\sigma), \tag{12}$$

which remains bounded by $\frac{1}{3}$ and 1 for all spectral frequencies σ.

Non-negativity: The last result shows that if the initial condition c_i^n is nonnegative, then the predicted concentration c_i^{n+1} will always be non-negative.

Accuracy: The amplification behavior of the numerical method may be contrasted with the amplification factor ξ_{analy} derived for an analytic solution of (2) found by substituting $c(x_i, t^n) = Ce^{\beta t^n}e^{i\sigma x_i}$ into (2) with D constant and $v = 0$, and solving for $c(x_i, t^{n+1})/c(x_i, t^n)$:

$$\xi_{\text{analy}} = e^{-D\sigma^2 \Delta t}. \tag{13}$$

The explicit three-point formula (2) is unconditionally stable, though its accuracy at higher spectral frequencies is diminished.

Different behavior will arise for other values of P. For example, the two-point formula is stable, yet will support oscillatory solutions.

Relationship to Finite Difference Methods: In real applications, the point "A" will not necessarily lie on a grid point, and an interpolation based on the neighboring grid values must be employed. If A is constrained to be less than or equal to Δx (through suitable choice of the time step), then the formula (11) can be reduced to

$$c_i^{n+1} \approx \gamma c_{i-1}^n + (1 - 2\gamma)c_i^n + \gamma c_{i+1}^n, \tag{14}$$

where $\gamma = \sqrt{D\Delta t/12\Delta x^2}$, assuming piecewise linear interpolation is used to estimate $c_{i\pm A}^n$ between nodes i and $i \pm 1$. This expression involves a differencing among nodes $i-1$, i, and $i+1$, and is similar in form to the explicit, $\mathcal{O}(\Delta t, \Delta x^2)$ finite difference formula,

$$c_i^{n+1} = \lambda c_{i-1}^n + (1 - 2\lambda)c_i^n + \lambda c_{i+1}^n, \tag{15}$$

where $\lambda = D\Delta t/\Delta x^2$. The finite difference expression gives rise to unstable results for γ or $\lambda > 1/2$. However, the quadrature formula will actually change to a stable five-node formula when the "A" becomes greater than Δx, *i.e.*,, when γ exceeds $1/6\sqrt{2}$.

More on Accuracy: Although stability is preserved through this adaptivity, the capacity to resolve fine-scale behavior diminishes as the Gauss point spread "$2A$" or the time step increases. This will occur because a three-Gauss point formula will fail to pick off variability of $c(x_i, t^n)$ occuring at wavelengths that are small relative to the quadrature point separation. In other words, the form of $c(x_i, t^n)$ may not be as "well behaved" as the Hermite quadrature formula assumes. Hence, the minimum feature length resolvable is $\max(2\Delta x, 2A)$. Using more Gauss points, reverting to a Riemann sum, or constraining the size of the time step may all be useful in getting around this issue. Each alternative has important computational implications.

Diffusion–Reaction in an Infinite Spatial Domain

To solve reactive forms of (2) with $r \neq 0$, $v = 0$, and $D = 1$ we use the explicit method of fractional steps, as indicated earlier, to advance solutions over discrete time steps. This entails diffusing mass (in the absence of reaction) using a differencing formula like (14), and then reacting mass (in the absence of diffusion) using an integration method tailored to the reaction time scale and time step magnitude. We have used several different integrators for the reactions, including a variety of Euler, predictor-corrector, and Runge-Kutta schemes; each has advantages and disadvantages.

Real Ginzburg-Landau Equation in One Dimension: As an example application of the method, we consider the one-dimensional Ginzburg-Landau equation,

$$\frac{\partial c}{\partial t} - \frac{\partial^2 c}{\partial x^2} = c - c^3. \tag{16}$$

This diffusion-reaction equation has a stronger nonlinearity in the reaction term than the Fisher equation, in which the reaction is quadratic. If c is initially bounded with $0 \leq c \leq 1$, the solution will grow towards an equilibrium value of 1 and propagate in directions of lower concentration. For the specific initial condition

$$c(x) = (1 + e^{x/\sqrt{2}})^{-1}, \tag{17}$$

the analytical solution [5] is a steady traveling wave $c(x - St)$ moving in the positive x direction at a speed of $S = 3/\sqrt{2}$.

This solution has also been shown to be sensitive to perturbations in the initial and far field conditions [5]. Now examine the response of the solution to a slight modification of the far-field initial condition. Analogous to Reitz [8], we use a set of initial data that decay exponentially ahead of the wave, but are otherwise equal to (17) on or behind it. McKean [6] shows that such an initial wave should evolve into another one moving at a constant speed different from that resulting from (17). Although the differences between this initial data and the original condition are extremely small the solution dynamics are quite distinct and illustrate the sensitivity and nonlinearity embodied within (16).

A one-dimensional PTT model was applied to the region $-20 \leq x \leq 180$ with a uniform grid spacing $\Delta x = 0.5$. The initial condition (17) evaluated at grid points. Because this is an advection-free problem with constant diffusion, the trace-back portion of the algorithm is trivial and the quadrature portion employs identical Gauss points (relative to the trace-back point). The leftmost $ceiling(Co)$ cells were always specified with $c = 1$ and we use $c = 0$ for $x > 180$ to avoid boundary issues in this example. A time step of $\Delta t = 0.1$ was used to advance the simulation. Following the diffusion step, concentrations $\Delta c_R(x_g)$ were evaluated using either an Euler or fourth-order Runge Kutta integration method. Since this system is stable and has equilibrium points at 0 and 1, concentrations and their changes are always positive. Hence, new mass is added to the system; the largest change c could make over $\Delta t = 0.1$ using an Euler scheme is 0.03849, so no stiffness problems were anticipated. Simulations using our method for the initial condition (17) and for perturbed initial conditions once again show excellent matches with analytical results.

Additional Topics and Directions

"Pass The Trash" has been developed in this paper for one-dimensional spatial domains. The methods used to implement the method have been highlighted in this paper. PTT has several very interesting properties and has performed well in demonstration problems (only one of which was discussed here).

Several important issues have not been discussed here.

1. Extension of the method to two and three dimensions has already occurred; this will be reported elsewhere. It is important to note that products of Gauss-Hermite quadrature rules are not appropriate if one wishes to pre-serve non-negativity and unconditional stability. A detailed development of a suitable quadrature scheme will appear in the abovereferenced report.

2. Variable coefficients, whether scalar, vector, or tensor, need to be accomodated. A formal development based on local approximations has been completed and one-dimensional testing using smoothly varying coefficients have yielded very good results. The two- and three-dimensional extensions have occurred, but testing has been limited to embedded one-dimensional problems. Complex fields will require careful tracking of the quadrature points across jumps, or large changes, in the velocity field.

3. This paper has focussed on infinite spatial domains, although the computational example occurred on a finite domain and hence had boundaries. As in many other characteristics-, adjoint-, or particle-based methods, the treatment of boundaries needs significant attention. A formal development has occurred, but advances remain necessary before this problem can be called solved.

4. Finally, there is the issue of computational efficiency and complexity. The non-negative, unconditionally stable properties of the method come at the expense of off-node evaluations of the solution variable to compute solutions on-node. Many methods are now using tracking methods, but this method has taken this idea farther than others.

Acknowledgements

DED's work was supported by the National Science Foundation under Grant No. ASC-9100226; the U.S. Government has certain rights in this material. Any opinions, findings, and conclusions or recommendations expressed in this material are those of the authors and do not necessarily reflect the views of the NSF. AFBT's work was conducted under the auspices of the U. S. Department of Energy by Lawrence Livermore National Laboratory under contract W-7405-Eng-48 and supported by the Subsurface Science Program of the Office of Health and Ecological Research of the US DOE.

References

[1] Carnahan, B., H. A. Luther, and J. O. Wilkes, *Applied Numerical Methods*, (Wiley, New York, 1969).

[2] M. A. Celia, T. F. Russell, I. Herrera, and R. E. Ewing, An Eulerian-Lagrangian localized adjoint method for the advection-diffusion equation, *Advances in Water Resources*, **14**, 187 (1990).

[3] J. Douglas, Jr. and T. F. Russell, Numerical methods for convection-dominated diffusion problems based on combining the method of characteristics with finite element or finite difference procedures, *SIAM J. Num. Anal.*, **19**, 871 (1982).

[4] Garven, G. and R. A. Freeze, Theoretical analysis of the role of groundwater flow in the genesis of stratabound ore deposits, I. Mathematical and numerical model, *Am. J. Sci.* **284**, 1085–1124 (1984).

[5] P. Kaliappan, An exact solution for travelling waves of $u_t = Du_{xx} + u - u^k$, *Physica D*, **11**, 368 (1984).

[6] McKean, H. P., Application of Brownian Motion to the Equation of Kolmogorov-Petrovskii-Piskunov, *Comm. Pure Appl. Math.*, **28**, 323–331 (1975).

[7] Oran, E. S. and J. P. Boris, *Numerical Simulation of Reactive Flow* (Elsevier, Amsterdam, 1987).

[8] Reitz, R. D., A study of numerical methods for reaction-diffusion equations, *SIAM J. Sci. Stat. Comput.*, **2**, 95–106 (1981).

[9] Tompson, A. F. B. and D. E. Dougherty, Particle-grid methods of reactive flows in porous media with application to Fisher's equation, *Applied Mathematical Modeling*, **16**, 374:383 (1992).

[10] Tompson, A. F. B. and L. W. Gelhar, Numerical simulation of solute transport in randomly heterogeneous porous media, *Wat. Resour. Res.*, **26**, 2541–2562 (1990).

[11] Yeh, G. T. and V. S. Tripathi, A critical evaluation of recent developments in hydrogeochemical transport models of reactive multichemical components, *Wat. Resour. Res.*, **25**, 93–108 (1989).

[12] Yanenko, N. N., *The Method of Fractional Steps*, (Springer Verlag, New York, 1971).

NUMERICAL INTEGRATION METHODS FOR THE DUAL POROSITY MODEL IN SORBING POROUS MEDIA

GIUSEPPE GAMBOLATI[†], CLAUDIO GALLO[‡] and CLAUDIO PANICONI[‡]

[†] *Dept. of Mathematical Methods for Applied Sciences, University of Padua, Italy*

[‡] *CRS4, Via Nazario Sauro 10, 09123 Cagliari, Italy*

Three numerical integration-in-time methods are presented for the Galerkin solution of a dual porosity model describing the transport of reactive contaminants through a sorbing porous medium characterized by intra-aggregate diffusion. In the coupled approach the model is solved simultaneously for the mobile and immobile region concentrations c_m and c_{im} over the nodes of a finite element grid, while in the other two approaches c_m and c_{im} are obtained separately with a decoupled procedure by which the per time step CPU cost is reduced approximately by a factor of 3. The decoupled approach relying on an integro-differential formulation, recently developed by the authors, appears to be relatively less robust as it requires a smaller time integration step in the limiting case when the diffusion process is so rapid that the solution to the dual porosity model approaches the solution to the classical transport equation with retardation.

INTRODUCTION

The "dual porosity model" is frequently used to describe the movement of reacting solutes in both saturated and unsaturated porous media. This model relies on a sorption mechanism controlled by the so-called non-local equilibrium assumption (non-LEA), according to which the fluid phase is divided into a mobile and an immobile solution with a diffusive mass transfer driven by the difference in concentration between the mobile and immobile regions [*Coats and Smith, 1964*]. Non-LEA conditions hold whenever sorption onto solid grains of contaminant transported by the groundwater flow field is a slow process compared to other processes (diffusion, dispersion, advection) affecting the contaminant concentration, so that chemical equilibrium cannot be established between the sorbent and the solution phase [*Valocchi, 1985*]. In this case the rate of mass transfer from the mobile solution toward the immobile sorbing region can influence significantly the distribution and fate of the pollutant and lead to breakthrough profiles exhibiting a pronounced "tailing effect" [*Giddings, 1963*], which is easily accounted for by the dual porosity model. The dual porosity model can be further enhanced by introducing additional realistic processes, including biodegradation or radioactive decay, and instantaneous mass exchange controlled by a linear or nonlinear adsorption isotherm in both the mobile and

A. Peters et al. (eds.), Computational Methods in Water Resources X, 621–628.
© *1994 Kluwer Academic Publishers. Printed in the Netherlands.*

immobile regions [*van Genuchten and Wierenga, 1976*]. The dual porosity model may be numerically solved using a coupled approach wherein the mobile and immobile concentrations, c_m and c_{im} respectively, are treated separately and simultaneously, or a decoupled approach wherein first the kinetic equation is solved for c_{im}, and then c_{im} is substituted into the transport equation, which is solved for c_m. Decoupling can be obtained with a special time stepping scheme in the integration of the coupled model [*Leismann et al., 1988*], or with an integro-differential approach [*Gambolati et al., 1993*] where the convolution integral resulting from the analytical integration of the kinetic equation is substituted into the transport equation.

In this paper we first present the three solution approaches for a dual porosity model which also incorporates decay and instantaneous sorption. We then discuss the limiting solution of the non-LEA model when the mass transfer coefficient becomes sufficiently large that the classical LEA transport equation for c_m applies. Finally, we present some preliminary numerical results and a comparison of the performance (CPU time and storage) of the three different approaches.

DUAL POROSITY MODEL

Nonequilibrium contaminant transport may be mathematically described using a dual porosity model wherein an unsaturated, aggregated porous medium is subdivided into five regions [*van Genuchten and Wierenga, 1976*]: (1) the air phase; (2) the region containing the *mobile* water phase, located in the largest pores; (3) the region containing the *immobile* water phase. Solute transfer into this region occurs by diffusion from region 2; (4) the *dynamic* soil region, located around the mobile water region. Solute transfer is by instantaneous sorption with region 2; (5) the *stagnant* soil region, located around the immobile water region. Solute transfer is by instantaneous sorption with region 3. Under these assumptions, the general equations of the dual porosity model are [*Gambolati et al., 1993; 1994*]:

$$\frac{\partial}{\partial x_i}\left(D_{ij}\frac{\partial c_m}{\partial x_j}\right) - v_i\frac{\partial c_m}{\partial x_i} = T_m\frac{\partial c_m}{\partial t} + T_{im}\frac{\partial c_{im}}{\partial t} + \lambda\left(T_m c_m + T_{im}c_{im}\right)$$
$$+ q(c_m - c^*) - f \tag{1}$$

$$T_{im}\frac{\partial c_{im}}{\partial t} = \alpha(c_m - c_{im}) - \lambda T_{im}c_{im} \tag{2}$$

where the indices i, j on the left hand side of (1) denote summation over the coordinate dimensions; c_m is the concentration of the dissolved constituent in the mobile water region; c_{im} is the concentration in the immobile water region; $D_{ij} = n_m S_{w_m}\tilde{D}_{ij}$; n_m is the porosity of the mobile water region (ratio of the volume of voids in the mobile region to the total volume); S_{w_m} is the water saturation in the mobile water region; $\tilde{D}_{ij}$ is the dispersion tensor; v_i is the Darcy velocity; $T_m = n_m S_{w_m} R_m$; $T_{im} = n_{im} R_{im}$; n_{im} is the porosity of the immobile water region; $R_m = 1 + (\rho_s F k_{d_m})/(n_m S_{w_m})$ is the retardation factor for the dynamic water and soil regions; $R_{im} = 1 + (\rho_s(1 - F)k_{d_{im}})/n_{im}$ is the retardation factor for the stagnant water and soil regions; $\rho_s = (1 - n)\gamma_s$ is the bulk soil density; $n = n_m + n_{im}$ is the total porosity; γ_s is the density of solid grains; F is the fraction of adsorption sites which are in the dynamic soil region; k_{d_m} is the distribution coefficient in the linear Freundlich isotherm describing the instantaneous sorption between the mobile water and soil regions; $k_{d_{im}}$ is the linear isotherm distribution coefficient for the immobile water and soil regions; λ is the radioactive or biodegradation decay constant; $q(x_j) = \partial v_i / \partial x_i$ is the distributed source or sink (volumetric flow rate per unit volume); c^* is the concentration of the contaminant injected or withdrawn with the fluid source or sink; f is the distributed mass rate of the contaminant per unit volume; α is the mass transfer coefficient for the diffusion process between the mobile and immobile water regions; x_i is the ith Cartesian coordinate; t is time.

If α becomes infinitely large, i.e., the mass exchange between the mobile and the immobile region tends to be instantaneous, equation (2) implies $c_{im} = c_m$ and equation (1) reduces to the classical LEA transport equation:

$$\frac{\partial}{\partial x_i}\left(D_{ij}\frac{\partial c_m}{\partial x_j}\right) - v_i\frac{\partial c_m}{\partial x_i} = (T_m + T_{im})\left(\frac{\partial c_m}{\partial t} + \lambda c_m\right) + q(c_m - c^*) - f \quad (3)$$

In addition, in fully saturated aquifer systems ($S_{w_m} = 1$) with $k_{d_m} = k_{d_{im}} = k_d$, equation (3) simplifies to:

$$\frac{\partial}{\partial x_i}\left(D_{ij}\frac{\partial c_m}{\partial x_j}\right) - v_i\frac{\partial c_m}{\partial x_i} = R\left(\frac{\partial c_m}{\partial t} + \lambda c_m\right) + q(c_m - c^*) - f \quad (4)$$

where $R/n = 1 + [(1-n)\gamma_s k_d/n]$ is the retardation factor for the porous medium.

Using equation (2), we can write equation (1) as:

$$\frac{\partial}{\partial x_i}\left(D_{ij}\frac{\partial c_m}{\partial x_j}\right) - v_i\frac{\partial c_m}{\partial x_i} = T_m\frac{\partial c_m}{\partial t} + \alpha(c_m - c_{im}) + \lambda T_m c_m + q(c_m - c^*) - f \quad (5)$$

The solution to equations (5) and (2), with suitable boundary and initial conditions, can be obtained by three different approaches, summarized below.

SOLUTION BY A COUPLED SCHEME

Equations (5) and (2) are integrated by finite elements in space and finite differences in time using the Galerkin formulation and a weighted time stepping scheme. The solutions c_m and c_{im} are expressed as linear combinations of N known basis functions $N_j(x_i)$ with time-dependent Galerkin coefficients $c_j^{(m)}(t)$ and $c_j^{(im)}(t)$: $c_m = \sum_{j=1}^N c_j^{(m)}(t)N_j(x_i); \; c_{im} = \sum_{j=1}^N c_j^{(im)}(t)N_j(x_i)$. Substituting these expressions into equations (5) and (2), we get the residuals $L_1(x_i, t)$ and $L_2(x_i, t)$ which are both orthogonalized against each basis function $N_j(x_i)$, $j = 1, \ldots, N$ over the flow domain Ω. A weak formulation is used by applying Green's first identity to the dispersive component of equation (5). Denoting by $c^{(m)}$ and $c^{(im)}$ the N-size vectors containing the Galerkin coefficients (which coincide with the nodal mobile and immobile concentration if, as is usually the case, piecewise basis functions are adopted), the following algebraic system is obtained:

$$
\left[\nu_1 (A + B + \widetilde{E} + F)_{k+\nu_1} + \frac{1}{\Delta t_k} \widetilde{G}_{k+\nu_1} \right] c_{k+1}^{(m)} - \nu_1 R c_{k+1}^{(im)} =
$$

$$
\left[\frac{1}{\Delta t_k} \widetilde{G}_{k+\nu_1} - \nu_{11} (A + B + \widetilde{E} + F)_{k+\nu_1} \right] c_k^{(m)} + \nu_{11} R c_k^{(im)} - r_{k+\nu_1}^* \tag{6a}
$$

$$
\left[\frac{1}{\Delta t_k} G^* + \nu_2 R^* \right] c_{k+1}^{(im)} - \nu_2 R c_{k+1}^{(m)} = \left[\frac{1}{\Delta t_k} G^* - \nu_{22} R^* \right] c_k^{(im)} + \nu_{22} R c_k^{(m)} \tag{6b}
$$

where ν_1 and ν_2 ($0 \leq \nu_1, \nu_2 \leq 1$) are weighting parameters; $\nu_{11} = 1 - \nu_1$; $\nu_{22} = 1 - \nu_2$; k indicates time level; $A, B,$ and $\widetilde{G}$ are the stiffness, convection, and capacity matrices, respectively; $\widetilde{E}, F, R,$ and R^* are capacity-type matrices arising from the c_m and c_{im} terms on the right hand sides of equations (5) and (2) and from the convective component of Cauchy boundary conditions; G^* is the capacity matrix for equation (2); r^* contains source/sink terms, Neumann boundary conditions, and the total solute flux across the Cauchy boundary. Equations (6a) and (6b) represent a nonsymmetric system of $2N$ coupled equations for the nodal mobile and immobile region concentrations. The system is solved by a conjugate gradient-like method specifically designed for nonsymmetric systems.

DECOUPLED SOLUTION: ALGEBRAIC SUBSTITUTION METHOD

Following *Leismann et al.* [1988], we first apply a weighted time difference approximation to the coupled system (5) and (2) to obtain

$$
\frac{\partial}{\partial x_i} \left(D_{ij}^{k+\nu_1} \frac{\partial}{\partial x_j} \left[\nu_1 c_m^{k+1} + \nu_{11} c_m^k \right] \right) - v_i^{k+\nu_1} \frac{\partial}{\partial x_i} \left[\nu_1 c_m^{k+1} + \nu_{11} c_m^k \right] =
$$

$$
T_m^{k+\nu_1} \frac{c_m^{k+1} - c_m^k}{\Delta t_k} + \alpha \left[\nu_1 c_m^{k+1} + \nu_{11} c_m^k \right] - \alpha \left[\nu_1 c_{im}^{k+1} + \nu_{11} c_{im}^k \right]
$$

$$
+ \left[\lambda T_m^{k+\nu_1} + q^{k+\nu_1} \right] \left[\nu_1 c_m^{k+1} + \nu_{11} c_m^k \right] - q^{k+\nu_1} c^{*k+\nu_1} - f^{k+\nu_1} \tag{7a}
$$

$$
T_{im} \frac{c_{im}^{k+1} - c_{im}^k}{\Delta t_k} = \alpha \left[\nu_2 c_m^{k+1} + \nu_{22} c_m^k \right] - (\alpha + \lambda T_{im}) \left[\nu_2 c_{im}^{k+1} + \nu_{22} c_{im}^k \right] \tag{7b}
$$

We solve equation (7b) for c_{im}^{k+1}:

$$
c_{im}^{k+1} = \frac{T_{im} c_{im}^k + \alpha \Delta t_k \left[\nu_2 c_m^{k+1} + \nu_{22} c_m^k \right] - \Delta t_k (\alpha + \lambda T_{im}) \nu_{22} c_{im}^k}{\nu_2 \Delta t_k (\alpha + \lambda T_{im}) + T_{im}} \tag{8}
$$

Substituting equation (8) into (7a) and applying the Galerkin approach we obtain a decoupled system in the N unknown mobile region concentrations $c_{k+1}^{(m)}$:

$$
\left[\nu_1 (A + B + \widetilde{E} + F)_{k+\nu_1} + \frac{1}{\Delta t_k} \widetilde{G}_{k+\nu_1} - \nu_2 E^* \right] c_{k+1}^{(m)} =
$$

$$
\left[\frac{1}{\Delta t_k} \widetilde{G}_{k+\nu_1} - \nu_{11} (A + B + \widetilde{E} + F)_{k+\nu_1} - \nu_{22} E^* \right] c_k^{(m)} + E^{**} c_k^{(im)} - r_{k+\nu_1}^* \tag{9}
$$

where E^* and E^{**} are capacity-type matrices involving the terms $\left(\nu_1\alpha^2\Delta t_k\right)/\delta$ and $\alpha\left[T_{im} + \Delta t_k\left(\alpha + \lambda T_{im}\right)\left(\nu_2 - \nu_1\right)\right]/\delta$, respectively, with $\delta = \nu_2\Delta t_k\left(\alpha + \lambda T_{im}\right) + T_{im}$.

DECOUPLED SOLUTION: INTEGRO-DIFFERENTIAL APPROACH

Equation (2) is a linear first order ordinary differential equation which may be integrated analytically. Setting $\beta = \alpha/T_{im}$ and assuming $c_{im} = 0$ at time $t = 0$, the solution to (2) is $c_{im} = \beta e^{-(\beta+\lambda)t}\int_0^t e^{(\beta+\lambda)\tau}c_m d\tau$. Substituting this solution into (5) leads to an integro-differential equation for the mobile region concentration:

$$\frac{\partial}{\partial x_i}\left(D_{ij}\frac{\partial c_m}{\partial x_j}\right) - v_i\frac{\partial c_m}{\partial x_i} = T_m\frac{\partial c_m}{\partial t} + (\alpha + \lambda T_m + q)c_m - (qc^* + f)$$

$$- \alpha\beta e^{-(\beta+\lambda)t}\int_0^t e^{(\beta+\lambda)\tau}c_m(\tau)d\tau \qquad (10)$$

Equation (10) is integrated in space by the Galerkin approach. The time derivative is approximated by a weighted ($0 \leq \nu \leq 1$) time stepping scheme. The convolution integral is performed either applying the mean integral theorem, as is done in *Gambolati et al.* [1993], or the trapezoidal rule. In the former case the convolution integral adds the following term to the left hand side coefficient matrix $\nu\left(A + B + \widetilde{E} + F\right)_{k+\nu} + \widetilde{G}_{k+\nu}/\Delta t_k$: $M_1 = -\nu\alpha\beta(1 - e^{-(\beta+\lambda)\Delta t_k})S/(2\beta + 2\lambda)$ where S is the basic capacity matrix with coefficient $s_{ij} = \int_\Omega N_iN_j dV$. In the case of trapezoidal rule integration, the additional term appearing in the left hand side matrix is $M_2 = -\nu\alpha\beta\Delta t_k S/2$.

Note that $M_1 \to M_2$ in the limit as $\Delta t_k \to 0$. Also note that, since $\widetilde{E}$ contains the term αS, the left hand side coefficient matrix differs from the classical (LEA) convection-dispersion-reaction equation by the introduction of the term $M = \nu\alpha S + M_1$ for the scheme based on the mean integral theorem, and $M = \nu\alpha S + M_2$ for the scheme based on the trapezoidal rule. In the limiting case when $\Delta t_k \to 0$, both schemes lead to $M_{\Delta t_k \to 0} = \nu\alpha S$. For $\Delta t_k < 2/\beta$, or for any Δt_k value when M is formed using M_1, the symmetric matrix M is positive definite and the magnitude of its nonzero coefficients, which are all positive, grows as Δt_k becomes smaller and M tends to its limiting value $\nu\alpha S$. Therefore M increases the importance of the diagonal coefficients of the dispersive-convective matrix, thereby accelerating convergence of the conjugate gradient-like solvers. In other words the integro-differential approach is particularly suited to enhance the performance of iterative solvers.

NUMERICAL TEST PROBLEM

The three different approaches for the solution of the nodal dual porosity model have been analyzed using a one-dimensional sample problem adapted from *van Genuchten and Wierenga* [1976] and involving the transport of a reactive solute through a semi-infinite porous column subjected to a pulse input. Boundary conditions are of Cauchy type at the inlet ($x_3 = 0$): $vc_m - D\partial c_m/\partial x_3 = vc_0$ for $T < 3$ and 0 for $T \geq 3$, where $T = vt/(nL)$ is the effluent pore volume and $L = 30$ cm is the effluent point in the column where we wish to evaluate c_m. For numerical purposes the finite length of the column was taken to be 200 cm. At $x_3 = 200$ the Dirichlet condition $c_m = 0$ is prescribed. The model results at $x_3 = L$ are analyzed and plotted against the relative concentration c_m/c_0. The 1-D case was simulated with a 2-D triangular grid, using a 3-node discretization in the direction orthogonal to v. The parameter values used for this preliminary series of simulations are: $v = 10$ cm/day, $n = 0.4$, $\gamma_s = 2.8889$ g/cm^3, $\lambda = 0$, $k_{d_m} = k_{d_{im}} = 0.5$ cm^3/g, $D = 30$ cm^2/day, $n_m/n = 0.65$, $F = 0.4$, $\Delta x_3 = 0.2$ cm, and $\nu_1 = \nu_2 = 1/2$. Note that with this set of values $\Delta t = 1.2\Delta T$.

BEHAVIOUR OF DUAL POROSITY MODEL FOR LARGE α VALUES

As was pointed out earlier the solution of the dual porosity model tends to the solution of the LEA transport equation (3) as $\alpha \rightarrow \infty$. To check whether the non-LEA solution is close to the LEA solution we calculate

$$\epsilon = \frac{1}{k_{tot}} \left[\sum_{k=1}^{k_{tot}} \frac{(c_m^k - c_{lea}^k)^2}{c_{max}^2} \right]^{1/2} \tag{11}$$

where k_{tot} is the total number of time steps in the simulation, c_m^k is the mobile non-LEA concentration at the fixed node located at $x_3 = L$, c_{lea}^k is the corresponding LEA concentration obtained from equation (3), and c_{max} is the maximum value of c_{lea} over all time steps.

Figure 1 shows how the three non-LEA numerical models approach the LEA solution as α increases, for various time integration steps Δt. Note in Figures 1a and 1b that ϵ progressively decreases as α becomes larger, independently of the magnitude of the time step. Hence both the coupled and algebraic substitution approaches capture the LEA solution at a sufficiently "large" α, value irrespective of Δt. By contrast, Figure 1c emphasizes that for the integro-differential approach, Δt must not be too large in order to capture the LEA solution at large α. In other words, there appears to be an upper bound on the product $\alpha \Delta t$ beyond which the integro-differential model does not yield a reliable limiting LEA solution. However, it should be observed that the ϵ value at which the integro-differential solution begins to deviate significantly from the LEA solution is smaller than 10^{-2}, even for the largest Δt values. Hence it may be acceptable to assume that the corresponding α value is large enough that the sorption process is practically an instantaneous one, and thus equation (3) can be used in lieu of the system (1)-(2).

CPU COMPARISON OF THREE INTEGRATION METHODS

In Figure 2 the CPU performance of the three dual porosity models is compared using different time step sizes (2a), and different grid resolutions (2b). The simulations were run on an IBM RISC System/6000 model 560 workstation. The coupled approach, which requires the assembly and solution of a linear system of $2N$ equations, is clearly the most expensive of the three solution approaches. The algebraic substitution model is slightly more efficient than the integro-differential model, both of which

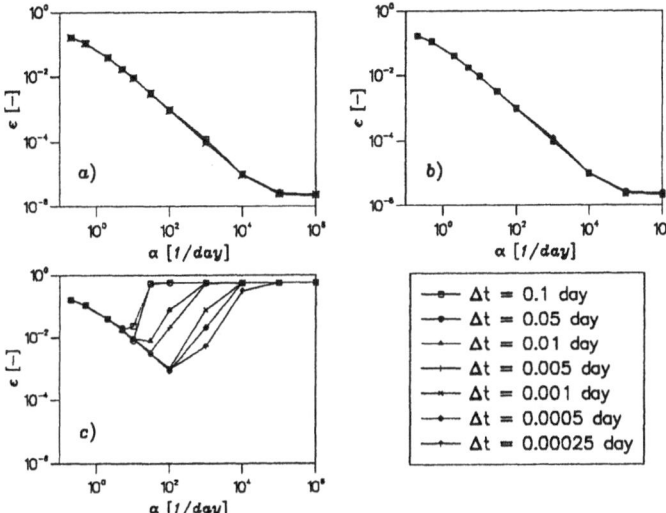

Figure 1. Difference between the solutions from the dual porosity and the LEA models using a) the coupled approach, b) the algebraic substitution approach, and c) the integro-differential approach.

produce linear systems of size N. For a given grid and time step size, the integro-differential and algebraic substitution models are approximately three times faster than the coupled model, and require about 2.5 times less computer storage.

CONCLUSIONS

We have presented three numerical methods for the integration in time of the dual porosity model describing the movement of a contaminant through a sorbing porous medium characterized by instantaneous sorption and by intra-aggregate diffusion controlled by a first order kinetic. Preliminary results show that all approaches yield the solution of the classical LEA transport equation when the mass transfer coefficient α is sufficiently large. However, the results obtained with the integro-differential method deviate from the correct LEA solution if the product between α and the time step Δt becomes too large. The CPU cost of the coupled integration approach is about three times larger than for the other two methods, and computer storage is about 2.5 times more. The algebraic substitution method appears to be quite attractive since it is as robust as the coupled approach and as computationally efficient as the integro-differential method. Further research is under way on the numerical accuracy and stability of the three methods.

Acknowledgments. This work has been supported by the Italian CNR, Gruppo Nazionale per la Difesa dalle Catastrofi Idrogeologiche, linea di Ricerca n. 4, by Fondi Ministeriali 40%, and by the Sardinia Regional Authorities.

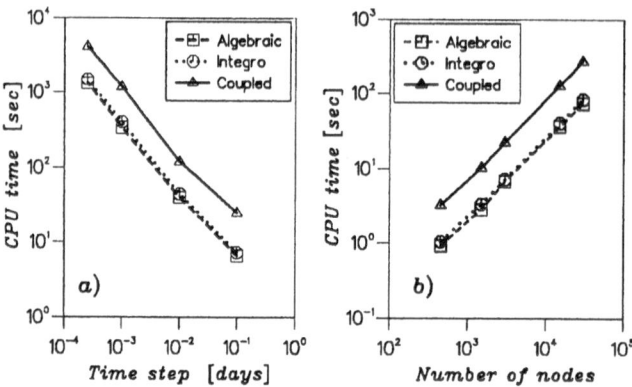

Figure 2. CPU comparison of the three non-LEA models as a function of a) time step size and b) problem size.

REFERENCES

Coats, K. H. and B. D. Smith, Dead-end pore volumes and dispersion in porous media, *Soc. Pet. Eng. J.* 4, 73–84, 1964.

Gambolati, G., C. Paniconi and M. Putti, Mass transfer analysis in sorbing porous media by an integro-differential approach. In: Wang, S. S. Y. (ed.) *Advances in Hydro-Science and -Engineering, Volume I, Part B.* The University of Mississippi, University, MS, pp 1819–1828, 1993.

Gambolati, G., G. Pini, M. Putti and C. Paniconi, Finite element modeling of the transport of reactive contaminants in variably saturated soils with LEA and non-LEA sorption. In: Zannetti, P. (ed.) *Environmental Modelling II.* Computational Mechanics Publications, Southampton, UK, 1994.

Giddings, J. C., Kinetic origin of tailing in chromatography, *Anal. Chem.* 35, 1999–2002, 1963.

Leismann, H. M., B. Herrling and V. Krenn, A quick algorithm for the dead-end pore concept for modeling large-scale propagation processes in groundwater. In: Celia, M. A., L. Ferrand, C. A. Brebbia, W. G. Gray and G. F. Pinder (eds.) *Proc. VII International Conference on Computational Methods in Water Resources.* CMP Elsevier, Amsterdam, pp 275–280, 1988.

Valocchi, A. J., Validity of the local equilibrium assumption for modeling sorbing solute transport through homogeneous soils, *Water Resour. Res.* 21(6), 808–820, 1985.

van Genuchten, M. T. and P. J. Wierenga, Mass transfer studies in sorbing porous media: 1. Analytical solutions, *Soil Sci. Soc. Amer. J.* 40(4), 473–480, 1976.

MATHEMATICAL MODELLING OF SOLUTE TRANSPORT AFFECTED BY SUBSURFACE GEOCHEMICAL REACTIONS.

KHALIL GHABAEE[1] , DAVID M. BURLEY, AND CHARLES D. CURTIS[2]
The University of Sheffield, School of Mathematics and Statistics,
P O Box 597, Sheffield, S3 7RH
England.

ABSTRACT

Development of transport models that incorporate the interactions between migrating pore fluids and permeable media, is fundamental to an understanding of processes such as ore deposition, the chemical diagenesis of the sediment and the migration of toxic chemical and radioactive wastes. A numerical model has been developed, which simulates transient two-dimensional coupled flow and chemical reaction within the porous medium of any pre-defined geometry and initial porosity and permeability distribution. The numerical model developed exploits advantageous features of characteristic and moving point methods to reduce the 'numerical diffusion' caused by the application of conventional numerical methods to the solution of advection dominated chemical kinetic equations. Porosity modification due to geochemical reaction is coupled with permeability, using various derived correlations, which are believed to capture the essential features of the real systems. The developed geochemical flow models are applied to a number of simulation experiments. The initial values of system parameters such as reaction rate parameters, diffusion coefficient, spatial porosity and permeability distribution and imposed flow rate are chosen so that they predict the chemical diagenesis pattern consistent, at least qualitatively, with those observed in the nature, but in a reasonable computer time.

INTRODUCTION

There is substantial theoretical, experimental, and petrological evidence that pore water migrating through sediments within evolving basin systems, interact with the sediment and modify their porosity and permeability accordingly (Hays, 1979; Berner, 1980; Parker and Sellwood, 1981; Ortoleva, et al, 1987).
Net dissolution of a mineral in pore water will generally depend on the degree of undersaturation, the rate of detachment of ions or molecules from the mineral surface and the rate of transport of these away from the site - this in turn being a function of diffusion rate and fluid migration rate. If any subsurface dissolution reactions are transport controlled (rate limited by the transport of solutes away from the site of dissolution rather than by surface reaction control), both porosity and permeability must change in a systematic way. There will be positive feedback and region of high permeability will be further enhanced. We have set out to develop a model, which allows the consequences of such feedback to be

---

[1] Now at AEA Technology, Consultancy Services, Dorset, England.
[2] Now at Department of Geology, University of Manchester, Manchester, England

A. Peters et al. (eds.), Computational Methods in Water Resources X, 629–636.
© 1994 *Kluwer Academic Publishers. Printed in the Netherlands.*

explored and, potentially at least, to predict the spatial style of permeability contrast enhancement. The present model is kinetic rather than based on local equilibrium assumptions, which only hold if reaction is reversible and fast in comparison with flow rate (Robin, 1983; Bahr and Robin, 1987). It uses the advantageous features of finite difference, characteristic and moving point methods to reduce the numerical diffusion associated with the solution of advection dominated advection-diffusion-reaction equations.

MATHEMATICAL FORMULATION

The conservation of mass of water flowing through a non-deformable porous medium in a Cartesian coordinates system is expressed as (Bear and Beachmat, 1991):

$$-\nabla.(\rho_w \phi \bar{u}) = \frac{\partial(\rho_w \phi)}{\partial t} \tag{1}$$

where $\bar{u}$ is the superficial velocity vector, ρ_w is the density of water and ϕ is the effective porosity of medium. For incompressible water (constant water density), Equation (1) becomes:

$$-\nabla.(\phi \bar{u}) = \frac{\partial \phi}{\partial t} \tag{2}$$

If the flow of fluid in porous media of interest obeys Darcy's law, the superficial velocity, $\bar{u}$ is given by:

$$\bar{u} = -\frac{k}{\mu}\nabla p \tag{3}$$

where k is the permeability of medium, p is the pressure and μ the viscosity of the fluid, assumed constant. Substitution of Equation (3) into Equation (2), gives the pressure equation:

$$\nabla.\left(-\phi \frac{k}{\mu}\nabla p\right) = \frac{\partial \phi}{\partial t} \tag{4}$$

Let c be the concentration of solute per unit volume of solution, then the conservation of mass for the solute transport can be expressed as:

$$\frac{\partial(c\phi)}{\partial t} = \nabla.(c\phi \bar{u}) + \nabla.(D\phi \nabla c) + \left[\frac{\partial(c\phi)}{\partial t}\right]_{chem} \tag{5}$$

where D is the mass diffusivity coefficient. The first and second term on the right hand side of Equation (5) are the solute flux due to advection and diffusion, respectively. The last term describes the effect of chemical reactions between minerals and water; the transport limited reactions for dissolution or precipitation are usually expressed as a function of departure from equilibrium (Lerman, 1979; Berner, 1980):

$$\left[\frac{\partial(c\phi)}{\partial t}\right]_{dissolution} = k_r A\phi(c_{eq} - c) \qquad\qquad c < c_{eq} \tag{6}$$

$$\left[\frac{\partial(c\phi)}{\partial t}\right]_{precipitation} = k_r A\phi(c_{eq} - c) \qquad\qquad c > c_{eq} \tag{7}$$

where A is the reactive solid surface area per unit volume, k_r is the rate parameter of dimensions of velocity, and c_{eq} is the equilibrium concentration.

The rate of change of porosity can be related in the absence of compaction to the rate equation by the following equation:

$$\left[\frac{\partial(c\phi)}{\partial t}\right]_{chem} = \left[\frac{\partial c_s}{\partial t}\right]_{chem} = \rho_s \frac{\partial \phi}{\partial t} \tag{8}$$

where $c_s = (1 - \phi)\rho_s$ is the concentration of solid per unit volume of total sediment and ρ_s is the average density of solid, assumed constant. From Equation (6 or 7), Equation (8) becomes:

$$\frac{\partial \phi}{\partial t} = \frac{\phi}{\rho_s} k_r A(c_{eq} - c) \qquad (9)$$

In view of Equations (2,8), solute transport equation (Equation 5), after some arrangement, becomes:

$$\phi \frac{\partial c}{\partial t} = \phi \overline{u} \nabla . c + \nabla .(D\phi\nabla c) + k_r A\phi(c_{eq} - c) \qquad (10)$$

Surface area and permeability modification

As the moving fluid reacts with the solid, the surface area of minerals is modified and this in turn affects the overall geochemical reaction (Lerman, 1979). There are several possible ways to be account for this. Two are considered here:
a)   Precipitation of cement in sandstone or dissolution of uniformly cemented sandstone or carbonate rock. In the latter case, dissolution occurs at all surfaces. For this situation, the "spherical-grain onion skin" model has been utilised, which for an array of rhombohedrally packed, single-size spheres with different coatings of cement provides a relationship between surface area and porosity as:

$$A = \sigma\phi^{2/3}d^2 \qquad (11)$$

where σ is a proportionality constant (= 3.86), and d is the initial grain diameter.
b)   Dissolution in non-uniform composition framework- supported sandstone. In this case the integrity of the medium is maintained by inert framework grains whilst reactive grains dissolve. Then, supposing that the numbers of grains per unit volume of rock for reactive and non-reactive minerals are N_a and N_b respectively, we have:

$$A = \sigma N_a^{1/3}(\phi_f - \phi)^{2/3} \qquad (12)$$

where ϕ_f is porosity of rock without the reactive mineral grains and σ is proportionality constant (= 4.84).
Corresponding to the change of porosity is a change in permeability. But there is no single, simple correlation between these two variables. Numerous measurements from natural sandstones document its general nature (Parker and Sellwood, 1981) and this can be reasonably approximated for the case of uniformly cemented rock, case (a) above, using "spherical-grain onion skin model", as:

$$k = \beta d^2 \phi^{5/3}(\phi - 0.127)^{10/3} \qquad (13)$$

where β is a constant parameter, for permeability in millidarcy (md), d, the initial grain diameter (mm) and porosity in fraction, is (= 2.9 10^9). The factor (0.127) is introduced in Equation (13) because when cement is just thick enough to close off a pore throat, the permeability becomes zero as porosity reduces to 12.7%.
For the case of non-uniform framework composition, the relationship between porosity and permeability is approximated by the Fair-Hatch relationship:

$$k = \beta\phi^5 \left\{ (\phi_f - \phi)^{2/5} N_a^{1/5} + (1 - \phi_f)^{2/5} N_b^{1/5} \right\}^{-2} \qquad (14)$$

where β is a constant parameter, for permeability in millidarcy (md), N_a and N_b in cm^{-3}, and porosity in fraction, has a value of 8.55 10^8.

NUMERICAL SIMULATION

The space of a two-dimensional domain Ω representing the porous medium domain of unit thickness is partitioned into a network of (n x m) identical control elements $\Omega_{ij}, i = 1, n$; j=1,m). The surrounding boundaries of Ω are assumed impermeable except at an injection point, where the fluid is continuously injected at constant pressure $p = p_i$, and outflow point, where pressure $p = 0$ is assumed at all times. The porous medium of domain is assumed initially to be completely saturated with a fluid containing a solute in equilibrium with the reactive mineral. Since departure from equilibrium is the necessary condition for reaction, the porous medium is then injected with a fluid undersaturated or supersaturated with respect to the reactive mineral. In solving the pressure Equations (4), the finite difference analogue of the equation is derived and solved implicitly for pressure and velocity distributions as the porosity and permeability of the medium are modified due to geochemical reactions, using Equations (9, 13 and 14) For the solute transport equation (Equation 10), a numerical method is developed that incorporates an improve version of the characteristic and moving point method.

In order to incorporate the heterogeneity with respect to permeability and porosity, the discretized network element (n x m), is divided into n layers. From the range of porosity values commonly observed in petroleum or water reservoirs, a porosity value ϕ_i is assigned to each of these layers. The spatial variability of porosity is obtained by a single dimensionless scale factor, α_{ij}. Using a normal distribution with a small standard deviation, (n x m) values of α_{ij} corresponding to the (n x m) elements are generated. These values of scale factors are used to calculate porosity values at the centre of elements from the allocated porosity of layers by $\phi_{ij} = \alpha_{ij}\phi_i$. The permeability of each element is then calculated using the chosen permeability and porosity relationship (Equation 13 or 14).

Characteristic method

The principle motivation for using the characteristic as compared to the finite difference method is that the former approximates the physics of the problem more closely and therefore it is expected to lead to a gain in accuracy (Douglas and Russel, 1982). Considering the transport Equation (10) in two-dimensional form, we have:

$$\phi\frac{\partial c}{\partial t} + u\frac{\partial c}{\partial x} + v\frac{\partial c}{\partial y} = \nabla.(D\phi\nabla c) + k_r A\phi(c_{eq} - c) \qquad (15)$$

Let the vector function $\overline{\psi}$ represents the coefficients u, v and ϕ of transport Equation (15) in the rectangular Cartesian coordinates. The unit vector in the direction of $\overline{\psi}$ is given by:

$$\overline{\eta} = \left[\frac{\phi}{|\psi|}, \frac{u}{|\psi|}, \frac{v}{|\psi|}\right] , \qquad |\psi| = \sqrt{(u^2 + v^2 + \phi^2)} \qquad (16)$$

Then the characteristic direction associated with operator $\phi\frac{\partial c}{\partial t} + u\frac{\partial c}{\partial x} + v\frac{\partial c}{\partial y}$ is given by:

$$|\psi|\frac{\partial c}{\partial \overline{\eta}} = |\psi|\left\{\overline{\eta}.\left[\frac{\phi}{|\psi|}, \frac{u}{|\psi|}, \frac{v}{|\psi|}\right]\right\} = \nabla.(D\phi\nabla c) + k_r A\phi(c_{eq} - c) \qquad (17)$$

To evaluate the spatial derivatives following the fluid stream, we take (x⁻,y⁻) as the position of characteristic curve at time t and (x, y) at time t + Δt.. Then x⁻ and y⁻ are approximated by: $$x^- = x - \frac{u\Delta t}{\phi} \quad \text{and} \quad y^- = y - \frac{v\Delta t}{\phi}$$ (18)

Substituting the Equation (18) into Equation (17), we obtain:

$$|\psi| \frac{\partial c}{\partial \eta} = |\psi| \frac{c^{r+1}(x,y) - c^r(x^-, y^-)}{\sqrt{\{(x-x^-)^2 + (y-y^-)^2 + \Delta t^2\}}}$$ (19)

$$= \phi \frac{c^{r+1}(x,y) - c^r(x^-, y^-)}{\Delta t} = \nabla.(D\phi\nabla c) + k_r A\phi(c_{eq} - c)$$

In Equation (19), $c^{r+1}(x,y)$ is the concentration at time step $r+1$ at the fixed mesh point (x,y). The value $c^r(x^-, y^-)$ is the concentration at time step r, not at the point (x,y) but at the point (x^-, y^-) on the characteristic curve. By advection the point (x^-, y^-) will move to the point (x^-, y^-) over the time interval allotted and this will carry with it the concentration at the (x^-, y^-). Therefore, we first evaluate $c^r(x^-, y^-)$ neglecting diffusion and chemical reaction and then evaluate the right hand side of Equation (19) over the time interval Δt.

Moving points algorithm

In the method of characteristics described above, the computational mesh is a fixed mesh. If this is used without a proper choice of values of step size (Δx, Δy) and time interval (Δt), it may lead to smearing of the concentration front due to numerical diffusion. To alleviate this, the moving point method based on the method of characteristics is used. The moving point method developed by Farmer and Norman (1986) appears to be too complicated for practical use. The present numerical scheme is based on the idea put forward by Pietlicki and Archer (1987), although it differs conceptually from it. Suppose for every fixed mesh grid (i,j), there is a moving counter part (I,J). Let (x,y) and (x^-, y^-) represent the position of fixed and moving points respectively, in a two-dimensional co-ordinate system. At the start of movement the moving point (I,J) coincides with its counterpart fixed mesh point (i,j), i.e. $(x^-, y^-) = (x,y)$. The moving point is then allowed to move with the characteristic velocity, $\bar{u}$ in the upstream direction. The position of the moving point is determined by $M > 1$ applications of the Equation (18) with a time step of $\Delta \tau = \Delta t/M$. The subdivision of time step into M sub-time steps increases the accuracy, as compared to a using single time step, when the velocity and porosity vary significantly over the mesh spacing. The values of velocity, porosity and concentration at the moving point, are determined by a piecewise two-dimensional linear interpolation from the fixed mesh values using three upwind points. The procedure of updating the positions, velocities, porosities and concentrations at the moving points with time, is continued until:

$$F = \frac{\Delta x_{ij}}{\Delta x} + \frac{\Delta y_{ij}}{\Delta y} = 1 - \varepsilon \qquad \varepsilon \leq 0.01$$ (21)

where $\Delta x_{I,J} = |x^- - x|$ and $\Delta y_{I,J} = |y^- - y|$ are the distances of moving point (I,J) from the fixed mesh point (i,j) and Δx and Δy are the step lengths in the fixed mesh. If the criterion is satisfied for a point then the algorithm is completed for this point, and the grid moving point is returned to its original fixed mesh position. The nodal concentration value, which remains frozen at the fixed mesh concentration until F =1-ε, is updated to the moving concentration $c^r(x^-, y^-)$. The criterion (F =1-ϵ) is used to avoid an excessive movement due to advection. This is found to capture the concentration front in a more satisfactory manner. The exact values of ϵ and M depends on the extent of reduction in numerical diffusion.

Piecewise-linear interpolation used above in determining the moving point concentration is proved to be accurate for the case of strong domination of advection over diffusion. When advection is dominated by diffusion or roughly equivalent to it, the moving point concentration must be calculated using quadratic interpolation from the fixed mesh concentrations, as opposed to linear interpolation. Once the advection effect on the solute concentration has been dealt with, the values of $c^{r+1}(x, y)$ due to reaction and diffusion are then determined by Equation (19) using the central difference approximation of fixed mesh concentrations.

RESULTS AND DISCUSSION

The developed geochemical flow models are applied to a number of simulation experiments. The initial values of system parameters such as reaction rate parameters, diffusion coefficient, spatial porosity and permeability distribution and imposed flow rate are chosen so that they predict the chemical diagenesis pattern consistent, at least qualitatively, with those observed in the nature, but in a reasonable computer time. Simulated cases that are presented in this paper to show its capabilities and robustness are as follow:

Case 1: Solute transport with no mineral reactions and zero mixing.
 This situation with no diffusion and reaction is used to test the ability of model to reduce the numerical dispersion to a minimum. In the first instance the fluids can be thought of as either immiscible or miscible but with zero diffusive mixing across solute concentration discontinuities. For this problem, where a fluid of one composition is displaced by one of a different composition, the numerical model should transport the injected fluid without any smearing of concentration profile due to false diffusion, and any fingering of the displacing fluid should have a sharp front. These properties, which are depicted by Figure 1, are captured by the present model. Figure 1 illustrates a time series of solute concentration within the model system. The more dilute solution fingers through the medium, as would be expected from the inlet/outlet geometry. With time, all the higher concentration fluid would be displaced through the "edges" of the outlet.

Case 2: Diffusion and advection with reaction between fluid and porous medium.
This simulation experiment represents the combined effect of advection, diffusion and interaction between fluid and mineral on the solute concentration through a porous medium. This is seen in Figure 2 for the case of a single phase solid, which dissolves, when the initial fluid, which is saturated with respect to the solid ($c/c_{eq} = 1$), is displaced by an undersaturated one ($c/c_{eq} = 0.5$). This figure illustrates that as the injected undersaturated fluid flow imposes disequilibrium in the porous medium, it reacts with minerals and establishes a moving dissolution front. Because of the initial larger porosity in some regions, the local flux is faster there, as permeability is an increasing function of porosity. An increase in the flow of undersaturated fluid means an increase in the local porosity and permeability due to dissolution, leading to the local focusing of the flow. Focus of

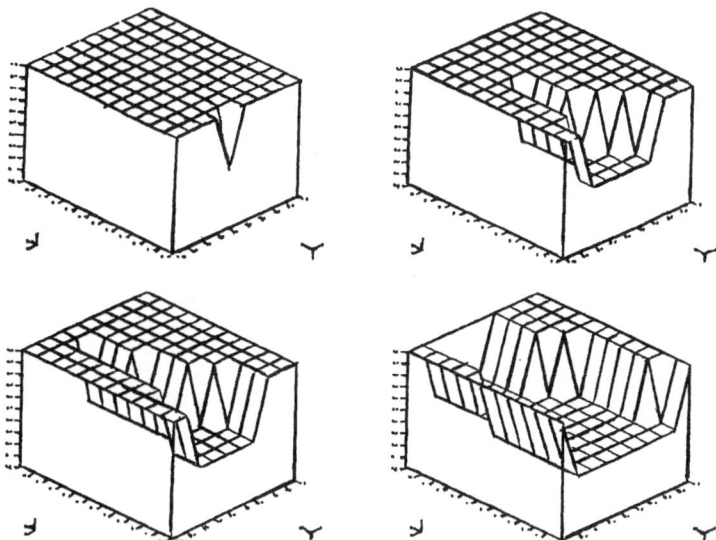

Figure 1. Simulated solute distribution - zero diffusive mixing (vertical axis solute concentration, fluid entry centre right): at times t = 0, t =20 hrs, t = 29 hrs, and t =70 hr

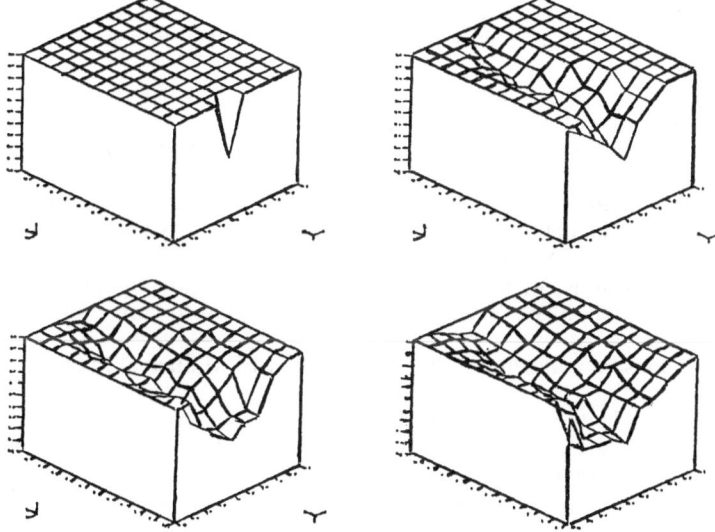

Figure 2. Simulated solute distribution - mineral dissolution with advection and diffusive mixing (vertical axis solute concentration, fluid entry centre right): at times t = 0, t =20 hrs, t = 29 hrs, and t =70 hr

undersaturated solute bearing reactive water to higher permeability zone, leaves other regions, specially the dissolution front, unaffected. Channel development is obvious, although the maximum porosity reached is still only 30%. This together with the fact that dissolution increases the solute concentration, reduces the channelling speed through the porous medium, leading to a shorter length finger and a delay in the breakthrough. This effect will persist until, in the simplest case, all the solid is removed sequentially from unit volumes adjacent to the fluid inlet.

CONCLUSIONS

A true two-dimensional mathematical model of fluid flow has been developed, which accommodates chemical reaction between fluid and porous medium by a first order kinetic model relating dissolution or precipitation to surface area and degree of undersaturation or supersaturation. Subsequent porosity and permeability modification are investigated. The numerical model developed incorporates an improved version of characteristic and moving point method, which reduces the numerical diffusion caused by the application of conventional numerical methods to the solution of advection dominated chemical kinetic equations. Using a three-dimensional graphic output format, the temporal changes in the solute distribution due to mineral reactions are monitored.

ACKNOWLEDGEMENTS

This work was performed as part of the Fluid Processes Special Topic Programme sponsored by the Natural Environment Research Council.

REFERENCES

Bahr, J.M., and Robin, J., 1987, Direct comparison of kinetics and local equilibrium formulation of solute transport effected by surface reaction, Water Resour. Res., 23(2), 438-452.
Bear, J. and Bachmat, Y., (1991), Introduction to modelling of transport phenomena in porous media, Kluwer Acadamic Publisher, The netherlands.
Berner, R.A., 1980, Early diagenesis : a theoretical approach, Princeton University Press, Princeton, N.J.
Douglas, Jr. J., and Russell, T.F., 1982, Numerical methods for convection dominated diffusion problems based on combining the method of characteristics with finite element or finite difference procedures, SIAM J. Numer. Anal. 19(5), 871-885 .
Farmer, C.L., and Norman, R.A., 1986, The implementation of moving point methods for fluid dynamics II, (Morton, K.W., and Bains, M.J., Eds.), Oxford University Press, 635-644 .
Hayes,J.B., 1979, Sandstone diagenesis - the hole truth., SEPM Special Publication, 26, 127-139 .
Lerman, A., 1979, Geochemical processes., Wiley Interscience, New York.
Pietlicki, R., and Archer, J.S., 1987, Novel scheme for control of numerical diffusion in reservoir simulation., Chem. Eng. Res. Des., 65, 107-112
Parker, A., and Sellwood, 1981, Sediment Diagenesis, D. Reidel Publishing Company, Holland.
Ortoleva, P., Chadam, J., Merino, E., and Sen, A., 1987, Geochemical Self-Organisation II: The Reactive-Infiltration Instability, Am. Jour. Sci, V 287, 1008-1040.
Robin, J., 1983, Transport of reacting solute in porous media : relation between mathematical nature of problem formulation and chemical nature of reaction, Water Resour. Res., 19(5), 1231-1252 .

REACTIVE TRANSPORT IN HETEROGENEOUS MEDIA

W. K. H. KINZELBACH and W. SCHÄFER
Institute of Environmental Physics
Heidelberg University
Im Neuenheimer Feld 366, D-69120 Heidelberg
Germany

In order to be able to extrapolate reactive pollutant transport behaviour from the laboratory experiment to the aquifer field scale it is necessary to combine the deterministic lab results with the stochastics of the field situation into effective macroscopic properties. Effective time scales for chemical reactions such as adsorption and degradation are no longer determined by the time scales of processes on the molecular scale only but also by rate limiting steps arising from the variable accessibility of reactive parts of the medium. For a bacterial degradation reaction it is suggested that an exchange coefficient between mobile water and reactive phase can parametrize subscale processes. The coefficient depends on the degree of heterogeneity of the medium.

INTRODUCTION

One of the major difficulties in describing and predicting reactive transport processes in aquifers is the usually unknown heterogeneity of natural sediments both with respect to hydraulic conductivity and chemical properties. In order to be able to extrapolate from the laboratory experiment to the field scale it is necessary to combine the deterministic lab results with the stochastics of the field situation into effective macroscopic properties. An observation which was made many times is that properties of soils show scaling behaviour. This is not only true for dispersion coefficients (for a review see Gelhar et al., 1992) which parametrize the subscale variability of hydraulic conductivity but also for bulk chemical properties such as cation exchange capacity, denitrification potential or organic carbon content (e.g. Murphy et al., 1992). The observation can at least partially be understood by noting that the availability of some chemical capacity depends on the accessibility to water flow of the site where it resides.

A second observation which relates to the theme is the long duration of cleanup of contaminated aquifers by biological or chemical in situ remediation methods. It is due to the fact that injected solutes do not reach all sites of residual pollutants equally well. While easily accessible channels may be cleaned several times over, other regions of the medium have not participated to any appreciable degree yet. Thus the overall rate of reaction is usually much slower than the one observed in the batch reactor.

The theoretical treatment of flow and tracer transport in heterogeneous aquifers goes

637

A. Peters et al. (eds.), Computational Methods in Water Resources X, 637–648.
© 1994 Kluwer Academic Publishers. Printed in the Netherlands.

back to the work of Gelhar and Axness (1983), Dagan (1984) and Matheron and de Marsily (1980). The major results of these investigations are that heterogeneity in hydraulic conductivity shows in solute transport in the form of dispersion and the size of the dispersion coefficient can be obtained from the statistical properties of the conductivity distribution. The latter procedure, however, makes only sense if asymptotic behaviour is reached, that means that the length scale of the transport process is large compared to the correlation length of the hydraulic conductivity distribution of the medium.

SIMPLE EXAMPLES

Mechanisms of forming effective chemical properties can be illustrated by simple examples. Here a one dimensional medium with constant hydraulic conductivity and stochastically varying adsorption coefficient is considered. The adsorption is assumed fast with a linear adsorption isotherm. The transport process is described by the equation

$$R\frac{\partial c}{\partial t} + \vec{\nabla}(\vec{u}c) = \vec{\nabla}(D\vec{\nabla}c) \tag{1}$$

where the retardation factor R is a random variable in space. Let us neglect dispersive transport for a moment and let us assume that the medium is built of N blocks of equal length L which have retardation factor $R = R_o > 1$ with a probability p or retardation factor $R = 1$ with a probability (1-p) (fig. 1).

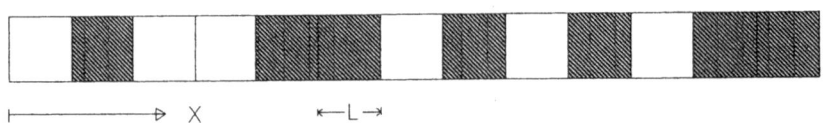

Fig. 1: Binary stochastic medium

A pulse traversing N-k unretarded and k retarded blocks on his way through the N block medium takes the time:

$$t = \frac{L(N-k)}{u} + \frac{kLR}{u} \tag{2}$$

There are 2^N possible realizations produced by this binary process. The probability distribution of arrival times of a pollutant pulse entering any realization of the medium at x=0 and time t=0 is

$$P(t,x) = \sum_{k=0}^{N} C_N^k p^k (1-p)^{N-k} \delta(t - \frac{L(N-k)}{u} - \frac{kLR}{u})$$ (3)

The arrival time distribution can be simplified with a method from Bouchaud and Georges (1988). This leads to the expectation value of the transport velocity

$$u_{eff} = \frac{u}{\overline{R}}$$ (4)

$$\text{with} \quad \overline{R} = pR_0 + (1-p)1$$

and a dispersion coefficient

$$D_{eff} = \frac{L}{2}p(1-p)\frac{L(R_0-1)^2}{\overline{R}^2}\frac{u}{\overline{R}}$$ (5)

From (4) it is seen that the average retardation factor is the arithmetic mean of the distribution. As in any single realization there was no diffusion nor dispersion this coefficient expresses the uncertainty of pollutant arrival. Interestingly the width of the distribution grows with the square root of time as in a normal diffusion process. The dispersion coefficient is, however, not a physical property of the single realization but rather a measure of the lack of information on a concrete realization. It may still be of interest, as in practical applications an ensemble average can also be a useful concept. The input of pesticides into groundwater under an agricultural area, for example, can be thought of as an ensemble mean of fluxes through a large number of vertical soil columns extending from groundsurface to the groundwater table.

Real physical diffusion/dispersion and its modification by heterogeneity of the retardation factor is investigated into by Chrisikopoulos et al. (1992) looking at a single medium with a periodically varying retardation coefficient. For a sinusoidal distribution they find that the diffusion coefficient is

$$D_{eff} = \frac{D}{\overline{R}}(1 + w \frac{(\frac{uL}{2\pi D})^2}{1 + (\frac{uL}{2\pi D})^2})$$ (6)

$$\text{with} \quad w = \frac{\overline{R^2} - \overline{R}^2}{\overline{R}^2}$$

while the effective retardation factor is the arithmetic average of the distribution. Metzger et al. (1994) show that starting out from the full transport equation (1) a dispersion coefficient can be derived through a straightforward perturbation theory approach which is essentially the sum of (5) and (6).

$$D_{eff} = \frac{D}{R}(1+w(1 - e^{-\frac{uL}{2D}}))+\frac{L}{2}w\frac{u}{R} \qquad (7)$$

It is clear that if the first term is smaller than the second one no ergodic dispersion is reached. In this case a physically more meaningful property of the medium itself is only obtained if a priori knowledge of the medium is introduced for conditioning of the ensemble over which the average is taken. If for example the average retardation coefficient of the medium is known by taking enough samples, term (4) must go to zero. Rajaram and Gelhar (1994) make the point that rather than averaging over the ensemble first and then taking the second moment the inverse order is more appropriate to arrive at the physically relevant effective dispersion coefficient.

For a one-dimensional medium with variable degradation rate the effective rate is obtained in a quite straightforward manner. Again a medium of the type of fig. 1 is considered, where the distributed property now is a first order reaction rate λ which varies statistically between values $\lambda = \lambda_o$ (probability p) and $\lambda = 0$ (probability 1-p) in blocks of length L. As the flow field is not influenced by the degradation rate, both the first and the second moment of the concentration distribution remain unaffected by the distribution of the degradation coefficient.

In purely advective transport the effective degradation rate is then given by

$$\lambda_{eff} = \lambda_0 p \qquad (8)$$

In order to observe this rate the asymptotic average situation must be reached before concentrations have dropped to zero i.e. the residence time in a block must be short compared to the time scale of degradation. In two dimensions the situation may be different if hydraulic conductivity varies statistically in space. Then the portion of the degradation capacity "seen" by the solute depends on the correlation between conductivity and degradation rate. This effect is discussed in a more realistic example.

MORE REALISTIC EXAMPLE

A numerical model of reactive transport is applied to a hypothetic case of in situ bioremediation. The model domain is a so-called "quarter-five-spot" 20 m x 20 m in size with an injection well in the upper left hand corner and a pumping well in the lower right hand corner. It is surrounded by impermeable boundaries. The organic contaminant is distributed between pore water and aquifer material according to a linear isotherm:

$$AOC = k_d \cdot DOC \qquad (9)$$

where AOC and DOC the adsorbed and dissolved organic carbon concentrations of the contaminant and k_d is the distribution coefficient.

The remediation consists of an injection of dissolved oxygen (DO). The microorganisms use the injected DO as electron acceptor and oxidize the electron donor DOC to carbon dioxide. DOC is also used to build up cell material. Most of the aquifer clean-up is performed through microbial oxidation, but a minor part of the contaminant is also extracted by the pumping well.

The model for simulating this process in two spatial dimensions considers the advective and dispersive transport of DO, DOC and a non-reactive tracer. The heterotrophic microorganisms are immobile and reside in a biophase. The degradation happens exclusively in the biophase, i. e. DO and DOC have to enter the biophase from the pore water prior to microbial consumption. The exchange between pore water and biophase may be either instantaneous or limited by a linear exchange term.

Microbial growth is simulated with the help of Monod-kinetics. The consumption of DOC and DO is coupled to microbial growth via yield coefficients and stoichiometric relations. This model is a simplified version of a denitrification model, which is presented in more details in Kinzelbach et al. (1991). In Schäfer and Kinzelbach (1992) the above mentioned model configuration was used to study the effects of heterogeneity on remediation time. Here we will focus on the effect of aquifer heterogeneity on the effective kinetics of clean-up as manifested by the breakthrough curve (BTC) of DO at the pumping well. This is done by comparing the behaviour of the heterogeneous medium to the behaviour of a homogeneous block, starting with the homogeneous case.

Homogeneous case

The model parameters for the homogenous case with instantaneous exchange are displayed in table 1.

Parameter	Value	Unit	Parameter	Value	Unit
hydraulic conductivity	43	m/d	effective porosity	0.15	-
distribution coefficient	0.8	l/kg	recharge/discharge	86.4	m^3/d
longitudinal dispersivity	0.25	m	DO concentration of recharge	50	mg/l
transverse dispersivity	0.05	m	spacing in x- and y-direction	0.5	m
aquifer thickness	30	m	time step	0.1	d

Table 1: Model aquifer parameters for the homogeneous case

The tracer has the same properties (e. g. injection rate, maximum concentration etc.) as DO but does not undergo any microbial reduction.

The BTCs at the pumping well for the non-reactive tracer and for DO are shown in fig.

2. After about 10 days the breakthrough of the tracer begins. It reaches the maximum (i.e. concentration at the injection well) after about 40 days. DO breakthrough begins later, because it can only arrive at the pumping well when the available DOC between injection and pumping well is used up. The area between the two curves is a measure for the amount of contaminant available for microbial oxidation.

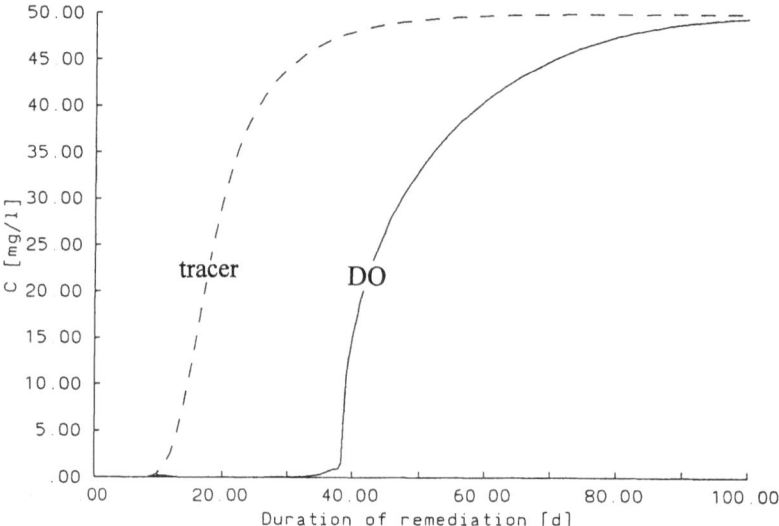

Fig. 2: BTCs for the non-reactive tracer and for DO in the pumping well for the homogeneous case with instantaneous exchange

The BTC for DO is not only delayed in comparison to the tracer, but also exhibits a different shape. It is initially steeper than the tracer BTC. The microorganisms develop in the transition zone where they have DO from injection and DOC from desorption at their disposal. The dispersive precursors, which are visible in the tracer movement, are reduced for DO by microbial activity which leads to DO-front sharpening. Later, the slope of the DO breakthrough becomes less steep. This effect is the result of the geometry of the model domain and the particular flow field that develops. The two opposing wells lead to increased velocities and DO supply along the diagonal with a less efficient supply along the boundaries where the contaminant is used up at a slower rate and represents a small but long lasting sink for DO. Thus the time to complete DO breakthrough is prolonged.

Heterogeneous cases

The two aquifer parameters thought to be the most important for this model setup were chosen as random variables, namely the hydraulic conductivity k_f and the distribution coefficient k_d. For both variables a log-normal distribution was assumed. The inital conditions for the contaminant distribution were chosen such that each model cell was assigned a constant concentration of DOC of 2 mg/l. Then the initial concentration of

AOC was determined from equation 9. This procedure controls the initial contaminant distribution via the k_d-distribution. The heterogeneous distributions for k_f and k_d were obtained by means of a turning bands generator (Mantoglou and Wilson, 1982). We will discuss three main cases for the heterogeneous aquifer:

-   Case 1: Moderately heterogeneous distribution for k_f with $\sigma_{\ln kf} = 1.15$, correlation length clx = 1 m and uniform k_d-distribution
-   Case 2: Strongly heterogeneous distribution for k_f with $\sigma_{\ln kf} = 1.8$, clx = 1 m and uniform distribution for k_d. This case tests the effect of increased heterogeneity expressed in terms of an increased standard deviation.
-   Case 3: Moderately heterogeneous distribution for k_f with $\sigma_{\ln kf} = 1.15$, clx = 1 m and linear negative correlation between k_f and k_d distributions. This case tests the effect of large amounts of contaminant in regions with low hydraulic conductivity.

For all three cases we selected an exponential variogram model. To guarantee comparability the k_d distributions were scaled by a factor such that in all realizations the inital contaminant mass in the aquifer was the same. All other model parameters remain the same as in the homogeneous case (tab. 1).

The model was run for ten realizations in each case. In figs. 3a - c the BTCs for the tracer and for DO (averaged over the ten realizations) are shown for all three cases. Both the BTC of the tracer and that of DO are less steep than in the homogeneous case, but the effect is more pronounced for DO breakthrough. For the tracer BTC this behaviour could be expected, as the heterogeneous distribution of k_f increases the variability of the velocity field in the aquifer. The variability leads to macro-dispersion which increases with increasing $\sigma_{\ln kf}$ (compare tracer BTCs in figs. 3a and 3b).

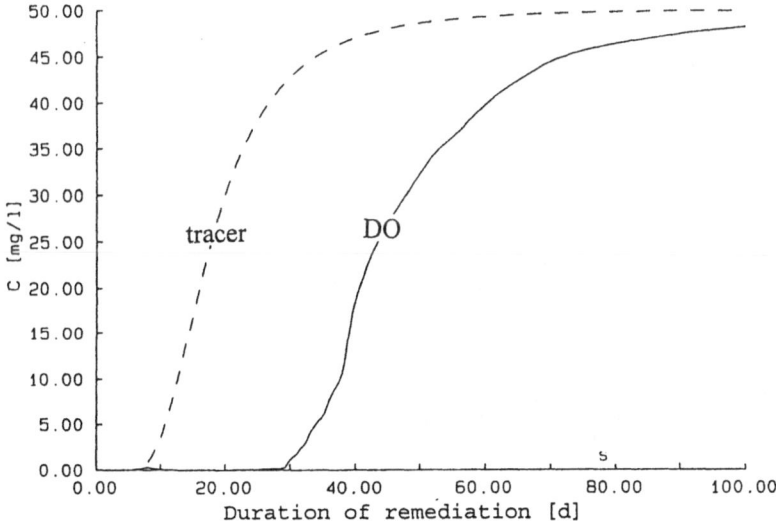

Fig. 3a: BTCs for the tracer and for DO averaged over ten realizations of case 1

For the reactive compounds DO and DOC the heterogeneity has the additional effect of reducing their mixing, which is the prerequisite for microbial degradation. In the homogeneous case this mixing is optimal, because DOC transport is retarded and degradation can happen rather uniformely at the front of the DO movement.

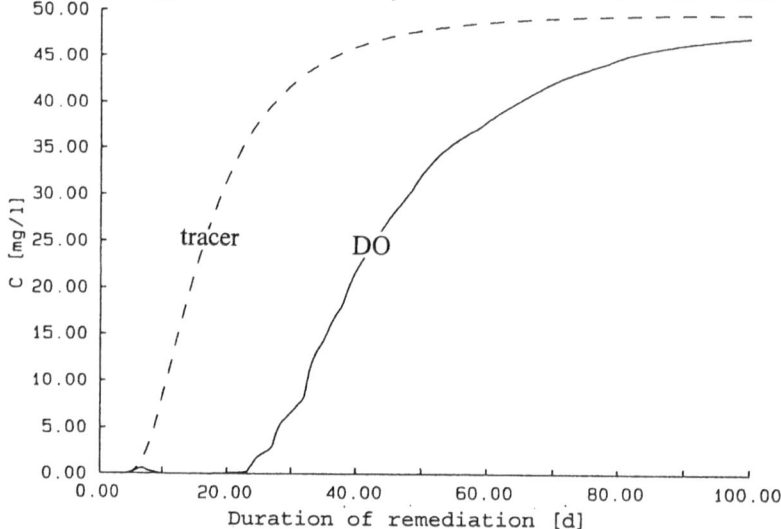

Fig. 3b: BTCs for the tracer and for DO averaged over ten realizations of case 2

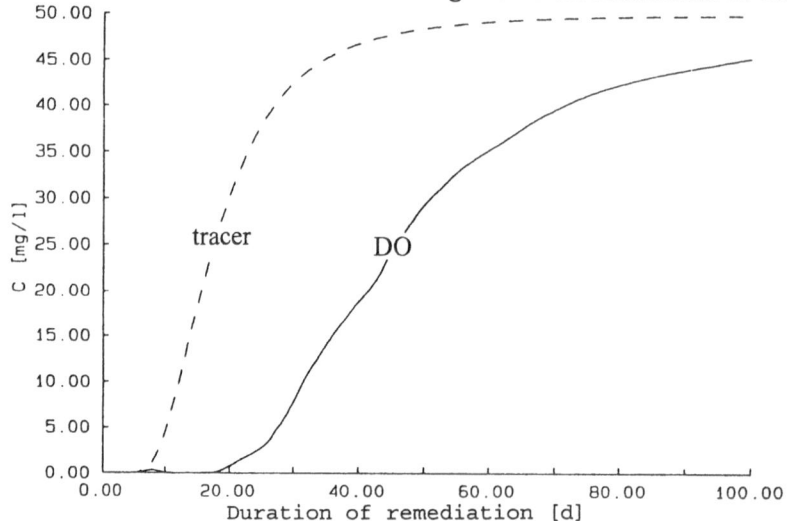

Fig. 3c: BTCs for the tracer and for DO averaged over ten realizations of case 3

In the case of a heterogenous k_f-distribution some pathways are better supplied with DO and are therefore cleaned up faster. Along these pathways DO breakthrough can happen earlier than in the homogeneous case. Regions with decreased supply of DO constitute longer lasting sinks for DO which in turn prolong the time to maximum DO arrival at the pumping well. On the whole, the decrease in steepness of the BTC is stronger for DO than for the tracer. This effect increases with increasing $\sigma_{\ln kf}$ (compare tracer and DO BTC in figs. 3a and 3b). It is still more pronounced, when the contaminant systematically resides in regions with low k_f (compare figs. 3a and 3c).

Equivalent homogeneous medium

According to theory the coefficient of macrodispersion increases with increasing heterogeneity of the k_f-distribution. That means that in order to reproduce the tracer BTC of the heterogeneous cases 1 and 2 in the homogenous aquifer, the value of the longitudinal dispersivity has to be increased from 0.25 m to 1 m and 3 m respectively. While the value of 0.25 m accounts for heterogeneity at a scale below the 0.5 m x 0.5 m resolution of the model, the increase represents the heterogeneity effect on the model scale. The ratio of longitudinal to transverse dispersivity was kept constant. But the increased macro-dispersivity alone cannot account for the heterogeneity effect on the BTC of DO. As explained before, dispersivity has only a reduced effect on DO transport due to front sharpening, while aquifer heterogeneity has a larger effect on DO than on the tracer. To accommodate heterogeneity effects on the BTC of DO, a parameter in addition to dispersivity has to be found.

We selected the exchange parameter α which controls the exchange of the reactive species between porewater and the biophase to parametrize heterogeneity. In the homogeneous case α was set infinite, so that the exchange between porewater and biophase was instantaneous. In this case the concentrations of the reactive species are the same in the pore water and inside the biophase, and biodegradation is not exchange-limited. A decreasing α means a decreasing exchange between the two phases. The microorganisms that grow exclusively inside the biophase consume DO (and DOC) and thereby deplete the biophase relative to the porewater. The exchange is driven by the concentration gradient, e.g. :

$$\frac{dDO_m}{dt} = -\alpha \cdot (DO_m - DO_{im}) \tag{10}$$

DO_m and DO_{im} denote DO in the pore water and in the biophase respectively. This approach is analogous to the dual porosity model of Coats and Smith (1964). Variations of this concept are frequently used to simulate non-equilibrium adsorption and desorption in the subsurface (e. g. Brusseau et al., 1989). In our case the choice of the dual porosity approach is motivated by the concept that the contaminant is primarily located in less permeable zones to which DO has rate limited access. One might argue that the tracer movement should consequently use the same concept. However, the rather small amount of immobile volume required has no impact on tracer movement as it is saturated in a short time. DO on the other hand is used up in the immobile zone over and over again. Thus the impact is large. When α in equation 10 becomes small DO arrives earlier at the pumping well and needs a longer time to reach maximum concentration (fig. 4).

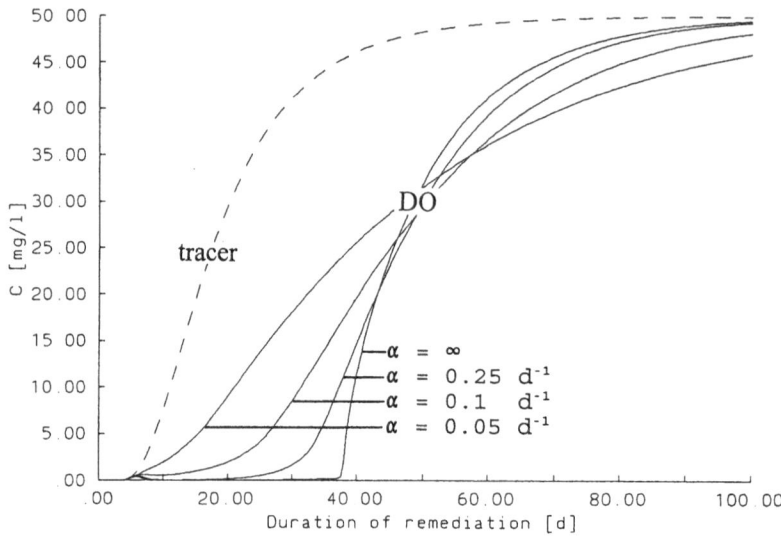

Fig. 4: BTCs for tracer and DO for different values of the exchange coefficient α

The effects of aquifer heterogeneity on the flow field and on DO and DOC concentrations for a single realization of case 3 can be seen in figs. 5a-c. In the most pervious channel DO has almost reached the pumping well, but one can also see less permeable regions where DO has not yet entered and consequently DOC is not yet oxidized. These regions act as sinks for DO over a prolonged period (fig. 5b and 5c). Table 2 lists the effective parameter values for the different cases.

	longitudinal dispersivity (m)	exchange coefficient (1/d)
Homogeneous case	0.25	∞
Heterogeneous case 1	1.5	0.25
Heterogeneous case 2	3.	0.16
Heterogeneous case 3	1.5	0.1

Table 2: Values for longitudinal dispersivity and exchange coefficient in the homogeneous case and fitted effective values necessary to reproduce the effects of the three heterogeneous cases under homogeneous assumptions.

Similar to dispersivity the exchange coefficient shows a dependence on the degree of heterogeneity of k_f (compare cases 1 and 2). But while the dispersivity of the non-reactive tracer is of course not influenced by the k_d-distribution, a negative correlation of k_f and k_d affects the exchange coefficient (compare cases 1 and 3). A negative correlation of the type used here may not be unrealistic. Even if it does not exist in the beginning of an aquifer contamination it may develop in aged spills when regions with larger hydraulic conductivity are already cleaned up and the contaminant still remains in less permeable areas.

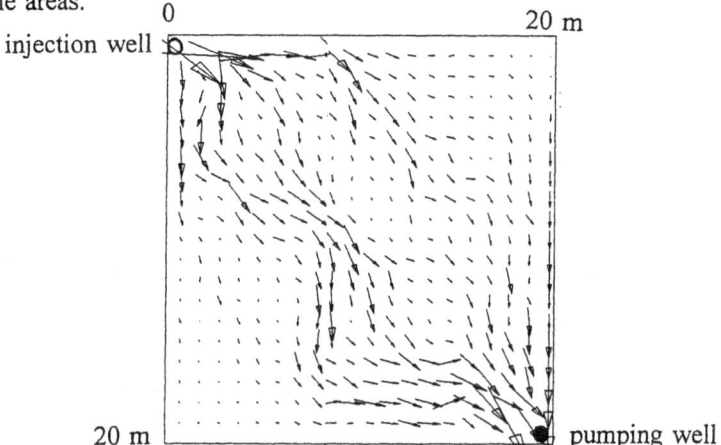

Fig. 5a: Velocity distribution for one realization of case 3.

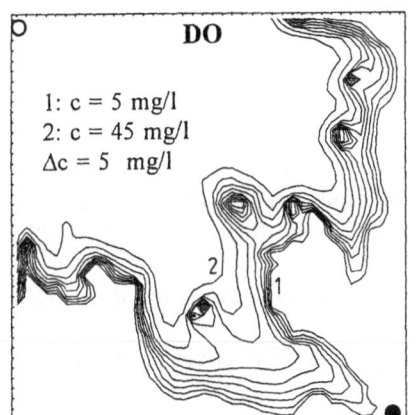

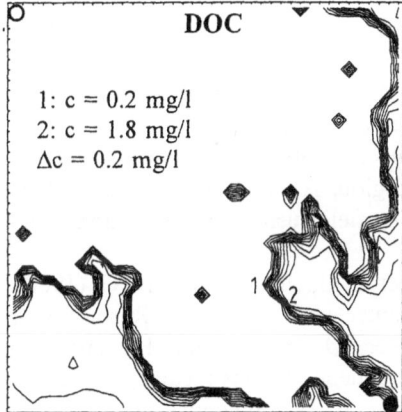

Fig. 5b and 5c: DO and DOC concentration for one realization of case 3 30 d after beginning of the remediation.

CONCLUSIONS

The simple examples show that effective parameters can be defined. But their definition requires comparability of realizations in order to obtain physically meaningful properties. Otherwise spurious properties result which accomodate only information uncertainty.

In a more realistic example a model of diffusion limited exchange between pore water and immobile biophase was able to reproduce BTCs for DO in heterogeneous aquifers in equivalent homogeneous media. While the dispersivity is the effective parameter that accommodates heterogeneity in a homogeneous model for a non-reactive tracer, the exchange coefficient between pore water and biophase can play this role for the reactive transport of DO. Ongoing research investigates in how far this approach is valid for the prediction of the degradation activity.

LITERATURE

Bouchaud, J. P. and Georges, A. (1988) "Un modèle simple de dispersion hydrodynamique", C. R. Acad. Sci. Paris 307, 1431-1436.

Brusseau, M. L., Jessup, R. E., and Rao, P. S. C. (1989) "Modeling of transport of solutes influenced by multiprocess nonequilibrium", Water Resour. Res. 25, 1971-1988.

Chrisikopoulos, C. V., Kitanidis, P. K., and Roberts, P. V. (1992) "Macrodispersion of sorbing solutes in heterogeneous porous formations with spatially periodic retardation factor and velocity field", Water Resour. Res. 28, 1517-1529.

Coats, K. H. and Smith, B. D. (1964) "Dead-end pore volume and dispersion in porous media", Soc. Petrol. Eng. J. 4, 73-84

Dagan, G. (1984) "Solute transport in heterogeneous porous formations", J. Fluid Mech. 145, 151-177.

Gelhar, L. W. and Axness, C. L. (1983) "Three-dimensional stochastic analysis of macrodispersion in aquifers", Water Resour. Res. 19, 161-180.

Gelhar, L. W., Welty, C., and Rehfeldt, K. R. (1992) "A critical review of data on field-scale dispersion in aquifers", Water Resour. Res. 28, 1955-1974.

Kinzelbach, W., Schäfer, W., and Herzer, J. (1991) "Numerical modeling of natural and enhanced denitrification processes in aquifers", Water Resour. Res. 27, 1123-1135.

Mantoglou, A. and Wilson, J. L. (1982) "The turning bands method for simulation of random fields using line generation by a spectral method", Water Resour. Res. 18, 1379-1394.

Matheron, G., and de Marsily, G. (1980) "Is transport in porous media always diffusive? A counter example", Water Resour. Res. 16, 901-917.

Metzger, D., Kinzelbach, H., and Kinzelbach, W. (1994) "Effective properties of heterogeneous porous media", in preparation.

Murphy, E. M., Schramke, J. A., Fredrickson, J. K., Bledsoe, H. W., Francis, A. J., Sklarew, D. S., and Linehan, J. C. (1992) "The influence of microbial activity and sedimentary organic carbon on the isotope geochemistry of the Middendorf aquifer", Water Resour. Res. 28, 723-740.

Rajaram, H. and Gelhar, L. W. (1994) "Plume-scale dependent dispersion in heterogeneous aquifers", Water Resour. Res. 29, 3249-3276

Schäfer, W. and Kinzelbach, W. (1992) "Stochastic modeling of in situ bioremediation in heterogeneous aquifers", J. Contaminant Hydrol. 10, 47-73.

6. FRACTURED POROUS MEDIA

CONTINUUM MODELING OF COUPLED THERMO-HYDRO-MECHANICAL PROCESSES IN FRACTURED ROCK

R. ABABOU*, A. MILLARD*, E. TREILLE*, M. DURIN*, F. PLAS**
*Commissariat à l'Energie Atomique, Centre d'Etudes de Saclay,
 DRN/DMT/SEMT, 91191 Gif-sur-Yvette Cedex, France.
**Agence Nationale pour la Gestion des Déchets Radioactifs,
 92260 Fontenay-aux-Roses, France.

ABSTRACT

An equivalent continuum approach for modeling coupled Thermo-Hydro-Mechanical (THM) processes in 2-D and 3-D fractured rock is presented. The equivalent, homogenized, model can be used for studying the stability of nuclear waste geologic repositories, e.g., the response of water-saturated fractured rock, to disturbances caused by excavation and heat production. The continuum approach leads to anisotropic hydro-mechanical equations, equivalent to Darcy's law and to Biot's poro-elastic equations, with smoothly variable tensorial coefficients. A hypothetical 2-D test problem, involving a complex network of thousands of fluid-filled joints, is used for illustration of the approach. The effect of homogenization scale on hydraulic conductivity is examined.

INTRODUCTION

Brittle rocks like granite are considered as a potential host rock for the underground storage of high level nuclear wastes. The overall response of the rock mass and of groundwater, due to excavation and to subsequent heat production by wastes, generally involves coupled Thermo-Hydro-Mechanical (THM) phenomena. It is important to model such phenomena in order to appraise the environmental safety of a mined geologic repository. In this paper, we present an effective approach for modeling coupled THM processes in water-filled fractured rock. The processes are modeled as if they were taking place in a continuum, without explicitly simulating individual joints.

It is assumed that natural fractures can be classified in two categories: (i) major fractures or faults extending through a large part of the domain; and (ii) fractures or joints of a lesser extent. The model presented in this paper concerns the second category, where the joints may have irregular or 'random' spatial distribution, and a broad spectrum of sizes and orientations. Major fractures or faults can be taken into account separately using a discrete approach. The advantage of the continuum approach is that it can be used for modeling coupled THM processes in the presence of many, variously oriented joints, while a discrete joints approach would become rapidly untractable as the number of joints and their geometrical complexity increases.

A. Peters et al. (eds.), Computational Methods in Water Resources X, 651–658.
© 1994 *Kluwer Academic Publishers. Printed in the Netherlands.*

Starting from a known rock matrix with a given distribution of joints, the constitutive laws of the equivalent continuum are obtained by a linear superposition approach, based on the methods developed by several authors, in particular *Snow [1]* for hydraulics, and *Oda [2]* for hydro-mechanics. The continuum equations are formulated for a 2-D or 3-D fractured rock made up of an elastic impervious matrix, and irregularly distributed, water-filled elastic joints. A hypothetical 2-D test problem involving thousands of joints is used for illustration. Here, we focus on certain essential features of the coupled continuum model, rather than on detailed simulation results.

BASIC ASSUMPTIONS

Dimensionality and Geometry : In 3-D space, the fractured rock is assumed to be made up of intact rock and planar joints or fractures, with known shapes, lengths, orientations, and apertures. The 2-D case arises when the fracture planes are all parallel to a given direction, e.g. horizontal. In this case, fractures may be represented by line segments in the plane orthogonal to all fractures, e.g. vertical. The assumption of plane strains can be used to model THM processes within the plane. This is the case for the 2-D model problem to be described further below (application section).

Thermal Processes : In general, THM processes are fully coupled. However, assuming that fluid velocities in the joints are sufficiently small, heat convection effects can be neglected. Moreover, heat conduction can be approximated based solely on intact matrix properties. With these assumptions, heat transport is only 'one-way' coupled to hydro-mechanical processes. The temperature field history can therefore be calculated independently, then injected into the hydro-mechanical model. Full thermal coupling, including heat convection and the effect of joints, is postponed to a later stage.

Hydro-Mechanical Processes : The mechanical behaviour of joints is assumed to be linear-elastic in compression, tension, shear. Nonlinear joint behaviour, e.g. through a Mohr-Coulomb elastoplastic model, may be considered later. The hydraulic behavior of joints is assumed to be governed by Poiseuille's law, which can be viewed as an approximation to the full Navier-Stokes equations (neglecting inertial terms, transient effects, and non-planar flow components within each joint). The intact rock matrix is a homogeneous, isotropic, elastic medium, satisfying Hooke's law. Finally, the matrix is assumed impervious given the low permeability of intact granite.

Local Constitutive Laws and Equations : For the rock matrix, we use Hooke's law $\sigma_{ij} = \lambda e_{kk} \delta_{ij} + 2\mu e_{ij}$, where λ and μ are the matrix Lamé coefficients ($\lambda \leftrightarrow$ compression, $\mu \leftrightarrow$ shear). Within each water-saturated fracture, we use Terzhagi's "effective" stress concept, namely: $\sigma'_{ij} = \sigma_{ij} + p\delta_{ij}$, where the convention of negative compressive stress is used. That is, (σ'_{ij}) is equal to total stress (σ_{ij}) *minus* negative fluid stress ($-p\delta_{ij}$). We then assume that each joint behaves elastically. Thus, starting from an equilibrium state, crack aperture 'a' varies linearly with normal effective stress (σ'_n) : $\delta\sigma'_n = K_n\delta a$. Similarly, crack length l varies linearly with shear effective stress (σ'_s) : $\delta\sigma'_s = K_s\delta l$. The normal and shear crack stiffness coefficients, K_n and K_s, are taken constant, independent of stress, same for all fractures ($K_n \approx 10^{11}$ Pa/m, $K_s \approx 10^{10}$ Pa/m). Finally, fracture flow obeys Poiseuille's law : $V = a^2(g/12v) J$, where V is the average velocity in the joint, 'g' the acceleration of gravity, 'v' the kinematic viscosity, and $J = -\nabla(p+\rho gz)/\rho g$ the hydraulic gradient.

EQUIVALENT CONTINUUM PROPERTIES

We now consider an arbitrary set composed of N fractures having various apertures, lengths, and orientations. Equivalent homogenized properties are determined based on a linear superposition approximation, which may be applied either to the whole domain, or more generally, to a subdomain. The size of the homogenization subdomain, or homogenization scale ℓ_H , may be larger or smaller than the Representative Elementary Volume (REV). The choice of ℓ_H , and its relation to REV size, may have an important effect on the interpretation of results obtained with the equivalent continuum model. This point will be discussed later.

Equivalent Darcy Law (Hydraulics)

Following Snow's superposition approach [1,2,3], a homogenized Darcy law is developed for quasi-steady flow in fractured rock. First, we use Poiseuille's *cubic law* to compute the discharge rate q of a single joint as a function of the hydraulic gradient **J** imposed at the scale of the domain. This yields $q = (a^3 g/12 v) \mathbf{J}'$, where $\mathbf{J}'$ is the projection of **J** on the fracture plane. If **n** is the unit vector normal to the fracture, then $\mathbf{J}' = \mathbf{J} - (\mathbf{J}.\mathbf{n}) \, \mathbf{n}$. Secondly, we superimpose the contributions of all fractures to the global flow, by arithmetic summation of the local q's, with **J** constant for all fractures. This yields a linear relation between the global or mean discharge rate **Q** (output), and the global hydraulic gradient **J** (input). The equivalent permeability tensor, K_{ij} , such that $Q_i = K_{ij} J_j$, is expressed in terms of fracture apertures and lengths, or alternatively, fracture porosities and specific areas :

$$(1) \quad K_{ij} = \sum_{f=1}^{N} K_f \left(\delta_{ij} - n_{i,f} \, n_{j,f} \right) \text{ with } K_f = \frac{g}{12 \, v} \frac{a_f^3}{\ell_f} \text{ or } K_f = \frac{4g}{12 \, v} \frac{\Phi_f^3}{\sigma_f^2} ,$$

where K_f is the directional conductivity of fracture number 'f. In these formulas, ' $n_{i,f}$ ' is the ith component of the unit vector normal to fracture 'f, a_f is fracture aperture, and ℓ_f is fracture length (2-D case). The first K_f formula can be found in *[1,2,3]*. More generally, the second formula gives K_f in terms fracture porosity Φ_f and fracture specific area σ_f (accounting for both fracture walls). The second formulation is applicable to 2-D as well as 3-D fracture sets *[3]*, including the case of arbitrarily shaped plane fractures. Remarkably, it generalizes the classical isotropic Kozeny-Carman formula. Finally, the equivalent hydraulic porosity is given by :

$$(2) \quad \Phi = \sum_{f=1}^{N} \Phi_f \ .$$

Equivalent Poro-Elastic Stiffness Coefficients (Mechanics & Hydro-Mechanics)

A strain-based superposition approach was developed in *[2]* to obtain equivalent hydromechanical laws for an elastic rock containing many cracks. The individual cracks or joints were assumed to behave elastically or quasilinearly under compression and shear *[2]*, and to satisfy Terzaghi's "effective stress" approximation *[4]*. Our implementation assumes linear elastic laws with constant coefficients. The mean strain, due to the imposed global stress tensor σ_{ij} , is calculated by linear superposition of the local displacements occuring throughout the intact rock matrix *and* the discrete joints, keeping the global stress constant. This leads to linear hydro-mechanical laws coupling solid stress and fluid pressure to solid strain and fluid strain (or fluid production), similar to the poro-elastic laws developed earlier by *Biot [5]*. The results can be summarized as follows. A relation between global variables e_{ij} , σ_{ij} , p , i.e. fractured rock strain, total stress, and fluid pressure in the joints, is obtained :

(3) $$e_{ij} = T'_{ijk\ell}\,\sigma_{k\ell} + B'_{ij}\,p$$

where T'_{ijkl} and B'_{ij} are equivalent homogenized coefficients, namely the total compliance coefficient (4th rank tensor), and the strain-pressure coupling coefficient (2nd rank tensor). Another relation can be derived from the same superposition method, based on fluid/solid mass conservation. This relation couples linearly 'fluid production' (ξ) to "effective" stress σ'_{ij} :

(4) $$\xi = \frac{1}{K_n\ell}\,F_{ij}\sigma'_{ij} \quad \text{with} \quad F_{ij} = \frac{1}{2}\sum_{f=1}^{N}\ell_f\,\sigma_f\,(n_i)_f(n_j)_f\ ,$$

Note that ξ represents the net variation of volume of fluid per unit volume of the deformable medium; it is related to the spherical strains of the solid ($e = e_{kk}/3$) and fluid ($\varepsilon = \varepsilon_{kk}/3$) by the simple conservation equation $\xi = -3\Phi(\varepsilon-e)$. The 2nd rank tensor F_{ij} is purely geometrical. Note that the 'prime' sign in equation (3) serves to distinguish the *strain vs. stress* formulation. The tensorial relations (3)-(4) need to be inverted in order to obtain suitable forms of the continuum model (for numerical reasons and for comparison purposes). The new unprimed coefficients involved in the *stress vs. strain* formulation below are *stiffness* coefficients, 'inverse' in some sense of the primed *compliance* coefficients.

Now, the 4th rank tensorial equation (3) is inverted by using appropriate tensor algebra, and exploiting the particular symmetry properties of the elastic tensors, for the general case of non-orthotropic media. The inverted relation is found to be of the form :

(5) $$\sigma_{ij} = T_{ijkl}\,e_{kl} - B_{ij}\,p$$

Comparing term by term equations (3) and (5), and taking also into account that $\mathbf{T}$ and $\mathbf{B}$ satisfy the required symmetry properties of elastic tensors, one obtains indeed the correct relation between a symmetric 4th rank tensor and its inverse :

(6) $$T'_{k\ell mn}\,T_{mnij} = \left(\delta_{ki}\,\delta_{\ell j} + \delta_{kj}\,\delta_{\ell i}\right)/2\ ,$$

Comparing again (3) and (5) yields a relation between Biot's coefficient $\mathbf{B}$ and its reciprocal ($\mathbf{B'}$) :

(7) $$B_{mn} = T_{mnij}\,B'_{ij} \quad \text{or} \quad T'_{k\ell mn}\,B_{mn} = B'_{k\ell}$$

Finally, inserting $\sigma'_{ij} = \sigma_{ij} + p\delta_{ij}$ in (4) and using previously established relations, gives also the equation coupling pressure to strains, or rather, to solid strain e and fluid production ξ :

(8) $$p = -G\left(B_{kl}\,e_{kl} - \xi\right)$$

Remarkably, eqs.(5) and (8) are of the same form as those of Biot's theory [5], provided the latter are properly generalized for fully anisotropic poroelasticity. More precisely, it can be shown as a consequence of linear superposition and conservation equations, that the "Biot" coefficient B_{ij} which appears in the stress/strain law (5), is identical to that appearing in the pressure/strain law (8). For fractured rock, the reciprocal Biot coefficient and the Biot modulus G are given by :

(9) $$B'_{ij} = F_{ij}/(K_n\ell)\ ,\quad G = \left\{B'_{ij}\left(\delta_{ij} - T_{ijk\ell}\,B'_{k\ell}\right)\right\}^{-1} = \left\{T'_{ijk\ell}\,B_{k\ell}\left(\delta_{ij} - B_{ij}\right)\right\}^{-1}$$

where l stands for mean fracture length. It will be interesting to specialize these relations to orthotropic and isotropic cases. In the isotropic case, let K_b be the volumetric or 'bulk' stiffness modulus of the drained medium, K_s that of the individual solid 'grains', and K_w that of water. Biot's coefficient is classically given as $\beta = 1 - K_b/K_s$, Biot's modulus as $G = [(\beta-\Phi)/K_s + \Phi/K_w]^{-1}$, and bulk stiffness is related to Lamé constants by $K_b = (\lambda+2\mu)/3$. Our tensorial relations for fractured rock constitute a generalization of such isotropic poro-elasticity formulas.

2-D APPLICATION : VARIABILITY, ANISOTROPY, & SCALE EFFECTS

For specific applications, an appropriate homogenization scale ℓ_H must be selected. When ℓ_H is chosen less than domain size, the equivalent coefficients and constitutive laws are non-unique, being spatially variable, and scale-dependent. Also, the 'Representative Elementary Volume' is worth investigating. We focus here on the equivalent hydraulic conductivity tensor.

Fractured Rock Data (2-D) : The 50 m × 50 m domain represents a 2-D vertical cross-section of a hypothetical rock formation. The rock mass contains 6580 fractures of various orientations, apertures, and lengths. *Figure 1* depicts the entire fracture network. Each fracture is represented as a thin segment, so that the various apertures are not being displayed. The geometric data of the network are : (i) fracture counter; (ii) (x,y) location of fracture midpoint; (iii) angle θ of the unit normal vector n; (iv) fracture aperture 'a'; and (v) fracture length 'ℓ'. Several statistical properties were analyzed, including probability distributions of 'a', 'ℓ', and 'θ". Fracture orientations were only roughly uniform (isotropic). A 'deficit' of fractures was observed for normal angles θ between $3\pi/4$ and π (fractures inclined at angles between $\pi/4$ and $\pi/2$ over horizontal axis x). Fracture lengths have positively skewed distribution, similar to lognormal, with a mean of 1.83 m and a coefficient of variation (CV) of 58%. Apertures are also positively skewed, but closer to an exponential distribution, with a mean of 4.57 μ (microns) and a CV of 86%. The equivalent porosity of the rock mass, by equation (2), is $\Phi \approx 2.2 \; 10^{-5}$. The mean fracture density is $\rho \approx 2.632$ fractures/m^2.

Hydraulic Conductivity Tensor at the Largest Scale (ℓ_H = 50m) : The K_{ij} tensor was computed at the scale of 50m x 50m, in the natural (x,y) system, using eq.(1). The principal components were then obtained by expressing K_{ij} (m/s) in the principal system (x*,y*) :

$$\left[K_{ij}{}^* \right] = \begin{vmatrix} +0.127E\text{-}08 & 0.000E\text{+}00 & 0.000E\text{+}00 \\ 0.000E\text{+}00 & +0.767E\text{-}09 & 0.000E\text{+}00 \\ 0.000E\text{+}00 & 0.000E\text{+}00 & +0.204E\text{-}08 \end{vmatrix}$$

with the convention that the first component is the largest, and the second is the smallest (the z component is irrelevant and therefore not concerned by this rule). The rotation angle from non-principal (x,y) to principal (x*,y*) system, is γ = -45.34 degrees. This gives the direction of largest conductivity, roughly orthogonal to the direction for which there is a deficit of fractures, as expected. The global anisotropy ratio is moderate, $A = K_{11}{}^*/K_{22}{}^* \approx 1.7$.

Anisotropy and Spatial Variability : The above calculations can be repeated at smaller scales, e.g. based on a partition of the 50m x 50m domain into subdomains. *Figure 2* depicts conductivity tensors obtained on 6.25m x 6.25m subdomains, corresponding to a partition into 8x8 squares. The 'local' K_{ij}'s are represented by their anisotropy ellipse, or directional conductivity ellipse, of major axis $(K_{11}{}^*)^{1/2}$ and minor axis $(K_{22}{}^*)^{1/2}$. There are 64 such ellipses. Their orientations and aspect ratios indicate the principal directions and the square-root anisotropy ratio of the local K_{ij}'s. The ellipses spatial variability is quite apparent. Some local anisotropy ratios are significantly larger than the global ratio A≈1.7.

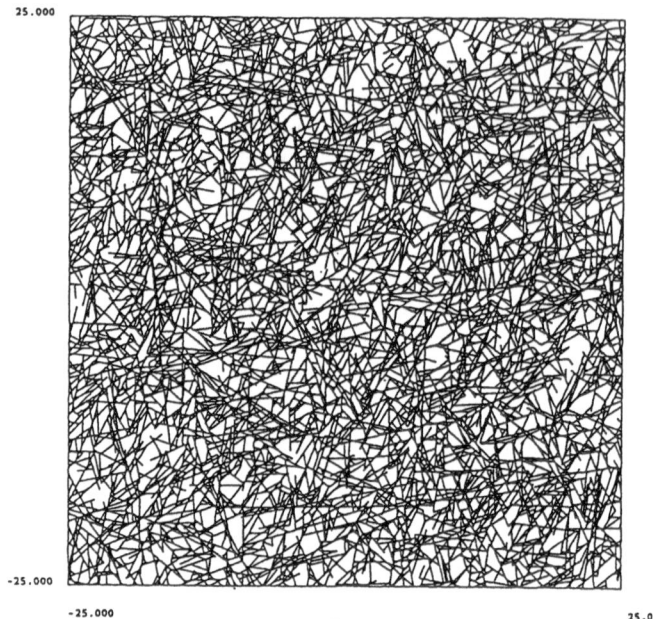

Figure 1 : Fracture network on the whole domain (50 m x 50 m)

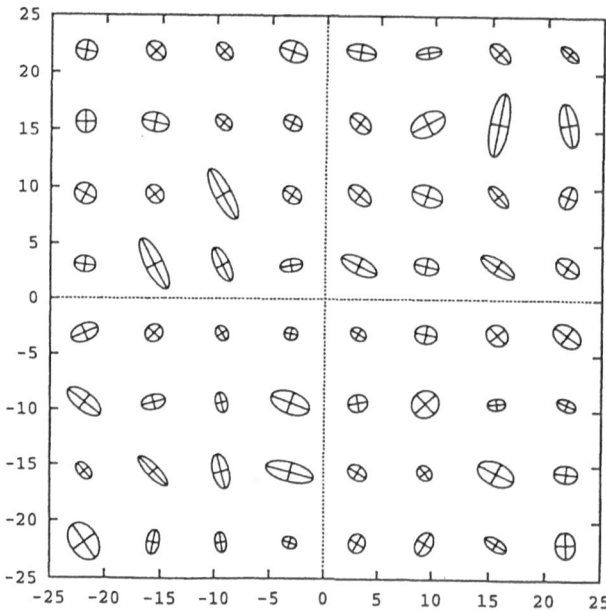

Figure 2 : Anisotropy ellipses, representing the spatial distribution of equivalent conductivity
for a homogenization scale of 6.25 m (partition into 8x8 subdomains)

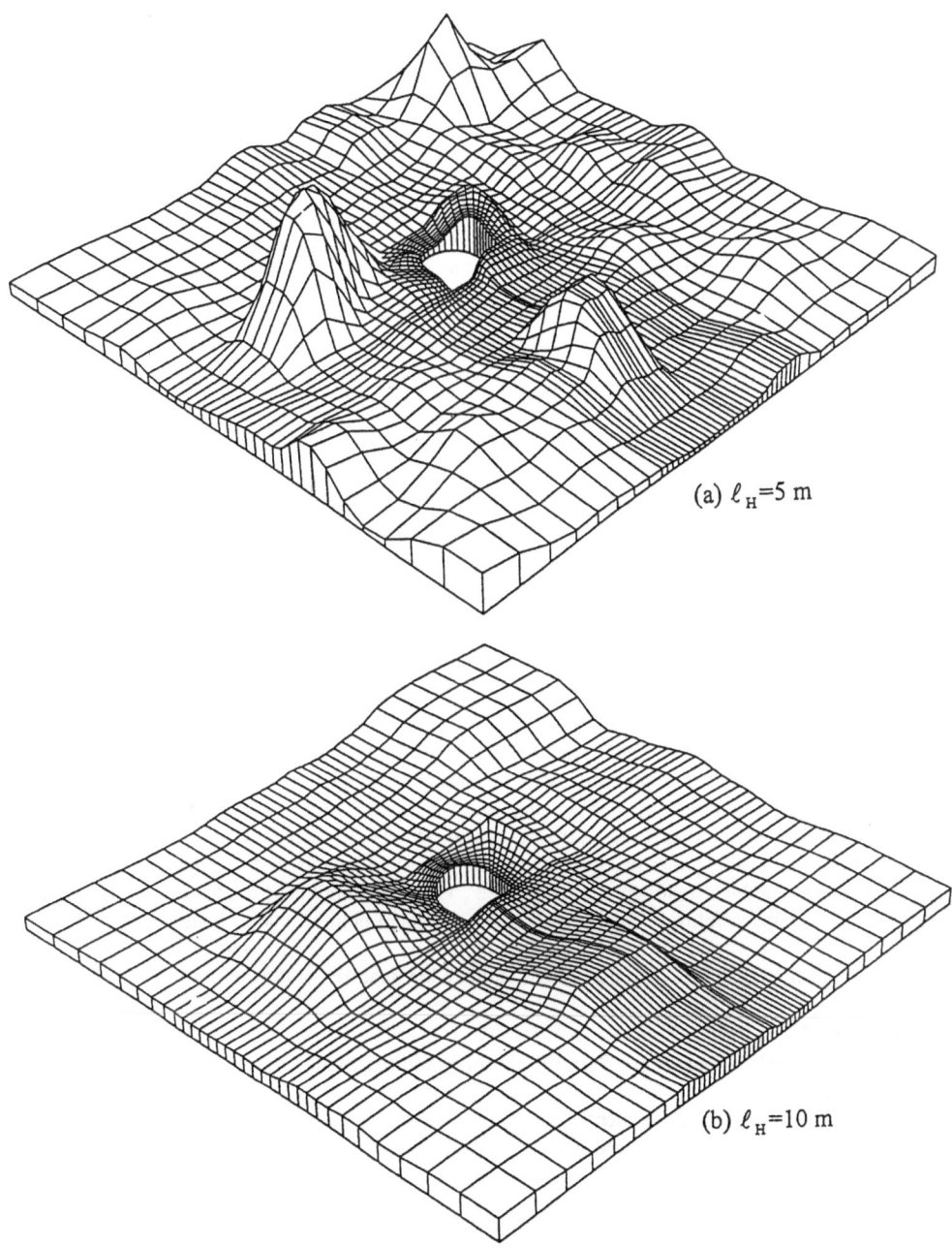

(a) ℓ_H=5 m

(b) ℓ_H=10 m

Figure 3 : Modulus, or root-mean-square value, of the equivalent principal conductivity components, as a function of space : (a) ℓ_H=5 m ; (b) ℓ_H=10 m

Scale Effects : Increasing homogenization scale l_H, or coarsening the subdomains, should smooth out spatial variability. Let us define a 'spherical' conductivity, $K^* = (K_{11}^{*2} + K_{22}^{*2})^{1/2}$, for each subdomain, and plot $K^*(x,y)$ using a standard plotting package. The results are shown in *Figure 3*, for $l_H = 5m$ and $l_H = 10m$. The 'tunnel' cavity in the middle of the domain gives the scale of inhomogeneity that must be resolved by the numerical mesh. Another analysis of the K_{ij}'s was developed based on nested subdomains of increasing sizes, growing from the center of the domain. Each K_{ii}^* was plotted versus l_H (*not shown here for lack of space*). The plots indicated an REV $\approx 25m$. Plotting instead A and γ versus l_H, suggests a smaller REV ≈ 15-$20m$. It should also be kept in mind that scale-dependence and REV size are relative to the property being studied.

SUMMARY AND OUTLOOK

Equivalent hydromechanical laws were obtained by linear superpositions of local fluxes and local strains, and then identified with anisotropic (non-orthotropic) versions of Darcy's law and Biot's theory of poro-elastic mixtures. The equivalent permeability tensor, stiffness tensor, and pressure coupling coefficients, were expressed in terms of (i) elastic properties of rock matrix, (ii) hydraulic and elastic properties of joints, and (iii) geometric properties of joints. The anisotropic 2nd rank tensors were given explicitly (the non-orthotropic 4th rank stiffness coefficient T_{ijkl}, not given here, is of similar form). The spatial variability and scale-dependence of Darcy's equivalent conductivity tensor was studied for a 2-D fractured rock containing thousands of joints.

New criteria and guidelines need to be developed for the 'optimal' choice of homogenization scale. Our model is being applied to a hypothetical case involving excavation of a gallery, and heating, in a 2-D fractured rock mass. The coupled THM equations are solved numerically for the equivalent anisotropic continuum, using 'CASTEM 2000', a general-purpose finite element code of the Commissariat à l'Energie Atomique. The primal variables are temperatures, displacements with quadratic basis functions, and pressure with linear basis functions. The elements are 6 node triangles and 8 node quadrangles. Time integration is done with an implicit one-step scheme. While preliminary simulation tests were restricted to elastic behavior, the homogenization approach described in this paper has been extended to model the sensitivity of material properties with respect to the ambient fields of strains, stresses, etc. Such nonlinear feed-back couplings will be taken into account in future. The possible effects of local nonlinear joint behavior should also be appraised. Comparisons of the continuum simulations with discrete joint or discrete element hydro-mechanical models, will help assess and validate key features of the continuum model.

REFERENCES
[1] Snow D.T. (1969). Anisotropic Permeability of Fractured Media. Water Resour. Res., 5(6), 1273-1289.
[2] Oda M. (1986). An Equivalent Continuum Model for Coupled Stress and Fluid Flow Analysis in Jointed Rock Masses. Water Resour. Res., 22(13), 1845-1856.
[3] Ababou R. (1991). Approaches to Large Scale Unsaturated Flow in Heterogeneous and Fractured Geologic Media. Report NUREG/CR-5743, U.S.NRC, Washington D.C.
[4] Terzaghi Von K. (1936). The Shearing Resistance of Saturated Soils and the Angle Between the Planes of Shear. First Int. Conf. Soil Mech., Vol.1, Harvard Univ., pp. 54-56.
[5] Biot M.A. (1956.a & b). Theory of Deformation of a Porous Viscoelastic Anisotropic Solid. J. Appl. Phys. 27, 459-467 (1956.a). General Solutions of the Equations of Elasticity and Consolidation for a Porous Material. J. Appl. Mech. 23, 91-96 (1956.b).

Dual-Continuum Modeling of Contaminant Transport in Fractured Formations

J. BIRKHÖLZER and G. ROUVÉ
Institute for Hydraulic Engineering and Water Resources Management
University of Technology, RWTH Aachen
Mies-van-der-Rohe-Straße 1, Aachen, 52056
Germany

The objective of the present paper is to discuss the performance of different dual-continuum models for simulating contaminant transport in fractured aquifers. In the first part of our study transport processes in fractured rock are simulated using a discrete representation of the fractures and the porous matrix blocks. In a second step the results of the discrete simulations are compared with the results obtained with either existing or new developed dual-continuum modeling approaches.

1 INTRODUCTION

Since the permeability of a fracture network is often substantially higher than the permeability of the host rock, the interconnected fractures play an important role in the transport of a contaminant in the groundwater. However, most of the capacity for storing a pollutant is provided by the pore system of the rock matrix. The concentration field in a fractured aquifer may significantly be influenced by the solute exchange between the fractures and the porous rock. Depending on the properties of the formation different physical processes may lead to the exchange of contaminants between the two pore systems: molecular diffusion, advective transport due to the local pressure gradient in rock blocks ("local" advection) and advective transport due to the regional pressure gradient in the aquifer ("regional" advection).

The simulation of solute transport in fractured porous formations is often performed with **dual-continuum** models. Central to those models is the assumption that the heterogeneous formation can be separated into two homogeneous media, one representing the fracture system with high conductivity, one representing the rock matrix with high storage capacity. Since both media are treated as different systems, the transport processes are described by two separate sets of equations coupled by a solute exchange term. In the general form, the socalled **dual-permeability** model, the global flow and transport processes take place in both media *(Teutsch, 1990; Gerke and van Genuchten, 1993)*. However, in most approaches it is assumed that the rock medium does not contribute to global transport since the hydraulic conductivity of the matrix is negligible (**dual-porosity** model).

Two formulations are available for the representation of concentration values in the matrix pore system: In the **quasi-steady** approach an averaging is made over the porous blocks (although large gradients may appear in the matrix as the fracture surface is approached). The solute exchange is assumed to be proportional to the average concentrations in the fracture and the rock medium. The **unsteady** formulation is more sophisti-

659

A. Peters et al. (eds.), Computational Methods in Water Resources X, 659–666.
© *1994 Kluwer Academic Publishers. Printed in the Netherlands.*

cated: A submodel is introduced representing the concentration values in single blocks by a one-dimensional equation. However, this formulation can only be used in combination with **dual-porosity** models since the assumption of one-dimensional transport is limited to cases when flow through the matrix blocks is negligible.

In the present paper different dual-continuum concepts and exchange approaches are tested for various medium properties. Special interest is focused on the **dual-permeability** modeling of formations with comparatively high conductivities in the rock matrix.

2   GOVERNING EQUATIONS

In the following section the dual-continuum problem shall be presented only for the transport part. As the numerical solution is similar but less complicated we shall assume that the flow field has already been solved.

Dual-Permeability Model

In the general form of a **dual-permeability** model the transport equation in the fracture (superscript F) and the rock (superscript M) continuum can be written as follows

$$n^F \frac{\partial C^F}{\partial t} + q_i^F \frac{\partial C^F}{\partial x_i} - \frac{\partial}{\partial x_i} D_{ij}^F \frac{\partial C^F}{\partial x_j} + W = 0 \tag{1}$$

$$\left(1 - n^F\right)\left(n^M \frac{\partial \overline{C}^M}{\partial t} + \overline{q}_i^M \frac{\partial \overline{C}^M}{\partial x_i} - \frac{\partial}{\partial x_i} \overline{D}_{ij}^M \frac{\partial \overline{C}^M}{\partial x_j}\right) - W = 0 \tag{2}$$

where C is the solute concentration [ML^{-3}], q_i is the DARCY-velocity [LT^{-1}], D_{ij} is the hydrodynamic dispersion tensor [L^2T^{-1}] and n is the porosity [ ]. Both media are assumed to behave as homogeneous continua, and therefore all values represent averages over the representative element volume (REV). The properties of the fracture domain are related to the unit volume of the whole continuum, whereas the properties of the rock matrix are related to the unit bulk volume of the rock. Note that the averaging over the REV is different from the averaging over the concentration profile in single rock blocks which is required for the quasisteady approach. The overbars in equation (2) indicate this second averaging for the rock domain. No such distinction is needed for the fracture continuum.

W is the solute exchange between the fractures and the porous rock per unit volume of the rock matrix [T^{-1}]. It comprises solute transfer by molecular diffusion, "local" advection and "regional" advection. The significance of each one of this processes depends on several medium properties; however, most important is the hydraulic conductivity of the porous matrix blocks. In tight, almost impermeable rock diffusion is dominating, whereas in formations with a comparatively high conductivity (i.e. sandstone, limestone) advective transport may play a significant role.

In a **quasi-steady** fomulation the diffusive transfer is given by

$$W^D = \left(1 - n^F\right)\Omega^2 \alpha \left(C^F - \overline{C}^M\right) \tag{3}$$

where Ω is the specific surface defined as the ratio between the surface and the volume of the rock blocks [L^{-1}]. The exchange coefficient α [L^2T^{-1}] is correlated to the diffusion coefficient, the matrix porosity and the geometry of the porous blocks. However, its physical significance is not well defined, except for simple geometries. In the general case, a shape factor is introduced which has to be calibrated. Then the exchange coefficient can be written as

$$\alpha = \beta n^M D_m^M \tag{4}$$

where D_m^M is the molecular diffusion coefficient [L^2T^{-1}].

Under transient flow conditions local pressure differences may force a fluid and solute transfer between fractures and rock. With a fluid exchange rate W^Q (which is known after solving the flow problem) the solute exchange term of "local" advection W^{AL} becomes

$$W^{AL} = \left|W^{QL}\right|\left(C^F - \overline{C}^M\right) \tag{5}$$

W^{AL} must only be considered in equation (1) or (2) when W^{QL} enters the medium.

The third form of solute transfer in dual-permeability formations is contributed by "regional" advection. The advective transport of a contaminant in the matrix into the direction of the "regional" gradient forces a mixing between the fracture flow and the porous matrix flow as transverse fractures are approached (see fig. 1). With a given fluid mixing rate W^{QR} the solute exchange of "regional" advection W^{AR} is defined by

$$W^{AR} = \left|W^{QR}\right|\left(C^F - \overline{C}^M\right) \tag{6}$$

W^{AR} must be considered in both equations (1) and (2) simultaneously since the mixing between fractures and rock takes place in both directions. In chapter 4.2 a new coupling term is presented to determine the fluid mixing rate W^{QR}.

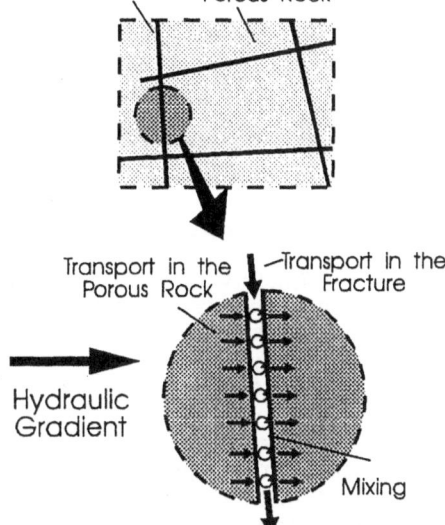

Fig. 1: Mixing between fracture flow and porous block flow

Dual Porosity Model

The system of coupled equations becomes much less complicated if flow in the matrix is negligible. In **dual-porosity** models equation (2) for the rock domain can be reduced to

$$\left(1 - n^F\right)n^M\frac{\partial\overline{C}^M}{\partial t} - W = 0 \tag{7}$$

Equation (7) can be lumped in the numerical scheme since all space derivatives are crossed out. The solute exchange term only comprises the diffusive exchange W^D (eq. (3)).

In the **unsteady** formulation equation (7) is replaced by a one-dimensional diffusion equation for representative rock blocks (see fig. 2)

$$n^M\frac{\partial C^M}{\partial t} - \frac{1}{A(s)}n^M D_m^M\frac{\partial}{\partial s}\left(A(s)\frac{\partial C^M}{\partial s}\right) = 0 \tag{8}$$

where $A(s)$ is the interface area for transport in the matrix blocks at a distance s from the block surface. Note that C^M is no longer averaged over the rock blocks. Boundary conditions are given by

$$C^M(s = 0) = C^F; \quad \frac{\partial C^M}{\partial s}(s = S) = 0 \tag{9}$$

After solving equation (8) the diffusive solute exchange is obtained by applying FICKs law at the interface between fractures and porous blocks

$$W^D = -\left(1 - n^F\right)\Omega n^M D_m^M \frac{\partial C^M}{\partial s}\bigg|_{s=0} \qquad (10)$$

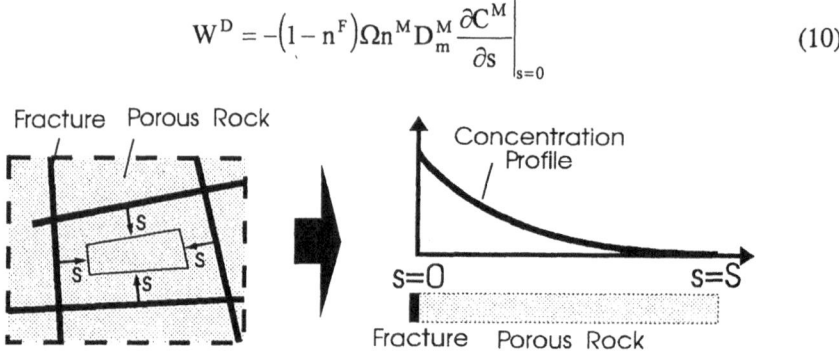

Fig. 2:   One-dimensional concentration profile in matrix blocks

3   BASIC CONCEPT OF THE NUMERICAL TRACER EXPERIMENTS

The methodology of comparing the results of discrete and continuum models implies a scale problem: A discrete representation of the fractures and the porous rock is only possible on small scales because of the enormous amount of data, whereas continuum models usually include large scales to meet the requirements related to the definition of REVs. Thus an appropriate concept is needed which allows for the performance of both discrete and continuums models. Figure 3 illustrates a schematic diagram of an idealized fractured formation with two sets of parallel equidistant fractures embedded in porous permeable matrix blocks with homogeneous and isotropic properties *(Berkowitz et al., 1988)*. The fluid is subject to steady flow in the positive x-direction. We refer to the migration of a contaminant disposed uniformly at x=0. The concentration at that location is $C=C_0=1$.

Although the conceptual model is highly idealized, it is coherent with the basic requirements needed for the present study: On one hand, it represents the tortuous flow paths through a natural fracture network which provide a large interface for mixing between the fractures and the porous blocks. On the other hand, the mixing in fracture intersections does not lead to dispersion. Thus it is possible to use the systems symmetry and only simulate the contaminant transport in a subdomain of the formation as shown in figure 3.

The discrete simulation runs were performed with a GALERKIN-Finite-Element-Code on a supercomputer SNI S-600/20. Due to the heterogeneity of the formation a refined discretization in space and time was needed to meet the Peclet and Courant criteria. Triangle elements were used for the porous matrix and one-dimensional line-elements for the fracture representation.

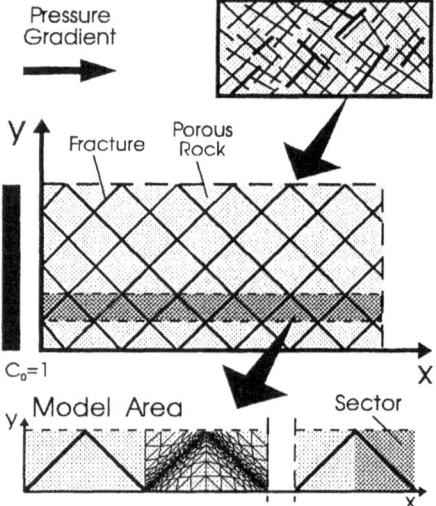

Fig. 3:   Conceptual Model with idealized fracture network

A maximum number of 100 sectors was considered with 1000 line elements, 20000 triangles and up to 30000 time steps. A sector is defined by the distance between two adjacent fracture intersections measured along the x-axis. The hydraulic gradient is imposed by Dirichlet-type boundary conditions on both sides of the model area; the solute disposal is represented by a Dirichlet-type boundary condition at x=0. Both continuous and time-dependent disposals are simulated. Initially the model area is not contaminated.

A wide range of medium properties was studied; however, only three simulations shall be presented here, considering different matrix conductivities (*Case A: 8.35 10⁻⁶ m/d, Case B: 8.3 10⁻³ m/d* and *Case C: 9.94 10⁻² m/d*). Table 1 provides the other parameters chosen for the fractures and the matrix blocks, respectively.

The dispersivities were not varied in our study: In the fractures the longitudinal dispersivity was set to 0.05 m; in the matrix a longitudinal dispersivity of 0.01 m and a transverse dispersivity of 0.001 m were chosen. The conductivity of single fractures was evaluated by applying the well-known Poisseuille equation for laminar flow between parallel plates. With an aperture of 10^{-4} m the conductivity of a single fracture is 703.0 m/d. A steady flow field was assumed with a hydraulic gradient of

Parameter	Present Value	Total Range Studied	Unit
Fracture			
Aperture	10^{-4}	$0.2\ 10^{-4}$ - 10^{-3}	m
Angle to x-axis	45	26.56 - 63.43	°
Fracture distance	0.707	0.447 - 0.894	m
Rock Matrix			
Porosity	0.05	0.01 - 0.25	-
Diffusion Coeff.	$2.0\ 10^{-5}$	$0.4\ 10^{-5}$ - 10^{-4}	m²/d
Size of Blocks	0.5	0.25 - 1.0	m²

Table 1: Medium Properties for the discrete simulation

0.01. Note that a continuously changing pressure boundary condition did not show any effect since the local pressure differences between fractures and rock vanished within very small time periods, due to the small storage coefficient in a saturated medium.

4   DUAL-CONTINUUM MODELING

The dual-continuum simulations are performed for the same model area as presented in figure 3. Due to the symmetry of the system, the resultant flow and transport processes in both continua are one-dimensional in x-direction. The equivalent properties of the two continua can easily be derived from the discrete parameters in table 1.

Two numerical solution techniques are used for the coupled dual-continuum problem: 1. The two continua are discretized and solved as different systems. The coupling may require iterative procedures. 2. The two continua are discretized simultaneously and directly solved in one system of equations. In our work the **dual-permeability** problem is solved with approach 2 to avoid iterative procedures. A Galerkin Finite-Element-Code was adopted to meet the special requirements of the coupling between fracture and rock continuum. For the **dual-porosity** problem it is appropriate to choose approach 1 since efficient direct solution techniques are available for both quasi-steady and unsteady formulations. See details in *Huyakorn et al. (1983)* or *Birkhölzer and Rouvé (1991)*.

4.1 Dual-Porosity Modeling

With a hydraulic conductivity as given in *Case A* the porous matrix is practically impermeable. Thus *Case A* is a typical application for a **dual-porosity** model. Figure 4 and 5 show breakthrough curves for *Case A* calculated at certain locations in the domain. The distance between the particular location and the tracer inlet shall be given by the number of sectors. Here distances of 2, 10, 30 and 90 sectors are considered. Only continuous disposal of a contaminant shall be presented.

All curves show the typical behaviour of dual-porosity media: In the beginning the contaminant build-up is very fast due to the large velocities in the fracture network. After a certain time the permanent diffusive loss from the fractures into the matrix pore system becomes obvious. The concentration build-up becomes slower and slower; the breakthrough curves are almost horizontal.

In figure 4 the results of the discrete simulation are compared with the **quasi-steady** dual-porosity formulation. However, the agreement between the discrete and the continuum model is not sufficient. The linear diffusive transfer term as given in equation (3) is not able to represent the steep gradients between fractures and porous rock, especially in the beginning of the contamination. Note that the shape factor β in equation (4) was carefully calibrated.

A much better representation of the time-dependent tracer migration is achieved with the **unsteady** formulation. The agreement between the discrete and the dual-porosity simulation results in figure 5 is almost perfect. The diffusion process in the matrix blocks can accurately be approximated by the one-dimensional equation (8).

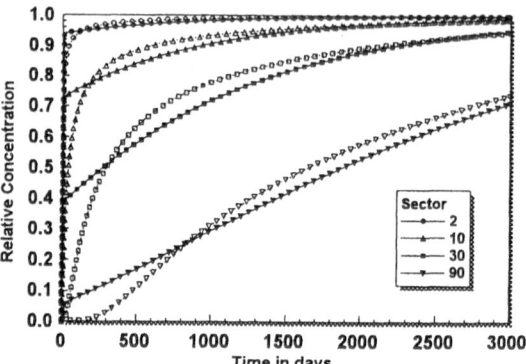

Fig. 4: Breakthrough curves for *Case 1*
Discrete model (line symbols) and **quasi-steady** dual-porosity model (filled symbols)

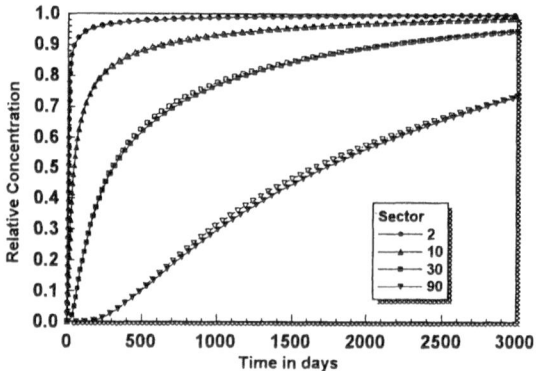

Fig. 5: Breakthrough curves for *Case 1*
Discrete model (line symbols) and **unsteady** dual-porosity model (filled symbols)

4.2 Dual-Permeability Modeling

In *Case B* the flow rate through the rock continuum provides about 8 % of the overall flow in the formation. The permeability of the porous matrix can not be neglected. Furthermore it has to be considered that the solute exchange between the two systems is based on diffusive as well as on advective transport. In *Case C* about 50 % of the fluid flows in the fracture and 50 % in the rock continuum. The mixing between the fracture and the porous block flow is so intense that the diffusive solute transfer is almost negligible. In the following paragraph a new coupling term is developed for the advective mixing between the fracture and the porous block flow.

In equation (5) a fluid mixing rate W^{QR} was introduced to derive the solute exchange term of "regional" advection. Figure 1 indicates that W^{QR} is correlated to the intensity and the direction of fluid flow in the matrix and to the geometry of the fracture network. If the majority of the fractures is parallel to the fluid flow in the matrix the mixing rate is small. Obviously only those parts of the fractures are relevant for the fluid mixing which are perpendicular to the the direction of flow in the matrix (see fig. 6). A new parameter, the **surface function** Ω_W, is introduced to describe the relevant part of the fracture surface

$$\Omega_W = \frac{1}{V^M} \sum A^F \frac{\overline{q}_i^M}{\left|\overline{q}_i^M\right|} n_i \qquad (11)$$

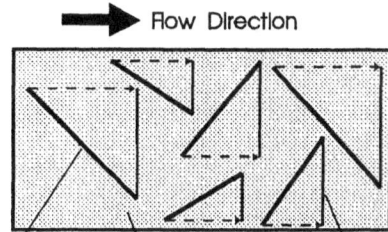
Flow Direction

The definition of Ω_W is similar to the definition of the specific surface, besides that it is a function of the flow direction rather than a constant. V^M is the bulk volume of the porous rock, A^F is the fracture surface and n_i is a unit vector perpendicular to A^F. Like other medium properties the relevant surface can be provided as a material parameter of the rock continuum.

Fracture   Porous Rock   Relevant Surface

Fig. 6:   Relevant fracture surface

For naturally fractured formations the surface function Ω_W can be estimated by Monte-Carlo-Studies using statistically generated fracture networks.

With a given function Ω_W the fluid mixing rate W^{QR} is easily derived as follows

$$W^{QR} = \left(1 - n^F\right)\left|q_i^M\right|\Omega^W \qquad (12)$$

As the flow field has already been solved, W^{QR} can directly be implemented into the solute transfer term of "regional" advection (see eq. (6)). For the conceptual model presented in chapter 3 only the value of Ω_W for the x-direction has to be considered. This value is given as 2.0 m^{-1}. With hydraulic conductivitites of $8.3 \cdot 10^{-3}$ m/d (*Case 2*) and $9.94 \cdot 10^{-2}$ m/d (*Case 3*) and a hydraulic gradient of 0.01 the mixing term W^{QR} becomes $1.66 \cdot 10^{-4}$ d^{-1} and $1.988 \cdot 10^{-3}$ d^{-1}, respectively.

With both the diffusive and the advective solute transfer term given as linear functions of the concentration values, the dual-permeability problem can directly be solved. Figure 7 and 8 show breakthrough curves for *Cases 2* and *3*, respectively. In both figures results of the discrete and the dual-permeability simulations are compared.

The additional advective mixing between the fracture flow and the porous block flow reduces the heterogeneity of the formation. In both figures the breakthrough curves start with a concentration build-up slower than in *Case 1* since a larger amount of solute is transferred from the fractures into the matrix (compare fig. 4 and 5). Due to the intense contaminant exchange, smaller time intervals are needed to contaminate the matrix. The tailing of the breakthrough curves is less significant.

In figure 7 differences occur between the discrete and the dual-permeability simulation, due to the fact that a quasi-steady formulation can not exactly represent the concentration profile in the porous blocks. However, the agreement is much better than in the quasi-steady dual-porosity simulation of *Case 1* (see fig. 4). It seems that the additional

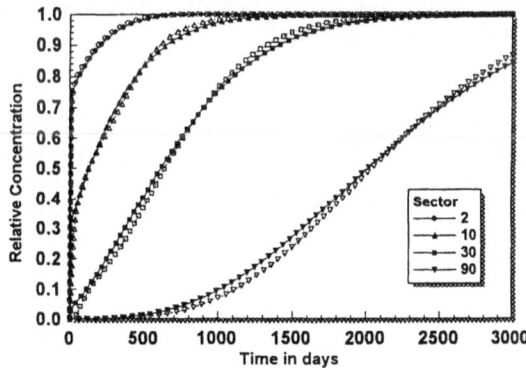

Fig. 7:   Breakthrough curves for *Case 2*
Discrete model (line symbols) and **quasi-steady** dual-permeability model (filled symbols)

"advective" solute exchange improves the performance of quasi-steady models.

In *Case 3* the agreement between the discrete and the dual-permeability simulation is almost perfect. The solute exchange of "regional" advection is accurately described by the new coupling term presented in the preceding section. The diffusive exchange term is not needed in this case since molecular diffusion in the matrix is negligible compared to the effect of advection.

Note that the breakthrough curves in figure 8 have the well-known S-shape typical for ideal porous media. In that case a classical continuum

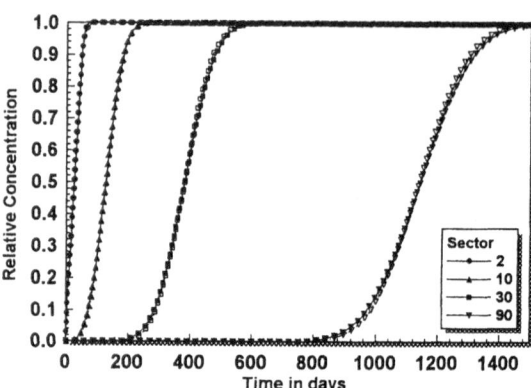

Fig. 8: Breakthrough curves for *Case 3* Discrete model (line symbols) and **quasi-steady** dual-permeability model (filled symbols)

approach seems sufficient for representing the heterogeneous system. However, treating fractures and porous rock as one single continuum requires the calibration of equivalent medium parameters. With a dual-continuum model the equivalent parameters can directly be determined from the fracture and matrix properties.

5   CONCLUSION

Dual-continuum models are widely used to simulate the flow and transport processes in fractured porous formations. In our study we simulate the migration of solutes in fracture-rock-systems by discrete models and compare the results with different dual-continuum approaches. It is shown that for almost impermeable rock the unsteady dual-porosity model leads to very good results. Quasi-steady formulations are less accurate and should not be applied when the main interest is focused on the exact description of the transient response in the matrix. If advective flow in the matrix is relevant a better performance of quasi-steady models is obtained. A new coupling term is introduced for the mixing between the fluid flow and the porous matrix flow which allows for an accurate description of the advective mass transfer.

6   REFERENCES

Berkowitz, B., Bear, J., Braester, C. (1988) "Continuum Models for Contaminant Transport in Fractured Porous Formations", Water Resour. Res., 24, 8, 1225 -1236.

Birkhölzer, J. and Rouvé, G. (1991) "Solute Transport in Fractured Porous Rock", Proc. 7th Int. Congress Rock Mechanics, Aachen, September 1991.

Gerke, H.H. and Van Genuchten, M.T. (1993) "A Dual-Porosity Model for Simulating the Preferential Movement of Water and Solutes in Structured Porous Media", Water Resour. Res., 29, 2, 305 - 319.

Huyakorn, P.S., Lester, B.H.; Mercer, J.W. (1983) "An Efficient Finite Element Technique for Modeling Transport in Fractured Porous Media: 1. Single Species Transport", Water Resour. Res., 19, 3, 841-854.

Teutsch, G. (1990) "An Extended Double-Porosity Concept as a Practical Modeling Approach for Karstified Terranes", Int. Symp. Field Seminar on Hydrogeological Processes in Karst Terranes, 7. - 17. Oct. 1990, Antalya, Turkey.

MONTE CARLO SIMULATION OF SUSPENSION FLOWS IN UNCONSOLIDATED POROUS MEDIA

V.N. BURGANOS, C.A. PARASKEVA and A.C. PAYATAKES
Institute of Chemical Engineering and High Temperature Chemical Processes, and
Department of Chemical Engineering, University of Patras
Patras 26500
Greece

Flow of underground water through unconsolidated porous rocks and deposition of suspended matter on the internal surface are simulated using Monte Carlo techniques. The structure of the rock is represented by regular and distorted 3-D networks of pores with converging-diverging geometry. The hydraulic resistance of each pore is computed by solving the creeping flow equation with a collocation technique. The microscopic pressure profile is calculated using the standard network solution. Trajectories of a large number of non-Brownian test particles across the pore network are computed that may lead to either escape through the working sample or deposition at some point on the internal surface. The selection of the pathway through the network is made in a stochastic manner that respects fully the local population balance. Local and global deposition rates are computed using both regular and distorted networks in the down-, horizontal, and upflow filtration modes. It is found that the relative particle capture efficiency of the three flow modes depends strongly on the pore size distribution and the superficial velocity used, whereas distortion of a regular cubic network improves the upflow mode of operation considerably.

INTRODUCTION

Naturally occurring and externally driven flow of underground water through porous formations is of particular importance in several engineering and scientific disciplines, such as water resource evaluation, aquifer rehabilitation, enhanced oil recovery, hydrogeology, agriculture etc. The characteristics of this flow that are of primary interest in relevant studies are the permeability and evolution of permeability, the rate of suspended matter deposition during flow, the distribution of regions of intensive collection throughout the porous medium, and the effluent clarity in the main and lateral flow directions.

There is a large number of theoretical works that attempt to model the flow of water through porous solids as flow of a dilute suspension through a pore network or through the interstitial space in a grain assemblage. It is beyond the scope of this work to review these approaches which are discussed in detail elsewhere (Tien and Payatakes, 1979; Tien, 1989; Sahimi et al., 1990). We will rather focus our attention on the "particle trajectory" methods which have proved capable of elucidating the basic transport and capture mechanisms that underly particle motion and physicochemical interaction with the collector surface.

In a series of recent publications, the authors have presented trajectory-based simulators of motion and deposition of non-Brownian particles in 3-D networks of pores with converging-diverging geometry. A 3-D particle trajectory analysis in sinusoidally

A. Peters et al. (eds.), Computational Methods in Water Resources X, 667–674.
© 1994 Kluwer Academic Publishers. Printed in the Netherlands.

shaped pores has been developed by Paraskeva et al. (1991), and used for the calculation of the local deposition rate. The network simulator developed by Burganos et al. (1992, 1993) revealed that unit cells not aligned with the macroscopic flow direction can be responsible for a large portion of deposition in the pore walls thanks to the low particle velocities that develop there combined with the action of gravity towards the lower unit cell surface. That simulator involved solution of the population balance equation written around each node of the network and computation of the particle collection efficiency of each unit cell. Alternatively, the authors developed recently (Burganos et al., 1994a) a Monte Carlo technique which, in essence, provides a stochastic solution to the population balance equation, and stochastic estimates of the collection rate in the unit cells. Horizontal, up-, and downflow modes were examined in that work. However, the results of those works were restricted to regular cubic networks with one principal direction parallel to the gravity direction. Consequently, upflow operation proved of very poor performance because of the practically nil removal efficiency of vertical upflow pores (Burganos et al., 1994b). In the present work, the highlights of this novel Monte Carlo simulator are presented and the significance of network distortion is investigated in the horizontal, up-, and downflow modes.

DESCRIPTION OF THE METHODOLOGY

The void space of an unconsolidated porous medium is represented here by a regular or distorted cubic network of unit cells with sinusoidal shape (Figure 1). Each unit cell is characterized by its length (l), constriction radius (r_c), and mouth radii (R_1, R_2). The coordination number of the network and the constriction and mouth size distributions remain constant upon distortion of a regular network. Structural periodicity in directions perpendicular to the average flow direction is imposed so that the working sample becomes equivalent to an infinitely large pore network.

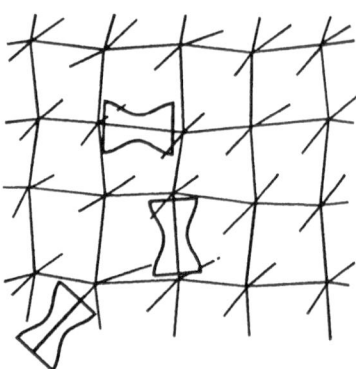

Figure 1. Sample of distorted cubic network of pores with converging-diverging geometry

Non-Brownian test particles are injected into the pore network through the entrance face. The probability of selecting an entrance pore for the initiation of the particle trajectory is proportional to the rate of particle transport through that pore. This selection procedure is implemented through a Monte Carlo routine that first calculates the particle transport rate through each entrance pore and then uses a pseudorandom number generator for the weighted selection of a pore. The particle transport rate in each pore is calculated using a force balance on the test particle at the pore mouth. Gravitational, hydrodynamic, London-van der Waals (including retardation) and double ionic layer forces are considered acting on the test particles in this work. However, in this part of the simulator the last two surface forces affect negligibly the particle entrance into a pore and the particle velocity is practically determined from the balance of gravity and drag forces.

The entrance position on the pore mouth is selected next. To this end, a uniform deviate in $(0,1)$, d_u, is generated and the radial coordinate, r_{in}, of the particle entrance position is calculated numerically from the equation

$$d_u = \int_0^{r_{in}} u(r)\, r\, dr \Big/ \int_0^{R-a_p} u(r)\, r\, dr$$

where a_p is the particle radius and $u(r)$ is the particle velocity at the position with radial coordinate r. The distribution of the angular coordinate, θ, of the particle entrance is uniform over $[0,2\pi)$.

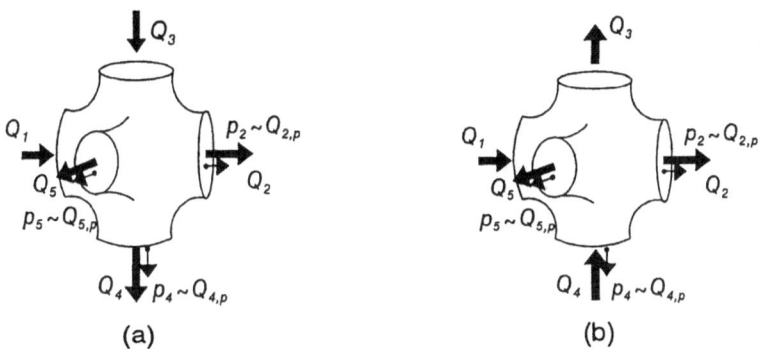

(a)   (b)

Figure 2. Pore selection procedure in (a) downflow and (b) upflow segments of a cubic network. p_i: probability of selecting the i-th pore; Q_i: flow rate in the i-th pore; $Q_{i,p}$: particle entrance rate in the i-th pore.

Once the particle has entered the pore, it translates and rotates within the pore space following, in general, a 3-D trajectory. This is accomplished by integrating the trajectory equation that results from consideration of local force and torque balances. Details of this procedure, which uses a fourth order Runge-Kutta routine for the numerical integration of the resulting ordinary differential equations, are given in Paraskeva et al. (1991) and will not be repeated here. If a particle touches the pore wall, it is assumed that instantaneous capture occurs and the trajectory is terminated. If the particle arrives at the exit mouth of the pore, it selects one of the adjacent pore segments to resume its travel. This pore selection is also implemented through a Monte Carlo procedure that weights the selection according to the local particle transport rate. Care is exercised in upflow cells where particles can actually move downwards (that is, in the direction opposite to the flow direction) in regions of low drag, or even stagnate. Thus, double entrance-double exit phenomena may develop (Burganos et al., 1994b) which complicate the trajectory calculations and the pore selection procedure. Figure 2 shows schematically the weighting used in the pore selection procedure upon arrival of a test particle at a downflow (a) and an upflow (b) pore network segment. The entrance position at the mouth of the selected pore is selected through the procedure described above for the entrance pores (weighting according to the local particle transport rate). Particle trajectory calculations are then performed within the pore space until capture or escape takes place and so on. If the particle manages to escape through the exit face of the working sample, it is considered as transmitted and a new test particle is examined. Figure 3 shows the algorithm used in the simulator.

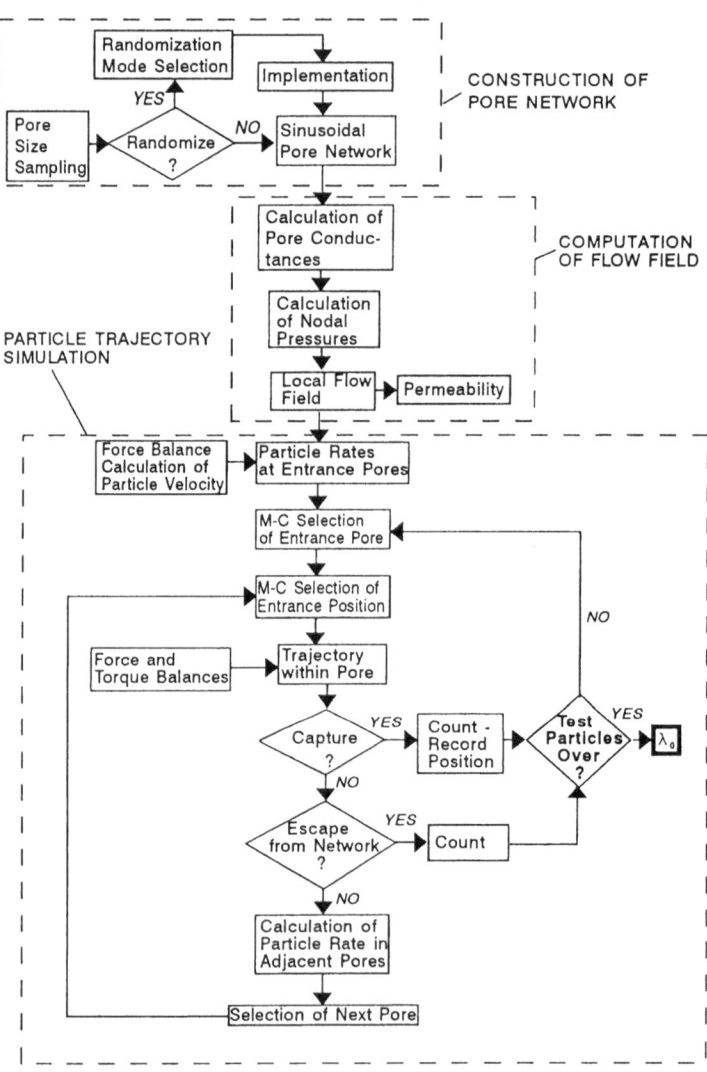

Figure 3. Algorithm used in the simulator.

RESULTS AND DISCUSSION

Figure 4 shows the dependence of the impacted fraction in down- and upflow vertical unit cells (that is, the number fraction of particles that enter the unit cell per unit of time and deposit on the pore surface) on the cell inclination. (Typical values of the main physicochemical parameters are used in this work. The interested reader is referred to any of our previous publications.) Vertical unit cells ($\phi=90°$) can collect some suspended matter only if operated in the downflow mode. However, deviation from the vertical orientation by ~20° or more favors the upflow mode yielding higher deposited fraction than in the downflow mode (Burganos et al., 1994b).

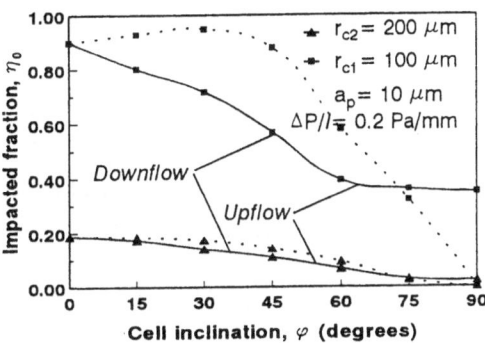

Figure 4. Impacted fraction vs. cell inclination during down- and upflow operation. Variation with the constriction size.

At the network scale, we can use the Monte Carlo simulator described in the previous section to investigate the distortion effects on the overall network collection efficiency quantified by the filtration or retention coefficient, λ. At the initial stages of suspension flow through a porous medium, the retention coefficient can be obtained from

$$\lambda_0 = -\frac{1}{L} \ln \frac{C_{eff}}{C_0} = -\frac{1}{L} \ln f_{esc}$$

where C_0, C_{eff} are the influent, effluent stream concentrations, respectively, and L is the distance between entrance and exit faces. The value of λ_0 can be obtained from our simulations by calculating the number fraction of test particles that escape through the exit face of the sample, f_{esc}.

Figure 5a shows the dependence of the retention coefficient during horizontal and downflow of suspensions through regular unimodal pore networks on the superficial velocity. Horizontal flow results in higher collection efficiency over the entire range of typical superficial velocity values. The upflow mode yields retention coefficient values that are negligibly small and do not appear in the graph. However, inclination of all the vertical cells while retaining the horizontal pores intact, favors deposition in the upflow mode yielding for small superficial velocity values, retention coefficients higher from both during horizontal and downflow operation, (Figure 5b).

Results for networks with nonuniform constriction size are presented next. Figure 6a shows the retention coefficient values obtained from numerical experiments on a fixed network realization with discrete bimodal constriction size distribution with varying number of test particles. Note that more than ~ 5,000 particles are needed for reliable statistics. For constant number of test particles, Figure 6b shows that networks of at least 20 unit cells in each main direction must be used for statistically accurate retention

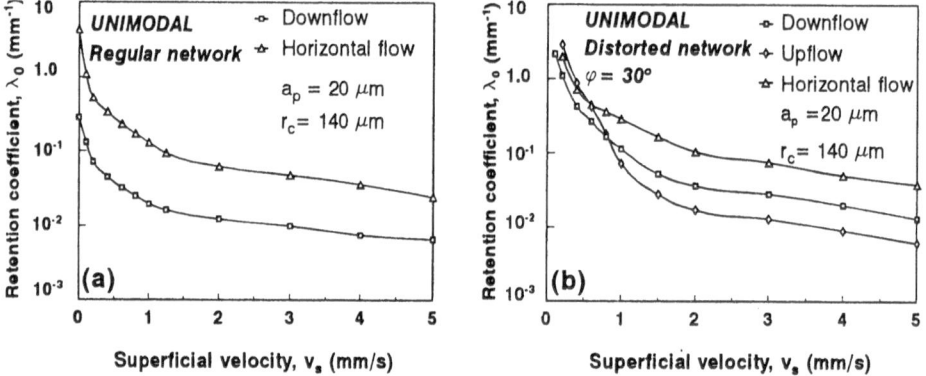

Figure 5. Dependence of the retention efficiency on the superficial velocity for (a) regular and (b) distorted unimodal cubic networks.

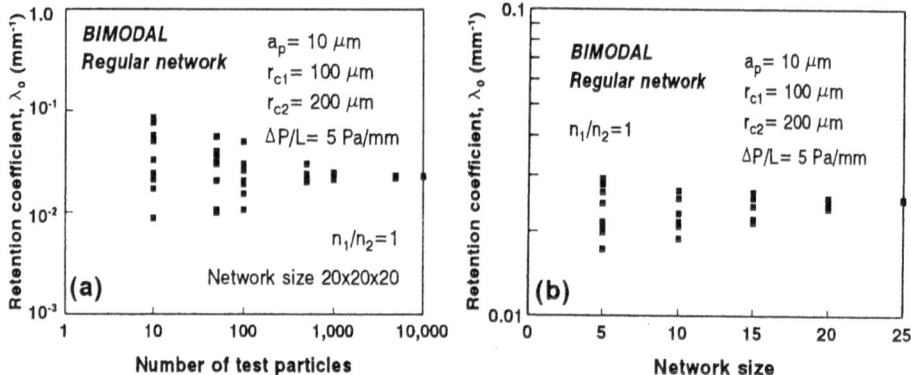

Figure 6. Retention efficiency of downflow networks with discrete bimodal constriction size distribution vs. (a) number of test particles used in the simulation (fixed network realization) and (b) network size (sufficiently large number of test particles in all cases). Population ratio of narrow to wide cells $n_1/n_2=1$.

coefficient estimates. These observations served as guidelines for the construction of Figure 7. It is seen there that distortion of an upflow regular cubic network with discrete bimodal distribution of the constriction size through inclination of vertical pores to $\varphi=30°$ causes considerable increase in the filter coefficient over the entire range of the number fraction of wide cells. This increase attains its maximal value at the lower unimodal limit as is also predicted by the results presented in Figure 4.

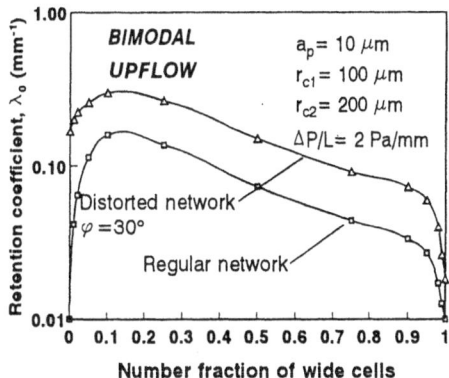

Figure 7. Dependence of the retention efficiency of upflow networks with discrete bimodal constriction size distribution on the number fraction of wide cells. Comparison of regular and distorted network cases.

CONCLUSIONS

Flow of dilute suspensions in unconsolidated porous media was simulated using a three-stage Monte Carlo procedure. First, a regular or distorted cubic network of pores with converging-diverging geometry is constructed with mouth and constriction sizes sampled from prescribed distributions. Then, the hydraulic conductance of each unit cell is calculated under creeping flow conditions using a collocation procedure. The nodal pressure values are obtained from a standard network solution. Finally, particle trajectories in the pore space are computed using Monte Carlo techniques for the selection of the particle pathway through the network and for the selection of the particle entrance position at pore mouths. In both selection procedures the decision was weighted according to the local particle transport rate. The particle trajectory within unit cells was entirely deterministic once the entrance position was selected and was dictated by the local force and torque balance equations. It was found that distortion of a cubic network can lead to highly increased particle collection rates during suspension flow in any of the horizontal, up-, and downflow modes. However, combination of small superficial velocities and network skeleton distortion favors the upflow operation in particular, rendering the upflow capture efficiency higher (for particles with density larger than that of the fluid) than that of both horizontal and downflow modes. As a general conclusion, it must be noted that microscopic structure details related not only to pore size and geometry but also to pore orientation are of primary importance in suspension flow and particle deposition studies and deserve further theoretical and experimental investigation.

NOTATION

a_p suspended particle radius
C_0, C_{eff} influent and effluent stream concentrations
f_{esc} number fraction of test particles that escape through the exit face of the sample
L   length of the bed

p_i probability of selecting the i-th pore on an intersection
l length of a unit cell
Q_i flow rate in pore i
$Q_{i,p}$ particle entrance rate in pore i
R_1,R_2 mouth radii of the unit cell
r radial cylindrical coordinate
r_c constriction radius of the unit cell
$u(r)$ particle velocity at position with radial coordinate r
v_s superficial velocity

Greek Letters

ΔP pressure drop

η_0 impacted fraction

θ angular coordinate

λ_0 retention coefficient

ϕ angle formed by the axis of a unit cell and the horizontal

REFERENCES

Burganos,V.N., Paraskeva, C.A. and Payatakes, A.C. (1992) "Three-dimensional trajectory analysis and network simulation of deep bed filtration", J. Colloid Interf. Sci., **148**, 167-181

Burganos, V.N., Paraskeva, C.A. and Payatakes, A.C. (1993) "Parametric study of particle deposition in sinusoidal pores of arbitrary orientation", J. Colloid Interf. Sci., **158**, 466-475

Burganos, V.N., Paraskeva, C.A. and Payatakes, A.C. (1994a) "Monte Carlo network simulation of horizontal, upflow, and downflow depth filtration", AIChE J., accepted

Burganos, V.N., Paraskeva, C.A. and Payatakes, A.C. (1994b) "Motion and deposition of non-Brownian particles in upflow collectors", Sep. Technol., **4**, 47-53

Paraskeva, C.A., Burganos, V.N. and Payatakes, A.C. (1991) "Three-Dimensional trajectory analysis of particle deposition in constricted tubes", Chem. Eng. Comm., **108**, 23-48

Tien, C. and Payatakes, A.C., "Advances in deep bed filtration", AIChE J., **25**, 737-759

Tien, C. (1989) Granular Filtration of Aerosols and Hydrosols, Butterworths series in Chemical Engineering, Boston

Sahimi, M., Gavalas, G.R. and Tsotsis T.T. (1990) "Statistical and continuum models of fluid-solid reactions in porous media", Chem. Eng. Sci., **45**, 1443-1502

FLOW TO A WELL IN A FRACTURED AQUIFER : INFLUENCE OF THE UNSATURATED ZONE

K.S.HARI PRASAD, M.S.MOHAN KUMAR and M.SEKHAR
Department of Civil Engineering, Indian Institute of Science,
Bangalore 560 012
INDIA.

Flow to a well in a fractured aquifer using double porosity approach is studied by considering the unsaturated zone in the weathered zone (unconfined block). The effect of the thickness of the unsaturated zone and the unsaturated parameters on drawdowns in the fracture and the water table (block) are discussed. It is found that neglecting the unsaturated zone in the block resulted in lesser fracture and water table drawdowns. This implies an underestimation in the specific yield parameter and an associated underestimation in ground water resource in fractured aquifer.

INTRODUCTION

Ground water flow and storage in fractured aquifers has always been a matter of great interest and importance to hydrogeologists in India, as well as world over from both ground water quantity and quality related concerns. It is generally acknowledged that theories developed for a homogeneous porous medium may not be applicable for a heterogeneous and anisotropic fractured rock aquifer system. It is recognized that flow through such a system is very significantly influenced by the fracture characteristics. It must be noted that inherently, the problem of understanding flow and storage in fractured rock aquifers is very complex. There is the heterogeneity associated with the mechanical discontinuity resulting from the presence of fractures. The rock matrix or blocks surrounding the fractures is generally not impermeable. Though its permeability may be very small, flow transfer takes place between the fractures and blocks. The fractured formations also generally display anisotropic characteristics.

A proper conceptual model is a prerequisite for assessment of ground water resources in fractured areas. Studies of fluid flow in fractured rock mass where fissure flow is augmented by contributions from blocks, have generally adopted the double porosity concept proposed by Barenblatt et al. (1960). In these models, the fracture rock mass is assumed to consist of two interacting overlapping continua, a

A. Peters et al. (eds.), Computational Methods in Water Resources X, 675–682.
© 1994 Kluwer Academic Publishers. Printed in the Netherlands.

continuum of low permeability, primary porosity blocks and a continuum of high permeability, secondary porosity fissures. Double porosity models differ in the approach to the treatment of the block to fracture leakage. Based on the spatial variation of hydraulic head gradient in the block, two theories namely, pseudo-steady state flow approach and transient flow approach were considered. Moench (1984) unified the above theories of flow to a well in double porosity reservoir by using the concept of fracture skin. Recently, Sekhar et al. (1993) have considered the effect of water table in the weathered zone on the hard rock aquifers using the conceptual similarity between the double porosity and the leaky aquifer models.

All the above idealisations restrict attention to flow in the saturated zone, assuming that flow in the unsaturated zone has little effect on flow in the saturated zone. Recent studies of Narasimhan and Zhu (1993) and Nwankwor et al. (1992) show that drainage processes above water table significantly affect the response in an unconfined aquifer. The present paper studies the effect of the unsaturated zone, above the moving water table, on the flow behaviour towards a well in a fractured aquifer. A coupled saturated-unsaturated model applicable for fractured aquifers is required for both ground water resource assessment and contaminant movement studies.

CONCEPTUAL MODEL AND MATHEMATICAL FORMULATION

Fig.1 presents a parallel fracture double porosity system with well fully penetrating the fracture aquifer overlain by a rock matrix block. If the weathered zone above the the fractured aquifer is looked upon as a block, which is influenced by surface processes such as infiltration, evaporation and leakage to the fracture zone, a saturated-unsaturated model for the block is necessary. The flow in the fracture is assumed radial and the flow in the block is assumed vertical.

The governing equations for flow in fracture and block due to pumping from a well, ignoring well bore storage are as follows. In the following discussions, the terminology fracture or aquifer, block or aquitard are used interchangeably.

Fracture :

$$T_r \left(\frac{\partial^2 h}{\partial r^2} + \frac{1}{r} \frac{\partial h}{\partial r} \right) = S_t \frac{\partial h}{\partial t} - K'_r K' \left. \frac{\partial h'}{\partial z} \right|_{z=0} \tag{1}$$

$$\begin{aligned} t = 0 \; &: \; h = h_o \\ t > 0 \; &: \; r \to \infty \quad \frac{\partial h}{\partial r} = 0 \\ &\quad lim_{r \to 0} r \frac{\partial h}{\partial r} = \frac{Q}{2\pi T_r} \end{aligned} \tag{2}$$

Block :

$$\frac{\partial}{\partial z}\left[K'_r K' \frac{\partial}{\partial z}(\psi + z)\right] = \left(\frac{d\theta}{d\psi} + \frac{\theta}{\phi}S'_s\right)\frac{\partial \psi}{\partial t} \tag{3}$$

$$
\begin{aligned}
t = 0 \quad &: \quad \psi(z) = h_o - z \\
t > 0 \quad &: \quad z = 0 \,, \quad \psi = h \\
&\qquad z = L \,, \quad \frac{\partial \psi}{\partial z} + 1 = 0
\end{aligned}
\tag{4}
$$

In eqns. (1)-(4) h is the piezometric head in the fracture, h' is the piezometric head in the block, ψ is the pressure head in the block, θ is the volumetric moisture content in the block, z is the vertical coordinate taken positive upwards, r is the radial coordinate in the equivalent isotropic domain, t is the time. T_r and S_t are the equivalent transmissivity and storage coefficient of the fracture respectively. K', K'_r, S'_s, ϕ are the vertical saturated conductivity, relative hydraulic conductivity, specific storage and porosity of the of the block respectively. L is the thickness of the block, h_o is the prepumping head and Q is the pumping discharge.

To solve eqns. (1) and (3) two constitutive relationships between (i)pressure head and moisture content, (ii) conductivity and moisture content, in the block have to provided. The empirical equation suggested by van Genuchten (1980) is used for pressure head and moisture content relationship. The equation is

$$
\begin{aligned}
\Theta \;&=\; \left[\frac{1}{1 + \|\alpha\psi\|^n}\right]^m \quad \text{for } \psi \le 0 \\
&=\; 1 \qquad\qquad\qquad \text{for } \psi > 0
\end{aligned}
\tag{5}
$$

where α and n are unsaturated soil parameters with $m = 1 - \frac{1}{n}$ and Θ is effective saturation, defined as

$$\Theta = \frac{\theta - \theta_r}{\theta_s - \theta_r} \tag{6}$$

where θ_s and θ_r are saturated moisture content and residual moisture content of the block respectively.

Mualem's (1976) equation is used to calculate the relative hydraulic conductivity K'_r

$$K'_r = \Theta^{1/2} \left[1 - \left(1 - \Theta^{1/m} \right)^m \right]^2 \tag{7}$$

The above equations are nondimensionalised to facilitate comparison with earlier solutions by choosing the length scale as h_o and the time scale as $\frac{S_t h_o^2}{T_r}$.

NUMERICAL SCHEME

The fracture and block equations, eqns.(1) and (3) are coupled due to the presence of leakage term in the fracture eqn.1 and the continuity requirement at the block fracture interface, eqn.4. At the current simulation time, these equations can be solved iteratively, by solving in each iteration, the heads in the fracture at all the nodes in the radial direction and heads in the block at all the nodes in the vertical direction, column by column. It is to be noted that an iterative procedure is required for solving the system of equations due to the nonlinearities in saturated-un saturated model in the block eqns.(3) and (4) for each outer iteration. The non-linear system of equations in the block are solved using Picard's scheme Paniconi et al. (1992).

A finite element scheme leads to a tridiagonal system of equations for the fracture as well as for each column in the block which can be solved conveniently by double sweep method or Thomas algorithm.

RESULTS AND DISCUSSION

Influence of the unsaturated zone is studied by plotting fracture drawdowns (s) and water table drawdowns (s') in the block for various cases with nondimensional drawdown ($s / \dfrac{Q}{4\pi T_r}$, $s' / \dfrac{Q}{4\pi T_r}$) versus nondimensional time (T/R^2) where T and R are chosen as $\dfrac{t T_r}{S_t h_o^2}$, $\dfrac{r}{h_o}$ respectively. In the following discussion the values of parameters T_r, S_t, K', S'_s are used which correspond to the values obtained in earlier studies in hard rock aquifers of Vedavati river basin.

a. Effect of thickness of unsaturated zone

Two thicknesses of unsaturated zone in the block are considered, with one case having no initial unsaturated zone, while the second case considers a 5-m thick unsaturated zone. Fig.2 shows smaller drawdowns in fracture and water table for the case of higher thickness of unsaturated zone. However the effect of the thickness of the unsaturated zone is felt only at later time in the case if fracture drawdowns. The smaller drawdowns associated with a higher thickness of unsaturated zone can be attributed to the larger amount of drainable water available in the unsaturated zone at any instant of time.

b. Effect of soil parameters α, n

Here the effect of variation in the soil parameters α and n on the fracture and water table drawdowns are considered with five meter thick initial unsaturated zone. Fig.3 shows that the effect of α is negligible on fracture drawdowns while significant effect can be seen on wate table drawdowns with a higher drawdown for low value of α. Physically, a lower value of α represents a higher thickness of capillary fringe in the unsaturated zone which in turn retains more drainable water and hence larger drawdowns. Fig.4 shows that the effect of n is found to be negligible on the fractured drawdowns with less pronounced effect on water table drawdowns. A higher water table drawdown is observed for the case of low n values, which physically correspond to a soil type having slow drainage mechanism.

Influenc∊ of unsaturated zone

To study the influence of unsaturated zone, two models are considered. Model 1 considers a saturated-unsaturated model in the block. Model 2 considers a saturated model(Sekhar et al. , 1993) with a moving water table in the block, which assumes that the parameter S'_y accounts for the drainage processes above the water table and is approximately equal to the difference between saturated moisture content and the residual moisture content (θ_s - θ_r). For purposes of comparison with Model 2, Model 1 is run with no initial unsaturated zone. It is observed that the deviation between Model 1 and Model 2 is small in the case of fracture drawdowns, while it is significantly large in case of water table drawdowns with higher drawdowns obtained with Model 1. A higher drawdown can be simulated with Model 2 using a smaller S'_y.

This implies that a saturated model (Model 2) in comparision with saturated-unsaturated model (Model 1) underestimates the parameter S'_y to obtain field drawdowns. However, the error in the underestimation of S'_y is less when there is sufficient thickness of unsaturated zone (Fig.2) or soil having smaller thickness of capillary fringe (Fig.3) or a soil having faster drainage (Fig.4).

It is clear that processes discussed above can have significant influence on the estimation of specific yield S'_y. A lower estimate of S'_y would underestimate the ground water resources in fractured aquifers.

REFERENCES

1. Barenblatt, G. E., Zheltov, I. P., & Kochina, I. N., (1960), Basic concepts in the theory of seepage of homogeneous liquids in fissured rocks, J. Applied Mathematical Methods (USSR), 24, 1286-1303.

2. Moench, A. F., (1984), Double-porosity models for a fissured groundwater reservoir with fracture skin, Water Resour. Res., 20, 831-846.

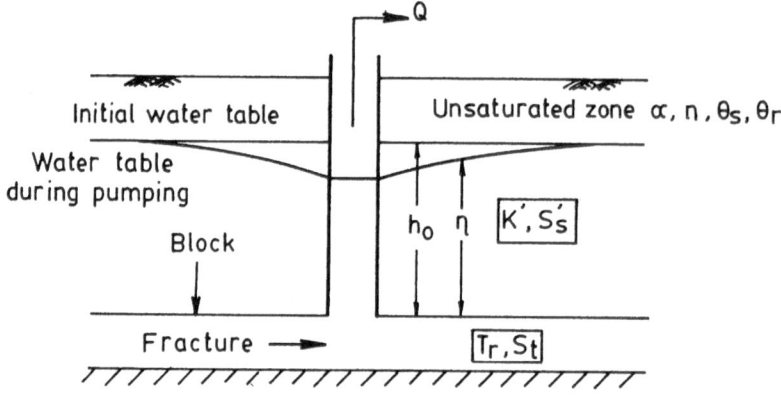

FIG.1-SCHEMATIC DIAGRAM OF SATURATED-UNSATURATED
FRACTURED AQUIFER MODEL

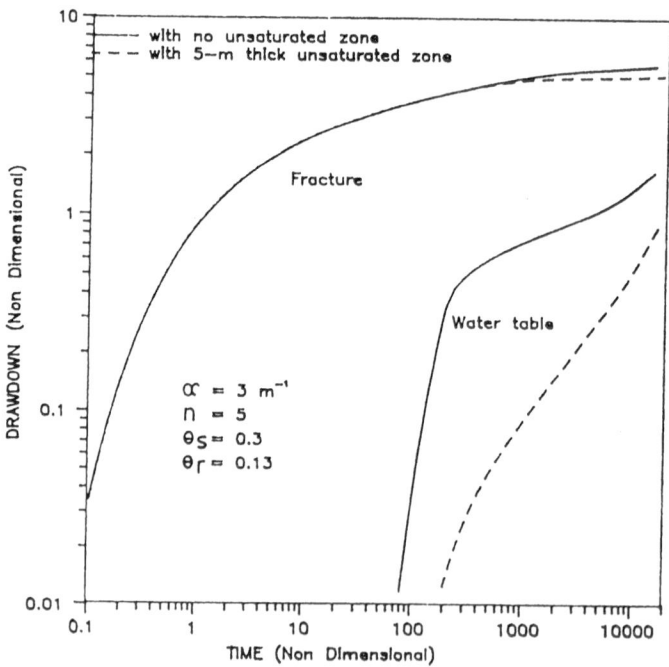

Fig.2 Effect of Thickness of Unsaturated Zone

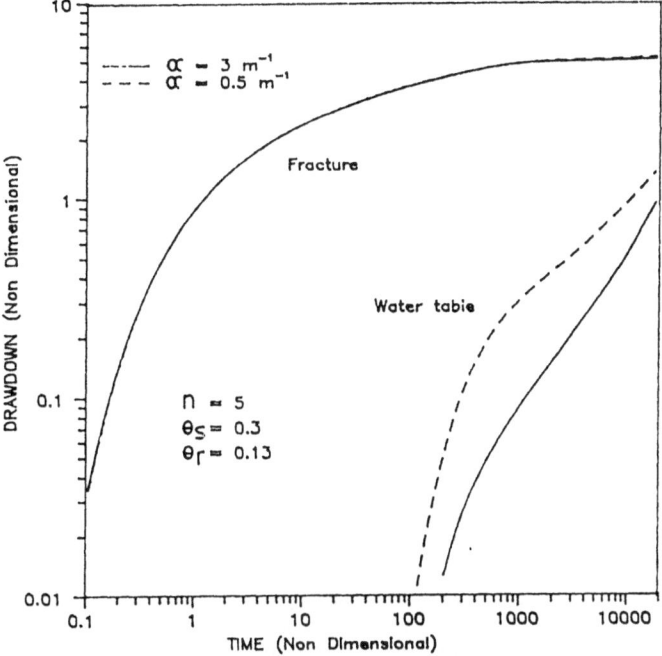

Fig.3 Effect of Unsaturated Parameter, α

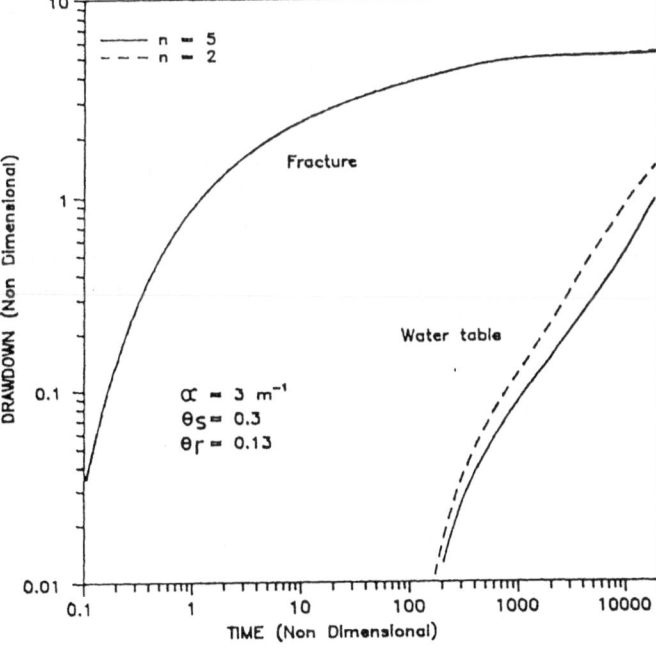

Fig.4 Effect of Unsaturated Parameter, n

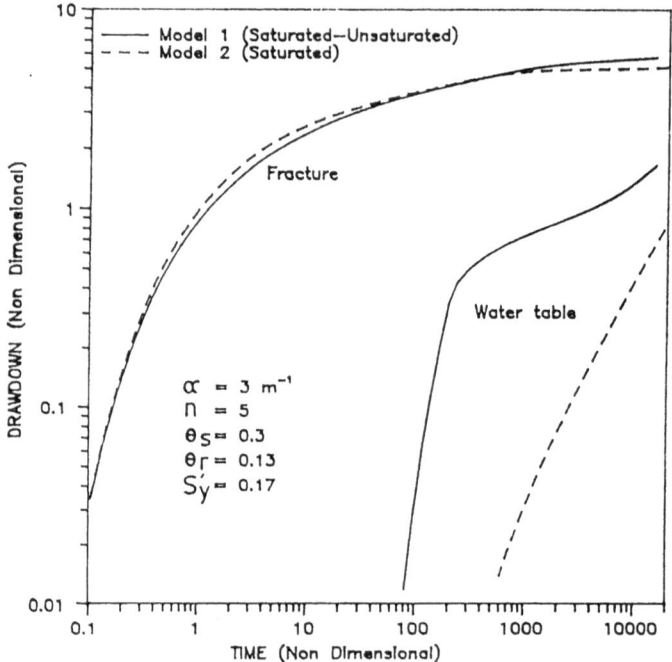

Fig.5 Comparison of Drawdowns for Model 1 and Model 2

3. Sekhar, M., Mohan Kumar, M. S. and Sridharan, K., (1993), A leaky aquifer model for hard rock aquifers, Journal of Applied Hydrogeology, (in press)

4. Narasimhan, T. N. and Ming Zhu, (1993), Transient flow of water to a well in an unconfined aquifer: Applicability of some conceptual models, Water Resourc. Res., 29(1), 179-191.

5. Nwankwor, G. I., Gillham, R. W., Van der Kamp, G. and Akindunni, F. F., (1992), Unsaturated and saturated flow in response to pumping of an unconfined aquifer: Field evidence of delayed drainage, Ground Water., 30(5), 690-700.

6. van Genuchten, M.Th.,(1980), A closed-form equation for predicting the hydraulic conductivity of unsaturated soils, Soil Sci. Soc. Amer., 44, 892-898.

7. Mualem, Y., (1976), A new model for predicting the hydraulic conductivity of unsaturated porous media, Water Resour. Res., 12, 513-522.

8. Paniconi, C., Aldama, A.A. and Wood, E.F., (1991), Numerical evaluation of iterative and noniterative methods for the solution of the nonlinear Richards equation, Water Resour. Res., 27(6), 1147-1163.

SIMULATION OF FLOW AND TRANSPORT IN FRACTURED POROUS MEDIA WITH COUPLING FROM FEM AND BEM

H. SHAO, R. HELMIG [1] and W. ZIELKE
Institute of Fluid Mechanics
University Hannover, Appelstraße 9a, 30167 Hannover
Germany

Modelling of flow and transport in fractured porous media is of increasing interest in the engineering and environmental fields. Since the simulation of such large and complex systems leads to high computational efforts with conventional numerical domain methods, the development of an optimized numerical method is necessary. Based on the discrete approximation of the geological structures, a numerical model will be presented which combines finite element approximation for linear or nonlinear flow and advective and dispersive transport in discrete fractures and boundary element approximation for linear flow and diffusive transport in the matrix.

INTRODUCTION

Problems involving fluid flow and contamination transport in fractured porous media have gained rapidly increasing attention, due to the necessity of evaluating the suitability of geologie sites for nuclear waste repositories, and of protecting the groundwater resources. This problem includes the comprehensive physical phenomena in a complex geological system, which are beening simulated by diverse numerical models with the development of the high-quality computer and numerical simulation techniques.

The powerful numerical methods developed in the last thirty years, may be devided between domain and boundary formulations. Domain methods, e.g. finite element method (FEM), offer powerful attributes in dealing with domain characters, for example, the complex nonlinear flow or advective transport, but the extensive meshing

---

[1] Institut für Wasserbau, Uni. Stuttgart

A. Peters et al. (eds.), Computational Methods in Water Resources X, 683–691.
© 1994 Kluwer Academic Publishers. Printed in the Netherlands.

within the domain introduces additional internal degrees of unknowns which are sometimes not required in some cases of engineering. On the other hand the boundary
solution, e.g. boundary element method (BEM), is very suitable to such problem in
which the ratio of the volume of the geometry to its surface is very large. In this
case less node points are necessary to be computed. However, this method has only
a restricted application to the existence of the fundamental solution of the partial
differential equations.

In the fractured system the fluid flow is normally treated as nonlinear flow because
of the high permeability in the fracture and correspondently the advective transport
is dominant. Both of them should be modelled by the FEM. And in the rock matrix
due to the relative lower permeability, the flow can be observed linearly and the
advection may be neglected. This procedure could be simulated by BEM. In this
case the coupling of FE-formulation and BE-formulation offer the potential of using
each of the different numerical procedures in the environment to which they are best
suited. The difficulty in dealing with time dependent problems with BEM occures in
the simulation of the transient flow and diffusive transport. Owing to the technique
SR-BEM (Secondary Reduction Boundary Element Method) proposed by Aral, M.
M., and Tang, Y. [1], the satisfied results of the simulation was received.

The objective of this paper is to present the development, verification of the coupling
of FE-formulation and BE-formulation for simulation of two dimensional fluid flow
and solute transport in a fractured system.

GOVERNING EQUATION

A generalized constitutive relationship for flow in porous and fractured media may
be represented by extended Darcy's law

$$v = - \mathbf{K}(\operatorname{grad} h)^{\alpha-1} \cdot \operatorname{grad} h \tag{1}$$

and continuity equation

$$S_0 \frac{\partial h}{\partial t} + \operatorname{div} \cdot (v) = q. \tag{2}$$

where v is the Darcy's flow velocity, $\operatorname{grad} h$ is the driving hydraulic gradient, and
$\mathbf{K}(\operatorname{grad} h)^{\alpha-1}$ is the hydraulic conductivity. The nonlinearity will be represented
with α. Wenn α is equal to 1 the flow will follow the laminar law and the small
Reynold's number has been reached. When α is not equal to 1 the nonlinearity is
displayed experimentaly for turbulent and inertial effects

The conservative solute transport processes in the fractured system can be described by two coupled equations, one for the fracture where the advection is dominant and the other for the porous matrix where the diffusion is dominant. The continuity of mass flux and concentration along the interface will be realized through source term. The differential equation for the fracture can be represented as

$$n\rho\frac{\partial c}{\partial t} + n\rho\bar{v}_a \cdot \mathbf{grad}\, c - \mathbf{div} \cdot (\underline{D}^h \cdot n\rho\, \mathbf{grad}\, c) + n\rho qc = 0, \tag{3}$$

where $v_a = \frac{v}{n}$ is the distance velocity calculated from the normal velocity, which has been modified in the flow equation and shall be evaluated before the calculation of the transport, devided with the effective porosity. c is the concentration of solute expressed in terms of mass per unit volume of solution. The hydrodynamic dispersion coefficient $\underline{D}^h$ is further expressed as $\underline{D}^h = \underline{D}^D + \underline{D}^m$. $\underline{D}^D$ is the dispersive coefficient and $\underline{D}^m$ is the molecular diffusion coefficient in water. The differential equation for the porous matrix can be represented as

$$n\rho\frac{\partial c}{\partial t} - \mathbf{div} \cdot (\underline{D}^h \cdot n\rho\, \mathbf{grad}\, c) + n\rho qc = 0, \tag{4}$$

the effective diffusion coefficient is defined here as $\underline{D}^h = \tau\underline{D}^m$, where τ is the matrix tortuosity.

The correspendence initial and boundary conditions enclose the global equation system which may be difficult to solve analytically, when the geometry and boundary conditions are complex.

NUMERICAL METHODS

In this work the standard FEM is applied to the nonlinear fluid flow in the fracture, while the BEM approximates to the linear flow in the porous media. As the form of the flow equation in some sense is the same as the diffusion equation without the advection term, we will not discuss about that further here, instead our attention will be focused on the treatment of the solute transport in the fracture and in the porous media. In this work the solute transport with dominate advection will be approximated by FEM, namely the Taylor-Galerkin FEM. And the transport with dominate diffusion in the rock matrix will be approximated by BEM, namely the special techniques SR-BEM. In the following both of them will be discussed.

Finite Element Formulation

Since transport in the fracture domain is dominated by advection, the numerical solution of the standard Galerkin FEM may exhibit unacceptable oscillation, even if

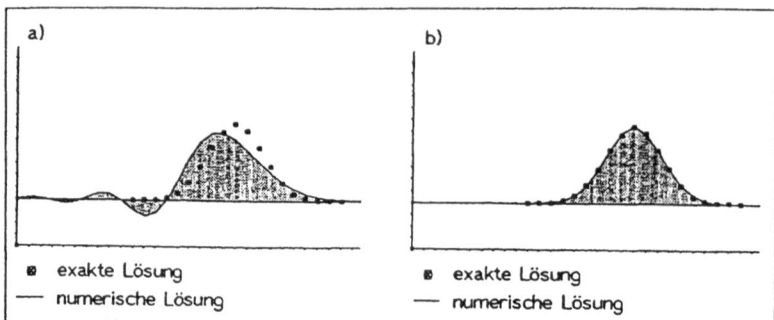

Figure 1: Numerical transport results with Pe=50 a) without error term b) with error term

the Peclet's number is smaller than 10. To overcome this problem the Taylor-Galerkin FE-formulation developed by Gärtner, S. (1987) [5] and Kröhn, K.-P. (1990) [7] has been used in this work. The idea of this techniques is that, with the help of the analysis of the consistancy between the Taylor-series-expansion of the solution c up to the third order in time and in space and the difference equation an additional error term has been added to the FE-equation. This error term considered as an artificial diffusion term in the original transport equation is presented in detail in the works [5], [7]. This procedure offers a stable solution with high Peclet's number. Figure 1. shows a calculation of the concentration with the standard Galerkin and Taylor-Galerkin method under the condition of a Peclet's number of 50 in the fracture.

Boundary Element Formulation

The fluid flow and transport equation in the fractured system is a time-dependent problem, in which some difficulties occur with the BEM. After the comparison of different techniques, for example, time-dependent fundamental solutions [2], dual-reciprocity BEM [3] and SR-BEM [1], SR-BEM have been selected here to simulate the flow with relatively small permeability or transport with a relatively small diffusion coefficient in the porous media. Although this method is not sparely in computer time, especially by the treatment of the domain integral, it offers the acceptable results of the computation.

As was described in [1] SR-BEM uses the time independent fundamental function

$$c^* = \frac{1}{2D^h\pi} \ln(\frac{1}{r}) \tag{5}$$

with $r = \sqrt{\frac{(x-\xi)^2}{D_x^h} + \frac{(y-\eta)^2}{D_y^h}}$ and $D^h = \sqrt{D_x^h D_y^h}$ as the weighting function. Applying

Green's second identity and the general idea of the BEM the differential equation will be reduced in matrix form as

$$[H]\{c\} + [G]\{q_c\} = \int_\Omega c^* \frac{\partial c}{\partial \tau} d\Omega. \tag{6}$$

The temporal derivative on the right-hand side of eq. (6) can now be handeled using a secondary interpolation to reduce it to a boundary only form

$$\frac{\partial c}{\partial \tau} = f(r_j)\beta_j(\tau), \qquad\qquad j = 1, 2, ..., n, \tag{7}$$

where $f(r_j)$ is a function defined for any field point in terms of Euclidean distance that can be defined experimentally

$$f(r_j) = \exp\left(-\frac{r_j}{r_0}\right) \qquad , \qquad j = 1, 2, ..., n \tag{8}$$

with $r_0 = max_j(r_j)$ and β_j is a time dependent unknown parameters. Inverting the equation of eq. (7) $\{\beta\}$ can be expressed as

$$\{\beta\} = [E]^{-1}\{\frac{\partial c}{\partial \tau}\}. \tag{9}$$

Thus one may write the right hand side of eq. (6) as

$$\begin{aligned}
\int_\Omega c^* \frac{\partial c}{\partial \tau} d\Omega &= \sum_{j=1}^{n_k} \beta_j(\tau) \int_\Omega c^* f(r_j) d\Omega \\
&= \sum_{j=1}^{n_k} \beta_j(\tau) [\sum_i^{n_e} \int_{\Omega_i^e} c^* f(r_j) d\Omega] \\
&= [M]\{\beta\} = [M][E]^{-1}\{\frac{\partial c}{\partial \tau}\}.
\end{aligned} \tag{10}$$

This is the SR-BEM description on the transport equation in the porous media only defined on the boundary. The temporal derivative may be integrated using standard finite difference procedure

$$\frac{\partial c}{\partial \tau} = \frac{1}{\Delta t}(c^{t+\Delta t} - c^t). \tag{11}$$

And other terms may be linearly interpolated as

$$c(x, y, z; t) = \Theta_c c^{t+\Delta t} + (1 - \Theta_c)c^t$$
$$q_c(x, y, z; t) = \Theta_{\partial c} q_c^{t+\Delta t} + (1 - \Theta_{\partial c})q_c^t. \tag{12}$$

where the explicit scheme for the derivative of the concentration in the normal direction, e.g. $\Theta_c = 0$ and the implicit scheme for the concentration itself $\Theta_c = 1$ have been used in this studies.

Coupling Schema

As has been mentioned above, the total geometry system may be devided into two parts, one is the fracture and the other is the rock matrix, both of them could be observed as a coupled system. In the coupled system there are different physical phenomena, which could be simulated by different numerical methods. From the view point of the numerical treatment, no particular problems arise in coupling boundary and domain methods if the corresponding conditions are introduced in the interface of the fracture and porous media, namely on every point of the interface the conditions, compatibility and equilibrium should be followed.

The collocation process of the BEM introduces an unsymmetric global matrix in the subdomain of porous media. It loses a great advantage of the symmetry of the system matrix when it combines with the subdomain of the fracture which offers a symmetric matrix through FE-formulation. Of course this is a great shortcoming of the coupling techniques, Because it can not use fast iterative solver, e.g. CG-method. If the symmetry of the matrix is made through the minimization of the error square, then the accurate will be influenced in some cases. On the other hand, the Gelerkin method of boundary element can overcome the disadvantage, but it is difficult to integrate the complex kernal function. The unsymmetric system matrix has been solved by sky-line solver in this work.

NUMERICAL EXAMPLE

A typical example analysed by Tang [10], about the contaniment transport, where advection is dominant in the fracture and diffusion is dominant in the matrix,

is illustrated with coupling of FE and BE method. Kröhn [7] calculated the example with combined FEM, and made a comparison with analytical solution from [10]. This example describes the case of a thin rigid fracture situated in a saturated porous matrix. The groundwater velocity in the fracture is assumed constant 0.25 m/day, and a contaminant source of constant strengh exists at the origin of the fracture, at the beginning to calculate 100 (Fig. 2).

During the modelling, the matrix has been approximated with the boundary elements and the fracture with one dimensional linear finite elements. A subdomain technique

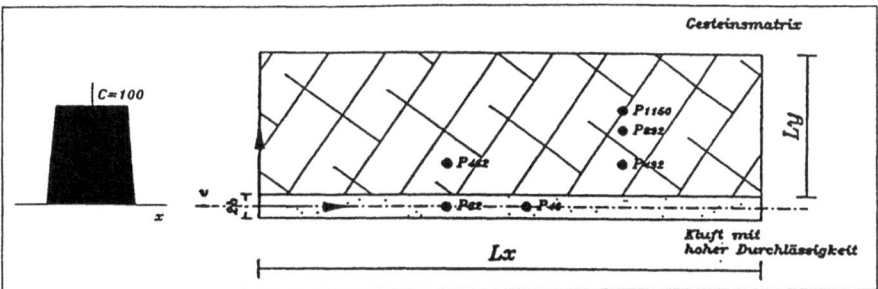

Figure 2: System geometry

has been used to display the domain character in the diffusion transport in the matrix. If the subdomain is so fine as the finite cell, then it seems better to use the FEM, and lose the advantage of the BEM.

Compared with the coupled FEM [7], the matrix was approximated with 2 dimensional finite elements and the fracture with one dimensional elements.

Fig. 3 illustrates the concentration of the solute in the matrix for the beginning, after 3 days, 5, 7, 10 and 15 days. Where the dot lines are the results from the coupling method, while the dashed lines are the results from coupled FE model. A very good agreement between the two methods in the middle of the figure can been seen. And a relatively poor agreement exits near the front of the concentration and near the wall of left side. It is because the curves in the middle is somehow flat, and is very sharp at the two ends. In this case if a very fine discretization of the great 'macro' element (subdomain) in the matrix is used, this error could be avoided.

Fig. 4 shows the breakthrough curves at the points 32 and 46 (see Fig. 2). A small numerical oszillation from both methods implied that, the simulation of pure advection in the fracture may be influenced by the effect from the diffusion from the matrix as a source of solute to the fracture.

CONCLUSION

A coupling model of FEM and BEM has been developed to simulate the fluid flow and solute transport in the complex geologic fractured system. Where the BEM is used to describe the physical phenomena (linear flow law and diffusion transport) in the rock matrix and FEM is to used to describe the physical phenomen (non-linear flow law and advection-dispersion transport) in the fracture. In unsteady situation the SR-BEM has been used so that the solute transport can also be simulated successfully. The calculation of the example shows that the accuracy has been arrived compared with the FEM. The problem is that, although a lot of unknowns have been reduced,

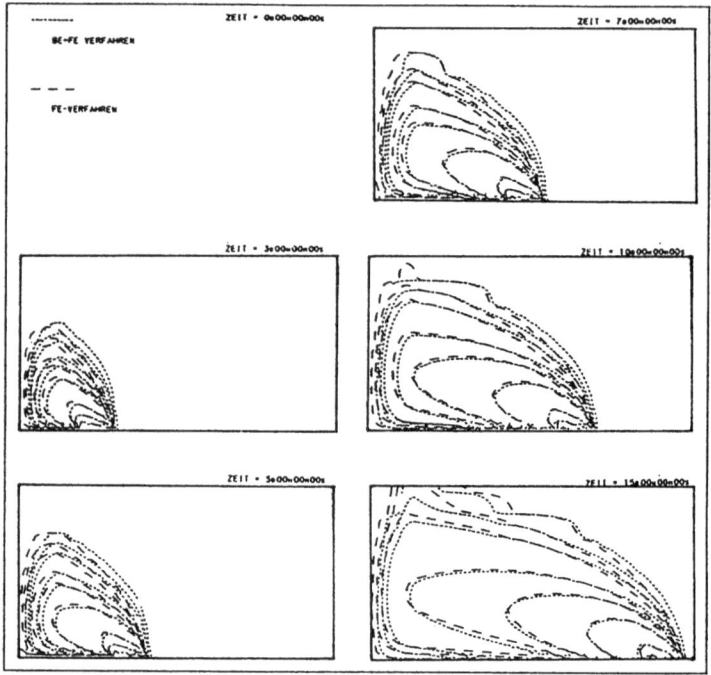

Figure 3: Comparison of the concentration between FEM and coupling of FEM and BEM in the rock matrix

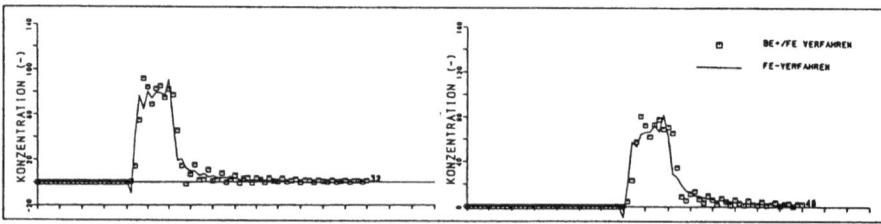

Figure 4: Breakthrough curves in the fracture

the computering time is still as high as that by FEM. The causes stem from the calculation of unsymmetric matrix, the treatment of the domain integration of the equation (10) and the computation of values at the internal points. However, the great advantage is the reduction of the node points. It will be more sensible when a three dimensional problem will be modelled.

References

[1] Aral, M. M. and Y. Tang (1988) *A New Boundary Element Formulation for Time-Dependent Confined and Unconfined Aquifer Problems*, Water Resources Research, Vol. 24. No. 6. pp. 831-842.

[2] Brebbia, C. A. and D. Nardini (1986) *Solution of Parabolic and Hyperbolic Time Dependent Problems Using Boundary Elements*, Comp. & Maths. with Appls. ol. 12B No. 5/6 pp. 1061-1072.

[3] Brebbia, C. A. and L. C. Wrobel (1979) *The boundary element method for steady state and transient heat conduction*, Numerical Methods in Thermal Problems, ed. R. W. Lewis and K. Morgan, Pineridge Press, Swansea, Wales.

[4] Elsworth, D. (1987) *A Boundary Element-Finite Element Procedure For Porous and Fractured Media Flow*, Water Resources Research, Vol. 23, No.4, pp. 551-560.

[5] Gärtner, S. (1987) *Zur diskreten Approximationn kontinuumsmechanischer Bilanz-gleichungen*, Bericht Nr. 24, Institut für Strömungsmechanik und Elektronisches Rechnen im Bauwesen, Universität Hannover.

[6] Liggett, J. A., P. L-F. Liu (1979) *Unsteady Flow in Confined Aquifers: A Comparison of Two Boundary Integral Methods*, Water Resources Research, Vol.15, No.4, pp. 861-866.

[7] Kröhn, K-P. (1991) *Simulation von Transportvorgängen im klüftigen Gestein mit der Methode der Finiten Elemente*, Bericht Nr. 29, Institut für Strömungsmechnik und Elektronisches Rechnen im Bauwesen, Universität Hannover.

[8] Rizzo, F. J., and D. J. Shippy (1970) *A method of solution for certain problems of transient heat conduction*, AIAAJ., 8(11), pp. 2004-2009.

[9] Shao, H. (1994) *Simulation von Strömungs- und Transportvorgängen im geklüftigen Gestein mit der Kopplung von FEM und BEM*, to be present, Institut für Strömungsmechnik und Elektronisches Rechnen im Bauwesen, Universität Hannover.

[10] Tang, D. H. , E. O. Frind and E. A. Sudicky (1981) *Contaminant Transport in Fractured Porous Media: Analytical Solution for Single Fracture*, Water Resources Research, Vol. 17, No. 3, pp. 555-564.

A STUDY OF MULTIPHASE FLOW IN FRACTURED POROUS MEDIA USING A MICROSCALE LATTICE BOLTZMANN APPROACH

W.E. Soll[‡], K.E. Eggert[‡], D.W. Grunau[‡], and A. L. Schafer-Perini[†]
[‡]Earth and Environmental Sciences[+]
Los Alamos National Laboratory
Los Alamos, NM, 87545
USA

[†]E. G. & G. Idaho, Inc.
2251 North Blvd.
Idaho Falls, ID, 83401
USA

The lattice Boltzmann technique has been shown to be an efficient and reliable approach to modeling single- and multi- fluid flow in porous media systems. The flexibility of this approach in discretizing the pore/solid space means it is particularly well suited to capturing fluid behavior, fluid-fluid interactions, and fluid-solid interactions at the scale of the individual pores. Such flexibility readily lends itself to studying processes occurring at physical interfaces, such as between a fracture and the surrounding porous matrix. Here we present pore-level simulations of fluid flow through a fracture embedded in an unsaturated matrix. Simulations are run on the massively parallel Connection Machine 5 (CM-5) using the two-fluid, two-dimensional lattice Boltzmann flow simulator developed at Los Alamos National Laboratory. We look at the effect of pressure gradients and initial matrix saturation on infiltration into the matrix and fluid flow along the fracture.

INTRODUCTION

The hydrogeologic environment at Los Alamos presents a challenge to current groundwater modeling capabilities. The topography consists of mesa tops and canyon bottoms. Beneath the mesas lies a thick (~260m) vadose zone made up of layers of volcanic tuff. This tuff is moderately permeable, 0.1-1 darcy, has a very low moisture content, typically about 5%, and very high capillary retention. Fractures, mostly vertical, are present in many of the tuff layers. The climate is semi-arid, with precipitation concentrated in the summer months, often in the form of short duration, high intensity events. Concerns about the relationship between precipitation on these mesa tops (some of which contain waste sites) and the deep aquifer have called attention to the limitations of our understanding of such systems. Does water enter the fracture and travel rapidly to the saturated zone? Or is water in the fracture imbibed into the matrix so fast that fractures do not provide fast flow paths? Similar concerns regarding groundwater flow in unsaturated, fractured media exist at other sites as

A. Peters et al. (eds.), Computational Methods in Water Resources X, 693–700.
© 1994 Kluwer Academic Publishers. Printed in the Netherlands.

well (e.g., Yucca Mountain, NV; Idaho National Engineering Laboratory, ID).

To model this problem computationally it is necessary to have a host of information about the system. We must have information about the porous matrix: its porosity, permeability, relative permeability, and saturation. We must have information about the fractures: location, aperture, extent, permeability, saturation, surface condition. Finally, we need to know something about how the fracture and matrix communicate, and how the water actually moves down the fracture. The difficulty is that much of this information is not available. In particular, our understanding of how the fracture and matrix interact is still quite limited. Similar concerns apply to saturated systems, in modeling diffusion processes between fracture and matrix.

There are a few different approaches currently being applied to modeling flow and transport in fractures. The most common approach models the system as a single, effective porous medium. The system is modeled as a homogeneous porous medium with constitutive characteristics that are some average of the fracture and matrix values. This approach has been shown to be fairly accurate for saturated systems with frequent, regular, and closely space fractures. However, this approach does not explicitly address fracture-matrix interaction. Another approach treats the system as a network of fractures. In this approach it is commonly assumed that all flow occurs in the fractures, and that the matrix is effectively impermeable. Some models assume that the fractures have perfectly parallel sides, and are uniformly spaced through the matrix. If the water is then assumed to be transported as a slug, semi-analytic solutions are available for predicting how far it will travel down the fracture, and how far it will penetrate into the matrix. If we choose not to make so many assumptions, we might use a numerical model that simulates the system as a dual porosity (DP) or dual porosity/dual permeability (DPDP) medium. In DP models the matrix acts solely as storage, with all flow occurring in the fractures The DPDP model allows for flow in both the matrix and the fractures. The major deficiency with the existing DP and DPDP models are that, currently, the fractures are characterized as a high permeability porous medium because of the lack of detail about how fluids are transported in fractures. This results in a model that is actually only a layered porous media system.

Regardless of which of the computational approaches are chosen from the above selection, many simplifying assumptions are made that allow the user to pretend the system is very much like a simple porous medium. Details about the fracture-matrix interface are lost. Weathering of fracture surface, or flow only along fracture walls only are ignored. Are these models truly capturing the nature of flow and transport in fractured systems? It is hard to know because of the lack of supporting experimental information. However, there is an alternative approach to obtaining answers to the many questions that have been posed in these first few paragraphs. The alternative is to study the systems at the micro-scale, through existing numerical techniques that apply fundamental principles of flow. Through these approaches, we can "observe" how the fluids distribute themselves, and apply what is learned toward improving the characterization of the physical domain and fluid-pore space interactions in the continuum scale models.

The lattice gas (LG) / lattice Boltzmann (LB) technique has been shown to reproduce correct hydrodynamics in complex porous media through modeling of "first principles" of physics. The LB/LG method discretizes the pore space at the microscopic level, and uses simple, local, parallel operations to move particles through the lattice such that the macroscopic, averaged properties of the system reproduce the physical phenomena that satisfy the partial differential equations of interest. In the case of subsurface flow the PDEs of interest are the Navier-Stokes equations. The natural parallelism of the computational process allows rapid and accurate computation of flow fields in arbitrarily complex pore spaces utilizing massively parallel computing technology.

The ability of the LB/LG methods to handle arbitrary boundaries and multiple fluids makes them natural candidates for studying the fracture-porous matrix system described in the introductory paragraphs. Because of the nature of the LB method, it is not necessary to make any modifications to handle cases of fracture flow and fracture-matrix interactions. The proper flow physics are naturally handled in both the large volume fractures and the tortuous porous matrix, and no arbitrary boundary is imposed between the fracture and matrix.

History of lattice gas / lattice Boltzmann methods

The basic lattice gas automata model (Frisch, Hasslacher, and Pomeau, 1986; Frisch et al., 1987) consists of numerous identical particles moving on a regular lattice (a triangular lattice in 2-D and a face-centered hypercubic (FCHC) lattice in 3-D). Each particle has the same mass, momentum, and kinetic energy. Particles may occupy only the links of the lattice, and have a unit velocity corresponding to the direction of the occupied lattice link. An exclusion rule, used to minimize storage requirements, states that only one particle may occupy any given link. These rules result in a binary system, in which each lattice link is occupied (1) or empty (0). At each time step there are two stages in the process of updating the system state: advection and collision. In the advection process, a particle moves from its present site to the nearest neighbor site in the direction of the velocity vector. In the collision process, particles at each site interact, and particle momentums are redistributed at the same site such that particle number and total momentum are locally conserved. The lattice Boltzmann approach is an extension of the LG approach that uses real numbers to represent particle distributions instead of binary values.

In either LG or LB continuum scale flow fields (i.e. velocity or pressure) are obtained through averaging over a number of the lattice sites. The LB approach eliminates the need for spatial and temporal averaging that is necessary for obtaining accurate velocity and pressure fields in the LG model. Another important advantage of the continuum description of the LB method as compared to the Boolean LG models is that in the LB approach much of the statistical noise that arises in the system is eliminated.

The basic LB approach will not be described in any more detail here, as it would take more space than we are allowed, and is thoroughly described in a number of excellent references. The reader is directed to the original LG derivations referenced above, as well as to Grunau (1993) for a good discussion of both LG and LB method derivations. McNamara and

Zanetti (1988) is the original publication on development of the LB approach from lattice gas methods. Gunstensen et al. (1991) and Grunau et al. (1993) offer excellent descriptions of LB models for two-fluid systems. Soll et al. (1994, in this proceedings) present a general discussion of the latest capabilities of the LANL LB model used for this work, and cover some of the other areas to which this model is being applied.

FRACTURE FLOW SIMULATIONS

In this section, simulations are presented from a number of different simple fracture configurations. The generic system is meant to approximate infiltration of water into a fracture and subsequent downward movement through both the fracture and the porous matrix, where the medium is unsaturated. Fluid enters into the fracture from the left side of the domain, but is restricted from entering the matrix on the inflow face. Outflow on the right side of the domain is open across the entire domain height. The flow is driven by a uniform pressure difference across the system. The entering fluid is strongly wetting , and has a higher viscosity than the resident fluid, which, by comparison, is non-wetting. Gravity is neglected in these simulations, although the LB code is capable of accommodating gravitational forces. In all of the figures the solid is shown in black, the wetting fluid (WF) is shown in white, and the non-wetting fluid (NWF) is shown in gray. The infiltrating fluid in all of the following simulations is the wetting fluid and always advances from left to right.

The simplest fracture is a pair of parallel, impermeable plates. This model is Poiseuille flow and, though not very realistic, provides an easy means of verifying proper model response. Figure 1 shows a step from a simulation of two-fluid flow between parallel plates. We observe the expected curved front. Figure 2 shows the corresponding velocity profile. When two immiscible fluids flow simultaneously, side by side, the velocity profile changes to reflect the difference in viscosities. The dotted line shows the actual velocity profile across the channel width overlaid on each of the individual velocity distributions.

Similar simulations have been run for "fractures" with parallel, impermeable walls into which a wall roughness has been introduced. Figure 4 shows the fluid distribution in this rough-walled system. Note that the front retains the curved profile and much of the symmetry at the leading edge, where the flow front is far enough from the wall not to be influenced by the roughness. However, the flow is somewhat retarded with respect to the smooth walled system (Fig. 1), and residual NWF occupies some of the roughness volume behind the advancing WF front.

The next level of complexity of pore space is having a porous matrix bounding the parallel fracture walls. In this configuration the porous matrix is from a digitization of a micromodel. A 512x512 subsection is taken from the micromodel space, and a parallel "fracture" is introduced down the middle of the domain. The fracture width is 32 lattice sites, sufficiently wide to balance the relatively large pore widths in the micromodel and produce a fracture with greater permeability than that of the pore space. For the initial analysis of fracture-matrix interactions we chose to run the simulations in a 2-D pore space to reduce the computational burden and allow easy visualization. 3-D systems will be

Fig. 1. Wetting fluid (white) displacing non-wetting fluid (gray) between a parallel plate "fracture."

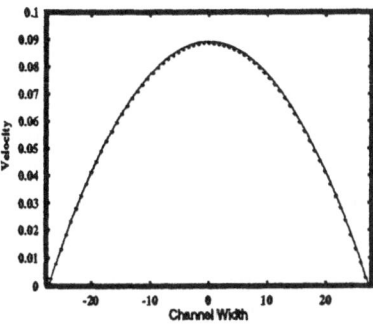

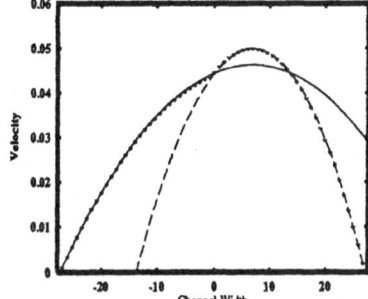

Fig. 2. Velocity profile for a single fluid flowing between parallel plates.

Fig. 3. Velocity profile for flow between parallel plates with two concurrently flowing fluids.

Fig. 4. Wetting fluid (white) displacing non-wetting fluid (gray) between a rough walled parallel "fracture".

addressed in future work.

Simulations were run to compare infiltration into the matrix from the fracture for different forcing pressures across the fracture length. Figure 5 shows a comparison of two systems in which the pressure gradient in Figure 5a is twice that in Figure 5b. Both figures represent the fluid distributions after the same number of time steps. The WF has infiltrated significantly farther into the matrix in Figure 5a as in Figure 5b. A similar WF saturation as in Figure 5a for the higher forcing pressure is obtained for the lower forcing pressure after slightly more than twice the number of time steps, suggesting that the matrix infiltration is largely being governed by the flow rate.

The model was also used to analyze the effect of different initial fluid saturations in the matrix on fluid movement between fracture and matrix. Initial conditions with non-zero WF saturation are generated by assigning a saturation fraction, which the code distributes homogeneously through the pore space. The system is then allowed to equilibrate under no external forces. During equilibration the fluids separate due to interfacial forces, and the wetting fluid migrates to the solid surface. The equilibrium state then provides the initial fluid distribution for the flow simulations.

Figure 6 shows a comparison of infiltration at two different initial matric wetting-fluid saturations. The system in Figure 6a started with an initial WF saturation of zero, while the system in Figure 6b had an initial matric WF saturation of approximately 0.40. The displacement in these simulations is viscous-dominated as a result of choosing a relatively large driving pressure gradient. In this comparison the WF in the initially NWF saturated system (Fig. 6a) has penetrated farther into the matrix near the fracture entrance than that in the partially WF saturated system (Fig. 6b). However, the WF has penetrated more rapidly down the fracture in Figure 6b than in Fig 6a., and the WF has penetrated into the matrix along more of the length of the fracture. A simulation run with an initial WF saturation of 0.10 followed the same trend. The difference in the WF penetration into the matrix is partly a result of reduced capillary pressure and relative permeability in the matrix at the higher WF saturation. A second factor in the reduced matrix penetration in Figure 6b is the change in effective pressure drop across the fracture in the partially saturated system as the WF flow volume in the fracture increases.

Note that Figure 6a and Figure 5a are from the same simulation at 5000 and 10000 time steps, respectively. Comparing these two figures shows that, once the WF crosses the entire fracture, the distance along the fracture that is saturated by WF increases slowly, while the distance the WF has penetrated the matrix near the entrance grows measurably. The increase in WF saturation in the matrix is small relative to the WF flux through the fracture.

THE NEXT STEP

We have presented preliminary results of a study to characterize flow through a porous system composed of fractures and porous matrix. The data presented here were limited to some of the simplest pore space configurations. The next step is to move to more realistic

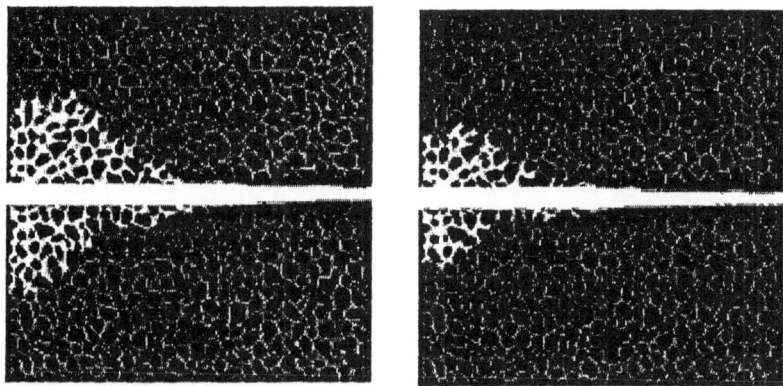

Fig. 5. Effect of pressure gradient on fracture-matrix flow. White is wetting fluid. Gray is non-wetting fluid. Gradient in (a) is twice as large as (b).

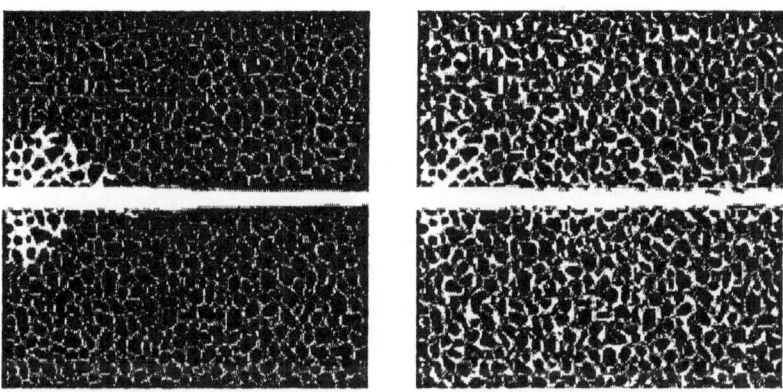

Fig. 6. Effect of initial wetting fluid matrix saturation on fracture-matrix flow. Initial matrix saturation of wetting fluid (a) is 0.0, (b) is 0.50.

physical systems. To do this we must be able to generate such systems. We intend to apply a pore space generation code, that has been developed by the authors, to generate 3-D pore spaces with more complex fractures. We also hope to obtain high resolution digital images of a fractured porous medium. Systems with both single and multiple fractures will be considered. Simulation results will be compared against available analytic or semi-analytic solutions, as well as any available experimental data.

ACKNOWLEDGEMENTS

This work has been supported by the Los Alamos National Laboratory which is operated by the University of California for the U.S. Department of Energy. The authors wish to acknowledge the Advanced Computing Laboratory of Los Alamos National Laboratory, Los Alamos, NM 87545. This work was performed on computing resources located at this facility. The authors also thank S.Y. Chen for code concept development.

REFERENCES

Chen, S.Y., G.D. Doolen, K. Eggert, D. Grunau, and E.Y. Loh. (1991). "Lattice Gas Simulations of One and Two-Phase Fluid Flows Using the Connection Machine-2," in A.S. Alves (ed.), Discrete Models of Fluid Dynamics, World Scientific. pp. 232-249.

Frisch, U., B. Hasslacher, and Y. Pomeau (1986). "Lattice-Gas Automata for the Navier-Stokes Equation". Physical Review Letters, 56, 1505-1508.

Frisch, U., D. d'Humieres, B. Hasslacher, P. Lallemand, Y. Pomeau, and JP. Rivet. (1987). "Lattice Gas Hydrodynamics in Two and Three Dimensions". Complex Syst, 1. pp. 649-707.

Grunau, D.W. (1993). Lattice Methods for Modeling Hydrodynamics. PhD Dissertation, Colorado State University, Dept. of Math.

Grunau, DW., SY. Chen, and K. Eggert (1993). "A Lattice Boltzmann Model for Multi-phase Fluid Flows". Phys. Fluids A.,5(10). pp. 2557-2562.

Gunstensen, A.K., D.H. Rothman, S. Zaleski, and G. Zanetti. (1991). "Lattice Boltzmann Model of Immiscible Fluids". Phys. Rev. A., 43(8). pp. 4320-4327.

McNamara, G.R., and G. Zanetti (1988). "Use of the Boltzmann Equation to Simulate Lattice-Gas Automata". Phys. Rev. Lett., 61(20). pp. 2332-2335.

7. PARAMETER ESTIMATION

A numerical method to characterize the averaging process invoked by a slug test

ROGER BECKIE AND BIN WANG
Department of Geological Sciences, University of British Columbia
Vancouver, B.C., V6T 1Z4, rbeckie@geology.ubc.ca

We conceptualize the slug test measurement process to be a linear transformation from a hydraulic conductivity field defined on a smaller scale, to a measured hydraulic conductivity field. We present a methodology to determine the kernel, or filter function, of the linear transformation. The filter function characterizes the averaging volume of the slug test, and the nature of the averaging process. We utilize a 2-d numerical flow model to simulate slug tests in heterogeneous media and spectral - image analysis techniques to estimate the empirical filter function invoked by the slug test method. We illustrate the effect of storage and anisotropy on the filter function.

Introduction

Aquifer tests, such as pump tests and slug tests, are used to determine the hydraulic conductivity and storage of porous media at a sampling location. In heterogeneous systems, the parameter values that result from these two tests are not directly comparable, because the methods sample different volumes of the subsurface. To be able to relate parameters estimated from different tests to each other, it is important to quantify the averaging affected by the measurement process and to characterize the volume of porous media that a measured parameter represents.

Most researchers define the range of the slug test, and thereby the volume investigated by the slug test, as that distance at which the head perturbation caused by the slug test has dissipated to a specified fraction of the initial excess head at the well bore [2], [4], [7]. For example, [4] shows that this distance is proportional to a dimensionless storage parameter: $r/r_w \propto \sqrt{\pi\, r_c^2/2\pi r_w^2\, S}$, where r is the maximum distance traveled by a specified fraction of initial excess head over the duration of the slug test, r_w is the radius of the well screen, $\pi\, r_c^2$ is the well bore storage, r_c is inner radius of the well casing, and S is the storativity of the aquifer.

A. Peters et al. (eds.), Computational Methods in Water Resources X, 703–710.
© 1994 Kluwer Academic Publishers. Printed in the Netherlands.

As noted in [5], a problem with these approaches is that the fraction of the initial head chosen to define the range of the slug is arbitrary; *e.g.* 1 %, 5 % or 10% [4]. Most authors examine only homogeneous or idealized heterogeneous systems, for example, dual porosity, fractured media, or annular heterogeneities such as skin effects [2], [7], [9].

Harvey [5], characterizes the way a slug test averages the hydraulic conductivity proximal to the well bore in heterogeneous systems. He uses a numerical approach that explicitly considers the effects of heterogeneity and the inversion process. He characterizes the averaging process with a spatial power average of the form $K_{p,r} = \left[\frac{1}{V(r)} \int_{V(r)} K \, dV\right]^{1/p}$, where K is the smaller scale hydraulic conductivity, $V(r)$ is the volume of a cylinder of radius r about the well screen, p is the exponent in the power average ($p = 0$ requires a log transformation and corresponds to geometric averaging), and $K_{p,r}$ is the power law averaged hydraulic conductivity. Harvey [5] determines the optimal p and r statistically, from an ensemble of numerical slug tests performed in synthetic aquifers. The optimal p and r are those that minimize the sum of the squared differences between the set of pairs of $K_{p,r}$ determined from the power average of the small scale conductivity and the K_s inverted with the model of Cooper [3].

Our work is similar to [5]. We are interested in the relationship between the small scale conductivity field K and K_m, the conductivity measured by a conventional slug test. Although we examine the role that storage plays in determining the measured conductivity, we do not consider the effects of heterogeneous storage on the measurement process.

Conceptual model

We assume the slug test process to be a bounded linear mapping from a small scale hydraulic conductivity field to the hydraulic conductivity field measured by a slug test. We actually define the mapping $\mathcal{G}$ in terms of log conductivity fields as $\mathcal{G} : Y \longrightarrow Y_m$, where $Y = \ln K - \langle \ln K \rangle$, and $Y_m = \ln K_m - \langle \ln K_m \rangle$, where the angle brackets denote the spatial mean $1/A \int \ln K \, dA$. A general expression for such a linear mapping is

$$Y_m(x) = \int G(x - x') \, Y(x') \, dx' + N(x), \tag{1}$$

where we call G the filter function (or impulse response) and N is a zero mean Gaussian noise term, which we assume is uncorrelated with Y and G. The functional $\mathcal{G}$ is analogous to a spatial filter, where G is the filter function. The nature of the measurement process, and the sampling volume investigated by the slug test, can be inferred from G. The noise N can be interpreted as the sum of the model misfit and random truncation errors. It may not be Gaussian, or small, especially if the model (1) is a poor representation of the measurement process.

Admittedly, the measurement process may not be linear as given in (1). However, the linearity assumption is appealing because of the computational simplicity it affords, and the clear interpretation of G as a spatial filter. We thus proceed pragmatically and assume linearity.

In our spatial filtering model (1), each point in Y_m is a spatial average of Y, where the filter function G is the same everywhere. Accordingly, the filter function that we estimate can be considered to be the average of the individual mappings from Y to the each point in Y_m. The filter is thus defined with respect to the single realization Y and in a spatially averaged sense. Alternatively, Harvey [5] employs a Monte Carlo approach, where a single slug test is taken in each realization of the field Y, and the relationship between Y and Y_m is found by averaging across many realizations. Both definitions are valid, and at this initial stage, it is not clear to us which is more useful.

In the following, we explain how we generate Y_m from Y and then estimate G using (1).

Method

There are essentially four steps in our methodology: 1) generate a heterogeneous field K and define $Y = \ln K - \langle \ln K \rangle$; 2) in every gridblock of the 2-d heterogeneous field K, simulate the recovery of head in the bore hole in response to an instantaneous change of head in the well (thus we "probe" the field K at every point with a slug test); 3) fit the 2-d head recovery data with the 1-d model of [3] and thus determine K_m and $Y_m = \ln K_m - \langle \ln K_m \rangle$ in every gridblock; 4) estimate the transfer function (and thus the filter function) using spectral analysis methods and the fields Y_m and Y.

An appropriate mathematical model for the head h in a 2-d constant thickness aquifer, and in the well bore H, is

$$\nabla \cdot (K \cdot \nabla h) = S_s \frac{\partial h}{\partial t}, \qquad r \geq r_w, \tag{2}$$

$$\int_0^{2\pi} K \frac{\partial h}{\partial r} r d\theta = \pi r_c^2 \frac{dH}{dt}, \qquad r = r_w, \tag{3}$$

$$h(r_w, t) = H(t), \tag{4}$$

$$H = H_0, \qquad t = t_0, r = r_w \tag{5}$$

where K is the hydraulic conductivity, S_s is the specific storage, and H_0 is the initial excess head. We solve this set of equations using a standard 5-point block centered finite difference scheme [1]. We use harmonic averaging to define interblock conductances. The mathematical problem is posed on an infinite domain. To minimize the influence of a bounded domain, we use a method similar to [5] and surround an inner domain of $n_i \times n_i$ constant sized gridblocks with an outer rim of n_o, gridblocks in each direction. The outer gridblocks increase geometrically in size away from the edge of the inner domain.

We approximate the well boundary condition (3) and (4) by assigning the gridblock that contains the well a specific storage $S_s = 1$, and a conductivity 7 orders of magnitude greater than the geometric mean of the field. In effect, the entire gridblock functions as a well. We set $r_c = r_w$. We determine the radius of the well as that which yields the best match between a 2-d simulation in a homogeneous aquifer and the 1-d analytical results of [3]. We can reproduce the results of [3] almost exactly for specific storage $S_s < 0.15$, if we set the well radius to $r_w = 0.59 \, \Delta x$, where Δx is the length of the gridblock. The results that we present here use $r_w = 0.59 \, \Delta x$.

Harvey, [5], using a 3-d code and a different storage convention for the well bore, finds $r_w = 0.427 \, \Delta x$. Another effective radius can be defined by relating the area of the gridblock to that of the well, $Area = \pi r_w^2 = (\Delta x)^2$, or $r_w = 0.56 \, \Delta x$.

We define the slug test measured conductivity K_m and specific storage S_s to be those parameters of the 1-d model that minimize the squared error $\mathcal{E}(K_m, S_s)$ between head recovery computed with the 2-d model, H_{2-d}, and H_{1-d}, the head recovery calculated using the 1-d model of Cooper [3] with spatially constant conductivity K_m and specific storage S_s,

$$\mathcal{E}(K_m, S_s) = \sum_{i=1}^{n} \left[H_{2-d}(t_i) - H_{1-d}(t_i; K_m, S_s) \right]^2, \tag{6}$$

where the n timesteps are spaced evenly in log time. We use a steepest descent algorithm to find the K_m and S_s that minimize $\mathcal{E}$.

We estimate both conductivity and storage, because in practice, both parameters are unknown. Alternatively, one could fix the storage to the true value and estimate conductivity independently. Harvey [5] shows that the conductivity estimated alone with the storage fixed at its true value is higher than the conductivity that is estimated together with storage.

The last step in our method is the estimation of G from Y and Y_m. This is difficult to do using the convolution form of (1) directly. Assuming that the proper technical conditions are met, we can apply the Fourier transformation to (1),

$$\widehat{Y}_m(f) = \widehat{G}(f)\widehat{Y}(f) + \widehat{N}(f), \tag{7}$$

where the hat indicates the Fourier transformation and f is the spatial frequency. The function $\widehat{G}$ is known as the transfer function. It may seem appropriate to rearrange (7) and solve for the transfer function,

$$\widehat{G}(f) = \frac{\widehat{Y}_m(f)}{\widehat{Y}(f)} - \frac{\widehat{N}(f)}{\widehat{Y}(f)}. \tag{8}$$

The problem with this expression is the noise term, $\widehat{N}(f)/\widehat{Y}(f)$ which we can expect to be large at those frequencies where Y has little power.

As an alternative to (8), we employ a Wiener filtering technique from image analysis [6], [10]. The idea is to minimize the expected square error $E\left[\tilde{G} - G\right]^2$ between the estimated filter function $\tilde{G}$ and the true (unknown) filter function G. If both the noise N and the unknown filter function G are considered to be Gaussian stochastic processes, then a linear estimate of G in terms of Y_m can be shown to be optimal [6]. As noted by Sondhi, [10], the Gaussian hypothesis is not likely to be valid in most image analysis contexts, nor would we necessarily expect it to hold here. As a consequence, a linear estimator may be suboptimal in the mean squared error sense.

Under the Gaussian and linearity hypotheses, it can be shown that the minimum expected squared error estimate of $\widehat{G}$ is given by, [6],

$$\widehat{\tilde{G}}(f) = \frac{S_{YY_m}}{S_{YY} + \frac{S_{NN}}{S_{GG}}}, \tag{9}$$

where S_{YY}, S_{NN}, S_{GG} and S_{YY_m}, are the empirical power spectral densities of Y, the noise N, and G and the Y - Y_m cross spectral density respectively.

The last term in the denominator is crucial to obtain a reasonable estimate of G. This term grows large at those frequencies where G has little power; that is where S_{GG} is small. As a consequence, at those frequencies, the estimate of G is forced to zero. The net effect is to obtain a noise tolerant estimate of G. The nature of the noise term can be inferred from an analysis of residuals. Our initial results are calculated setting S_{NN}/S_{GG} to a constant.

We estimate the spectral densities using a periodogram with the multiple taper method of [8]. The tapers are constructed from prolate spheroidal sequences, and have better bias, variance and leakage characteristics than a standard single taper approach.

The image processing technique that we present here is not essential to characterize the averaging process invoked by a slug test. Indeed, assuming a spatial filter model (1), and given the two fields Y and Y_m, one can apply one of many methods to determine the filter function G. However, the spectral approach is particularly convenient because of the efficiency with which it can be implemented using the fast Fourier transform (FFT).

Example

The greyscale figures below show a 64×64 gridblock, (or $108\ r_w \times 108\ r_w$), inner domain of an isotropic small scale $\log 10(K)$ field and two measured $\log 10(K)$ fields. In this example, the outer rim of $n_o = 19$ gridblocks. Our tests show boundary effects are minimal on this mesh, even for slug tests taken on the edge of the inner domain. We generate the small scale field from a target Markovian spectrum. The realization has $\sigma_{\ln K} = 1.1$, and an approximate correlation length of $\lambda/r_w = 18$. We invert the 2-d slug test data with a 1-d model containing 100 gridblocks.

The histograms and greyscales show the smoothing affected by the slug test when the true aquifer specific storage is $S_s = 0.09$ and $S_s = 0.03$. As can be observed in the greyscales and the histograms, the smoothing increases with decreasing storage.

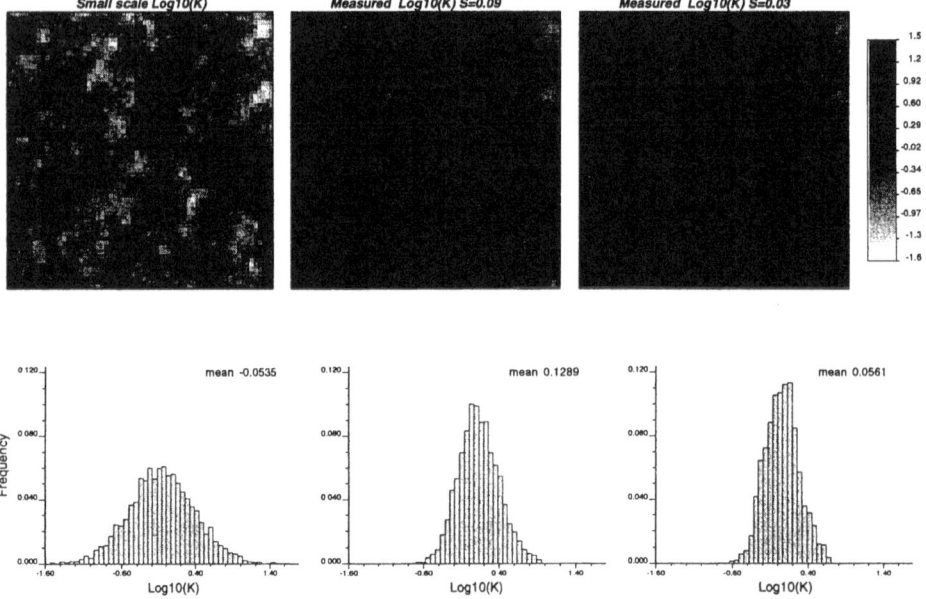

The following contour plots show the empirical filter functions $\tilde{G}$, where the center of the function corresponds to the well location. These functions show that the conductivity closest to the well receives the greatest weight in the averaging process. Although the filter functions for the two values of specific storage have similar support, in this example approximately $r/r_w \approx 13$ to the $G = 1$ contour line, the $S_s = 0.09$ filter weights the near well much more strongly than the $S_s = 0.03$.

The $G = 1$ contour demarks the area where G is well resolved. The spurious off center peaks are the effects of noise on the estimate. We can reduce the size of these off center areas by increasing the value we assign to S_{NN}/S_{GG}. However, we do this at the price of lost resolution. Therefore, we set the noise term to a value that gives us a reasonably smooth, well resolved estimate of G.

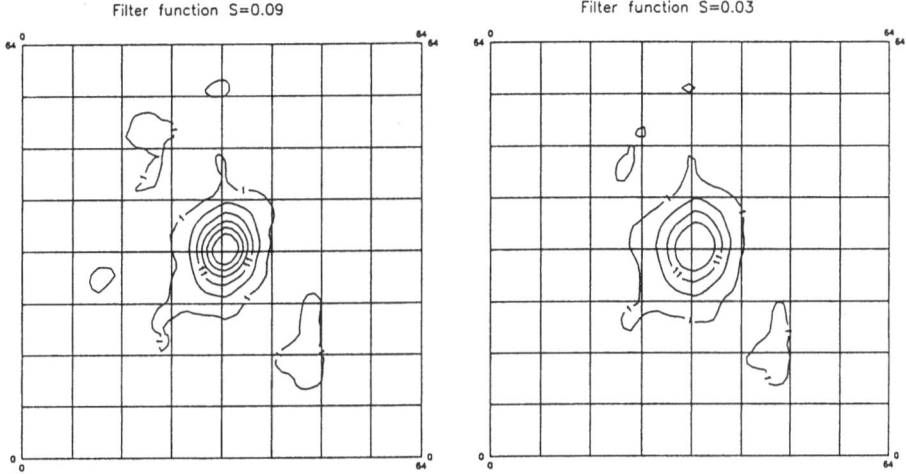

We demonstrate the methodology for an anisotropic field in the figure below. In this example, the anisotropy ratio is approximately $\lambda_y/\lambda_x = 5.0$, where $\lambda_y/r_w = 42$ and $\sigma_{\ln K} = 1.7$. The filter function shows a distinct anisotropy aligned with the small scale conductivity field. The value of the noise term that we use to estimate G is taken from the previous example. It may be somewhat too low, as can be observed by the increase in the number of spurious off center peaks.

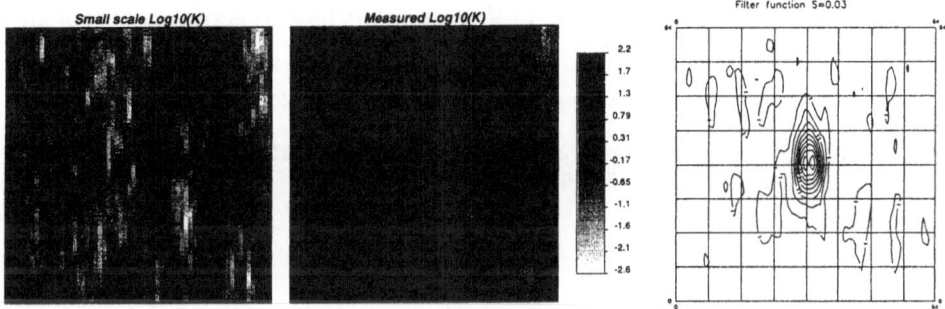

Acknowledgements

We benefited greatly from discussions with Charles Harvey of Stanford University, and Tom Clemo of UBC. This work is supported by an NSERC operating grant awarded R. Beckie.

References

[1]   K. AZIZ AND A. SETTARI, *Petroleum reservoir simulation*, Applied Science Publishers, 1979.

[2]   J. A. BARKER AND J. H. BLACK, *Slug tests in fissured aquiferss*, Water Resour. Res., 19 (1983), pp. 1558–1564.

[3]   H. H. COOPER, J. D. BREDEHOEFT AND S. S. PAPADOPOULOS, *Response of a finite diameter well to an instantaneous charge of water* , Water Resour. Res., 3 (1967), pp. 263–269.

[4]   D. GUYONNET, S. MISHRA AND J. MCCORD, *Evaluating the volume of porous medium investigated during slug tests*, Groundwater, 31 (1993), pp. 627–633.

[5]   C. F. HARVEY, *Interpreting parameter estimates obtained from slug tests in heterogeneous aquifers*, Applied Earth Sciences Dept., Stanford Univ., M.Sc. Thesis, 1992.

[6]   C. HELSTROM, *Image restoration by the method of least squares*, J. Opt. Soc. Am., 57 (1967), pp. 297–303.

[7]   K. KARASAKI, J. C. S. LONG AND P. A. WITHERSPOON, *Analytical models of slug tests*, Water Resour. Res., 24 (1988), pp. 115–126.

[8]   J. PARK, C. R. LINDBERG AND F. L. VERNON, *Multitaper spectral analysis of high frequency seismograms* , J. Geoph. Res., 92 (1987), pp. 12675–12684.

[9]   A. SAGEEV, *Slug test analysis*, Water Resour. Res., 22 (1986), pp. 1323–1333.

[10] M. M. SONDHI, *Image restoration: the removal of spatially invariant degradations*, Proc. IEEE, 60 (1972), pp. 842–853.

GROUNDWATER PARAMETER ESTIMATION WITH NON-DIAGONAL WEIGHTING MATRICES

L. R. BENTLEY
Department of Geology & Geophysics
University of Calgary
2500 University Dr. NW
Calgary, AB T2N 1N4, Canada

Groundwater flow simulators are well established, valuable engineering tools. Groundwater simulations are routinely used in water supply, dewatering and contamination studies. Optimization techniques are often used in the calibration of groundwater models. Optimizations are usually performed on a limited number of zoned parameters using diagonal weighting matrices. In this paper, optimization results obtained with complete parameterization (no zoning) will be presented and the benefits of using fully populated weighting matrices will be demonstrated.

BACKGROUND

The two dimensional, isotropic, steady state groundwater flow equation is,[1]

$$-\frac{\partial}{\partial x}\left(T\frac{\partial h}{\partial x}\right) - \frac{\partial}{\partial y}\left(T\frac{\partial h}{\partial y}\right) +$$

$$B(h - h_r) - I_f - \sum_{p=1}^{N_Q} Q_p \delta(\bar{x} - \bar{x}_p) = 0 , \qquad (1)$$

where h is hydraulic head, T is transmissivity, B and h_r are leakage parameters, I_f is infiltration, Q_p is the pumping rate for well p (positive in), N_Q is the number of pumping wells and $\delta(\bar{x} - \bar{x}_p)$ is the Dirac delta function. In addition, appropriate boundary conditions must be specified to complete the description of the problem. The application of discretization theory leads to a system of algebraic equations that approximates equation (1). These equations can be represented as a matrix expression,

$$\mathbf{A}H = f , \qquad (2)$$

where $\mathbf{A}$ is a matrix that is a function of the discretized system parameters such as

A. Peters et al. (eds.), Computational Methods in Water Resources X, 711–718.
© 1994 Kluwer Academic Publishers. Printed in the Netherlands.

T, H is a vector of head values at the nodes and f is a forcing vector which contains boundary conditions and system forcings.

The effectiveness of groundwater simulations is limited by the modelor's ability to establish realistic, spatially detailed estimates of the model parameters. Hydrogeologic parameters are inferred from aquifer tests, laboratory tests, or by comparison with values obtained in other areas. The amount and the spatial distribution of test data are always limited, and the test results contain errors. Hence, the values and the spatial distribution of the parameters are never known with complete certainty.

Parameter identification techniques are often used to improve the inprecise original estimates of the model parameters. The general procedure is to adjust the original parameter estimates to obtain model results that better match a set of field measurements of the hydraulic heads. One approach is to solve the minimization problem,[2]

$$\underset{\text{w.r.t. } H \,:\, P}{\text{MIN}} \; (\hat{h}(H) - h^*)^t \mathbf{W}_M (\hat{h}(H) - h^*) + (P - P^*)^t \mathbf{W}_P (P - P^*) \;,$$

$$\text{Subject To} \quad R_G = 0 \;, \tag{3}$$

where $\hat{h}(H)$ is a vector of trial function values at the measurement points, H is a vector of hydraulic heads at the nodes, h^* is a vector of the measurement values, $\mathbf{W}_M$ and $\mathbf{W}_P$ are weighting matrices, P is a vector of parameters, P^* is a vector of original estimates and R_G is a vector containing the discretized governing equations. The objective is to find a set of heads that is close to the field measured values and a set of parameters that is close to the original estimates, subject to the condition that two sets honor the discretized governing equations.

The penalty can be used to solve equation (3)[2],

$$\underset{\text{w.r.t. } H \,:\, P}{\text{MIN}} \; (\hat{h}(H) - h^*)^t \mathbf{W}_M (\hat{h}(H) - h^*) + (P - P^*)^t \mathbf{W}_P (P - P^*)$$

$$+ \, R_G^t \mathbf{W}_G R_G \;, \tag{4}$$

where $\mathbf{W}_G = \lambda_G^2 \mathbf{I}$ is a diagonal weighting matrix. Solution of equation (4) gives an updated set of model parameters P and an approximate solution to the groundwater flow equation H.

Another approach is to apply the reduced gradient method to equation (3),

$$\underset{\text{w.r.t. } P}{\text{MIN}} \; (\hat{h}(P) - h^*)^t \mathbf{W}_M (\hat{h}(P) - h^*) + (P - P^*)^t \mathbf{W}_P (P - P^*) \;. \tag{5}$$

Solution of equation (5) gives an updated set of parameters which can then be used in equation (2) to solve for a new set of heads. This procedure is analagous to the expressions derived from the maximum likelihood method [3,4,5] and from regression theory[6,7].

The approaches of equations (4) and (5) require the specification of the weighting matrices W_M and W_P. For optimal results these matrices should be the inverse of the covariance matrix for head measurment errors and the inverse of the covariance matrix for the original parameter estimation errors, respectively. In many cases the parameter of interest is the natural log-transmissivity (ln-T). Since ln-T is generally correlated in space, it is reasonable to assume that errors in the original estimates of ln-T will be correlated as well, and, in general, the weighting matrix W_P will be non-diagonal. The weighting matrices have two basic effects. Original parameter estimates and head measurements that are reliable will have associated small error variance values. Since inverting a small variance value leads to a large value in the weighting matrix, the more reliable data will be more strongly weighted than the less reliable data. If the weighting matrix is not diagonal, it will also have a spatial filtering or smoothing effect. Inclusion of the regularization residual portion of the objective function improves the conditioning of the objective function, which in turn leads to more rapid convergence of the solution algorithm. Most reported results have been obtained with diagonal regularization weighting matrices. In fact, only diagonal weighting matrices are incorporated in Modflowp, the USGS parameter estimation program.

Most reported parameter estimation results rely on zoning parameters into large areas of equal value. Zoning is used to reduce the number of decision variables and to avoid problems associated with over-parameterization. In effect, zoning establishes absolute correlation of the original transmissivity estimate errors for all elements within a zone. By necessity, zoning destroys the description of the smaller scale heterogeneities. Zoning is a fundamental model decision which affects the spatial distribution of the parameters. Generally, zoning takes place before the optimization step. It is difficult to determine whether the spatial extent of the zones is appropriately defined or not.

Oscillatory solutions are one of the problems associated with over-parameterized systems. It has been shown that the filtering effects of non-diagonal weighting matrices can reduce oscillations and substantially improve solutions[8]. Non-diagonal weighting matrices can be used to mimic the effects of zoning.

In this report, results are presented in which the ln-T of all elements are decision variables. Solutions are compared which are computed using different regularization residual weighting matrices. Optimization will be performed using the Penalty Method[2].

EXAMPLES

The simulated Walker Lake transmissivity field[9] was used for the following tests. The field has realistic features such as spatial connectivity of extreme values and nested scales of heterogeneity. The original field was subdivided into a 300 by 260 grid with cell dimension of 1 m by 1 m. The ln-T field has a maximum integral range of correlation of 34 m and a minimum integral range of 15 m with the direction of maximum correlation being N14°W. A new grid of 15 by 13 elements with cell dimensions of 20 m by 20 m was constructed. The block transmissivities of the new grid are the geometric mean of the original grid transmissivities contained within each block. The new grid has an average ln-T of 4.01 and a ln-T variance of 0.629. Transmissivities are plotted in Figure 1A. Note the high transmissivity channel that runs southward from the northern boundary and then bends towards the eastern boundary. Areas of lower transmissivity separate the channel from the relatively high transmissivities in the northeast and southwest corners.

A flow simulation was run with constant head boundary conditions of $H = 3$ m and $H = 0$ m for the northern and the southern boundaries, respectively (Figure 1A). The eastern and western boundaries are no-flow. The head field undulates around the linear decline in head with maximum departures on the order of 0.4 m. Simulated measurements were created by taking the forward solution values and corrupting them with uncorrelated Gaussian noise of standard deviation 0.01.

All elements were assigned ln-T values of 4.00 and new parameter estimates were computed using the penalty method of equation (4). Every other non-Dirichlet node was used as a measurement point. Consequently, there were 195 ln-T values to estimate using 98 measurement values. In all cases head measurement weights were 10^4 times the identity matrix. Performance indicators are the average sum of squares of the differences between the computed and forward solution heads at all locations in the domain (SSE(H)) and the average sum of squares of the difference between the computed ln-T and the true ln-T fields (SSE(Y)). Results are shown in Table 1. The original estimate of the ln-T field has an SSE(Y) of 0.63, and the quality of the improvement in the ln-T estimate can be seen by comparing the calculated SSE(Y) values to the original value. A perfect solution would have SSE(Y) and SSE(H) equal to zero.

In Example 1, the regularization residual weights were 1.26 times the inverse of a correlation matrix which was constructed using the integral ranges of the Walker Lake field and an anisotropic exponential correlation function. This represents a statistically based weighting matrix which plausibly could be derived from geological observations. The weighting matrix will be designated "best" since it is superior to the weighting matrices of the other examples. It does not imply that this weighting

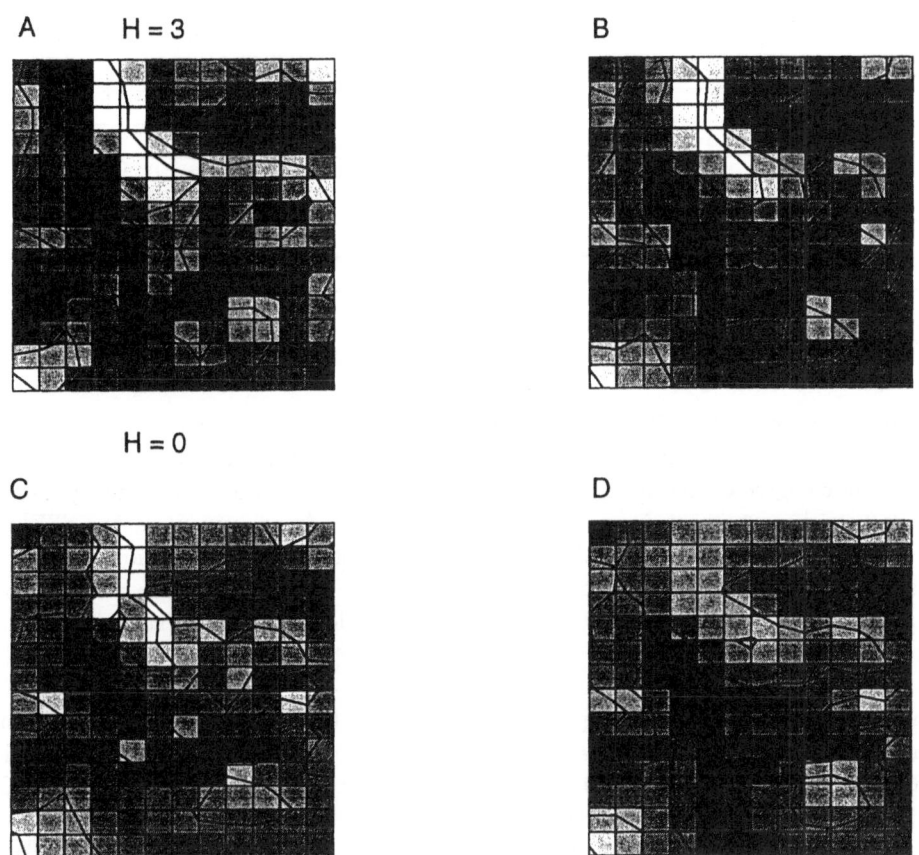

Figure 1: Log-Transmissivity Fields. A. Original field, B. Example 1 Results , C. Example 2 Results, D. Example 3 Results. Ln-T values run from 5.6 (light areas) to 2.2 (dark areas). Contour interval is 1.0.

EXAMPLE	Weighting	No. Meas. Pts.	SSE(H)	SSE(Y)
Original Estimate	-	-	-	0.63
Example 1	Best $\mathbf{W}_P$	98	3.66×10^{-4}	0.18
Example 2	Diagonal $\mathbf{W}_P$	98	6.31×10^{-4}	0.24
Example 3	$\mathbf{W}_P$ from $\lambda_C = 60.$	98	3.14×10^{-4}	0.27
Example 4	$100 \times$ Best	98	1.81×10^{-3}	0.31
Example 5	$100 \times$ Diagonal	98	2.74×10^{-3}	0.35
Example 6	$100 \times \lambda_C = 60.$	98	2.00×10^{-3}	0.38

Table 1: Performace Indicators.

matrix is correct or that it is the best of all possibilities. The algorithm produces an excellent result, and Figure 1B shows that all of the major features have been captured with a fair degree of detail. The results in Table 1 demonstrate that the hydraulic heads have been accurately matched and a significant improvement in the SSE(Y) has been obtained. Although the solution is excellent, the computational effort is extreme, and it took 112 iterations for the solver to converge. The poor conditioning of the objective function is due to the small magnitude of the regularization weights.

The weighting matrix of Example 2 was 1.26 times the identity matrix. The convergence rate of the algorithm was much slower than when the full weighting matrix was used. The results after 200 iterations are presented in Table 1 and Figure 1C. Although the convergence criterion had not been met, it appears that the solution is close to optimal. Comparing the results to those of Example 1, it is seen that not using the best weighting matrix leads to a twenty-eight percent reduction in the accuracy of the ln-T estimates, and the detail of the ln-T field is not as well represented. The solution is not as smooth as the original ln-T field. The tendency towards spatial oscillations is indicative of over-parameterization.

The weighting matrix of Example 3 was constructed by multiplying 1.26 times the inverse of a correlation matrix constructed assuming isotropic correlation with an integral range of 60 m. The weighting matrix overstates the correlation range of the ln-T field and ignores the anisotropy of the true field. With this weighting matrix, the algorithm converged after 37 iterations. Results are presented in Table 1 and Figure 1D. The SSE(H) is the lowest of the three tests, and, although the SSE(Y) is the highest of the three tests, it is comparable that of the diagonal weighting case of Example 2. The smoothing effect of the non-diagonal weighting is evident in Figure 1D.

The conditioning of the objective function can be improved at the expense of the accuracy of the solution by increasing the magnitude of the regularization weighting. The increased regularization weight will cause the solution to stay closer to the original estimates than they should. Examples 4, 5 and 6 were computed with a regularization weighting matrices of one hundred times those of Examples 1, 2 and 3, respectively. In these cases, the solutions converged after 21, 21 and 20 iterations, respectively. The solutions are significantly poorer than those of the first three examples. The high and low transmissivity areas of the solutions have been correctly identified, but the computed values are too low in the high areas and not low enough in the low areas, reflecting the over-emphasis on the reliability of the original estimates. Although the solutions are not optimal, they are a significant improvements over the original estimates, and could be used to identify areas for zoning. It is also seen that the solution computed with the best structure of the regularization weighting matrix is superior to those which used other structures for the weighting matrices, just as in the first three examples.

CONCLUSION

A regularization weighting matrix must be defined in parameter estimation algorithms. Optimal solutions are obtained when the weighting matrix is the inverse of the covariance of the errors of the original parameter estimates. In general the matrix is fully populated, but most reported estimations are performed with diagonal weighting matrices. In this report, it has been demonstrated that the use of correctly specified fully populated weighting matrices can significantly improve the parameter estimates. The use of fully populated weighting matrices also allows the solution of problems that would otherwise suffer from severe difficulties due to over-parameterization[8]. Consequently, spatially detailed parameter estimation can be attempted, avoiding the problems associated with zoning. If zoning is eventually used, spatially detailed parameter estimates can aid in defining zone boundaries.

It seems plausible that correlation structures of different geological environments can be established. Future research should be directed at developing regularization weighting matrices from geological information. Such a procedure would constitute one method for incorporating soft geological and geophysical information into quantitative parameter identification.

A dense spatial network of head information is necessary for estimating detailed spatial variations in parameters. Research should be directed towards developing inovative ways for determining spatially detailed hydraulic head estimates such as, for example, the cokriging of surface elevation and head measurements.[10]

One difficulty associated with spatially detailed parameter estimates is the associated large computational effort. Improvements in computational efficiency are required in order for these techniques to become operationally practical. It may be possible to apply domain decomposition techniques and iteratively solve a series of smaller optimization problems.

ACKNOWLEDGMENTS

Support for this work was provided by the Natural Sciences and Engineering Research Council of Canada Operating Grant OP0122023.

REFERENCES

1. Pinder, G. F. and W. G. Gray (1977) FINITE ELEMENT SIMULATION IN SURFACE AND SUBSURFACE HYDROLOGY, Academic Press, New York.
2. Bentley, L. R. (1993) "Least squares solution and calibration of steady state groundwater flow systems", ADV. WAT. RES., 16, 137-148.
3. Carrera, J. and S. P. Neuman (1986) "Estimation of aquifer parameters under transient and steady state conditions: 1. maximum likelihood method incorporating prior information", WAT. RESOUR. RES., 22, 199-210.
4. Carrera, J. and S. P. Neuman (1986) "Estimation of aquifer parameters under transient and steady state conditions: 2. uniqueness, stability, and solution algorithms", WAT. RESOUR. RES., 22, 211-227.
5. Carrera, J. and S. P. Neuman (1986) "Estimation of aquifer parameters under transient and steady state conditions: 3. application to synthetic and field data", WAT. RESOUR. RES., 22, 228-242.
6. Cooley, R. L. (1982) "Incorporation of prior information on parameters into nonlinear regression groundwater flow models 1. theory", WAT. RESOUR. RES., 18, 965-976.
7. Cooley, R. L. (1983) "Incorporation of prior information on parameters into nonlinear regression groundwater flow models 2. applications", WAT. RESOUR. RES., 19, 662-676.
8. Bentley, L. R. (in press) "Solving and calibrating groundwater flow systems with the penalty method", STOCHASTIC DIFFERENTIAL EQUATIONS WITH APPLICATIONS IN HYDROLOGY (proc. Int. Conf. on Stoch. and Stat. Meth. in Hydr. and Env. Eng., June 21-23, 1993), Kluwer Academic, The Netherlands.
9. Desbarats, A. J. and R. M. Srivastava (1991) "Geostatistical characterization of groundwater flow parameters in a simulated aquifer", WAT. RESOUR. RES., 27, 687-698.
10. Hoeksema, R. J., R. B. Clapp, A. L. Thomas, A. E. Hunley, N. D. Farrow and K. C. Dearstone (1989) "Cokriging model for estimation of water table elevation", WAT. RESOUR. RES., 25, 429-438.

MODIFICATION OF DAGAN'S NUMERICAL METHOD FOR SLUG AND PACKER TEST INTERPRETATION

K.D. Cole[1] and V. A. Zlotnik[2]
University of Nebraska-Lincoln
[1]Department of Mechanical Engineering, 255 WSEC, Lincoln, NE, 68588, USA
[2]Department of Geology, 214 Bessey Hall, Lincoln, NE, 68588, USA

Slug and double packer tests are among the most common techniques used to determine hydraulic conductivities in-situ. Accurate data interpretation is complex due to a wide variety of factors: casing and screen configurations; the aquifer properties (anisotropy and heterogeneity); boundary conditions; and, skin effect.

Currently three models are used for estimation of the relationship between well discharge and hydraulic head at the well screen. Bouwer and Rice (1976) used finite difference electric analog modeling for a wide variety of geometries. Dagan's (1978) version of boundary element method does not require extensive numerical computations, however it is limited to screen-length-to-well-radius ratios larger than 50. Widdowson et. al (1990) removed the limitation on well radius using the finite element method.

In this paper Dagan's method was modified to account for a finite radius of the well. The boundary value problem is reduced to a Fredholm integral equation of the first kind, which allows for numerical determination of the spatial non-uniform distribution of sources at the surface of the well screen. The velocity and well discharge can be calculated accurately from only a few nodes using Green's functions appropriate for finite well radius. Numerical experiments proved good accuracy of the method for determination of the velocity distribution at the well screen. The method is quite competitive with finite elements for arbitrary screen-length-to-well-radius ratios.

INTRODUCTION

There is a vast literature on computational aspects of slug and packer test interpretation. The major problem in simulation of this common well testing technique is treatment of mixed-type boundary conditions. The only available closed-form analytical solution is only valid for simple test configurations (Boast and Kirkham, 1971). However, the computational effort to use it is prohibitive (Dagan, 1978).

In practice there is a need for a computational tool for test interpretation, applied to a wide variety of test configurations including estimation of local and

A. Peters et al. (eds.), Computational Methods in Water Resources X, 719–726.
© 1994 Kluwer Academic Publishers. Printed in the Netherlands.

integral flow characteristics. Universal numerical methods include finite differences (Bouwer and Rice, 1976) and finite elements (Braester and Thunvik; 1984 Widdowson et al., 1990). These methods are computationally intensive in treatment of mixed boundary value problems. Dagan's (1978) semi-analytical technique reduces the elliptic boundary value problem to a singular integral equation to obtain near-analytic accuracy with some of the geometric flexibility of finite-element methods. Dagan's (1978) method is limited to screen-length-to-well-radius larger than 50. A similar method of singular integral equations is available for parabolic problems (Hayashi et al., 1987) with similar advantages. Both Dagan (1978) and Hayashi et al. (1987) methods can be considered to be boundary element methods.

The objectives of this paper are: (1) extension of Dagan's (1978) method to arbitrary configuration of tested well, (2) demonstration of effectiveness for velocity evaluation; (3) comparison of extended method with results of Dagan (1978) and Widdowson et al. (1990).

STATEMENT OF PROBLEM

Slug and packer test interpretations are based on analysis of the relationship between well discharge Q and water level in a riser pipe y at measurement time t (refer to Figure 1). Consider a well of radius r_w penetrating an aquifer with saturated thickness D. A screen of length L is located at depth H from the water table in the unconfined aquifer (or from upper confining boundary in the confined aquifer) with hydraulic conductivity K. The integral characteristic of well hydraulics is a dimensionless shape factor P defined by discharge Q from the well casing, $P \sim Q/(2\pi KLy)$.

The shape factor P has the physical meaning of discharge per unit drawdown per unit screen length and unit hydraulic conductivity. It depends on the configuration of the tested well. In case of anisotropic aquifer P also depends on the ratio of horizontal hydraulic conductivity to vertical hydraulic conductivity. In this paper the isotropic aquifer is considered since extension to anisotropic conditions is straightforward (Zlotnik, 1994).

Since elastic properties of the aquifer can be safely neglected in most circumstances for simulation of slug or packer tests (Dagan, 1978; Hayashi et al., 1987; Widdowson et al., 1990), the boundary value problem for hydraulic head $h(r,z)$ with radial variable r and vertical coordinate z increasing downward from the water table or upper confining layer is:

$$\frac{1}{r}\frac{\partial}{\partial r}\left(r\frac{\partial h}{\partial r}\right) + \frac{\partial^2 h}{\partial z^2} \sim 0, \quad r_w < r < \infty, \ 0 < z < D \tag{1}$$

The condition on the impermeable aquifer base is $\partial h(r,D)/\partial z = 0$, for $r_w < r < \infty$, and the condition at the well screen is $h(r_w,z) = y$, for $H-L < z < H$.   The condition of zero groundwater flow velocity far from the well is $\partial h(\infty,z)/\partial r = 0$ for $0 < z < D$.

Slug and packer tests have different boundary conditions at the well face. For a slug test (Figure 1A), the well casing is impermeable to flow above and below the screened section. The slug test problem has no-flow boundary conditions above and below the well screen at the well radius: $\partial h(r_w,z)/\partial z = 0$; for $0 < z < H-L$, $H < z < D$.

For a packer test (Figure 1B), the well casing is perforated above and below the sealed sections allowing water exchange between the aquifer and cased borehole. Two packers with lengths ϵ seal off the tested well section from the upper or lower well sections. The packer test problem has a constant head condition above and below the packers at the well radius: $h(r_w,z) = 0$, for $0 < z < H-L-\epsilon$, $H+\epsilon < z < D$. No-flow conditions exist on contact surfaces between the packers and well casing, $\partial h(r_w,z)/\partial r = 0$, for $H-L-\epsilon < z < H-L$, $H < z < H+\epsilon$.

Finally, the condition on the upper aquifer boundary is needed. For an unconfined aquifer, a constant head condition is used at the water table (Bouwer and Rice, 1976; Dagan, 1978; Widdowson et al., 1990), so that $h(r,0)=0$, for $r_w < r < \infty$ whereas for a confined aquifer a no-flow condition is used $\partial h(r,0)/\partial r = 0$ for $r_w < r < \infty$.

The hydraulic head is a function of variables r,z and seven parameters $(y, r_w, D, L, H, \epsilon, K)$. Then discharge is determined from the local horizontal velocity $V(r,z)$

$$V(r,z) = -K \frac{\partial h(r,z)}{\partial r}, \quad Q(y,r_w,D,L,H,\epsilon,K) = 2\pi r_w \int_{H-L}^{H} V(r,z)\, dz \qquad (2)$$

According to dimensional analysis of this boundary value problem, the dimensionless shape factor P depends only on four dimensionless arguments (Zlotnik, 1994), $P = P(r_w/L, h/L, d/L, \epsilon/L)$.   The problem is to find the local horizontal velocity $V(r,z)$ and integral flow characteristic P.

NUMERICAL METHOD

Dagan's (1978) numerical method uses the Green's function G to express the solution for hydraulic head through unknown velocity function $V(r_w,z)$ in a Fredholm equation of the first kind.

$$h(r,z) = \frac{2\pi r_w}{K} \int_0^D V(r_w,z')\ G(r,z,r_w,z')\, dz' \qquad (3)$$

Let z_j, $j = 0,1,...N$; $z_0 = 0$, $z_N = D$ be nodes of a nonuniform spacial grid on z-axis with steps $L_j = z_j - z_{j-1}$, $j = 1,2,...N$, and $z_{j-0.5} = 0.5(z_j + z_{j-1})$, $j = 1,2,...N$. Grid nodes must also include all singular points: top and bottom of tested screen for slug test and top and bottom of each packer for packer test. Then equation (3) can be rewritten as

$$h(r,z_{i-0.5}) = \frac{2\pi r_w}{K} \sum_{j=1}^{N} \int_{z_{j-1}}^{z_j} V(r_w,z') \; G(r,z_{i-0.5},r_w,z')dz' \qquad (4)$$

Using discrete notation for hydraulic head and piece-wise constant approximation of velocity values at each grid step at the well screen: $H_i = h(r_w,z_{i-0.5})$, $V(r_w,z') \approx v_j$, $z \in (z_j,z_{j-1})$, $i,j = 1,2,...N$ one obtains a full system of linear algebraic equations for v_j, $j = 1,2,...N$ with matrix g_{ij} found from the Green's function

$$H_i = \sum_{j=1}^{N} v_j \, g_{ij}, \quad g_{ij} = \frac{2\pi r_w}{K} \int_{z_{j-1}}^{z_j} G(r_w,z_{i-0.5},r_w,z')dz', \; i,j = 1,2,...N \qquad (5)$$

The system (5) involves only nodes located in screen zones with water exchange between well and aquifer where head values H_i are known. Solution of this system provides $V(r_w,z)$ at well screen. Then from (2) and (3)-(5) one finds velocity

$$V(r,z) = -\sum_{j=1}^{N} v_j \, \tilde{g}_j(r,z), \quad \tilde{g}_j(r,z) = 2\pi r_w \int_{z_{j-1}}^{z_j} \frac{\partial G(r,z,r_w,z')}{\partial r} dz' \qquad (6)$$

and integral discharge for slug test (which may differ slightly for packer test)

$$Q = 2\pi r_w \sum_{i=1}^{N} v_i L_i \qquad (7)$$

The last two equations constitute a solution of the problem. This method is valid for arbitrary L/r_w ratio. For $L/r_w > 50$, the asymptotic Green's function (AGF) used by Dagan (1978) is valid. The suggested extension of this method is based on the exact Green's function (EGF). A useful collection of Green's functions for various boundary value problems is available (Beck et al., 1992).

VERIFICATION OF METHOD AND NUMERICAL RESULTS

Effectiveness of the method is demonstrated for the case of slug and packer tests in unconfined aquifers with a constant head at the water table.

The *slug test* problem, defined earlier, has corresponding EGF given below where K_0 and K_1 are modified Bessel's functions, and $\beta_n = \pi(n-0.5)$.

$$G(r,z,r_w,z') - \frac{1}{\pi} \sum_{n-1}^{\infty} \frac{\sin(\beta_n z/D) \sin(\beta_n z'/D)}{\beta_n r_w} \frac{K_0(\beta_n r/D)}{K_1(\beta_n r_w/D)} \tag{8}$$

Calculation of g_{ij} and $\tilde{g}_j$ in equations (5) and (6) is straightforward.

To verify the method, the shape factor for different L/r_w (Figure 2) and velocity distribution at the screen surface (Figure 3) was computed using the AGF and EGF methods. For the well screen located close to the water table ($H/L=1.25$) and far away from the water table ($H/L=16.0$), the shape factor is almost identical in both methods for $50 < L/r_w < 100$. This gives confidence in the EGF-extension of Dagan's (1978) AGF method for $L/r_w < 50$.

Figure 3 indicates that the EGF accurately delineates the velocity profile near the well screen including singularities at the top and bottom of screen and removes oscillations inherent to AGF. Still, the relative error of the integral characteristic P of the AGF method is only 2%.

The finite element method was also used for verification of the EGF method. Braester and Thunvik (1984) have indicated a match between Dagan's (1978) and finite element results. Their Figure 2 displays overestimation of P of less than 1%. Widdowson et al. (1990) published a comparison with Dagan's (1978) AGF method for shape factor and velocities, though error bounds and information on discretization were not reported. Their finite element method noticeably overestimated horizontal velocities almost uniformly along the screen (see their Figure 4), thus overestimating shape factor P. The EGF-based recalculation of shape factor for the wide range of L/r_w is shown in Figure 4. It is obvious that finite element method overestimates the shape factor within bound 9%. Thus accuracy of both AGF and EGF methods is on the order of 1%. For figures (2)-(4), N=20 uniformly-spaced nodes were placed on the well screen.

Packer test simulations used additional nodes at the well radius above and below the packers which allow exchange of water between the casing and the aquifer. The shape factor as a function of screen length is given in Figure 5 for packer lengths $\varepsilon/r_w=1.0$ and 10.0. The shape factor is slightly larger for short packers. Figure 6 demonstrates that the present method produces accurate velocity profiles, including singularities at both edges of each packer and inversion of the flow direction above and below each packer. For Figure 5, N = 14 uniformly-spaced nodes were used, 4 above the packers, 6 between the packers, and 4 below the packers. For Figure 6 N = 36 nodes were used, 12 above, 12 between, and 12 below the packers. The shape factor P is accurate for as few as N = 12 nodes but more nodes are needed for finding smooth velocity profiles. In calculating g_{ij} the infinite series was truncated so that the ratio of the last retained term to the truncated series was less than 0.0001. Depending on the geometry about 6000 terms were required for each g_{ij} for the EGF method.

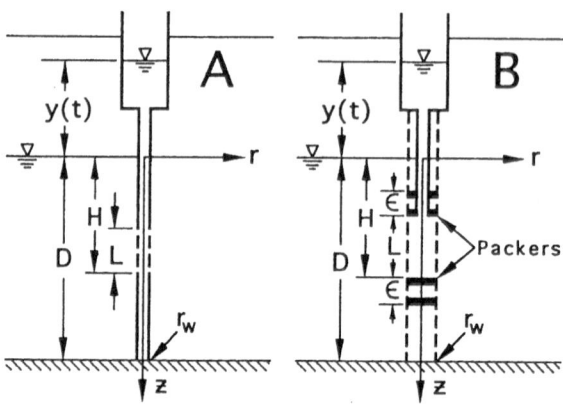

Figure 1. Test configuration is unconfined aquifer: (A) slug test; (B) packer test.

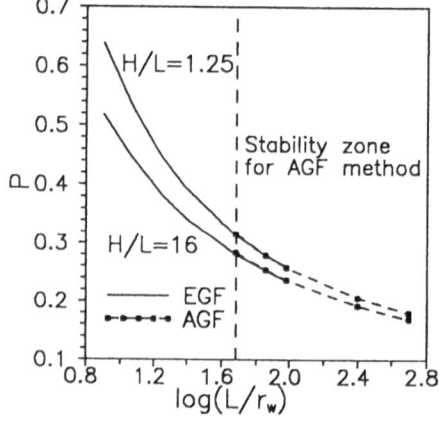

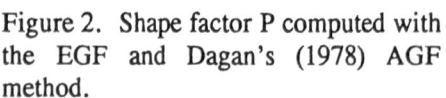

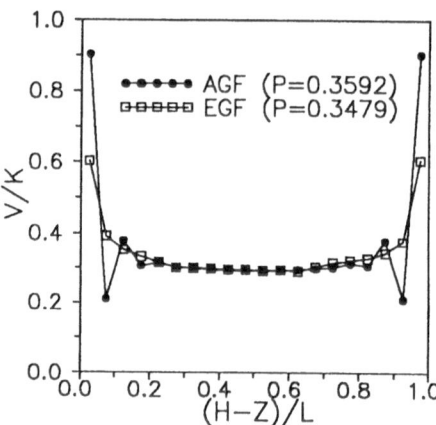

Figure 2. Shape factor P computed with the EGF and Dagan's (1978) AGF method.

Figure 3. Velocity profiles $V(r_w, z)$ for slug test ($H/L = 16$, $L/r_w = 24$).

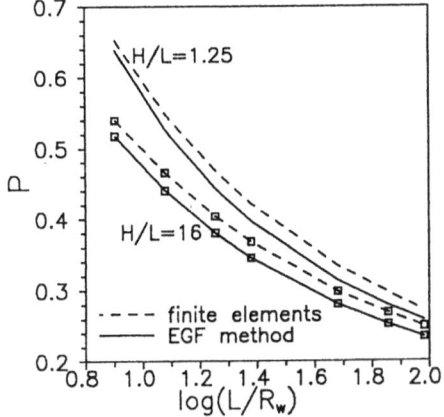

Figure 4. Shape factor P for the slug test in a comparison between the EGF method and results of Widdowson et al. (1990).

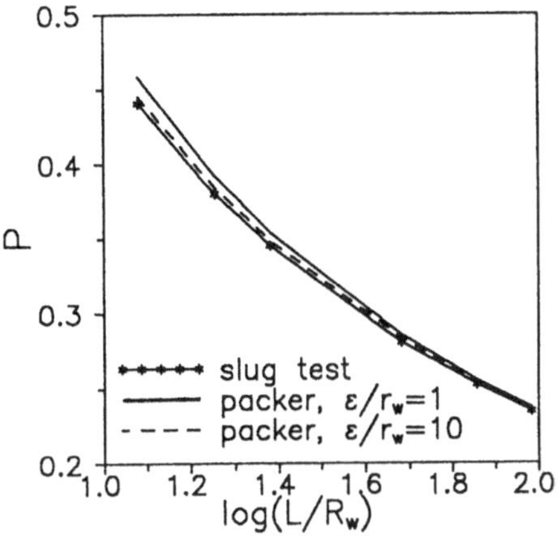

Figure 5. Shape factor P from the EGF method for slug test and packer tests with different packer lengths (H/L = 16).

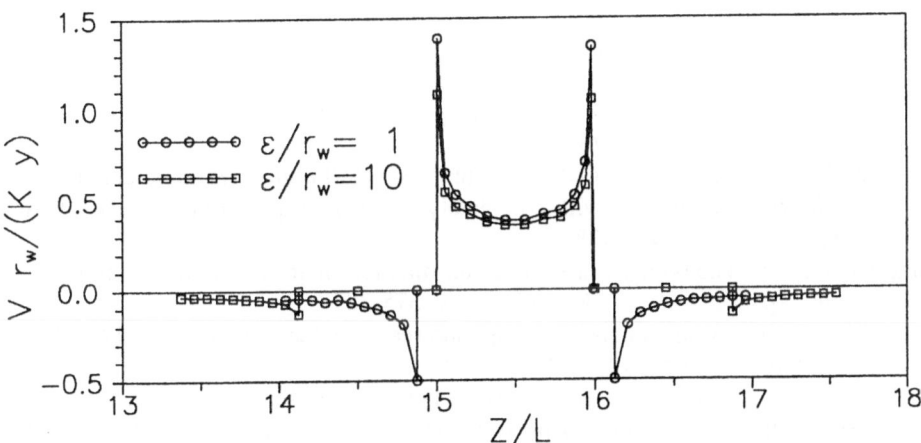

Figure 6. Dimensionless velocity profile $V(r_w,z)$ r_w $/(K\ y)$ for packer tests with different packer lengths computed with the EGF method (D/L = 25.0, H/L = 16.0, $L/r_w = 12.0$).

CONCLUSIONS

An extension of the semi-analytical method of Dagan's (1978) for solution of mixed-type boundary value problems common in slug and packer well testing interpretation is presented. The method is based on exact Green's functions to generate a Fredholm equation of the first kind. The accurate distribution of velocity (with singularities) and dimensionless shape factor P are obtained using only few nodes in discretized solution of the integral equation. Results are uniformly valid for all L/r_w ratios. For $L/r_w > 50$ the asymptotic Green's function of Dagan can be used. In the intermediate zone $50 < L/r_w < 100$ the discrepancy between methods is 1%. For $L/r_w < 50$ only exact Green's function can be used. Since the exact analytical Green's functions for any slug and packer test configurations are available, the numerical method is robust. Comparison with finite element solution shows that the finite element solution overestimates the shape factor and local velocities on the order of 1-9%.

ACKNOWLEDGEMENTS

This work is partially supported by the Water Center of the University of Nebraska-Lincoln and an annual Federal matching grant as authorized by the section 104 of the Water Resources Research Act, administered by U.S. Geological Survey.

REFERENCES

Beck, J. V., Cole, K. D., Haji-Shiekh, A., Litkouhi, B. (1992) Heat Conduction Using Green's Functions, Hemisphere, New York.
Boast, C. W., and Kirkham, D. (1971) "Auger hole seepage theory", Soil Sci. Soc. Amer. Proc. 35, 365-373.
Bouwer, H. and Rice, R. (1976) "A slug test for determining hydraulic conductivity of unconfined aquifers with completely or partially penetrating wells", Water Resour. Res. 12, 423-428.
Braester, C., and Thunvik, R. (1984) "Determination of formation permeability by double-packer tests", J. Hydrology 72, 375-389.
Dagan, G. (1978) "A note on packer, slug, and recovery tests in unconfined aquifers", Water Resour. Res. 14, 929-934.
Hayashi, K., Ito, T., and Abe, H. (1987) "A new method for the determination of in situ hydraulic properties by pressure pulse tests and application to the Higashi Hachimantai geothermal field", J. Geophys. Res. 92(B9), 9168-9174.
Widdowson, M.A, Molz, F.J., and Melville, J.G. (1990) "An analysis technique for multilevel and partially penetrating slug test data", Ground Water 28, 937-945.
Zlotnik, V. (1994) "Interpretation of slug and packer tests in anisotropic aquifers", Ground Water 32, in press.

PARAMETER ESTIMATION WITH DATA-DRIVEN ZONATION

M.J. EPPSTEIN and D.E. DOUGHERTY
Research Center for Groundwater Remediation Design
Department of Civil and Environmental Engineering
University of Vermont
Burlington, VT 05405
USA

1 INTRODUCTION

Parameter estimates are considered by many to be an "Achilles' heel" of numerical modeling of natural porous media. Among the important parameters for which estimates are needed are material property descriptors (e.g., hydraulic conductivity, storativity, distribution coefficient), source/sink terms (e.g., infiltration, history of contaminant releases, biomass density), and importantly—though rarely considered explicitly—the parameter structure for a given situation. Over the years many investigators have attended to the problems of devising parameter estimation schemes and of determining parameter values (see [11]). Typically, the parameter structure is assumed to be known and of low dimensionality. The method proposed a decade ago by Sun and Yeh [7] remains one of the few published quantitative approaches to automatic parameter structure detection of hydrological parameters.

Kalman filtering [5] has been widely used in control systems engineering as a means of recursively calibrating state variables based on time-series data for linear systems, and recently progress has been made in applying this technique to improve groundwater state variables in field scale domains (e.g. [2, 3, 12]). Extended Kalman Filtering (EKF) [1] is a non-linear adaptation of the technique. Early research efforts utilizing EKF for parameter estimation in groundwater systems showed promise (e.g. [10]). The computational unwieldiness of EKF has so far limited its use in further attempts at groundwater parameter estimation to low-dimensionality parameters with prespecified structure (e.g. [6, 8]), despite the fact that improper zonation can lead to significant errors in parameter estimates [7].

In this paper, we present early results from a new technique under development for automatic simultaneous estimation of parameter structure and values. The method is based on a combination of EKF and a divisive cluster analysis algorithm which progressively reduces the dimensionality of the parameter structure (and thus also the computational burden of the EKF) with each timestep. We illustrate the technique by estimating heterogeneous zoned and unzoned transmissivity fields from steady-state flow domains using a sequence of noisy head measurements at a limited number of

727

A. Peters et al. (eds.), Computational Methods in Water Resources X, 727–734.
© 1994 Kluwer Academic Publishers. Printed in the Netherlands.

observation wells. Test problems show that the integrated method achieves reductions in state prediction errors comparable to EKF alone, and is computationally feasible on much larger problems.

2   PARAMETER ESTIMATION ALGORITHM

2.1   Extended Kalman Filtering

Time advancement in a transient linear groundwater flow model can be expressed in state-space notation as

$$h^{n+1} = T^n h^n + L^n f, \tag{1}$$

where T and L are transition matrices incorporating the effects of system parameters, h is the state vector of heads, and f is the forcing function vector.

In EKF, the state space and transition matrices are augmented to include both state variables and parameters (p) to be estimated (such as zone log-transmissivities):

$$x = \left\{ \begin{array}{c} h \\ p \end{array} \right\}. \tag{2}$$

EKF proceeds in a two-step fashion for each timestep; a time update of state and state covariance, followed by a measurement update of state and state covariance. In the notation below, $(-)$ and $(+)$ are used to indicate values before and after measurement update, respectively.

2.1.1   EKF Time Update

The time update of the state x is performed deterministically by the numerical method of choice. A time update of the augmented state covariance B is performed as shown below, where Q is the unaugmented covariance matrix of the normally distributed system noise. These updates may be expressed as

$$x^{n+1}(-) = \Phi^n x^n(+) + \Lambda^n f \tag{3}$$

$$B^{n+1}(-) = C^n B^n(+) C^{nT} + W, \tag{4}$$

where the augmented matrices are

$$\Phi^{n+1} = \left[\begin{array}{cc} T^n & 0 \\ 0 & I \end{array} \right], \Lambda^n = \left[\begin{array}{cc} L^n & 0 \\ 0 & 0 \end{array} \right], C^n = \left[\begin{array}{cc} T^n & D^n \\ 0 & I \end{array} \right], W = \left[\begin{array}{cc} Q & 0 \\ 0 & 0 \end{array} \right], \tag{5}$$

and where

$$D^n = \frac{\partial}{\partial p}[T(p)h + L(p)f]\Big|_n \tag{6}$$

is the sensistivity (Jacobian) of the time update of the true (unaugmented) state with respect to parameter values.

2.1.2   EKF Measurement Update

Augmented state and state covariance matrices are stochastically adjusted in accordance with whatever measurements are available for the new timestep. The impact of the measurements is "filtered" via the so-called "Kalman gain matrix", which dynamically weights the impact of the new measurements depending on the relative uncertainty in the measurements versus in the state at each timestep. The resulting update is optimal in the sense that the trace of the state covariance matrix is minimized.

Allowing measurements to be taken at discrete times, assuming no direct observations of parameters (for discussion only, this is easily relaxed), and that measurements are corrupted with white noise with covariance R, we write

$$Z^{n+1} = H^{n+1}x^{n+1} + V^{n+1}, \tag{7}$$

where $H = [M \quad 0]$ is the augmented form of the observation vector M, and V has covariance

$$Y = \begin{bmatrix} R & 0 \\ 0 & 0 \end{bmatrix}. \tag{8}$$

Measurements are then used to update x and B by computing the augmented Kalman gain matrix G:

$$G^{n+1} = B^{n+1}(-)H^{n+1T}[H^{n+1}B^{n+1}(-)H^{n+1T} + Y]^{-1} \tag{9}$$

$$x^{n+1}(+) = x^{n+1}(-) + G^{N+1}[x^{n+1} - H^{n+1}x^{n+1}(-)] \tag{10}$$

$$B^{n+1}(+) = [I - B^{n+1}(-)H^{n+1}]. \tag{11}$$

EKF is computationally much more demanding than linear KF, since several large matrices (e.g., the Transition (Φ and Λ), Jacobian (D) and Kalman gain (G) matrices) must be recalculated at each timestep, due to their nonlinear dependence on the parameters p. Furthermore, the augmentation of n state variables with m parameters increases the size of each matrix by $m(m+2n)$, with concomitant increases in the time to perform operations on them.

One appealing feature of the EKF algorithm is that it naturally incorporates measurements whose frequency, location, and uncertainty varies with time. Several thorough descriptions of EKF exist in the literature (e.g. [1]).

2.2   Automatic Zonation through Cluster Analysis

The computational requirements of EKF are decreased by a reduction in the dimensionality of the parameter space. Unlike previous studies which have assumed parameter structure ([6, 8]), we determine structure through a progressive automatic zonation procedure, wherein the dimensionality of the parameter structure is dynamically determined and reduced (if possible) with successive timesteps.

The zonation proceeds as follows. Each element in the domain is initialized at time 0 to be a separate transmissivity zone. For each timestep, a time update and EKF measurement update are performed on all log-transmissivity zones. Following the EKF update, tranmissivity zones are coalesced, if possible, into larger zones according to an iterative partitional clustering algorithm [4].

The choice of appropriate cluster tolerance criteria [4] greatly affects the effectiveness of the clustering algorithm. We have experimented with several types of criteria. The best results have been achieved using a simple test based on a maximum allowable range of of transmissivities allowed in each cluster. Tolerance is initially set low to avoid incorrectly "locking" zones together at early timesteps before enough information is present to correctly determine zone boundaries. Tolerance is then allowed to increase gradually with subsequent timesteps.

3   EXPERIMENTS

3.1   Numerical Model Description

Each experiment was conducted by specifying a reference problem (with a solution), taking a time series of spatially distributed samples from the reference solution with measurement noise, and applying our methods to estimate heads and transmissivities. The reference problems are steady flow in an isotropic confined aquifer. Flow occurs in a rectangular domain, with heads (h) specified at two opposing sides, and no normal flow conditions on the other two sides. Storativity (S) is considered known. A Galerkin finite element method with backward Euler timestepping and bilinear rectangular elements is used to model the problem. Test problems were scaled so that $\Delta x = \Delta y = 10$, used $S = 0.001$, had Dirichlet conditions $h = 10$ and $h = 0$, and timestep $\Delta t = 1$.

In order to minimize boundary effects in the small test domains, boundary elements on Dirichlet boundaries were forced to take on the transmissivity values of adjacent internal elements after each EKF had been performed, before the cluster analysis.

Covariance matrices were constructed by assuming an exponential model of correlation with distance, with a correlation length of 4 elements for nodal head and initial elemental transmissivity values. Transmissivity zones formed by clustering are assumed to be uncorrelated. Variance of head state variables was initialized to 0.2, of log-transmissivity zones to 0.5, and of head measurements to 0.05.

3.2   Test Cases Considered

Four test cases are presented. The true transmissivity fields and locations of monitoring wells are shown in Figure 1 (leftmost graphs). In all cases, true transmissivity values ranged between 1 and 10, and the initial estimate of transmissivity was set to 5 for all elements. Cluster tolerance for timestep 1 was set to 0.2, and

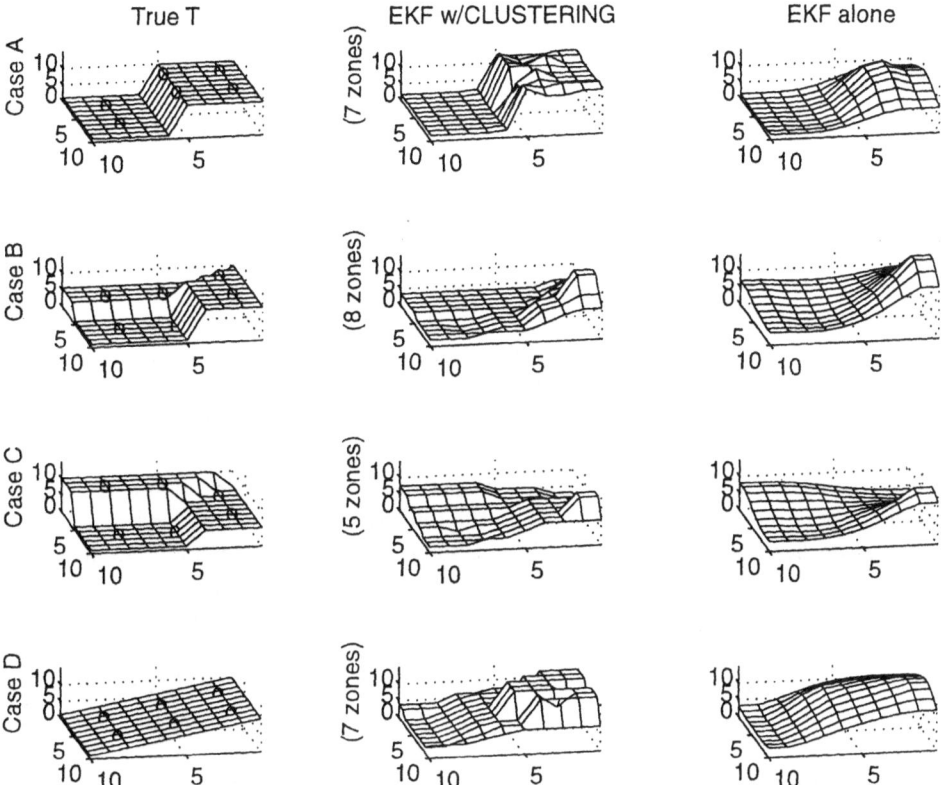

Figure 1: True and estimated elemental transmissivity values are shown for four test domains. The true transmissivity fields are shown in the left column, with monitoring well locations indicated. Transmissivity estimates after five timesteps starting from an initial transmissivity estimate of 5 across the domain, for EKF with clustering (final number of zones is indicated), and for EKF alone, are shown in the middle and right columns, respectively. In each case, flow is predominantly right to left.

was allowed to increase by 0.2 each timestep to a final value of 1.0. Simulations were run for 5 timesteps, with head measurements simulated by adding white noise to true head values at the six observation well locations.

Transmissivity estimates at the end of 5 timesteps of representative simulations are shown in Figure 1 for EKF with clustering (middle column of graphs) and for EKF alone (rightmost graphs). For Case A, a simple 2-zoned transmissivity field with zone boundary normal to flow, EKF with clustering was able to correctly detect the zone boundary. Clustering reduced the number of free zones at each timestep. This progressive zonation is typified by the Case A data, in which the number of zones dropped as follows: 100, 39, 26, 14, 10, 7. Zone boundaries which run parallel to flow are more difficult to detect (Cases B and C); the use of concentration measurements may improve zone detection in such situations. In Case D, the continuous ramp transmissivity field, clustering tends to form a step field of estimated transmissivity zones. Boundary effects in these small domains are evident in all simulations.

Nodal head errors (based on the heads predicted from the modified transmissivity estimates) and elemental transmissivity estimation errors were quantified as the sum of the squared errors as compared to the true system. Since the EKF conditions estimates on head measurements, it is not surprising that significant reductions in head error were achieved in all tests cases. Interestingly, the addition of clustering, although reducing the degrees of freedom, has little impact on head error reduction. An average of 96% of the head prediction error was removed by EKF alone in the four test cases (Figure 2a), as compared to an average of 95% for EKF with clustering (Figure 2b). Most of the head error reduction occurred during the first three timesteps.

Sum of squared transmissivity errors responds variably to the algorithm. In three out of the four test cases, EKF alone actually increased transmissivity errors after 5 timesteps (Figure 2c), whereas EKF with clustering reduced the error in transmissivity estimates in three of the four cases (Figure 2d). The ramp tranmissivity (Case D) yielded the highest transmissivity errors by both algorithms. The magnitude of the sum of squared transmissivity errors is greatly influenced by the initial estimate of transmissivity. This metric is a crude indicator of effectiveness and is a poor measure of accuracy in parameter structure detection.

The integrated algorithm offers significant reduction in computation time over EKF alone. The higher the cluster tolerance is set, the greater the reduction in dimensionality and resultant increases in speed, and the greater the risk of incorrect zonation and increased error. Efforts to quantify the speed-up afforded by our algorithm, as well as methods for objective determination of optimal cluster tolerance, are underway. Additional algorithm improvements are foreseen.

4   CONCLUSIONS

A new method for parameter estimation and structure detection shows promise

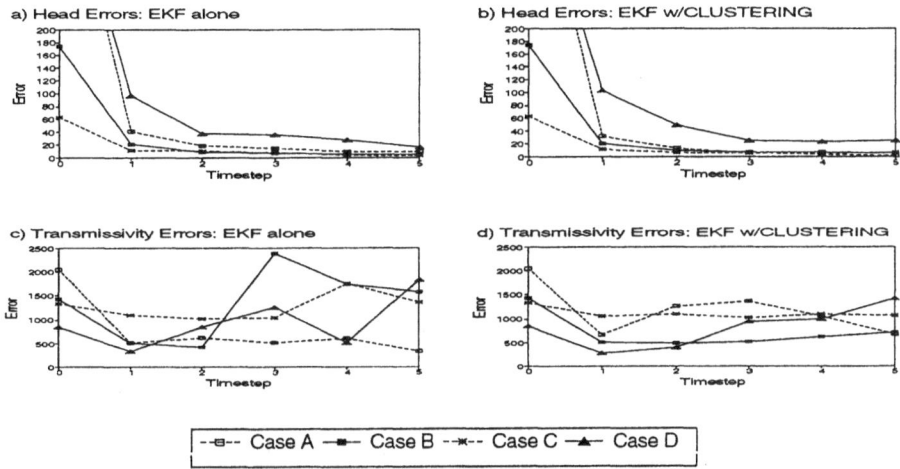

Figure 2: Reduction in sum of squared nodal head errors (a and b) and elemental transmissivity errors (c and d) after 5 timesteps for four test cases using EKF alone (a and c) and EKF with clustering (b and d).

for estimating heterogeneous transmissivity values and structures based on sparse head measurements. The method capitalizes on the recursive calibration abilities of Extended Kalman Filtering, while significantly reducing the time required by this computationally intensive algorithm through a progressive reduction in parameter dimensionality based on a divisive cluster analysis algorithm. Results from test problems show that the technique can afford reductions in predictive head error comparable to EKF alone, and clustering may actually improve transmissivity estimates relative to EKF alone.

We are investigating several additional potential improvements to the integrated estimation algorithm. Among these are the incorporation of other geostatistical data to aid in the zonation procedure, dynamic determination of cluster tolerance criteria, conditioning on solute concentration measurements, and parallelization of the EKF algorithm. It is hoped that such improvements will enable researchers to apply EKF to realistically sized problems for estimation of the myriad parameters required for accurate forecasts of groundwater systems.

ACKNOWLEDGMENTS

The research reported in this paper was supported in part by National Science Foundation Grant ASC-9100226; the Government may retain some rights to this work. M.J.E.'s participation in this project was made possible in part by a grant from the AAUW Educational Foundation.

REFERENCES

[1] Gelb, A., editor (1974), *Applied Optimal Estimation*, M.I.T. Press, MA.

[2] Graham, W.D. and McLaughlin, D.B. (1991) "A Stochastic Model of Solute Transport in Groundwater: Application to the Borden, Ontario, Tracer Test", *Water Resources Research*, **27**(6):1345-1359.

[3] Graham, W.D. and Tankersley, C.D. (1993) "Forecasting Piezometric Head Levels in the Floridian Aquifer: A Kalman Filtering Approach", *Water Resources Research*, **29**(11):3791-3800.

[4] Jain, A.K. and Dubes, R.C. (1988) *Algorithms for Clustering Data*, Prentice Hall, N.J.

[5] Kalman, R.E. (1961) "A New Approach to Linear Filtering and Prediction Problems", *Journal of Basic Engineering*, **March 1961**:95-108.

[6] Lee, S. and Kitanidis, P.K. (1991) "Optimal Estimation and Scheduling in Aquifer Remediation With Incomplete Information", *Water Resources Research*, **27**(9):2203-2217.

[7] Sun, N. and Yeh, W.W-G. (1985) "Identification of Parameter Structure in Groundwater Inverse Problem", *Water Resources Research*, **21**(6):869-883.

[8] Van Geer, F.C. and Van Der Kloet, P. (1985) "Two Algorithms for Parameter Estimation in Groundwater Flow Problems" *Journal of Hydrology*, **77**:361-378.

[9] Wang, H.F. and Anderson, M.P. (1982) *Introduction to Groundwater Modeling: Finite Difference and Finite Element Methods*, W.H. Freeman and Co., N.Y.

[10] Wilson, J., Kitanidis, P. and Dettinger, M. (1978) "State and Parameter Estimation in Groundwater Models", in *Applications of Kalman Filter to Hydrology, Hydraulics, and Water Resources*, Chiu, C., Editor, University of Pittsburgh, Pennsylvania.

[11] Yeh, W. W-G. (1986) "Review of Parameter Identification Procedures in Groundwater Hydrology: The Inverse Problem", *Water Resources Research*, **22**(2):95-108.

[12] Yangxiao, Z., Te Stroet, C.B.M., and Van Geer, F.C. (1991) "Using Kalman Filtering to Improve and Quantify the Uncertainty of Numerical Groundwater Simulations: 2. Applications to Monitoring Network Design", *Water Resources Research*, **27**(8):1995-2006.

HYDRAULIC TOMOGRAPHY

J. GOTTLIEB
Institute of Soil Mechanics and Rock Mechanics
Karlsruhe University
D-76128 Karlsruhe
Germany

ABSTRACT: A tomographical method, which allows the identification of spacially distributed hydraulic conductivities using 'point sources' and pore water pressure measurements in boreholes is presented.
We compare identifiability for transient and steady state conditions. The foreward problem is discritized in a FE algorithm, the ill - posed inverse problem is solved by a regularizing Gauss - Newton - like method, using underdetermined linearized equations.

INTRODUCTION

The knowledge of spacial distributions of hydraulic conductivity is of great interest in many geotechnical applications, as aquifer modelling, flow - mass transport problems and in - situ decontamination modelling.
In this paper, we describe a tomographical principle and its application to the identification problem under discussion. We use a transmission method by injecting water via 'point sources' in boreholes into the subsoil. In a second borehole observations of the pore water pressure response are made. We assume that the soil is water saturated. Then for the transient case we have the following set of parabolic partial differential equations

$$S(x)\partial_t h_i(x,t) - \nabla \cdot (k(x)\nabla h_i(x,t)) = Q_i(t), \qquad i = 1, \ldots, m, \qquad (1)$$

where $h_i(x,t)$ is the pore water pressure (water head), $S(x)$ the storage coefficient, $k(x)$ the hydraulic conductivity and $Q_i(t) = \delta_{x_i}(x)q_i(t)$ describe source terms, which are located at points x_i in the first borehole. The instationarity of h is caused by the storativity of the material soil which is in fact unknown, as well. For $S(x) = 0$ we come to the steady state equation

$$-\nabla \cdot (k(x)\nabla h_i(x)) = Q_i, \qquad i = 1, \ldots, m, \qquad (2)$$

A. Peters et al. (eds.), Computational Methods in Water Resources X, 735–741.
© 1994 Kluwer Academic Publishers. Printed in the Netherlands.

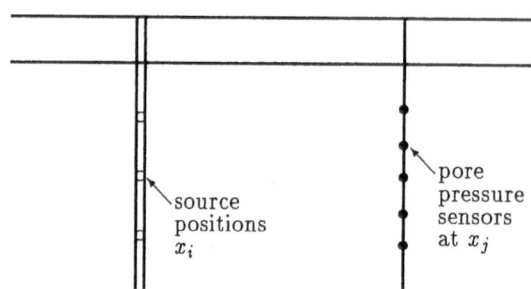

Figure 1: Hydraulic tomography: Experimental design.

which is related to the problem of electrical impedance tomography, see Calderón (1980), Barber and Brown (1984), Kohn and Vogelius (1984), Isaacson (1986), Dobson and Santosa (1993), for example.

Identification problems for equation (1) are also considered in many publications of groundwater mechanics, Carrera and Neumann (1986), Yeh (1986) and Sun and Yeh (1990). The main differences between our problem and those papers are

- the large amount of information we obtain by varying the position of sources and sinks and

- that we are mainly interested in the tomographical transmission principle, i.e. we want to investigate which information we need for reconstructing the conductivity distribution between boreholes from measurements outside the inaccessible regions.

The method presented here is called *hydraulic tomography*, see Gottlieb (1992), Gottlieb and Dietrich (1993) or Bohling (1993).
The higher the resolution required the more accurate the data should be. Because of the geometrical attenuation, the transmission distance can not be too large. Especially in 3-D problems the method works only in the range of some decameters.

The basic design consists of one borehole for the water source and another, which is endowed with pore water pressure sensors, as shown in figure 1. Configurations as described in figure 2 are possible, too. Here, we use a borehole with a point source of variable depth and a second borehole for the sink. Pore pressure observations are made over the whole depth of interest. This can be managed in special lances or boreholes.

In the following sections we discuss the mathematical model and its ill-posedness. We compare the identifiability under steady state and transient conditions and propose a numerical procedure, which gives u insight into the information content of our model and data. It shows the advantage of the transient conditions.

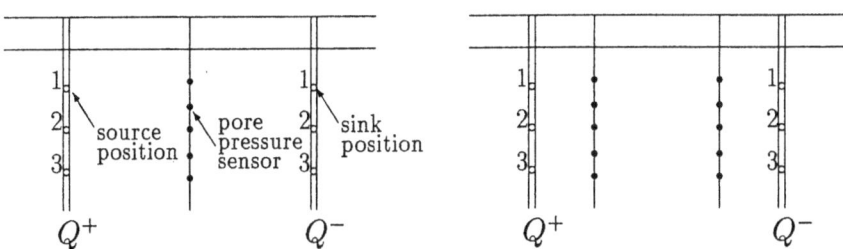

Figure 2: Hydraulic tomography. Experimental design with source and sink and one or two sensor lances.

THE MATHEMATICAL MODEL

We assume that boundary conditions are known, or the boundary is far away and without any influence, and the initial values $h_0(x) = h(x, 0)$ are given. Then for each i the equations (1) have a unique solution, if the coefficients $S(x)$ and $k(x)$ known.

The additional data for solving the inverse problem, i.e. the determination of the unknown coefficient $k(x)$ and $S(x)$ in (1), are pore water pressure measurements $z_{ij}(t) = \tilde{h}_i(x_j, t)$ at places x_j of the i-th source configuration Q_i. Should the occasion arise, we also make use of the pressure at the source. We define the response operator $\mathcal{R}_{k,S}$ as a map between the applied sources Q_i and the measurements $z_{ij}(t)$, i.e.

$$\mathcal{R}_{k,S}[Q_i(t)] = z_{ij}(t), \qquad j = 1, \ldots, n. \tag{3}$$

Let $H_1 = L^2(0, T) \times \ldots \times L^2(0, T)$ (m times) be a Hilbert space with inner product $(\cdot, \cdot)_1$ and $H_2 = L^2(0, T) \times \ldots \times L^2(0, T)$ (n times) a Hilbert space with inner product $(\cdot, \cdot)_2$. Then H_1 contains the domain of $\mathcal{R}_{k,S}$ and H_2 its image.

Analogously, under steady state conditions we have

$$\mathcal{R}_k[Q_i] = z_{ij}, \qquad j = 1, \ldots, n. \tag{4}$$

The stationary response operator maps from $H_1 = \mathbb{R}^m$ into $H_2 = \mathbb{R}^n$. In the following we want to employ the linearity of both response operators: Assume that m experiments with sources $Q_i(t)$ $(i = 1, \ldots, m)$ have been made, and we have the corresponding measured data. Then the superposition principle (and with respect to the time axis Duhamel's principle) gives additional synthetical data. So, we are able to extend the domain of definition of the response operators. This will be useful in the following two sections, where we discuss the reconstruction of k and S from $\mathcal{R}_{k,S}$ (k from $\mathcal{R}_k$ resp.). Both operators are compact.

Say, the pairs $\{k, S\}$ are elements of a normed spaces X. Then we have to invert a map $\mathcal{A}$ from X into the space of compact operators $\mathcal{K}(H_1, H_2)$, which map from H_1 into H_2:

$$\mathcal{A} : X \longmapsto \mathcal{K}(H_1, H_2). \tag{5}$$

It is known that this relation is nonlinear.

UNIQUENESS

In this section we want to refer to some uniqueness results. First, let us consider the steady state problem (2) in a domain $\Omega \subset I\!\!R^N$ ($N = 2, 3$)with zero Neumann boundary conditions. Instead of a finite number of experiments, let source positions be given at all boundary points $x \in \partial\Omega$. Assume further that measurements are made at the whole boundary. Then again by the superposition principle, the response operator is equivalently given by the so-called Neumann - Dirichlet operator, which maps all Neumann boundary data into corresponding Dirichlet boundary data on $\partial\Omega$:

$$\mathcal{R}_k[k\partial_N h|_{\partial\Omega}] = h|_{\partial\Omega}. \tag{6}$$

The uniqueness problem, i.e. to show that from $\mathcal{R}_k = \mathcal{R}_{k'}$ follows $k = k'$ was picked up by Calderón (1980). Sylvester and Uhlmann (1987) proved uniqueness for the 3-D case, and Nachman (1993) recently solved the more difficult 2-D case.

Now, we go to the transient problem. If we continuate the Neumann - Dirichlet map into the time coordinate, that is

$$\mathcal{R}_{k,S}[k\partial_N h|_{\partial\Omega \times [0,T]}] = h|_{\partial\Omega \times [0,T]}, \tag{7}$$

then we would come to a formally strongly overdetermined inverse problem.

Actually, there is a lot of redundancy in this map. Gottlieb (1994) has shown, that it is possible to recover the complete response operator $\mathcal{R}_{k,S}$ from just one pair of Neumann and Dirichlet boundary data

$$\{f, g\} = \{k\partial_N h|_{\partial\Omega \times [0,T]}, h|_{\partial\Omega \times [0,T]}\}.$$

Moreover, the input f can be chosen independently of k and S. Having the complete map $\mathcal{R}_{k,S}$ we are able to conclude that k and S are uniquely determined. This means that under transient conditions with observations at the complete boundary without any noise, just one experiment would be enough to reconstruct both coefficients S and k uniquely.

If observations are made not at the complete boundary of the domain of interest as in hydraulic tomography, we have no uniqueness for steady state conditions.

It is worth noting, that no boundary information can give uniqueness for anisotropic media.

AN ITERATIVE RECONSTRUCTION ALGORITHM

A popular method for dealing with nonlinear ill-posed problems is the Tikhonov regularization, see Engl et al. (1989). We want to go another way, based on the following considerations:

Nonlinear problems usually are solved by iterative methods, for example Newton-like methods. At each step of the iteration a linear problem has to be solved. If the nonlinear problem is ill-posed, then its linearizations are ill-posed as well. Therefore we also have to regularize the linear problems. In recent times, iterative regularization methods (like the conjugate gradient method) are better understood and they have some advantages in comparison to the classical Tykhonov regularization. Using such a method, we would come to an inner iteration for the linearized problem and an outer Newton iteration.

Let k^* be the true conductivity. For simplicity assume that $S = 1$ is given and only k^* is unknown.

The method that is proposed is based on minimizing the output least squares functional

$$J(k) = \|\mathcal{R}_{k^*} - \mathcal{R}_k\|_{\mathcal{K}} = \max_{\|Q\|_1=1} \|\mathcal{R}_{k^*} - \mathcal{R}_k\|_2, \tag{8}$$

$J(k)$ is just the first singular value of the difference of the measured response operator $\mathcal{R}_{k^*}$ and the calculated operator $\mathcal{R}_k$. The maximum in (8) is obtained for the first singular element Q_1.

The conductivity distributions k and k^* are called distinguishable with respect to an error level ε if $J(k) \geq \varepsilon$, Isaacson (1986).

For minimizing J, the Gauss - Newton method is applied. The restriction of data to only the first singular function acts as a regularization method. The linearizations in each Newton step lead to underdetermined problems for which the least square solution with minimum norm is calculated very rapidly.

An open problem is to find an optimal stopping rule. Until now the ε - criterion mentioned above is used ad hoc.

OPTIMAL PUMPING REGIMES

The optimal experimental design depends on the unknown conductivity k^* itself. Therefore one has to plan on the basis of an initial guess k^0. For getting an optimal pumping test regime it is proposed to proceed in the following way.

- Make an initial guess k^0.

- Decide, which geometrical design do you want (compare figure 1 and 2).

- Calculate the singular system of the response operator $\mathcal{R}_{k^0}$. This gives you a set of (synthetical) source configurations (optimal inputs), which are usually not managable. For example, you will get several sinks and sources in only one borehole at the same time.

- Select that optimal inputs, which corresponds to singular values greater then an error level.

- Find managable inputs, which gives you the optimal inputs by superposition.

We get an exact comparison of transient and steady state methods if we compare the number of singular values of the corresponding response operators, which are greater then an given error level.

REFERENCES

Alessandrini G. (1988) "Stable determination of conductivity by boundary measurements" Appl. Anal. 27, 153-172.

Barber, D.C. and Brown, B.H. (1984) Applied potential tomography J. Phys. E. Sci. Instrum. 17, 723-733.

Bohling, G. (1993) "Hydraulic Tomography in Two - Dimensional, Steady State Groundwater Flow" preprint, Kansas.

Calderón, A.P. (1980) "On an inverse boundary value problem", in W. Meyer and M. Raupp(eds.), Seminar on Numerical Analysis and its Application, Brazilian Math. Society, pp. 67-73.

Carrera, J. and Neuman, S.P. (1986) "Estimation of aquifer parameters under transient and steady-state conditions, 2. Uniqueness, stability and solution algorithms" Wat. Resour. Res. 22 (1986) 211-227.

Dobson, D.C. and Santosa, F. (1993) "An image enhancement technique for electrical impedance tomography" IMA Preprint Series No. 1145.

Engl, H.W., Kunisch, K. and Neubauer, A. (1989) "Convergence Rates for Tikhonov Regularization of Nonlinear Ill-Posed Problems" Inverse Problems 5, 523-540.

Gottlieb, J. (1992) "Hydraulische Tomographie: Identifikation räumlich verteilter Durchlässigkeiten im Untergrund" Deutsches Patentamt, P4236692.5, vom 30.10.1992.

Gottlieb, J. and Dietrich, P. (1993) "Identification of Permeability Distributions in Soil by Hydraulic Tomography" submitted.

Gottlieb, J. (1994) "Some problems in tomographical methods using potential fields" Proceedings of the Conference Methoden und Verfahren der mathmatischen Physik, Oberwolfach 1993. Peter Lang Verlag, Bern (to appear).

Isaacson, D. (1986) "Distinguishability of conductivities by electric current computed tomography" IEEE Trans. Med. Imag. 5, 91-95.

Kohn, R. and Vogelius, M. (1984) "Determining conductivity by boundary measurements" Comm. Pure Appl. Math. 37, 113-123.

Kravaris, C. and Seinfeld, J.H. (1985) "Identification of parameters in distributed parameter systems by regularization" SIAM J. Control. Optim. 23, 217-241.

Nachman, A. (1993) "Global uniqueness for a two-dimensional inverse boundary value problem" Dept. of Math. Preprint Series No. 19, Univ. of Rochester.

Nashed, M.Z. and Chen, X. (1993) "Convergence of Newton-like methods for singular operator equations using outer inverses" to appear in Numerische Mathematik.

Pidcock, M.K. at al. (1993) "POMPUS - A Fast Reconstruction Algorithm for Electrical Impedance Tomography" Proceedings of the 2. ECAPT conference, Karlsruhe 1993.

Sun, N.-Z. and Yeh, W.W.-G. (1990) "Coupled inverse problems in groundwater modelling, 2. Identifiability and experimental design" Water Resour. Res. 26, 2527-2540.

Sylvester, J. and Uhlmann, G. (1987) "A global uniqueness theorem for an inverse boundary value problem" Ann. Math. 125, 153-169.

Yeh, W.W.-G. (1986) "Review of parameter identification procedures in groundwater hydrology: the inverse problem" Water Resour. Res. 22, 95-108.

Yorkey, T.J., Webster, J.G. and Tompkins W.J. (1987) "Comparing reconstruction algorithms for electrical impedance tomography" IEEE Trans. Biomed. Eng. 34, 843-852.

A New Optimization Strategy for Global Inverse Solution of Hydrologic Models

Vijai K. Gupta and Soroosh Sorooshian
Department of Hydrology and Water Resources
The University of Arizona
Tucson, Arizona 85721
USA

Hydrologic models are only as reliable as model assumptions, inputs, and parameter estimates. While the use of measurements is gaining in importance, experience has shown that as a practical matter, virtually all models require calibration of at least some parameters. However, commonly used calibration (inversion) strategies are typically unable to find the "globally" optimal parameter estimates. Research has shown that the function response surface of hydrologic models can have hundreds, if not thousands of local solutions. Further, there can be sizeable regions of the parameter space that appear to give roughly equivalent results. A general global optimization strategy called the Shuffled Complex Evolution (SCE-UA) method, developed at The University of Arizona, has shown promise as an effective and efficient strategy under the types of conditions described above. In particular, the SCE-UA method is the only strategy so far shown to be capable of consistently locating the global solution for the Sacramento conceptual-rainfall runoff model. However, the problem of poor model identifiability still remains. We conclude by discussing our belief that the classical approaches have outlived their usefulness and that a new conceptual paradigm based on multi-objective solution procedures is necessary for further progress in model calibration strategies to be made.

INTRODUCTION

In order for a hydrologic model to be useful for the solution of practical problems, it has to be calibrated. Research to-date (both by our research group and others) has primarily focused on two issues - the development of specialized techniques for handling the kinds of errors present in the measurement data [Sorooshian and Dracup, 1980; Sorooshian, 1981; Sorooshian et. al., 1982, 1983; James and Burges, 1982; Sefe and Boughton, 1982; Lemmer and Rao, 1983; Kuczera, 1983a,b; Ibbitt and Hutchinson, 1984; Sorooshian and Gupta, 1983], and the search for an optimization strategy that can reliably solve the parameter estimation problem [Dawdy and O'Donnell, 1965; Nash and Sutcliffe, 1970; Monro, 1971; Clarke,

A. Peters et al. (eds.), Computational Methods in Water Resources X, 743–751.
© 1994 Kluwer Academic Publishers. Printed in the Netherlands.

1973; Sorooshian, 1983; Isabel and Villeneuve, 1986; Wheater et al., 1986;
Hendrickson et al., 1988; among others]. Much of the research addressing data
measurement error was predicated on the hope that the use of sophisticated
statistical techniques would help to minimize the difficulty in obtaining unique and
realistic parameter estimates. This led to the development of extensions of Least
Squares techniques, Maximum Likelihood approaches and Bayesian methods [eg.
Williams and Yeh, 1983; Troutman, 1985a,b; Sorooshian and Dracup, 1980;
Sorooshian 1981; Kuczera, 1983a,b].   However, in spite of the technical
sophistication of these methods, conceptual hydrologic models have remained
remarkably difficult to calibrate. Duan, et al. [1992] conducted the first thorough
investigation into the reasons for these difficulties and concluded that the 5 major
characteristics complicating the optimization problem are:

(1) **Multiple regions of attraction:** The function response surface contains more
    than one main region into which the optimization strategy can descend.
(2) **Multiple minor local optima:** Each region of attraction contains hundreds,
    and possibly thousands, of local solutions.
(3) **Roughness:** The function response surface is rough and has discontinuous
    derivatives.
(4) **Sensitivity:** The function value displays poor and varying sensitivity to the
    parameters in the region of the optimum.
(5) **Shape:**   The response surface is non-convex, with long curved ridges
    indicating non-linear parameter interaction.

Given the existence of such phenomena in even the simpler conceptual hydrologic
models, it is not surprising that hydrologists have not reported much confidence in
the results of their attempts at model calibration. While characteristics 1, 4 and 5
have been recognized and discussed in the
literature, Duan et. al [1992] were the first
to report the existence of characteristics 2
and 3, which turn out to be the most
serious of the problems. Figure 1 shows
the number and locations of the local
optima in a 3-parameter subspace of a
simple six-parameter conceptual rainfall-
runoff model.   Each dot indicates the
location of a separate local optimum to
which a classical local search optimization
algorithm can converge.   Under such
conditions, the likelihood of successfully
reaching the global optimum by means of
a local search algorithm initiated at some
arbitrary point in the parameter space is
extremely low.

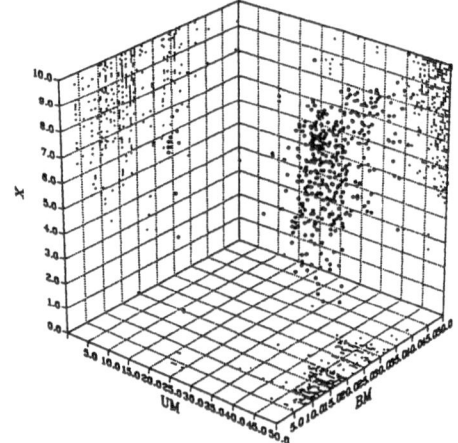

Figure 1: Locations of local optima in
a 3-parameter subspace

THE SHUFFLED COMPLEX EVOLUTION (SCE-UA) METHOD

In response to this problem, we have developed a general purpose global optimization algorithm called the Shuffled Complex Evolution (SCE-UA) method; the algorithm is specifically designed to perform well under the response surface conditions mentioned above, particularly the existence of several regions of attraction, each containing smaller-scale multiple local optima. Detailed descriptions and explanations of the method appear in Duan et al. [1992, 1993] (for computer code contact Dr. Q Duan, Hydrologic Research Lab., SSMC2 8th Floor, 1325 East West Highway, Silver Spring, MD 20910, USA). In brief, the SCE-UA method involves the initial selection of a "population" of points distributed randomly throughout the feasible parameter space. The population is partitioned into several "complexes", each complex consisting of $2n+1$ points, where n is the number of parameters to be optimized. Each complex is then allowed to "evolve" so as to independently search the parameter space in a manner that is based on an extension of the Simplex local-search algorithm. After a prescribed number of steps, the complexes are "shuffled" together and new complexes formed such that the information gained separately by each complex is shared. The evolution and shuffling procedures are repeated until prescribed convergence criteria are satisfied.

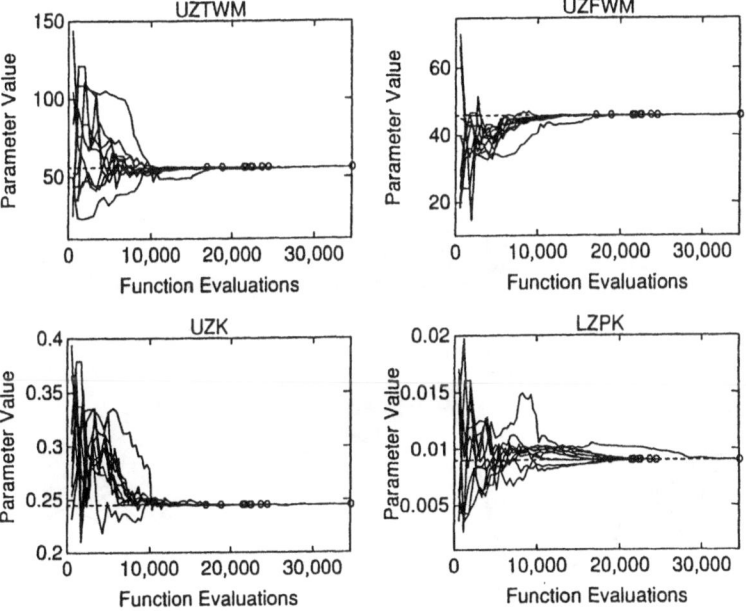

Figure 2: Convergence behavior of 4 SAC-SMA model parameters for 10 independent trials of the SCE-UA algorithm (synthetic data study)

The SCE-UA algorithm has been used to calibrate the Sacramento soil moisture accounting (SAC-SMA) riverflow forecasting model of the US National Weather Service. In the past, this model has proved to be very difficult to calibrate by hydrologists who do not have a great deal of experience with it. Therefore the development of reliable automated techniques for calibrating the model has been a priority issue in recent years. Calibration studies conducted, using both synthetic and real data, have shown that the SCE-UA method is a reliable calibration tool, being able to consistently recover the globally optimal parameter values when tested several times (Figure 2: reproduced from Sorooshian, et al., 1993].

In contrast, not one out of 100 independent trials of the Simplex local search algorithm [Nelder and Mead, 1965] was able to find the global optimum, even in the idealized conditions of a synthetic data study. In addition to being effective, the SCE-UA algorithm is remarkably efficient, being able to calibrate the model (using one water year of calibration data) in about one hour on a Sun Sparc 10 workstation; this amounts to approximately 20,000 function evaluations.

THE PROBLEM OF UNIQUENESS

Although the SCE-UA algorithm has the potential to overcome one of the serious

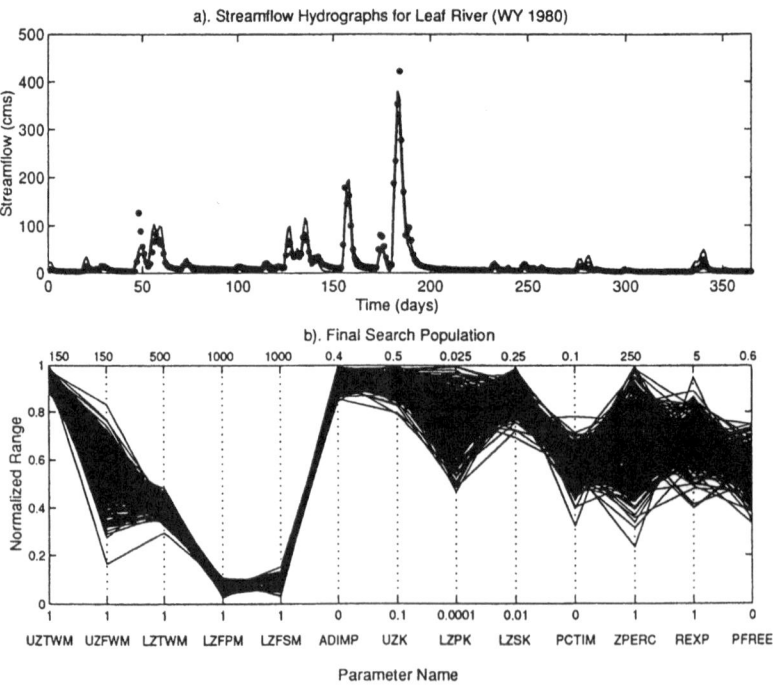

Figure 3: Results from a SCE-UA optimization run

hurdles that has been hindering more widespread use of conceptual rainfall-runoff models in operational settings, the model calibration problem is far from solved. A close investigation of the results provided by the algorithm during the evolution of its search has revealed that the problem of parameter uniqueness still remains to be addressed. At the end of each shuffling loop, the SCE-UA algorithm prints out the function value and parameter values for each of the points comprising its current search population. Figures 3 and 4 are graphical representations of some of these results. Plotted in Figure 3b are the 405 parameter locations constituting the search population of the SCE-UA algorithm at the end of an optimization run (calibration of the SAC-SMA model to one year of data from the Leaf River near Collins, Mississippi in the USA). Each line proceeding from across the graph represents one parameter location. Plotted in Figure 3a are the measured streamflow data (dots) and the upper and lower bounds on the simulated hydrographs obtained at all of the parameter locations (solid lines). Clearly the spread in parameter locations is still quite large, although the corresponding streamflow hydrographs are almost identical. Figure 4 shows (for a selected parameter, UZFWM) how the search progresses in terms of the function values for each of the points in the population. Clearly the response surface is quite flat, the function being largely insensitive to variations in the parameter values. Given that there are data measurement errors, model errors and that the calibration data are incomplete in its informativeness of the watershed, it would be foolhardy to confidently state that any one of these parameter locations is essentially superior to any other. This population of parameter locations therefore constitutes an "indifference" region from which the selection of a single "optimal" parameter set must necessarily be based on additional information, or must rely on a subjective decision by the hydrologist.

A NEW PARADIGM IS NECESSARY FOR MODEL CALIBRATION

We have come to believe that the classical approach to model calibration has been taken as far as it can go. It is clear that currently available calibration strategies are

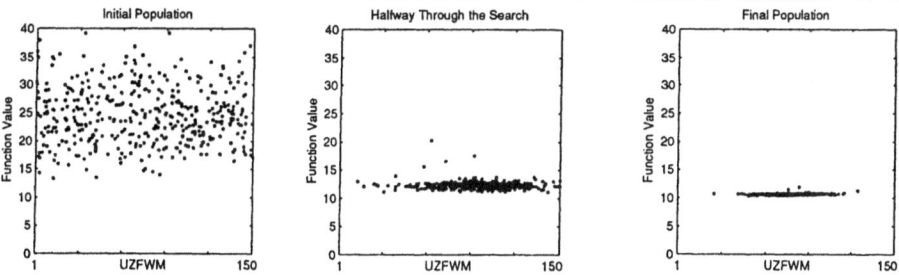

Figure 4: Function values displayed against parameter UZFWM at the beginning, middle and end of the SCE-UA optimization run

poorly suited to the task of calibrating conceptual hydrologic models. Moreover, these methods are proving to be woefully inadequate in the face of the emerging generation of distributed parameter hydrologic models. An entirely different conceptual paradigm is, therefore, necessary for further progress in model calibration methodologies to be made. We now believe that the philosophy behind the attempts to locate unique model parameters needs serious review, being that it is based on certain longstanding and questionable assumptions underlying traditional calibration procedures. In particular, the following issues require innovative attention.

(1) The structural errors arising from the fact that any model is only an (hopefully reasonable) approximation of reality cannot ignored, or treated as stochastic variables to be lumped into some output residual.
(2) The problem of model identification and calibration is inherently multi-objective, even in the case of only one output time series. The application of Least Squares and other statistical techniques (e.g. Maximum Likelihood and Bayesian) for model fitting is largely an attempt to bypass the difficulties inherent in multi-objective approaches.
(3) Errors in data, both input and output, cannot be ignored in many cases.
(4) There is a real need to be able to judge the reliability of a model, not as some overall approximate measure, but in terms of each model prediction.

When these factors are faced dead-on, it becomes apparent that there is no objective way in which a unique model solution can be obtained. Rather, the best that one can obtain using objective procedures is a model set, typically specifiable as a region of reduced uncertainty in the parameter estimate space (Figure 5). This model set

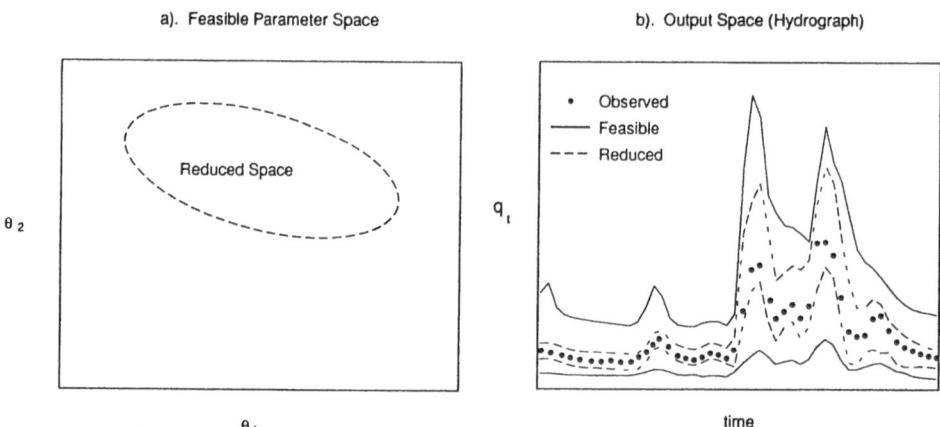

Figure 5: Conceptual representation indicating the relationship between the region of reduced parameter uncertainty and the ranges on the streamflow hydrograph

defines the "indifference region" or "non-dominated solution set" in which it is not possible to objectively select a specific model (parameter location) as being superior to any other model (parameter location), due to the uncertainties mentioned above. This uncertainty translates into an uncertainty in the model predictions (model is only capable of, at best, indicating the range in which the field observation will be observed). The size and properties of this model set and the sizes and properties of the uncertainty in the model predictions are the characteristics that define the adequacy or inadequacy of the model. Analysis of these features will provide insight into the manner in which the model needs to be improved and into the confidence that can be ascribed to model predictions. Therefore, objective calibration methods are required that can identify the non-dominated model set. This requires the innovative application of multi-objective theoretic concepts to the unique environment of hydrologic models.

Having identified the non-dominated model set, if sufficient computational power is available, the model user will be able to generate several plausible model forecasts or simulations. We believe that proper use of this new paradigm will result in model forecasts and simulations that consist of range forecasts indicating upper and lower limits and most probable values. However, it will often be necessary (for practical and computational reasons) for the model user to identify a specific model (parameter set) for forecasting or simulating single "most likely" values (e.g. for the streamflow hydrograph). In either case, graphical visualization techniques will be required, not only to display the decision variables, but also to facilitate the subjective procedure inherent in choosing a reduced model set (possibly a single model) from the non-dominated model set.

The research team at The University of Arizona is currently pursuing these avenues of investigation and we will be reporting our results in due course. If everything goes according to plan, we hope to present some preliminary findings at the conference venue.

REFERENCES

Clarke, R.T., (1973), A review of some mathematical models used in hydrology, with observations on their calibration and use, *Journal of Hydrology*, 19(1), 1-20.

Dawdy, D. R., and T. O'Donnell, (1965), Mathematical models of catchment behavior, *Journal of Hydraulics Division*, American Society of Civil Engineers, 91(HY4), 113-137.

Duan, Q., Gupta, V. K., and Sorooshian, S., (1992), Effective and efficient global optimization for conceptual rainfall-runoff models, *Water Resourources Research*, 28(4):1015-1031.

Duan, Q., Gupta, V. K., and Sorooshian, S., (1993), A shuffled complex evolution approach for effective and efficient global minimization, *Journal of Optimization Theory and Applications*, 76(3):501-521.

Hendrickson, J.D., S. Sorooshian, and L. Brazil, (1988), Comparison of Newton-type and Direct Search Algorithms for Calibration of Conceptual Rainfall-Runoff Models *Water Resources Research*, 24(5), 691-700.

Ibbitt, R.P., and P.D. Hutchinson, (1984), Model Parameter Consistency and Fitting Criteria, Proceedings of the IFAC 9th Triennial World Congress, Budapest, Hungary, vol 4, 153-157, July 2-6.

Isabel, D., and J.P. Villeneuve, (1986), Importance of the convergence criterion in the automatic calibration of hydrologic models, *Water Resources Research*, 22(10), 1367-1370.

James, L.D., and S.J. Burges, (1982), Selection, calibration and testing of hydrologic models, *in* Hydrologic Testing of Small Watersheds, edited by C.T. Haan, H.P. Johnson, and D.L. Brakensiek, 437-471, American Society of Agricultural Engineers, St. Joseph, Michigan.

Kuczera, G., (1983a), Improved parameter inference in catchment models. 1. Evaluating parameter uncertainty, *Water Resources Research* 19(5):1151-1162.

Kuczera, G., (1983b), Improved parameter inference in catchment models 2. Combining different kinds of hydrologic data and testing their compatibility, *Water Resources Research*, 19(5), 1163-1172.

Lemmer, H.R., and A.R. Rao, (1983), Critical Duration Analysis and Parameter Estimation in ILLUDAS, Water Resources Research Center, Tech. Report #153, Purdue University, West Lafayette, Indiana.

Monro, J.C., (1971), Direct search optimization in mathematical modeling and a watershed model application, NOAA Tech Memo NWS HYDRO-12, U.S. Dept. of Commerce, Silver Spring, Maryland.

Nash, J.E., and J.V. Sutcliffe, (1970), River flow forecasting through conceptual models, Part I -- A discussion of principles, *Journal of Hydrology*, 10(3), 282-290.

Nelder, J.A., and R. Mead, (1965), A simplex method for function minimization, *Computer Journal*, vol. 7, 308-313.

Sefe, F.T., and W.C. Boughton, (1982), Variation of model parameter values and sensitivity with type of objective function, *Journal of Hydrology (NZ)*, 21(1), 117-132.

Sorooshian, S., (1981), Parameter estimation of rainfall-runoff models with heteroscedastic streamflow errors: the non-informative data case, *Journal of Hydrology*, 52:127-138.

Sorooshian, S., (1983), Surface Water Hydrology: On-Line Estimation, *Reviews of Geophysics and Space Physics*, 21(3).

Sorooshian S. and Dracup J. A., (1980), Stochastic parameter estimation procedures for hydrologic rainfall-runoff models: correlated and heteroscedastic error cases, *Water Resources Research* 16(2):430-442.

Sorooshian, S., and V.K. Gupta, (1983), Automatic calibration of conceptual rainfall-runoff models: The question of parameter observability and uniqueness, *Water Resources Research*, 19(1), 251-259.

Sorooshian, S., V.K. Gupta, and J.L. Fulton, (1982), Parameter estimation of conceptual rainfall-runoff models assuming autocorrelated data errors--A case study, *in* V.P. Singh (Editor), Statistical Analysis of Rainfall and Runoff, Proceedings of the

International Symposium on Rainfall-Runoff Modeling, Mississippi State, Mississippi, May 18-21, 1981, WRP, Littleton, Colorado, 491-504, (refereed).

Sorooshian, S., V.K. Gupta, and J.L. Fulton, (1983), Evaluation of maximum likelihood parameter estimation techniques for conceptual rainfall-runoff models-- Influence of calibration data variability and length on model credibility, *Water Resources Research*, 19(1), 251-259.

Sorooshian, S., Q. Duan, and V.K. Gupta, (1993), Calibration of the SMA-NWSRFS conceptual rainfall-runoff model using global optimization, *Water Resources Research*, 29(4), 1185-1194.

Troutman, B.M., (1985a), Errors and parameter estimation in precipitation-runoff modeling 1: Theory., *Water Resources Research*, 21(8), 1195-1213.

Troutman, B.M., (1985b), Errors and parameter estimation in precipitation-runoff modeling 2: Case study, *Water Resources Research*, 21(8), 1195-1213.

Wheater, H.S., K.H. Bishop, and M.B. Beck, (1986), The identification of conceptual hydrological models for surface water acidification, *Hydrologic Processes, I,* 89-109.

Williams, B.J., and W. Yeh, (1983), Parameter estimation in rainfall-runoff models, *Journal of Hydrology*, 63, 373-393.

ADJOINT SENSITIVITY ANALYSIS OF STEADY-STATE SOLUTE TRANSPORT

L.J. HARTLEY
AEA Technology
Decommissioning & Waste Management
424.4 Harwell
Didcot, Oxfordshire, OX11 0RA
United Kingdom

When considering the safety of a deep underground repository for radioactive waste, performance measures of interest include radionuclide concentrations, radionuclide fluxes from the repository and in the surrounding geological formations, and rates of radionuclide discharge to the biosphere. In this study, the usefulness of adjoint sensitivity theory as a method for estimating the uncertainties in these performance measures due to parameter uncertainty is investigated. The adjoint equations and sensitivity coefficients for steady-state solute transport in a porous media are developed. The flow, transport and adjoint equations are solved using the Galerkin finite element method and a frontal solver. The new code developed was verified using a one-dimensional solute transport test case and a case of two-dimensional solute transport in a vertical cross-section which included three different strata. In the one-dimensional case the analytical adjoint states were accurately reproduced. In the two-dimensional case the sensitivity coefficients given by the adjoint method were in close agreement with those obtained from a set of explicit variant calculations.

INTRODUCTION

Numerical models of groundwater flow and radionuclide transport play an important role in the safety assessment of a deep underground repository for radioactive waste. Such models are used to predict the values of performance measures such as radionuclide concentrations, radionuclide fluxes from the repository and in the surrounding geological formations, and rates of radionuclide discharge to the biosphere. The uncertainty in these predictions due to uncertainties in the system parameters are often of interest in the assessment. The adjoint sensitivity method is a computationally efficient method of determining the sensitivities of scalar performance measures to the system parameters. Parameter uncertainties can then be estimated using the first-order second-moment technique.

The adjoint method has previously been used for a number of groundwater flow applications (e.g., Wilson and Andrews, 1985; Yeh and Sun, 1990). In this paper, the method is extended to the solute transport equation. Cases of steady-state transport of a non-sorbing tracer subject to convection, diffusion, and hydrodynamic dispersion are used to illustrate and verify the method. The first order derivatives of a performance measure with respect to each of the hydrogeological parameters (e.g. permeability and porosity) and transport parameters (e.g. diffusion and dispersion

A. Peters et al. (eds.), Computational Methods in Water Resources X, 753–760.
© 1994 Kluwer Academic Publishers. Printed in the Netherlands.

lengths) are taken to form the set of sensitivity coefficients. Equations are derived for two adjoint state variables arising from the pressure and concentration, respectively, and the sensitivity coefficients are expressed in terms of these. The AEA Technology groundwater flow and transport code NAMMU is used to solve the flow, transport and adjoint equations. This code uses the Galerkin finite element method and a frontal solver. A one-dimensional flow and transport problem which uses concentration at a point as the performance measure is used to verify the numerical results by comparing them with analytical solutions.

To verify and illustrate the adjoint method in a more realistic application, radionuclide transport in a two-dimensional vertical section through the geological strata underlying the Harwell site is also considered. Sensitivity coefficients for the discharge of radionuclide from a hypothetical repository to the surface are derived from the adjoint method and compared with those obtained from a set of explicit variant calculations.

SOLUTE TRANSPORT EQUATIONS

For a porous medium, the equations describing steady-state groundwater flow and the transport of a stable non-sorbing solute that does not affect the fluid density are

$$\frac{\partial}{\partial x_i}(\rho_l q_i) = 0 \quad i = 1, n \quad , \tag{1}$$

$$q_i = -\frac{k_{ij}}{\mu}\frac{\partial p}{\partial x_j} \quad i, j = 1, n \quad , \tag{2}$$

$$\frac{\partial}{\partial x_i}(q_i C) - \frac{\partial}{\partial x_i}\left(\phi D_{ij}\frac{\partial C}{\partial x_j}\right) = \phi s \quad i = 1, n \quad . \tag{3}$$

In this, n is the number of spatial dimensions, ρ_l is the fluid density ($\mathrm{kg\,m^{-3}}$), q_i is the specific discharge (Darcy velocity) ($\mathrm{m\,s^{-1}}$), k_{ij} is the permeability ($\mathrm{m^2}$), μ is the fluid viscosity ($\mathrm{kg\,m^{-1}s^{-1}}$), p is the residual pressure ($\mathrm{N\,m^{-2}}$), C is the solute concentration ($\mathrm{mol\,m^{-3}}$), ϕ is the rock porosity, D_{ij} is the hydrodynamic dispersion tensor ($\mathrm{m^2 s^{-1}}$), s is the solute source into the pore water ($\mathrm{mol\,m^{-3}s^{-1}}$), x_i is the cartesian coordinate (m). In (1)–(3) repeated indices indicate summation convention. The classical model for hydrodynamic dispersion is used in which D_{ij} is related to the pore water velocity $v_i = q_i/\phi$ by

$$D_{ij} = \frac{D_m}{\tau}\delta_{ij} + \alpha_T v \delta_{ij} + (\alpha_L - \alpha_T)\frac{v_i v_j}{v} \tag{4}$$

Here, D_m is the molecular diffusion ($\mathrm{m^2 s^{-1}}$), τ is the tortuosity, α_L and α_T are the longitudinal and transverse dispersion lengths (m), respectively, and $v = (v_i v_i)^{1/2}$.

The associated boundary conditions can be expressed in the form

$$p = \hat{p} \quad \text{on } \Gamma_1, \qquad \rho_l q_i n_i = \hat{q} \quad \text{on } \Gamma_2 \tag{5}$$

$$C = \hat{C} \quad \text{on } \Gamma_1', \qquad q_i n_i C - \phi D_{ij}\frac{\partial C}{\partial x_j}n_i = \hat{q}_C \quad \text{on } \Gamma_2' \tag{6}$$

Here $\hat{p}$ is a specified residual pressure on the boundary Γ_1, $\hat{q}$ is a specified groundwater flux normal to the boundary Γ_2, $\hat{C}$ is a specified solute concentration on the boundary Γ_1', and $\hat{q}_C$ is a specified solute flux normal to the boundary Γ_2' (Note that the flux term $\hat{q}_C$ in condition (6) can be expressed in such a way that it represents a specified advective or dispersive flux as well as a total flux).

PERFORMANCE MEASURES

Performance measures of interest related to solute concentration include:

1. a weighted average of solute concentrations at a set of points. This could be used to quantify the uncertainties in predictions of solute concentration in a sub-domain, e.g. one of the geological strata;

2. a weighted average difference between measured and predicted solute concentration. This is useful in model calibration exercises where measurements of solute concentration are available. The adjoint sensitivity method may then be used as part of an optimisation method to obtain ranges of parameters that best fit the measurements;

3. a weighted average of solute fluxes at a set of points. This could be used in an assessment to estimate uncertainties in the radionuclide flux out of a repository zone, or through surrounding geological formations;

4. integrated flux of solute from a surface. This could be used in an assessment to quantify uncertainties in the discharge of radionuclides to the environment.

The performance measure J is expressed as an integral over the spatial domain $\mathcal{D}$ of the groundwater flow and radionuclide transport:

$$J = \int_{\mathcal{D}} f\left(\{\alpha\}, p, C\right) d\mathcal{D}, \tag{7}$$

where $f(\{\alpha\}, p, C)$ is a function of the system parameters $\{\alpha\}$, pressure and radionuclide concentration. For use in a discrete numerical system the function f can be written as a sum over a discrete set of measure points $\underline{x} = \underline{x}_m$. The first and fourth of the above performance measures can be written, respectively, as

$$f_1 = \sum_{m=1}^{M} w_m C \delta(\underline{x} - \underline{x}_m), \quad \text{and} \quad f_4 = \sum_{m=1}^{M} w_m \left(q_i C - \phi D_{ij} \frac{\partial C}{\partial x_j} \right) n_i \delta(\underline{x} - \underline{x}_m) \tag{8}$$

Here $\delta(\underline{x})$ is the Dirac delta function. For the surface integral performance measure, the measure points $\underline{x} = \underline{x}_m$ and weights w_m are chosen to give an accurate evaluation of the integral and depend on the numerical discretization of the domain. For the concentration performance measure the weights might be set to give an average or total concentration in a sub-domain.

The marginal sensitivity of the performance measure J with respect to the system parameter α_k is defined as $dJ/d\alpha_k$ and from equation (7) can be written as

$$\frac{dJ}{d\alpha_k} = \int_{\mathcal{D}} \left\{ \frac{\partial f(\{\alpha\}, p, C)}{\partial \alpha_k} + \frac{\partial f(\{\alpha\}, p, C)}{\partial p} \frac{dp}{d\alpha_k} + \frac{\partial f(\{\alpha\}, p, C)}{\partial C} \frac{dC}{d\alpha_k} \right\} d\mathcal{D}, \tag{9}$$

ADJOINT EQUATIONS AND SENSITIVITY COEFFICIENTS

The adjoint equations for solute transport are derived as follows. Equations (3) and (4) are differentiated with respect to the system parameter α_k. An orthogonalization is then carried out by weighting these derivatives by differentiable functions over the domain, summing them together with equation (9), and applying Green's first and second identities (Cacuci *et al.*, 1980). The equations are formulated in terms of two adjoint state variables, θ and ψ, so that

$$\frac{\partial}{\partial x_i}\left[\frac{k_{ij}}{\mu}\left(\rho_l\frac{\partial\theta}{\partial x_j}+\frac{\partial\psi}{\partial x_j}C\right)+T_{ji}\frac{\partial\psi}{\partial x_j}\right]=-\frac{\partial f}{\partial p}\quad i,j=1,n \tag{10}$$

$$\frac{\partial}{\partial x_i}\left(q_i\psi\right)+\frac{\partial}{\partial x_i}\left(\phi D_{ij}\frac{\partial\psi}{\partial x_j}\right)=-\frac{\partial f}{\partial C}\quad i,j=1,n \tag{11}$$

Subject to boundary conditions

$$\theta=0\quad\text{on }\Gamma_1,\qquad\left[\frac{k_{ij}}{\mu}\left(\frac{\rho_l\partial\theta}{\partial x_j}+\frac{\partial\psi}{\partial x_j}C\right)+T_{ji}\frac{\partial\psi}{\partial x_j}\right]n_i=0\quad\text{on }\Gamma_2 \tag{12}$$

$$\psi=0\quad\text{on }\Gamma_1',\qquad\phi D_{ij}\frac{\partial\psi}{\partial x_j}n_i=0\quad\text{on }\Gamma_2' \tag{13}$$

The tensor T_{ij} in equation (10) is the derivative of the dispersive solute flux with respect to the pressure gradient

$$T_{ij}=\frac{\partial}{\partial(\partial p/\partial x_j)}\left(\phi D_{ik}\frac{\partial C}{\partial x_k}\right) \tag{14}$$

Note that these equations are coupled in such a way that they may be solved sequentially. Equation (11) can be solved first for ψ and then (10) can be solved for θ.

Having solved for θ and ψ the marginal sensitivities to each of the system parameters α_k can be obtained by evaluating

$$\frac{dJ}{d\alpha_k}=\int_{\mathcal{D}}\left[\frac{\partial f}{\partial\alpha_k}+\psi\frac{\partial(\phi s)}{\partial\alpha_k}-\frac{\partial(k_{ij}/\mu)}{\partial\alpha_k}\left(\rho_l\frac{\partial\theta}{\partial x_i}+\frac{\partial\psi}{\partial x_i}C\right)-\frac{\partial\psi}{\partial x_i}\frac{\partial(\phi D_{ij})}{\partial\alpha_k}\frac{\partial C}{\partial x_j}\right]d\mathcal{D}$$

$$-\int_{\Gamma_1}\left[\frac{k_{ij}}{\mu}\left(\rho_l\frac{\partial\theta}{\partial x_j}+\frac{\partial\psi}{\partial x_j}C\right)+T_{ji}\frac{\partial\psi}{\partial x_j}\right]n_i\frac{\partial\hat{p}}{\partial\alpha_k}d\Gamma_1+\int_{\Gamma_2}\theta\rho_l\frac{\partial\hat{q}}{\partial\alpha_k}d\Gamma_2 \tag{15}$$

$$-\int_{\Gamma_1'}\left[\phi D_{ij}\frac{\partial\psi}{\partial x_j}n_i\right]\frac{\partial\hat{C}}{\partial\alpha_k}d\Gamma_1'-\int_{\Gamma_2'}\psi\frac{\partial\hat{q}_C}{\partial\alpha_k}d\Gamma_2'.$$

The first term in the integral over the domain relates to the direct sensitivity effect, the second to the sensitivities to the source term, while the third and fourth terms represent indirect sensitivities to the flow and transport parameters. The four boundary integral terms relate to the sensitivities to specified values of pressure, to specified groundwater fluxes, to specified solute concentrations, and to specified fluxes of solute, respectively.

ONE-DIMENSIONAL TEST CASE

A simple one-dimensional flow and solute transport problem is used to illustrate and verify the numerical implementation of the adjoint sensitivity method. The problem is defined on a domain of length $L = 10,000$m with a permeability $k = 10^{-15}$m^2. Other parameters are $\phi = 0.5$, $D_m = 10^{-6}$m^{2}s^{-1}, $\tau = 1.0$, $\alpha_L = \alpha_T = 0.0m$, and $\mu = 10^{-3}$kg m^{-1}s^{-1}. The boundary conditions are $p = p_0 = 10^6$Nm^{-2} at $x = 0$, $p = 0$ at $x = L$, $C = 1$ at $x = 0$, and $C = 0$ at $x = L$. The solution to this problem is

$$p = \left(1 - \frac{x}{L}\right) p_0 \quad \text{and} \quad C = \left(e^{Vx/D_\phi} - e^{VL/D_\phi}\right)\left(1 - e^{VL/D_\phi}\right)^{-1} \tag{16}$$

where

$$V = \frac{k}{\mu}\frac{p_0}{L} \quad \text{and} \quad D_\phi = \phi D_m \tag{17}$$

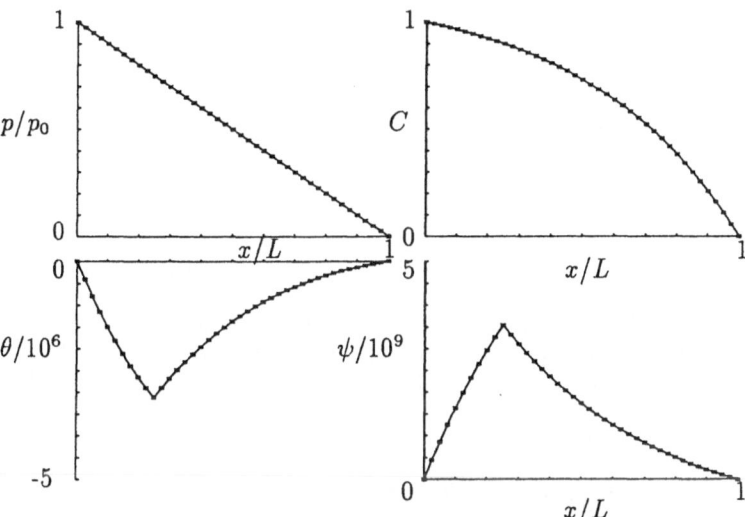

Figure 1: A comparison of the analytical solutions (solid line) and numerical solutions (points) of the flow, transport and adjoint equations. (a) Residual pressure. (b) Solute concentration. (c) θ adjoint state. (d) ψ adjoint state.

From equations (10) and (11) the corresponding adjoint states for a performance measure of concentration at the point $x = x_m$ are obtained by solving

$$\frac{d}{dx}\left(\rho_l \frac{d\theta}{dx} + C\frac{d\psi}{dx}\right) = 0 \tag{18}$$

$$V\frac{d\psi}{dx} + D_\phi\frac{d^2\psi}{dx^2} = -\delta(x - x_m) \tag{19}$$

The solutions for the adjoint states are

$$\theta = \begin{cases} \dfrac{-e^{Vx_m/D_\phi}}{\rho_l\left(1-e^{VL/D_\phi}\right)}\left[\dfrac{x}{D_\phi}\left(1-\dfrac{x_m}{L}\right) + \dfrac{\left(1-e^{-Vx/D_\phi}\right)\left(e^{V(L-x_m)/D_\phi}-1\right)}{V\left(1-e^{-VL/D_\phi}\right)}\right] & x < x_m \\[3em] \dfrac{-e^{Vx_m/D_\phi}}{\rho_l\left(1-e^{VL/D_\phi}\right)}\left[\dfrac{x_m}{D_\phi}\left(1-\dfrac{x}{L}\right) + \dfrac{\left(1-e^{-Vx_m/D_\phi}\right)\left(e^{V(L-x)/D_\phi}-1\right)}{V\left(1-e^{-VL/D_\phi}\right)}\right] & x \geq x_m \end{cases} \tag{20}$$

$$\psi = \begin{cases} \dfrac{1}{V}\left(e^{Vx_m/D_\phi}-e^{V(x_m-x)/D_\phi}\right)\left(1-e^{V(L-x_m)/D_\phi}\right)\Big/\left(1-e^{VL/D_\phi}\right) & x < x_m \\[2em] \dfrac{1}{V}\left(e^{Vx_m/D_\phi}-1\right)\left(1-e^{V(L-x)/D_\phi}\right)\Big/\left(1-e^{VL/D_\phi}\right) & x \geq x_m \end{cases} \tag{21}$$

These analytical solutions $(16, 17, 20, 21)$ are shown in Figures 1. The numerical results obtained using the finite element code NAMMU are also indicated. As can be seen the numerical results accurately reproduce the exact solutions.

TWO-DIMENSIONAL TEST CASE

A sensitivity analysis of a case of radionuclide transport in a two-dimensional vertical section was performed using the NAMMU code. The groundwater flow and transport model was derived from that used in the CEC PACOMA study (Winters *et al.*, 1990) of a hypothetical repository for intermediate level waste located in a clay layer below the Harwell site.

The idealised geological structure, physical dimensions and hypothetical repository location are shown in Figure 2a. The model consists of an upper chalk aquifer overlaying a clay aquitard, which in turn overlays a corallian aquifer. The finite element grid used for solving the flow, transport and adjoint equations is shown in Figure 2b. The boundary conditions for the flow and transport problems are also indicated in Figure 2b. A nominal radionuclide whose concentration is fixed at $C = 1$ in the repository region is considered.

Table 1: Hydrogeological parameters for the PACOMA flow and transport calculations.

Strata	k_{11}	k_{22}	ϕ	α_L	α_T	D_m	τ
chalk	$3.3\,10^{-13}$	$3.3\,10^{-13}$	0.37	82	8.2	$6\,10^{-6}$	100
clay	$1.3\,10^{-17}$	$8.2\,10^{-19}$	0.31	0	0.0	$6\,10^{-6}$	2.73
corallian	$5.2\,10^{-13}$	$5.2\,10^{-13}$	0.34	88	8.8	$6\,10^{-6}$	100

The hydrogeological properties for each layer are given in Table 1. It can be seen that this problem is defined in terms of 17 parameters and hence estimation of the sensitivities to the parameters by direct parameter sampling would require at least 17 additional computations of the flow and transport problems. The computed residual pressure and concentrations of the nominal radionuclide are shown in Figures 3. There are two pathways by which the radionuclide can reach the surface. Firstly, by diffusing down into the corallian aquifer, then advecting downstream before diffusing back up through the clay. Alternatively, it diffuses upwards through the clay to be dispersed to the surface of the chalk.

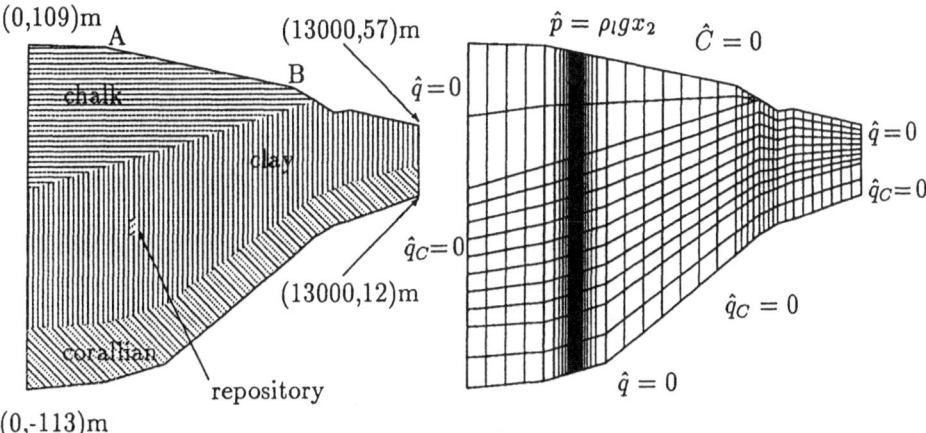

Figure 2: The PACOMA two-dimensional vertical section model. (a) Geological structure with the hypothetical repository location indicated. (b) The finite-element grid and boundary conditions.

As a demonstration of the adjoint method, an integral of the radionuclide flux out of the segment AB (shown in Figure 2a) of the top surface is used as the performance measure. All the sensitivity coefficients are computed by solving equations (10) and (11) once and evaluating the appropriate terms in the integral (15). Normalised sensitivities, that is $(dJ/d\alpha_k)/(J/\alpha_k)$, to the main parameters are given in Table 2. To verify these results the normalised sensitivities were also obtained by explicitly increasing each of the parameters in turn by 1% and resolving the flow and transport problems. As can be seen in Table 2, there is good agreement between the two sets of results.

Table 2: A comparison of normalised sensitivities as calculated by the adjoint and explicit variation methods.

Strata	k_{11}	k_{22}	ϕ	α_L	α_T
Adjoint normalised sensitivities					
chalk	$5.92\,10^{-4}$	$7.37\,10^{-5}$	$9.89\,10^{-8}$	$9.93\,10^{-4}$	$3.65\,10^{-3}$
clay	$2.51\,10^{-3}$	$-8.00\,10^{-1}$	$1.81\,10^{0}$	0.0	0.0
corallian	$-1.13\,10^{-2}$	$-2.22\,10^{-6}$	$1.41\,10^{-6}$	$-1.76\,10^{-3}$	$1.81\,10^{-3}$
Explicitly calculated normalised sensitivities for a 1% increase					
chalk	$8.05\,10^{-4}$	$7.20\,10^{-5}$	$1\,10^{-7}$	$9.75\,10^{-4}$	$3.85\,10^{-3}$
clay	$2.52\,10^{-3}$	$-7.96\,10^{-1}$	$1.81\,10^{0}$	0.0	0.0
corallian	$-1.12\,10^{-2}$	$-2.38\,10^{-6}$	$1.41\,10^{-6}$	$-1.75\,10^{-3}$	$1.79\,10^{-3}$

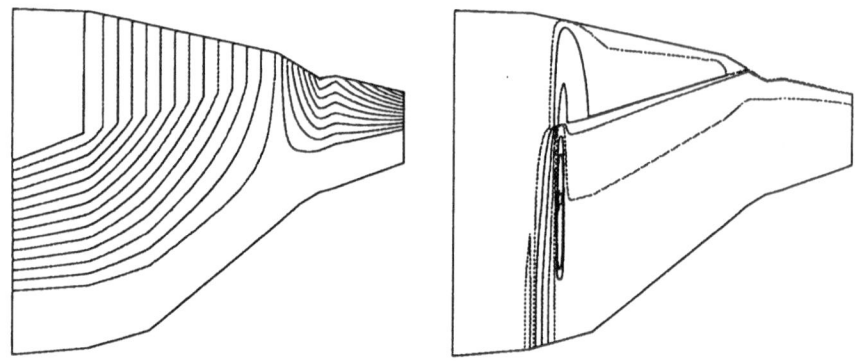

Figure 3: (a) Residual pressure, contours are plotted at an interval equivalent to 2m of head above Ordnance Datum from 58m to 108m. (b) Radionuclide concentration, contours at $C = 1, 0.3, 0.1, 1\,10^{-2}, 1\,10^{-3}, 1\,10^{-4}\,1\,10^{-5}, 1\,10^{-6}$.

DISCUSSION

Adjoint sensitivity theory has been successfully applied to the steady state solute transport equation. Once performance measures have been defined, the adjoint method facilitates the efficient computation of the uncertainties in these measures due to the parameter uncertainty. A numerical adjoint code can also provide sensitivities to parameters defined on individual elements or user-defined subregions. Such an ability can be used to identify the key regions of the system and the most important data to be obtained in future experimental observations (see Yeh and Sun, 1990).

The adjoint method should be used with caution when extrapolating the behaviour of a performance measure over large parameter ranges, because it only computes the first derivatives of the performance measure. Such first order extrapolation may be inaccurate if the performance measure depends on the parameters in a strongly nonlinear fashion. In the context of an assessment, adjoint sensitivity theory can provide a tool for rapidly ranking the importance of parameters according to the size of their contribution to the overall uncertainty.

ACKNOWLEDGEMENT This work is cofunded by the Nirex Safety Assessment Research Programme (NSARP) of UK Nirex Ltd. and by the Commission of the European Communities as part of its project 'Management and Storage of Radioactive Waste 1990 - 1994'.

REFERENCES
Cacuci, D.G., Weber, C.F., Oblow, E.M. and Marable, J.H. (1980) "Sensitivity Theory for General Systems of Nonlinear Equations", Nucl. Sci. Eng. 75, 88-110.
Sykes, J.F., Wilson, J.L. and Andrews, R.W. (1985) "Sensitivity Analysis for Steady State Groundwater Flow Using Adjoint Operators", Water Resour. Res. 21(3), 359-371.
Winters, K.H., Clark, C.M. and Jackson, C.P. (1990) "The UK Contribution to the CEC PACOMA Project: Far-Field Modelling of Radioactive Waste Disposal in Clay" UK Nirex report NSS/R 185.
Yeh, W.W.-G. and Sun, N.-Z. (1990) "Variational Analysis, Data Requirements, and Parameter Identification in a Leaky aquifer System" Water Resour. Res. 26(9), 1927-1938.

DEVELOPMENT AND APPLICATION OF MULTIPLE NONLINEAR SYSTEM TECHNIQUE FOR SUBSURFACE MODELS

BERNARD B. HSIEH
Hydraulics Laboratory
U.S. Army Engineer
Waterways Experiment Station
Vicksburg, MS 39180, USA

A newly developed multiple-input/single-output system with frequency domain basis and nonlinear decomposition processes has been used to incorporate the subsurface numerical models using the multiple frequency response function (MFRF). The developed procedures for integrating three important stages of modeling processes, namely, data recovery processes, the estimation of aquifer parameters, and the evaluation of model performance in verification, are presented.

INTRODUCTION

With recent advances in computer facilities and research development, conducting large-scale and long-term simulations for a multidimensional subsurface numerical model results in less time-consuming and more reliable solutions. However, before such models are used to address real-world problems, they must be verified. This requires field data at several interior points which often either do not exist or contain suspected errors, and proper estimation of aquifer parameters from very limited known state variables. In addition, the evaluation of model performance after verifying also needs to be addressed. A computational technique using either stresses or concentrations defined at the boundaries or as the source/sink terms of the groundwater flow model to estimate the response of the state variables in the same domain as the numerical model is proposed. A system response approach using the input/output relationships of the subsurface numerical model or field measurements to describe the dynamic behavior of subsurface flow phenomena can be used in this role. The technique is being used to aid in subsurface numerical modeling in three ways. First, missing data required for model verification can be reasonably estimated at interior and boundary locations. Secondly, a curve-matching procedure is proposed fit the

761

A. Peters et al. (eds.), Computational Methods in Water Resources X, 761–768.
© 1994 *Kluwer Academic Publishers. Printed in the Netherlands.*

modulus and phase transformations of the model to the corresponding transformations of the sample frequency response function (FRF). The goal is to find the optimal estimates for the aquifer parameters such that the theoretical FRF is in some sense closest to the sample FRF. Third, the system response coefficients, which are based on the ratio of numerical response system to measurement response system for each particular frequency band, are defined and used as an indication of model performance. These applications are illustrated by a numerical exercise using an area near Wichita, Kansas.

DEVELOPMENT OF NONLINEAR MULTIPLE SYSTEM

The behavior resulting from natural fluctuations in the subsurface can be prescribed by the partial differential equations of flow and transport. The solutions can be analytical or numerical approximation. Correspondingly, a multiple linear system with frequency domain approach also can transfer input/output from time domain to frequency domain using Fourier Transforms within a very small computational domain. The MFRFs, which are considered as the relative response strength of the system, are obtained from each frequency band through the transformation. Briefly, if defining N-dimensional transfer function vector $A(f)$, N-dimensional cross-power spectrum vector of the output with inputs $G_{xy}(f)$ and (NxN) matrix of the power and cross spectra of all the inputs $G_{xx}(f)$, the multiple linear system can be written in matrix notation as

$$G_{xy}(f) = G_{xx}(f) \cdot A(f) \qquad (1)$$

where

$$A(f) = [A_1(f),......,A_N(f)]^T \qquad (2)$$
$$G_{xy}(f) = [G_{1y}(f),.......,G_{Ny}(f)]^T \qquad (3)$$

$$G_{xx}(f) = \begin{bmatrix} G_{11}(f) \ G_{12}(f) \ ....G_{1N}(f) \\ G_{21}(f) \ G_{22}(f) \ ....G_{2N}(f) \\ . \qquad . \qquad . \\ . \qquad . \\ G_{N1}(f) \ G_{N2}(f) \ ....G_{NN}(f) \end{bmatrix} \qquad (4)$$

The MFRF $A(f)$ is solved by

$$A(f) = G_{xx}^{-1}(f) \cdot G_{xy}(f) \qquad (5)$$

The solutions to obtain the nonlinear MFRF through a multiple linear system is unlikely to occur in a real subsurface system. Two major barriers for solving a multiple nonlinear system with frequency domain basis are the group time delay and nonlinearity between each input series and output function. These difficulties can be eliminated by the processes of decomposition and shifting during the computation (Bendat 1990). In addtion, the correlated input series needs to be converted to mutual uncorrelated series. Hsieh (1993) developed the procedures of decomposing a mutiple nonlinear system as below. It can be coupled with a new developed software (MISOF4) for solving nonlinear multiple systems.

1. Define possible nonlinear finite memory system paths.
2. Decompose each path into nonlinear and linear components.
3. Insert zero-memory nonlinear system and a constant-parameter linear system to each nonlinear system with transformation.
4. Arrange the orders of the new parallel linear system.
5. Perform uncorrelated input transformation.
6. Solve each sublinear system.

To incorporate the system model with the numerical mode input/output structure, a moving response function with fixed input design (Figure1) for the system model is proposed. This makes each system output correspond to one computational cell from the numerical model. The advantage of this design is that even though the output cell of the point of interest is changed, the input functions still remain the same.

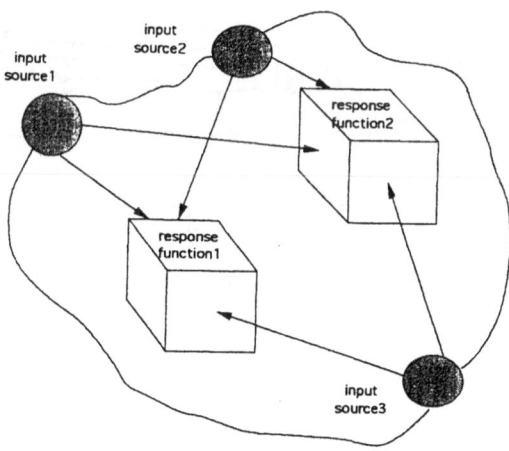

Figure 1. Moving response function with fixed inputs.

EXAMPLE OF GROUNDWATER MODELING STUDY

To demonstrate the application of technique to the subsurface environment , a numerical model of groundwater flow near Wichita, Kansas, from the information of Gelhar et al. (1974) was constructed. The study area is about 200 square miles with the Little Arkansas River on the northeast and east and the Arkansas River on the south and southwest. In this area, the water-bearing materials consist of unconsolidated deposits of sand, gravel, silt, and clay of the Pleistocene. The hydraulic gradient of the water table in the well field is toward the Little Arkansas River. The primary source of water for recharge to the well is local precipitation. Using the source code from Hromadka et al. (1985), a two-dimensional Galerkin finite-element groundwater flow model which computes on linear triangular elements to describe the study area was developed. The computational mesh includes 109 nodes and 202 elements (Figure 2). The boundary conditions on the flow domain are specified flux on the west side and specified head along the Little Arkansas River and the Arkansas River. From the literature (Williams and Lohman 1949), the average transmissivity of this area is of the order of 10,000 ft^2/day. The report estimated the storage coefficient for the wells were 0.33. Since the pumpage information was not available, the recharge was estimated mainly by the local precipitation. The model was first run by steady-state condition and then run with the transient condition.

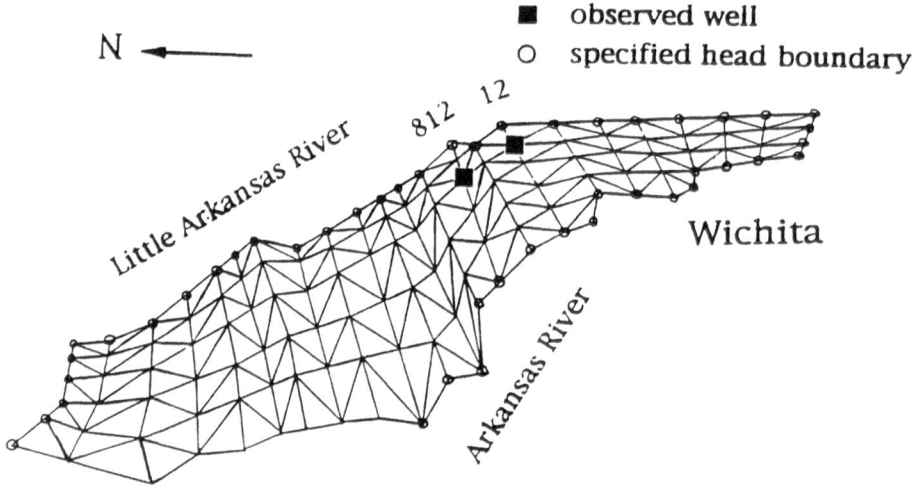

Figure 2. Computational mesh for a 2-D numerical model

Application of data recovery simulation

Due to the cost constraints, for obtaining the complete set of data to conduct the model verification is seldom occurred. The developed technique can use the response relationships between input and output series to fill needed gaps. With this approach, the transfer function is obtained using known input and output series with a common period. Frequency response functions of system parameters are then used with the remaining part of the known input series to simulate the output function. In this study, two groundwater table fluctuations (Figure 2) have been recorded since 1937. Well 12 and well 812 were used, with data collected from January 1938 to September 1971, and February 1937 to December 1962, respectively. The corresponding records of gauge height of the nearby stream and monthly precipitation at the city of Wichita, Kansas, were also prepared as the input series of system model. To verify the accuracy of system model prediction, all the records were divided into two portions. The first half was used to generate the MFRF. The second half was used to make the comparisons. For this consideration, the stream inflows and local precipitation were regarded as correlated input series while the groundwater table fluctuation was used as the system output. The computational procedures are processed by the decompositing and shifting steps. The transformation of natural log function for both input series recieved the best estimates. The MFRF obtained then takes inverse Fourier Transformation and multiples the second half of the input series to compute the simulated output. The comparisions between simulated and observed water table levels as an example are shown in Figure 3.

Application of aquifer parameter estimation

For the two-dimensional transient flow model, prime aquifer parameters for simulation of the head configuration are the storage coefficient and the transmissivity. Traditionally, the parameter estimation for a frequency domain approach uses a curve-matching process between theoretical and sample cross-spectral functions from the linear spectral theory of Fourier-Stieljes intergals and an inverse Fourier Transform. These were estimation criteria based upon modulus and phase transformations of the model and sample FRF. Gelhar et al. (1974) indicated that the matching procedure might appear to somewhat subjective. He suggested a more objective procedure such as a least squares fit. Ritzi , Sorooshian, and Gupta (1991) developed a complex vector estimation criterion and implemented it in a nonlinear, Gauss-Marquart optimization algorithm. However, when fitting the frequency response function, important frequencies of the process modeled needed to be included with weights of estimation criterion based on the variance distribution of sample transfer function.

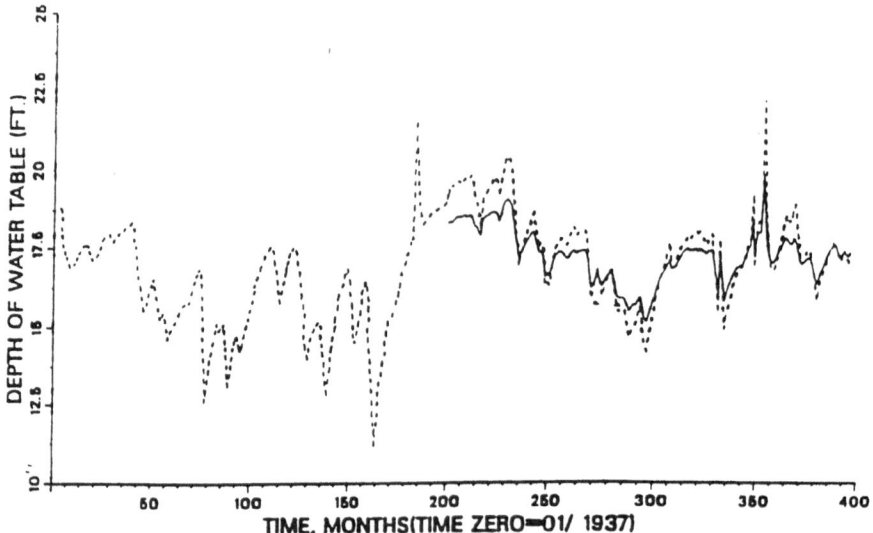

Figure 3.   Observed(---) versus simulated (——) water table
                levels for well 12.

Incorporating this concept and the developed nonlinear system solver, the new
object function as the estimation criterion (EC) to be minimized can be modified as

EC(q)= the sum of squared real errors and of squared
       imaginary errors between two frequency response   (6)
       functions with unit and variance weights.

The major procedures for estimating aquifer parameters are as follows:

1. Obtain theoretical and sample MFRF from a nonlinear approach.
2. Conduct curve-matching process to get initial estimate of parameters
3. Use nonlinear iteration process with above EC to compute final
    estimating parameters.

The final estimation of transmisivity was 4,600 ft^2/ day and the storage coefficient
was 0.304 for well 12..

Application of model performance analysis

In most studies of conducting subsurface models so far, the evaluation of model verification has consisted of simple statistical comparisions between model predictions and field observations. The comparable parameters are also often limited to simple variable evaluation. Performance criteria that involve system response and express actual physical contents are needed to identify how the verification has been done. Under these considerations, two response system were constructed. The first system presents a nonlinear system between input series and observed output. The MFRF in this system indicates the measurement response (MC_m). The computational response (MC_c), which correlates the numerical output and input series, is shown by the second system. The ratio of multiple coherence between these two systems for each frequency band is defined as the frequency response coefficients (RC) .

$$RC(f) = (MC_c(f)/MC_m(f)) \qquad\qquad (7)$$

The resulting response coefficients for well 12 were shown in Figure 4. It indicated that the model had better performance in the lower frequency than it had in the higher frequency. Since no other well record than these two locations including the pumping history were ready to use, the overall performance could not be fully evaluated. However, the main purpose of this study was to demonstrate the developed computational technique which can be incorporated with the processes of numerical subsurface modeling.

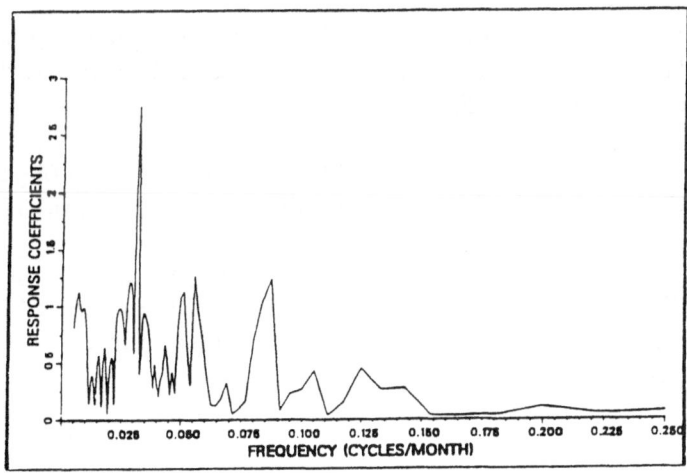

Figure 4.   Frequency response coefficients for a 2-D groundwater flow model in well 12.

CONCLUSIONS

The computational procedures for incorporating a nonlinear system analysis technique with the modeling processes of a subsurface numerical model have been demonstrated. To confirm the reliability of this developed technique, a subsurface numerical model verified by the data with more completely spatial distribution is required.

ACKNOWLEDGMENTS

Permission was granted by the Chief of Engineers to publish this information.

REFERENCES

Bendat, J. S. (1990) Nonlinear System Analysis and Identification from Random Data, John Wiley and Sons, New York.

Gelhar, L. W., Kuo, P. Y., Kwai, H. H., and Wilson, J. L. (1974) "Stochastic Modeling of Groundwater System", Report No. 189, Ralph M. Parson Laboratory for Water Resources and Hydrodynamics, Massachusetts Institute of Technology, Cambridge, MA.

Hromadka ,T. V., Durbin T. J., and Devries, J. J. (1985) Computer Methods in Water Resources, Lighthouse Publications, Mission Viejo, CA.

Hsieh, B. B. (1993) "A system analysis model of tidal hydrodynamic phenomena", in Sam S. Wang (ed.) , Advances in Hydro-Science and -Engineering, The University of Mississippi, MS, pp. 1677-1682.

Ritzi, R. W., Sorooshian S., and Gupta V. K (1991) "On the estimation of parameters for frequency domain models", Water Resources Research, 27(5), 873-882.

Williams, C C , and Lohman, S. W. (1949) "Geology and Ground-Water Resources of a Part of South-Central Kansas with Special Reference to the Wichita Municipal Water Supply", Bulletin 79, State Geological Survey of Kansas, F. Voiland, Jr., Topeka, KS.

THE APPLICATION OF AN INVERSE NUMERICAL MODEL FOR THE INTERPRETATION OF SINGLE OR MUTIPLE PUMPING TESTS.

L. LEBBE and W. DE BREUCK
Laboratory of Applied Geology and Hydrogeology.
Geological Institute, University of Ghent
Krijgslaan 281 (S8), Gent, B-9000
Belgium.

An inverse numerical model was developed for the interpretation of simple or multiple pumping tests in layered groundwater reservoirs. This model was obtained by the combination of a numerical model and a non-linear regression analysis. The numerical model is an axi-symmetric hybrid finite-difference finite-element model. In the non-linear regression analysis the logarithmic values of the hydraulic parameters are considered as well as of the calculated and observed drawdowns. With relative short-lasting pumping tests, the hydraulic resistance or the vertical conductivity of the semi-pervious layer can be derived with accuracies a slightly smaller than those obtained for the horizontal conductivities of the pumped layers. This is demonstrated by the interpretation of a double pumping test in a layered groundwater reservoir formed by quaternary sediments under Scheldt Valley (Belgium).

INTRODUCTION

In the framework of most hydrogeological problems in unconsolidated rocks consisting of an alternation of pervious and semi-pervious layers the accurate knowledge of the horizontal and vertical conductivity of respectively the pervious and the semi-pervious layers and their specific elastic storages is of considerable importance. Examples of such hydrogeological problems are the evolution of pollution plumes arround waste disposal sites, evolutions of watertables in the vicinity of water withdrawels and salt water intrusion problems. An accurate knowledge of these hydraulic parameters cannot be obtained from simple pumping tests and the interpretation of fragmented parts of the observed data, e.g. time-drawdown curves of different observation wells or distance-drawdown curves after a relative long period of pumping, by classical methods, where the representation of the groundwater reservoir is oversimplified. With relatively short-lasting multiple pumping tests, the hydraulic resistance or the vertical hydraulic conductivities of the semi-pervious layer can be derived with accurancies a slightly smaller than those for the horizontal hydraulic conductivities of the pumped layers. In these short-lasting multiple pumping tests every pervious layer is separately pumped while the drawdown is measured in the different layers. Pumping times extend usually not more than one or two days. All data are interpreted at the same time by means of an inverse numerical model that

A. Peters et al. (eds.), Computational Methods in Water Resources X, 769–776.
© 1994 Kluwer Academic Publishers. Printed in the Netherlands.

combines a numerical model with sensitivity analyses and a non-linear regression analysis. Here the application of this generalized interpretation method of simple or multiple pumping test is demostrated an a double pumping test in a layered groundwater reservoir formed by quaternary sediments under the Scheldt Valley (Belgium).

NUMERICAL MODEL

The applied numerical model is two-dimensional axi-symmetric. The model considers the groundwater reservoir to be subdivided in a number of homogeneous layers which are numbered from bottom to top (Fig. 1). Each layer is further subdivided in a number of concentric rings. The lowest layer, layer 1, is bounded below by an impervious boundary, and the uppermost layer is bounded above by the watertable. The horizontal flow and change in storage of each layer are characterized, respectively, by one value for the horizontal hydraulic conductivity and by one value of the specific elastic storage. The vertical flow between two layers is

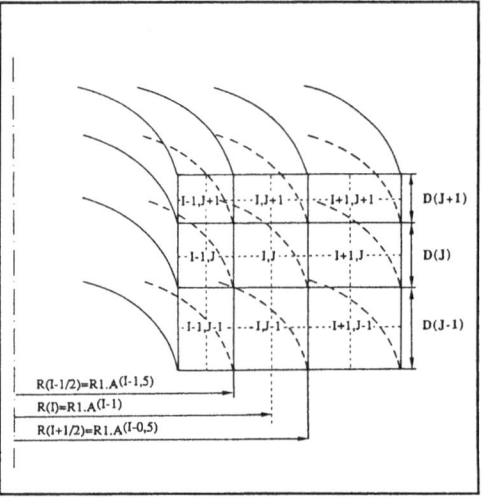

Fig. 1 The (I,J)-ring of the axial-symmetric grid.

governed by one value of hydraulic resistance between the layers. The hydraulic resistance is the thickness of a layer divided by its vertical conductivity. The amount of water delivered by a unit decline of the watertable is given by one value of the storage coefficient near the watertable. The drawdowns in the different layers at the different time steps are calculated with a hybrid finite-difference, finite-element model.

The input parameters which define the space grid are the number of layers, the number of rings per layer, the number of intervals, the initial time, T1, the initial radius, R1, and the factor A which determines the ratio of the outer and the inner radii of the successive rings (Fig. 1) and the ratio between the final and start time of the considered intervals. The initial radius is the inner radius of the smallest ring in the numerical model. It is assumed that between two successive nodal circles of ane layer the drawdown changes linearly with the logarithm of the distance from the pumped well. A detailed description of the numerical model is given in Lebbe (1988). In this work also the validation of the numerical model was demonstrated by the simulation of the models of Theis (1935), Jacob (1946), Hantush and Jacob (1955), Hantush (1960,1966) and the model of Boulton (1955,1963) as explained by Cooley (1971,1972) and Cooley and Case (1973).

NON-LINEAR REGRESSION ANALYSIS

After the schematization of the groundwater reservoir one has to estimate the initial values of the hydraulic parameters. With these values the model calculates the drawdowns at the same places and times where the observations took place. The differences between the logarithmic values of the observed and the calculated drawdowns for a certain parameter set are defined as residuals:

$$r = \log_{10}s^* - \log_{10}\hat{s} \qquad (1)$$

where r are the residuals,

    s^* the measured drawdowns,

    and $\hat{s}$ the calculated drawdowns.

The sensitivity J_{ij} of the drawdown s_i to the hydraulic parameter or group of parameters, P_j, are defined as follow:

$$J_{ij} = (\log_{10}\hat{s}_i(P_j \cdot sf) - \log_{10}\hat{s}_i) / \log_{10}(sf) \quad (2)$$

where sf is the sensitivity factor,

$\hat{s}_i$ is the calculated drawdown at the place and time of the i-th observation with the estimated values of the parameters for the first iteration or calculated values of the preceding iteration,

and $\hat{s}_i(P_j \cdot sf)$ is the calculated drawdown at the place and time of the i-th observation with the estimated values of the

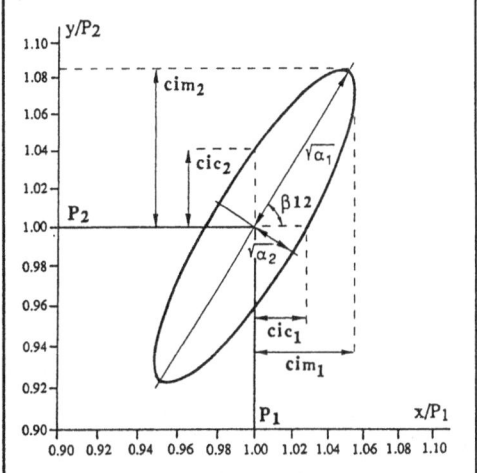

Fig. 2. Confidence intervals cim_j and cic_j for the case of two parameters

parameters with the exception of the value(s) of the j-th parameter or group of parameters whose estimated value(s) are multiplied by the sensitivity factor.

With the help of the residuals and the sensitivities the adjustment factors are calculated by means of the linearization method (Draper and Smith, 1981).

$$A = (J^T J)^{-1} J^T r \qquad (3)$$

where A is the vector of the logarithms of the adjustment factors of the different parameters.

The new estimated values of the parameters are obtained by multiplying the old ones by their corresponding adjustment factors or:

$$P_j^{n+1} = P_j^n \cdot 10^{A_{jn}} \qquad (4)$$

where P_j^n is value of the j-th parameter during the n-th iteration of inverse process,

    A_j^n is the logarithm in base 10 of adjustment factor of the j-th parameter deduced after the n-th iteration.

The algorithm is repeated until the adjustment factors become very small and the sum of the squares of the residuals reaches a minimum value.

When it is assumed that the residuals with their different weights approximate a normal distribution and that the drawdowns can be approximated as a linear function within the considered region then the joint probability distribution can be described by the mean and the covariance matrix of the parameters cov_p :

$$cov_p = \sigma_s^2 (J^T J)^{-1} \qquad (5)$$

where σ_s^2 can be estimated as $(\Sigma^n_{i=1}r^2_i)/(n-p)$ where n is the number of observations and p the number of parameters.

The marginal standard deviation sm_j of the j-th parameter can now be approximated as the square root of the j-th diagonal term of the covariance matrix. This standard deviation represents the variability when nothing is known about the other parameters. It may be a poor measure of the uncertainty when the estimates are correlated (Draper and Smith,1981 and Carrera and Neuman, 1986]. In Fig. 2 the confidence intervals cim_j and cic_j are given for the case of two parameters. A confidence interval cim_j, which is proportional to the marginal standard deviation, overestimates the variability. A better approximation of the actual variability is given by the conditional standard deviation. The confidence interval cic_j is proportional to this conditional standard deviation. It can be approximated by means of the eigenvalues and the eigenvectors of the covariance matrix as elaborated in Lebbe (1988):

$$\text{Sc}_j = \left(\sum_{k=1}^{p} \text{ß}_{jk}{}^2/\alpha_k \right)^{-1/2} \tag{6}$$

where Sc_j is the conditional standard deviation of the j-th parameter,
    α_k is the k-th eigenvalue of the covariance matrix,
    and ß_{jk} is the jk-th eigenvector of the covariance matrix.

With the aid of the marginal and conditional standard deviation the marginal and conditional confidence factor, $(1-\alpha)\%\text{Cfm}_j$ and $(1-\alpha)\%\text{Cfc}_j$, can be approximated as:

$$\log_{10}((1-\alpha)\%\text{Cfm}_j) = \text{sm}_j\sqrt{\text{pF}(\text{p,n-p,}1-\alpha)}$$
$$\log_{10}((1-\alpha)\%\text{Cfc}_j) = \text{sc}_j\sqrt{\text{pF}(\text{p,n-p,}1-\alpha)} \tag{7}$$

where $\text{F}(\text{p,n-p,}1-\alpha)$ is the F-distribution with p and n-p degrees of freedom and $1-\alpha$ the significance level. Because the standard deviation are located in the logarithmic parameter space the lower and upper limits of the marginal and conditional confidence interval can be found by dividing and multiplying the optimal values by their respective confidence factors.

INTERPRETATION OF DOUBLE PUMPING TEST

Lithostratigraphical cross-section and location of pumping and observation wells

The lithostratigraphical cross section (Fig. 3) is based on the description of core samples and on the results of geophysical borehole logs. In our particular case the resistivity measured with the short-normal device characterized the layering of the groundwater reservoir quite well. The reservoir is bounded below by a thick mainly clayey layer at a level of -9.5 (all levels are given in meters versus the Belgium Datum Level). Between -9.5 and -4 occurs a pervious layer which consists mainly of fine to medium fine sands with very thin layers of gravel and silt. Above this pervious layer lies a semi-pervious layer. This layer is composed of a sequence of fine layers (1 to 2 cm thick) silty fine sands, silt with fine sand and occasionally a peatbearing silt. Between the levels -2 and +3 occurs a second pervious layer which consists of gravelly sand with a small amount of fine silt layers which are seldom larger than one centimeter thick. The top of the groundwater reservoir is formed by a semi-pervious layer. This layer occurs between the levels +3 and +8. The lower part consists of silty fine sand. The top is composed of silt- and peatbearing alluvial clay.

The location of the two pumping wells and the observation wells are also shown in Fig. 1. The observation wells have a screen of one meter length and a diameter of 40 mm. Three observation wells are installed in the lower pervious layer, two in the middle semi-pervious layer, four in the upper pervious layer and finally two in the upper semi-pervious layer. All screens of the pumping wells and the observation wells are surrounded by calibrated sand. The annular space between the riser pipes and the semi-pervious layers are sealed with bentonitic clay.

Schematization of groundwater reservoir in numerical model

In the numerical model the groundwater reservoir is schematized in five layers. Layer 1 corresponds with the lower pervious layer, layer 2 with the middle semi-pervious

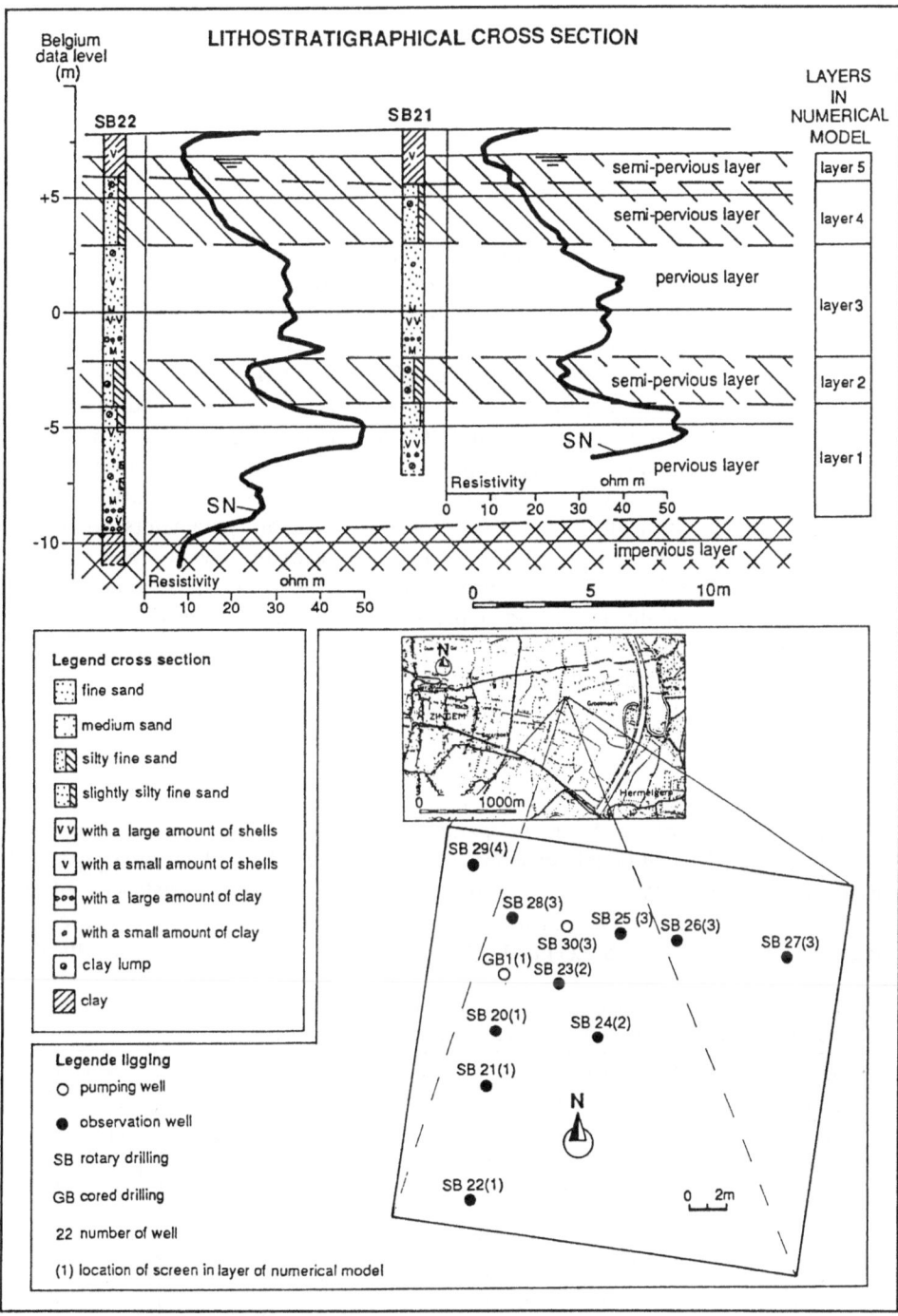

Fig. 3. Lithostratigraphical cross section, layers in numerical model and location of pumping and observation wells of double pumping test.

layer, layer 3 with the upper pervious layer and layer 4 and 5 with the upper pervious layer. The middle semi-pervious layer is only represented by one layer and was not further discretized in more than one layer because of his small thickness. The observation wells are located in the middle of the layers 1, 2, 3 and 4. The pumped wells has their screens over the entire thickness of layer 1 and 3.

Hydraulic parameters derived from the observed drawdown

Studying the sensitivities and the covariance matrices generated with the initial estimates of the parameters, the following hydraulic parameters can be deduced with the inverse model from all observed drawdowns of the double pumping test. Eleven hydraulic parameters or group of parameters can be deduced. It are the horizontal conductivities of the pervious layers $k_h(1)$ and $k_h(3)$, the specific elastic storage of the pervious layers $S_s(1)$ and $S_s(3)$, the hydraulic resistances of the lower and upper part of the middle semi-pervious layer, $c(1)$ and $c(2)$, and his specific elastic storage and his horizontal conductivity, $S_s(2)$ and $k_h(2)$, and the storage coefficient near the watertable or the specific yield S_0. The remaining parameters are classified in two groups of deducible parameters. The first group of deducible parameters comprises the specific storages of the upper semi-pervious layers or of the layers 4 and 5 of the numerical model, $S_s(4-5)$. The second group of deducible parameters comprises the hydraulic resistances of the lower and upper part of the upper semi-pervious layer, $c(3)$ and $c(4)$, and the horizontal conductivities of the layers 4 and 5 of the numerical model, $k_h(4)$ and $k_h(5)$, are included in this last group. It is assumed that the upper semi-pervious layer behaves as an isotropic one.

Interpretation

The estimated values of the hydraulic parameters are given in Table 1 along with their marginal and conditional standard deviations and their 98% marginal and conditional confidence factors. The accuracies of the hydraulic resistances of the semi-pervious layers are situated in the range of the accuracies of the specific elastic storages of the pumped layers $S_s(1)$ and $S_s(3)$. They are only a slightly smaller than

Table 1. Values of deduced hydraulic parameters along with their marginal and conditional standard deviation, sm and sc and 98% marginal and conditional confidence factors.

Hydraulic parameters	Units	Value	sm	sc	98%Cfm	98%Cfc
$k_h(3)$	m/d	$1,19.10^1$	0,0166	0,0095	1,1996	1,1097
$k_h(1)$	m/d	$1,83.10^1$	0,0167	0,0097	1,2008	1,1122
$S_s(3)$	m^{-1}	$6,26.10^{-5}$	0,0188	0,0144	1,2288	1,1710
$c(1)$	d	$6,87.10^1$	0,0266	0,0160	1,3385	1,1917
$c(2)$	d	$4,51.10^1$	0,0286	0,0191	1,3682	1,2329
$c(3),c(4)$ $k_h(4),k_h(5)$	d m/d	$4,50.10^1$ $5,07.10^{-2}$	0,0489	0,0210	1,7091	1,2588
$S_s(1)$	m^{-1}	$4,15.10^{-5}$	0,0357	0,0248	1,4789	1,3124
$S_s(4-5)$	m^{-1}	$9,04.10^{-5}$	0,0650	0,0323	2,0390	1,4248
$S_s(2)$	m^{-1}	$9,63.10^{-6}$	0,0729	0,0568	2,2234	1,8637
S_0	m^3/m^3	$1,44.10^{-2}$	0,0864	0,0715	2,5780	2,1895
$k_h(2)$	m/d	$4,71.10^{-1}$	0,1494	0,1167	5,1652	3,5935

the accurancies with which the horizontal conductivities of the pumped layers $k_h(1)$ and $k_h(3)$ can be deduced. The specific elastic storages of the semi-pervious layers $S_s(2)$ and $S_s(4\text{-}5)$ and the storages coefficient near the watertable S_0 can roughly be estimated from the measured drawdowns of the double pumping test. The conditional and marginal confidence intervals of the horizontal conductivity of the middle semi-pervious layer $k_h(2)$ are both larger than the square root of ten so that this parameter can hardly be considered to be deduced from the data of the double pumping test. The observed and calculated drawdowns are shown in Fig. 4. They correspond with the optimal parameter values. The calculated and observed drawdowns in the layers 1, 2 and 3 are very similar. The highest discrepancy occurs in the layer 4 during the pumping test on the upper pervious layer. The accurancies with which the hydraulic parameters can be derived are rather large because of the remaning residuals after the optimization The accurancies can further be ameliorated by the application of the biweighted least square method described by Wonnacott and Wonnacott (1985). In this method the observations obtain different weights depending of their residuals and of the interquartile range of these residuals (Lebbe et. al.,1992).

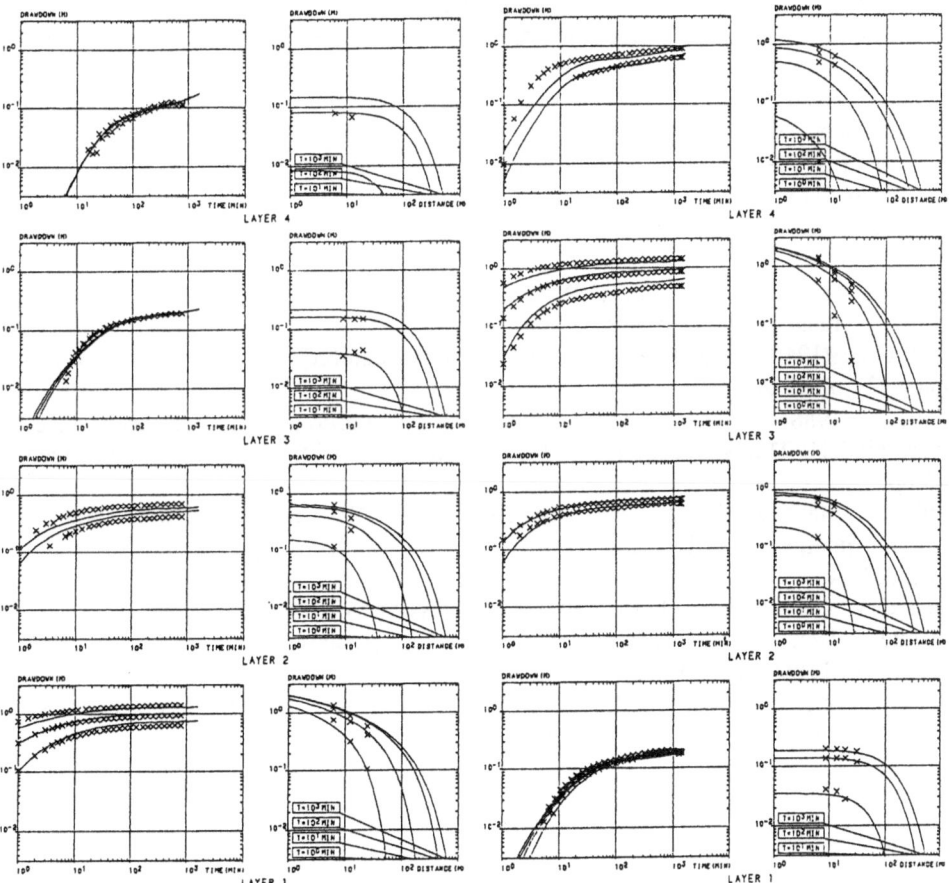

Fig. 4. Measured (x-signs) and calculated (solid curves) in time- and distance-drawdown graphs for the double pumping test (left two columns for pumping test in lower pervious layer, $Q(1)= 221\ \text{m}^3/\text{d}$, right two columns for pumping test in upper pervious layer, $Q(3)= 191\text{m}^3/\text{d}$).

Acknowledgement

The authors would like to thank the National Fund of Scientific Research (Belgium) under whose auspices the study was carried out. The pumping test was performed in the framework of a hydrogeological study which was financed by the Ministry of the Flemisch Community and the Water Company (T.M.V.W.).

References

Boulton,N.S. (1955) "Unsteady radial flow to a pumped well allowing for delayed yield from storage", IASH Assemblée Générale de Rome, Tome II, Publ. 37.

Boulton,N.S. (1963) "Analysis of data from non-equilibrium pumping test allowing for delayed yield from storage", Proc. Inst. Civ. Eng. 26,p.469-482.

Cooley,R.L. (1971) "A finite difference method for unsteady flow in variable saturated process media: application to a single pumping well", Water Resour. Res., 7 (6),p.1607-1625.

Cooley,R.L. (1972) "Numerical simulation of flow in an aquifer overlain by a water table aquitard", Water Resour. Res., 8 (4),pp.1046-1050.

Cooley,R.L. and Case,C.M. (1973) "Effect of a water table aquitard on drawdown in an underlying pumping aquifer", Water Resour. Res., 9 (2),p.434-447.

Draper,N.R. and Smith,H. (1981) Applied regression analysis, second edition. John Wiley & Sons, New York, 709 pp.

Hantush,M.S. (1960) "Modification of the theory of leaky aquifers", Journ. Geophys. Res., 65,p.3713-3725.

Hantush,M.S. (1966) "Analysis of data from pumping tests in anisotropic aquifers", Journ. Geophys. Res., 71,p.421-426.

Hantush,M.S. and Jacob,C.E. (1955) "Non-steady radial flow in an infinite leaky aquifer", Trans. Amer. Geophys. Union, 36,p.95-100.

Jacob,C.E. (1946) "Radial flow in a leaky artesian aquifer", Trans. Amer. Geophys. Union, 27,p.198-205.

Lebbe,L. (1988) "Uitvoering van pompproeven en interpretatie door middel van een invers model", Thesis Geagg. Hog. Onderw. Geologisch Instituut, Univ. Gent, Gent, 563 pp.

Lebbe,L., Mahauden,M. and De Breuck,W. (1992) "Execution of a triple pumping test and interpretation by an inverse numerical model", Applied Hydrogeology, 1(4),p.20-34.

Theis,C.V. (1935) "The relation between the lowering of the piezometric surface and the rate and duration of discharge of a well using groundwaterstorage", Am. Geophys. Union Trans., 16,p.519-524.

Wonnacott,R.J. and Wonnacott,T.H. (1985). Introductory Statistics, fourth edition. John Wiley and Sons, New York, 649pp.

ESTIMATING CONTINUOUS AQUIFER PROPERTIES FROM FIELD MEASUREMENTS: THE INVERSE PROBLEM FOR GROUNDWATER FLOW AND TRANSPORT

D.B. MCLAUGHLIN AND L.B. REID
Ralph M. Parsons Laboratory
Department of Civil and Environmental Engineering
Massachusetts Institute of Technology
Cambridge, Massachusetts 02139 USA

The inverse procedure described in this paper provides an efficient way to estimate spatially variable effective properties from scattered point measurements of related dependent variables. The procedure is illustrated with an example that considers the classic problem of estimating log transmissivity from noisy head measurements. In this case, the estimated log transmissivity function is derived from a Bayesian maximum *a posteriori* algorithm which relies on an efficient stochastic description of transmissivity variability. The transmissivity estimate is decomposed into a sum of basis functions with unknown coefficients. The basis functions are derived from a linearized approximation to the groundwater flow equation while the unknown coefficients are found with an iterative nonlinear search algorithm. Preliminary results from a synthetic test problem indicate that the algorithm can capture qualitative trends, although point estimation errors are large. Possible strategies for improving performance include the use of more accurate basis functions and the incorporation of solute concentration measurements which are more sensitive to transmissivity variations than the head measurements used in the example.

INTRODUCTION

Hydrogeological field investigations provide convincing evidence that soil properties such as hydraulic conductivity and related dependent variables such as hydraulic head, velocity, and solute concentration vary significantly over a range of spatial scales (Gelhar, 1993). In most modeling applications, the effects of small scale soil property fluctuations are accounted for implicitly, through the use of effective (or upscaled) parameters. Larger scale trends in these properties can be accounted for explicitly, by allowing the effective parameters to be functions of location. Effective parameters may be assumed to change only at the boundaries of identifiable geological units or, alternatively, allowed to vary continuously throughout the area of interest. In either case, these parameters are typically inferred from scattered field observations of piezometric head and/or solute concentration. Similar inference (or inverse) problems

A. Peters et al. (eds.), Computational Methods in Water Resources X, 777–784.
© 1994 Kluwer Academic Publishers. Printed in the Netherlands.

have been studied by many investigators in a number of different fields (see Ginn and Cushman, 1990, for a recent review of hydrologic applications).

In this paper we present a new approach to the groundwater inverse problem designed to deal with the nonlinearities inherent to the problem while remaining computationally efficient. This approach is based on Bayesian estimation theory and functional analysis concepts which are sufficiently general to be applied to both flow and transport problems. Efficiency is achieved by taking maximum advantage of closed form expressions for covariances which summarize, in an approximate way, physical relationships between the parameters to be estimated and the measurements available for estimation. The paper begins with a summary of the problem formulation and solution algorithm for a two-dimensional flow problem. This is followed by some preliminary results and a brief discussion of future research directions.

PROBLEM FORMULATION AND SOLUTION ALGORITHM

In order to focus on basic concepts we consider the problem of estimating a continuously varying effective transmissivity function from point measurements of piezometric head. We assume that the log transmissivity is a normally distributed random function with a known prior mean and covariance and that a two dimensional description of the flow problem is appropriate. The relationship between the log transmissivity and head is described by the following upscaled flow equation (Gelhar, 1993):

$$\nabla \cdot T\nabla h = \nabla \cdot e^{\alpha}\nabla h_l = 0 \qquad \mathbf{x} \in D \tag{1}$$

with the following boundary conditions:

$$h = h_b(\mathbf{x}) \qquad \mathbf{x} \in \partial D_d$$

$$-T\nabla h \cdot \mathbf{n} = -e^{\alpha}\nabla h \cdot \mathbf{n} = q_b(\mathbf{x}) \qquad \mathbf{x} \in \partial D_n$$

where h is the large-scale piezometric head, T is the effective transmissivity, and $\alpha = \ln T$ is the log transmissivity. The boundary head and flux functions, $h_b(\mathbf{x})$ and $q_b(\mathbf{x})$ are assumed to be known. An explicit solution to (1) may be written, at least formally, in terms of the flow equation Green's function $G(\mathbf{x}_i, \mathbf{x}'|\alpha)$, where $\mathbf{x}_i$ is some specified location:

$$h(\mathbf{x}_i) = \mathcal{F}_i(\alpha) = \int_{\partial D_d} \nabla_{x'}G(\mathbf{x}_i, \mathbf{x}'|\alpha) \cdot \mathbf{n}(\mathbf{x}')h_b(\mathbf{x}')\, d\mathbf{x}' + \int_{\partial D_n} G(\mathbf{x}_i, \mathbf{x}'|\alpha)q_b(\mathbf{x}')\, d\mathbf{x}' \tag{2}$$

This 'forward equation' defines a scalar functional $\mathcal{F}_i(\alpha)$ which maps $\alpha(\mathbf{x})$ to $h(\mathbf{x}_i)$.

Suppose that we have M noisy head measurements $z_1, z_2, \cdots, z_M$ taken at the points $x_1, x_2, \cdots, x_M$. These measurements may be related to $\alpha(\mathbf{x})$ with the following measurement equation:

$$\mathbf{z} = \mathcal{F}(\alpha) + \mathbf{v} \tag{3}$$

where $\mathcal{F}(\alpha)$ is an M vector composed of the $\mathcal{F}_i(\alpha)$'s evaluated at the measurement points and $\mathbf{v}$ is an M vector of random measurement errors. The measurement error term accounts for two distinct sources of uncertainty: 1) instrument and recording errors and 2) small-scale variability which influences the head measurement but is not captured in the upscaled flow equation. We assume here that $\mathbf{v}$ is independent of α and normally distributed with a mean of zero and a covariance matrix $\mathbf{C}_v$.

Our goal is to find a 'good' estimate of the random function $\alpha(\mathbf{x})$, given the measurement vector $\mathbf{z}$, the physical constraints embedded in the measurement equation, and prior statistical information about α and $\mathbf{v}$. This estimate should be well-posed and stable (i.e. it should yield unique solutions which are are not overly sensitive to small measurement errors) as well as consistent (i.e. it should converge to the true $\alpha(\mathbf{x})$ in the limit as the density of the head measurements increases). Unfortunately, these two requirements are contradictory for the problem outlined here (McLaughlin and Townley, 1994). Estimation techniques which successfully stabilize log transmissivity estimates (e.g. regularization techniques) diminish the influence of the measurements and generally yield estimates which are smoother than the true values, even when the number of measurements is large. Although the benefits provided by such smoothing usually outweigh the costs, the tradeoff needs to be recognized when selecting an inverse procedure.

In this paper we investigate an inverse approach which is typically referred to as maximum *a posteriori* (MAP) or Bayesian maximum likelihood estimation. When the unknown to be estimated is a vector the MAP estimator can be derived directly from Bayes theorem (Jazwinski, 1970). When the unknown is a function, as it is here, the classical derivation must be generalized. Generalized MAP theory characterizes the uncertain function $\alpha(\mathbf{x})$ by its prior mean $\bar{\alpha}(\mathbf{x})$ and its covariance $C_\alpha(\mathbf{x}, \mathbf{x}')$. The resulting MAP estimate of $\hat{\alpha}$ of α is defined as follows (Jazwinski, 1970; McLaughlin and Townley, 1994):

$\hat{\alpha}(\mathbf{x})$ is the $\alpha(\mathbf{x})$ which minimizes:

$$\mathcal{J}(\alpha) = [\mathbf{z} - \mathcal{F}(\alpha)]^T \mathbf{C}_v^{-1} [\mathbf{z} - \mathcal{F}(\alpha)] + \int_D \int_D [\alpha(\mathbf{x}) - \bar{\alpha}(\mathbf{x})] C_\alpha^{-1}(\mathbf{x}, \mathbf{x}') [\alpha(\mathbf{x}') - \bar{\alpha}(\mathbf{x}')] \, d\mathbf{x} d\mathbf{x}' \tag{4}$$

where $C_\alpha^{-1}(\mathbf{x}, \mathbf{x}')$ is an 'inverse covariance function' which is the functional counterpart to the inverse matrix (Tarantola, 1987). The first term on the right side of (4) penalizes 'lack-of-fit' while the second term penalizes deviations of the estimate from the prior mean. The relative weight of these two terms depends on the measurement error and log conductivity covariances.

The deterministic minimization problem posed in (4) can be formally solved with the methods of variational calculus. In practical applications, the problem must be discretized so that numerical solution methods can be used. Our approach to discretization is to project the infinite dimensional mean-removed log transmissivity function $\alpha'(\mathbf{x}) = \alpha(\mathbf{x}) - \bar{\alpha}(\mathbf{x})$ onto a finite dimensional space spanned by a set of M linearly independent basis functions. These basis functions are selected to provide a concise but physically plausible description of spatial variability. The basis function expansion may be written as:

$$\widehat{\alpha}'(\mathbf{x}) = \sum_{i=1}^{M} \beta_i \phi_i(\mathbf{x}) = \boldsymbol{\beta}^T \boldsymbol{\phi}(\mathbf{x}) \tag{5}$$

where $\boldsymbol{\phi}(\mathbf{x})$ is a vector of M basis functions $\phi_1(\mathbf{x}), \phi_2(\mathbf{x}), \cdots, \phi_M(\mathbf{x})$ and $\boldsymbol{\beta}$ is a corresponding vector of basis function coefficients. Each basis function-coefficient pair is associated with a particular measurement.

It seems reasonable to expect the basis function set selected for the problem discretization to reflect the influence of 1) the forward operator, 2) the sampling network, and 3) prior statistical information about $\alpha(\mathbf{x})$ and $\mathbf{v}$. A basis function set which meets these requirements can be identified from an analysis of the linear MAP estimation problem. When $\mathcal{F}(\alpha)$ is linear in α (or when it can be adequately approximated by a linear functional) the MAP estimate for the mean-removed parameter function $\alpha'(\mathbf{x})$ is equal to the conditional mean of $\alpha'(\mathbf{x})$ given $\mathbf{z}$ (Jazwinski, 1970). Moreover, this estimate has the same form as (5), with the basis function $\phi_i(\mathbf{x})$ given by:

$$\phi_i(\mathbf{x}) = C_{\alpha z_i}(\mathbf{x}) \tag{6}$$

where $C_{\alpha z_i}(\mathbf{x})$ is the covariance between $\alpha(\mathbf{x})$ and z_i (Dagan, 1989; Sec. 5.6).

Since the forward functional $\mathcal{F}(\alpha)$ defined by (1) and (2) is nonlinear, a basis function expansion which uses (6) for $\phi_i(\mathbf{x})$ does not, in fact, yield the true MAP solution. However, we still adopt the linearized basis function set for the nonlinear problem, since the linearized set satisfies the general requirements outlined above, although it is not strictly 'optimal'. In this nonlinear case the coefficient vector $\boldsymbol{\beta}$ is treated as an unknown and must be estimated. If (5) is substituted for $\alpha'(\mathbf{x})$ in (4), with $\phi_i(\mathbf{x})$ given by (6), the estimation problem becomes (Reid, 1994):

$$\widehat{\boldsymbol{\beta}} \text{ is the } \boldsymbol{\beta} \text{ which minimizes:}$$

$$J(\boldsymbol{\beta}) = [\mathbf{z} - \mathcal{F}(\bar{\alpha} + \boldsymbol{\beta}^T \boldsymbol{\phi})]^T \mathbf{C}_v^{-1} [\mathbf{z} - \mathcal{F}(\bar{\alpha} + \boldsymbol{\beta}^T \boldsymbol{\phi})] + \boldsymbol{\beta}^T \mathbf{C}_h \boldsymbol{\beta} \tag{7}$$

where $\mathbf{C}_h$ is a M by M matrix (the Gram matrix) with element (i, j) equal to the covariance $C_h(\mathbf{x}_i, \mathbf{x}_j)$ between the heads at the two measurement points $\mathbf{x}_i$ and $\mathbf{x}_j$. Note that this matrix is not inverted in (7) ! A closed form expression for the head covariance function $C_h(\mathbf{x}, \mathbf{x}')$ may be analytically derived from the linearization approximation introduced in (6) using Green's function techniques similar to those described in Dagan (1989, Sec. 3.7). The new minimization problem defined in (7) replaces the infinite dimensional unknown function $\alpha(\mathbf{x})$ by the finite dimensional unknown vector β. The minimizing coefficient estimate $\hat{\beta}$ may be found with an iterative optimization algorithm such as a quasi-Newton or gradient-based search (Tarantola, 1987). The corresponding log conductivity estimate $\hat{\alpha}(\mathbf{x})$ is then be constructed by substituting $\hat{\beta}$ into (5).

RESULTS FOR A SIMPLE EXAMPLE

The concepts introduced above can be conveniently illustrated with a model inverse problem based on synthetically generated data. Synthetic problems have the advantage of providing good diagnostic information, since the inverse estimate can be readily compared to a known true function. Such problems do not, of course, prove that the technique in question will work in a field setting. For this reason we consider the results shown here to be preliminary.

Our example problem is concerned with the estimation of a continuously variable random log transmissivity function from scattered measurements of piezometric head. The prior mean and covariance of the log transmissivity function and the measurement error are assumed to be known. The problem domain is a two-dimensional region of 450 by 150 m. discretized on a 27 by 15 cell finite difference grid. The log transmissivity spectrum is an isotropic Whittle A (Gelhar, 1993) with a correlation distance of 30 m. and a variance of 1.0. The geometric mean transmissivity is 1.0 m^2/day (giving a mean log transmissivity of 0.0). The head measurements are corrupted by uncorrelated synthetically generated normally distributed measurement errors with means of zero and standard deviations of 0.1 m. The boundaries are no-flux at the top and bottom of the domain, a specified head on the left, and a specified flux on the right. The boundary flux is 0.01 m^2/day, giving a mean hydraulic gradient from left to right of 0.01. The existence of a flux condition on the entire right boundary insures that the inverse problem is well-posed when viewed as a Cauchy problem with the head known perfectly everywhere in the domain (Dagan, 1989, Section 5.6). In practical applications where head information is available at only a finite number of points, regularization and/or prior information are required to maintain well-posedness.

Figure 1 shows a typical basis function obtained for the example problem. This function determines the influence of a head measurement located at (200,200) on the log transmissivity estimate at every point in the domain. The smooth variation of the basis function reflects the correlation imposed by the prior covariance of α and by the forward operator $\mathcal{F}_i(\alpha)$. As indicated in (5), the log transmissivity estimate is a linear combination of such basis functions, one for each measurement. Figure 2 shows the true (synthetically generated) log transmissivities and heads, the MAP estimates, and the estimation errors (differences between true and estimated values) for a typical log transmissivity replicate. The locations of the ten measurements used for estimation are indicated by the small dots on the horizontal plane. Although the estimation algorithm is able to capture most of the major qualitative features of the log transmissivity field, the point errors are quite large (in some cases as large as the original value). This reflects, to some extent, the absence of measurements in areas near prominent log transmissivity features. But it also reflects the inherent difficulty of estimating log transmissivities from head measurements which are relatively insensitive to point transmissivity values. Note that the estimation algorithm does a good job of reproducing the head field, which is only moderately heterogeneous, even in areas where the transmissivity estimates are poor.

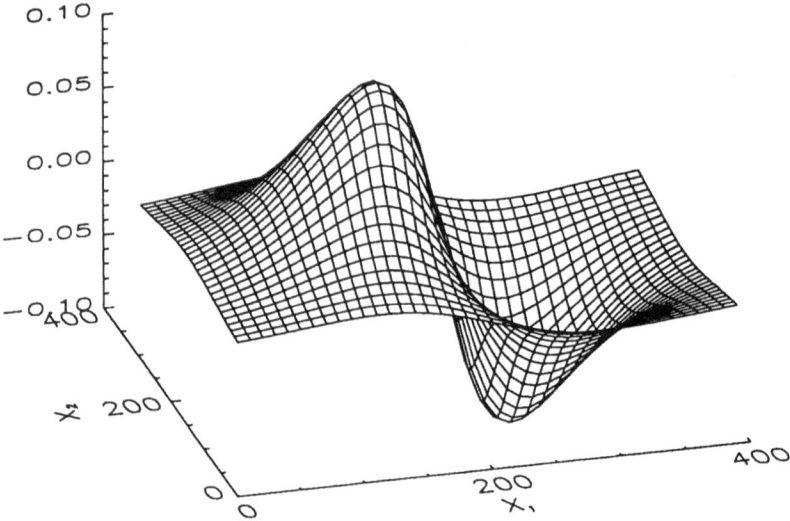

Figure 1: Typical basis function for the example problem.

The most obvious ways to improve the performance of the estimation algorithm are to use more head measurements and/or to include measurements of other variables (such as log transmissivity and solute concentration). The option of increasing the number of head measurements must be approached with caution since closely spaced measurements will be highly correlated. This leads to estimation equations which are algorithmically ill-conditioned, although mathematically non-singular. If too many measurements are added the near-singularity of the equations can lead to large numerical errors and poor estimates. Generally speaking, concentration measurements provide more useful information than additional head measurements for this problem.

It should be noted that the simplifying approximations used to derive the covariance/basis function $C_{\alpha z_i}(\mathbf{x})$ and the Gram matrix C_{hh} merit further examination.

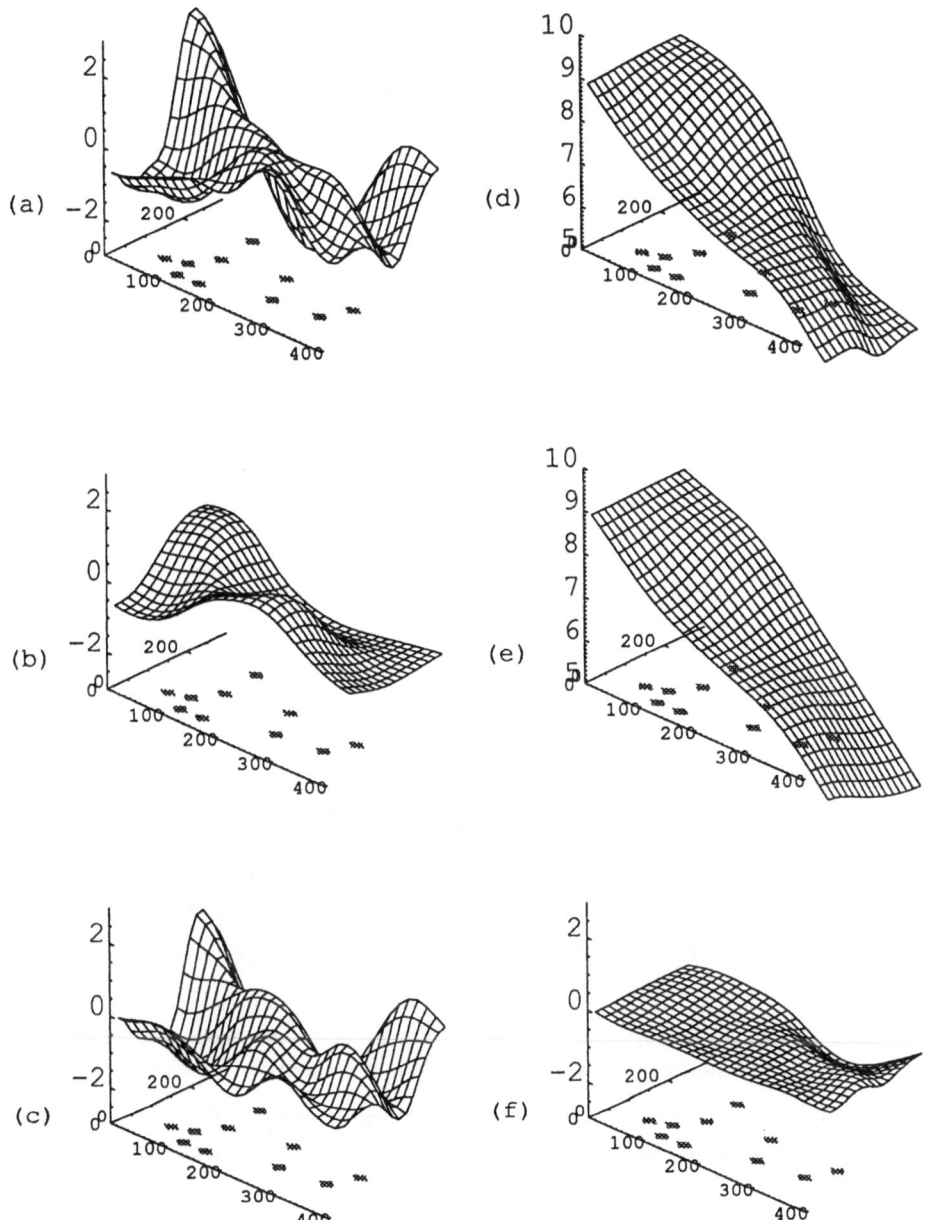

Figure 2: Example problem results: a) & d) true log transmissivity and head, b) & e) estimated log transmissivity and head, c) & f) log transmissivity and head errors. Dots show location of head measurements.

For example, the current version of our algorithm is based on the assumption that the mean hydraulic gradient points in the x_1 direction. This assumption could be refined after the first iteration to reflect new information provided by the head measurements. It is also possible that the estimates could be improved if cross-validation techniques were used to adjust the prior log transmissivity variance, which may not necessarily reflect the variability of any particular replicate (Wahba, 1989). We are currently studying both of these enhancements.

CONCLUSIONS

The simple example outlined above suggests that our approach to inverse estimation of continuous aquifer properties from discrete measurements is sufficiently promising to merit further investigation. This approach can be viewed as a nonlinear iterative extension to purely linear geostatistical approaches (Dagan, 1989, Section 5.6). In the future we intend to extend our approach to three-dimensions, add solute transport and sorption (so that solute and soil concentration measurements may be included), improve on the approximations used to derive the basis functions and Gram matrix, and investigate the possibility of using cross-validation methods. Further improvements and enhancements will be guided by field tests as well as synthetic experiments.

REFERENCES

Dagan (1989) Flow and Transport in Porous Formations, Springer-Verlag, Berlin.

Gelhar, L.G. (1993) Stochastic Subsurface Hydrology, Prentice Hall, New York.

Ginn, T.R. and J.H. Cushman (1990) Inverse methods for subsurface flow: A critical review of stochastic techniques, Stochastic Hydrology and Hydraulics, 4, 1-26.

Jazwinski, A.H.(1970) Stochastic Processes and Filtering Theory, Academic Press, New York.

McLaughlin, D. and Townley, L. (1994), A reassessment of the groundwater inverse problem, unpublished manuscript submitted to Water Resources Research.

Reid, L. (1994), Characterization of Groundwater Aquifers using Infinite-Dimensional Inverse Methods, unpublished PhD thesis, Dept. of Civil and Environ. Engr., Mass. Instit. of Tech.,Ccambridge, MA.

Tarantola, A. (1987) Inverse Problem Theory, Elsevier, Amsterdam.

Wahba, G. (1990) Spline Models for Observational Data, SIAM Publications, Philadelphia, PA.

INVERSE PARAMETER IDENTIFICATION OF SOIL HYDRAULIC PROPERTIES RESULTS OF A NEW SOIL COLUMN EXPERIMENT

NÜTZMANN, G.; MOSER*, H.; AND HANDKE, H.

Institute of Freshwater Ecology
Department of Hydrology
Rudower Chaussee 5
12484 Berlin, Germany

* Institute of Water Resources and Hydraulic Engineering
Technical University of Berlin
P.O. Box 100 320
10563 Berlin, Germany

A one step outflow experiment on a homogeneous sandy soil in a 150 cm column is considered. The cumulative outflow of the column is controlled by different fixed groundwater levels so, that a lower part of the soil remains saturated and a complete saturated-unsaturated flow regime is modeled. Hydraulic properties of the sand are assumed to be represented by van Genuchtens closed-form expressions. The parameters of this functions are evaluated by a simple optimization algorithm and a finite-element-code simulates the water movement in the saturated-unsaturated soil. Because the groundwater level is both the lower boundary of the column and a well defined inner point of the flow region (in a mathematical sense) with a constant hydraulic head, the conditions of identifiability are satisfied, the algorithm is stable and accurate results can be obtained.

INTRODUCTION

Knowledge of soil hydraulic properties is an essential requirement for prediction of flow and transport through the unsaturated zone. There are many laboratory and field methods to determine this highly nonlinear functions. A well-known laboratory test is the one-step outflow method (Kool et al., 1985; 1987; Kool and Parker 1988; Dam van et al. 1992). In recent years, there are many investigations about feasibility of determining water retention and conductivity functions from transient outflow experiments by numerical inversion of the initial boundary value problem. In such approaches one generally assumes a certain parametric model for the hydraulic properties. Unknown parameters are estimated by minimization deviations between observed and predicted cumulative outflow from the soil column.

As it was shown by Mous (1993), the non-uniqueness of the estimates is not due to a bad choice of the optimization algorithm, but it is merely a consequence of the structure of the model and the design of the experiment. This agrees with results of Hornung (1983), who gives mathematical conditions of identifiability, and of van Dam et al. (1992), who makes recommendations for improving the performance of outflow experiments.

Following that, the purpose of this paper is to discuss several one-step outflow experiments on a large soil column, where the initial fully saturated soil is only drained by a 'natural' groundwater table fluctuation, not by an 'artificial' pressure condition at the bottom of the column. Due to identifiability of the parameters in this experiments a simple estimation routine in connection with a numerical solution of the flow equation are used and the influence of the drainage gradient to the parameters α and n of the Mualem-Genuchten model is reported.

A. Peters et al. (eds.), Computational Methods in Water Resources X, 785–792.
© *1994 Kluwer Academic Publishers. Printed in the Netherlands.*

MATERIALS AND METHODS

The general layout of the experimental set up is depicted in Figure 1. A synthetic porous media is filled in a vertical column. The column is made of plexiglass with a thickness of 6 mm. The inner dimensions are length 1500 mm and diameter 206 mm. The upper part of the column is open and at the bottom is a inlet/outlet hole with a diameter of one inch. Below the inlet a 3-way valve is installed and connected with two constant level reservoirs. The discharge of the second reservoir is conducted to the flow measurement device. The material used for filling the column is quartz sand, with diameters ranging from 0.30 to 0.80 mm. In preparing the porous media a main criteria was to achieve a high and uniform compactation in order to avoid settlement or other structural changes and to provide porosity as homogeneous as possible. In a prein-vestigation the lowest and highest compactation of the porous medium was determined to $n_{min} = 0.40$ and $n_{max} = 0.48$ by means of a standard testing procedure for non-cohesive soils. Additionally the density of the solid particals was determined to $\rho_s = 2.60$ kg/m³. Due this the exact mass of sand was known to create a porous media with a constant porosity in a certain volume of the column. The packing of the column was done in sections of five centimeters and as the column was being filled, the model was being vibrated at high frequency by means of a vibrator attached to the underside of the steel-table, on which the column was erected to obtain a controlled porosity of $n = 0.40$. The permeability of the column was determined by means of classical Darcian Experiments to $k = 7 \cdot 10^{-10}$ m², and a tracer experiment was done to controll the effective porosity, the deviation from the above mentioned value was measured to $\pm 5\%$.

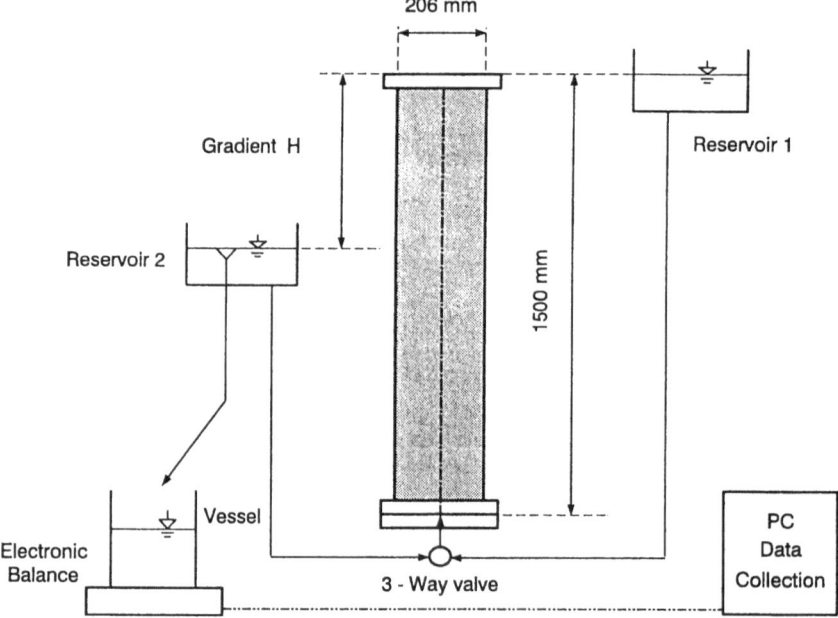

Fig. 1: Experimental layout

The initial conditions for the outflow experiments were a saturated column corrosponding to the water-level in reservoir number one. The reservoir number two is located at a certain position beneath the upper edge of the column. At the time point $t = 0$ the 3-way valve was switched over and at this initial time point the experiment was driven by the gradient between the water-levels of the reservoirs. The outflow of reservoir two was collected in a vessel, which was on an electronic balance. During the experiments the signal of the balance was recorded with time by a PC and the volumetric outlow as well as the cumulative volumetric outflow was stored.

MATHEMATICAL MODEL

The transient saturated-unsaturated water flow in a vertical soil column without sinks and sources is described by Richards-equation

$$C \frac{\partial h}{\partial t} + \frac{\partial}{\partial x}\left(K \frac{\partial h}{\partial x} \right) = 0 \tag{1}$$

where $h = \Psi + x$ is hydraulic head with Ψ the pressure head and x the vertical distance taking positive upwards, $C = d\theta/d\Psi$ is the water capacity with θ the volumetric water content, K is the hydraulic conductivity and t is the time. The appropriate initial and boundary conditions for the one-step outflow experiment are

$$h = h_0(x) \qquad t = 0, \qquad 0 \leq x \leq L \tag{2 a}$$

$$\partial h/\partial x = 1 \qquad t > 0, \qquad x = L \tag{2 b}$$

$$h = h_{GW} \qquad t > 0, \qquad x = 0 \tag{2 c}$$

where $x = 0$ is taken at the bottom of the column, $x = L$ is the top of the soil, and h_{GW} is the actual groundwater level in the column. The solution of Eq. (1) and (2) was obtained by a Galerkin finite element code SUNSOL of Nützmann (1991). Cumulative outflow $Q(t)$ is than calculated as

$$Q(t) = A \int_0^L \left[\theta(x,0) - \theta(x,t) \right] dx \tag{3}$$

where A is the cross-sectional area of the column, using trapezoidal rule.
In this study the hydraulic functions of an unsaturated porous medium are described by the commonly used Mualem-Genuchten-Model (Mualem, 1976; van Genuchten, 1980):

$$\Theta(\psi) = \frac{(\theta - \theta_r)}{(\theta_s - \theta_r)} = \left[\frac{1}{1 + (\alpha\psi)^n} \right]^m, \qquad m = 1 - 1/n \tag{4 a}$$

$$K(\Theta) = K_s \Theta^{0.5} \left[1 - \left(1 - \Theta^{1/m} \right)^m \right]^2 \tag{4 b}$$

where θ_s is the satured water content; θ_r is the residual water content; K_s is the saturated hydraulic conductivity and α, n and m are empirical parameters. This consistent set of parameter functions is well-suited to describe hydraulic properties of soils with a wide spectrum of pore size distribution, Durner (1991), like the sandy soil of this experiment. In the following parameter estimation examples we have measured K_s, θ_s and θ_r independently, and only the values of α and n are sought by solving the inverse problem. This is formulated as a nonlinear optimization problem, i.e. parameters are estimated by minimizing a suitable objective function which expresses the discrepancy between observed and predicted system response (Kool and Parker, 1988; Dam van et al. 1992). In the case of measurement of outflow values only, the objective function can be written as an ordinary least-square problem

$$Q(b) = \sum_{i=1}^{N} \left[w_i \left(Q_m(t_i) - Q_c(t_i, b) \right) \right]^2 \tag{5}$$

where b is the array of parameter values to be optimized, w_i is a weighting factor for outflow measurements (e.g. the inverse of measurement errors), Q_c is the predicted and Q_m is the measured outflow, N is the length of the data vector. Newton's and related methods for solving Eq. (5) lead to algorithms in which the so-called sensitivity matrix is required to calculate (Kool et al. 1988). The elements of the sensitivity matrix are the partial derivatives of $Q_c(t_i, b)$ with respect to the parameter b. As it was shown by Mous (1993), mathematical characteristics of the sensitivity matrix give information if the problem is identifiable or not. If the columns of these matrix are plotted versus time and the shapes are almost equal or are mirror images, than a linear dependence between these columns are given and this holds for a non-identifiability (Mous, 1993). In the most algorithms the sensitivity matrix is approximated simultaneously with the cumulative outflow Q_c. In the experiments considered here, this could be done analytically inserting Eq. (4a) into Eq. (3), determining the derivatives of Q_c with respect to α and n, and using the calculated hydraulic heads from the SUNSOL-simulation. In Fig. 2 exemplary the columns of the sensitivity matrix for the 0.20 m gradient experiment are plotted and their independence is evidently shown.

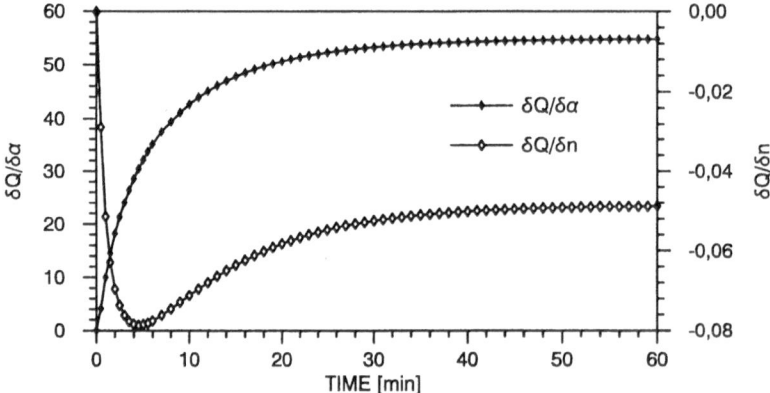

Fig. 2: Sensitivity analysis of the 0.20 m experiment

The minimization of the objective function (5) is now calculated with a simple gradient method.

The convergence criteria are given by the mean square error and the relative mean deviation between measured and predicted cumulative outflows of the soil.

RESULTS AND DISCUSSION

Several outflow experiments were carried out, with different groundwater fluctuations ranging from 0.20 m to 1.20 m, see Tab. 1, in order to measure the cumulative outflow. The data shown in table 1 is a selection of numerous experiments. This selection was done according to the aspect, that the first outflow experiments of each gradient differs in the total outflow mass from these experiments carried out afterwards at the same gradient. This difference in outflow can be explained by the fact the moisture content of the column does not reach again the fully saturation during rewetting.

Gradient	20	30	40	50	60	80	100	120
Time [min]	60	70	100	140	150	180	240	280
Mass [cm]	1,05	2,74	5,30	7,53	9,90	15,03	19,75	24,13

Tab. 1: Experimental Data

The data base shown in Tab. 1, was used to determine the hydraulic parameter functions of synthetic porous medium. A first estimation of the initial values α and n from Eq. (4) was based on van Dam et. al., (1992); these values $\alpha = 0.14$ and $n = 1.225$ were found by simulation runs and kept constant as initial conditions for every estimation cycle of the experiments. According to the increasing gradient, that means here the distance of the top of the column to the fixed groundwater level, the fits were done as described above. The criteria of convergence were given by the least squares and the relative mean square error.

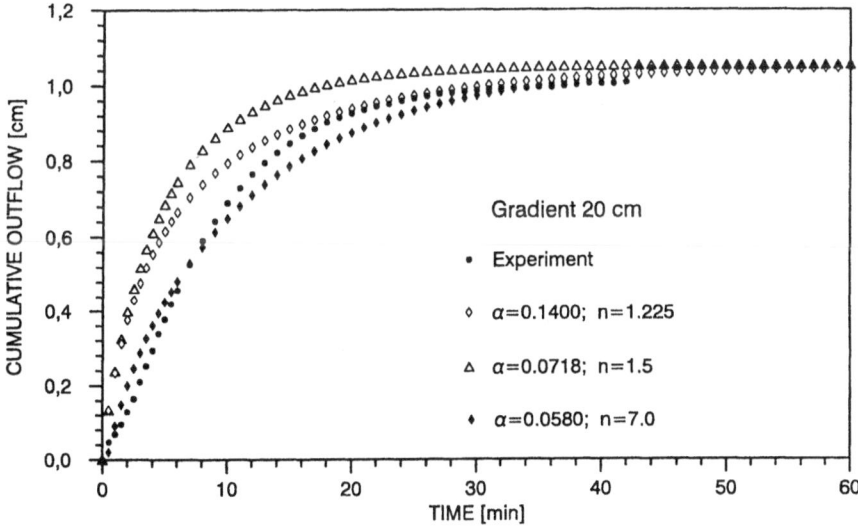

Fig. 3: Calculated and measued cumulative outflow of the 0.20 m experiment

In Fig. 3and 4 the results for the gradient 0.20 m and 1.20 m are depicted. Three different types of approximations to the experimental outflow curve can be observed. For small gradients the calculated curve overshoots the measured curve for the initial values of the parameters and turns down during minimization procedure (see Fig. 3). In contrast to this the simulated curves for large gradients show a different behaviour and reach the measured outflow curve from underestimated starting runs (see Fig. 4). For the medium gradients a combination of the above mentioned effects can be stated. It should be remarked, that the largest deviations between predicted and observed outlow curves are at the very beginning of the draining process, where the rate of change of the slope reaches a maximum.

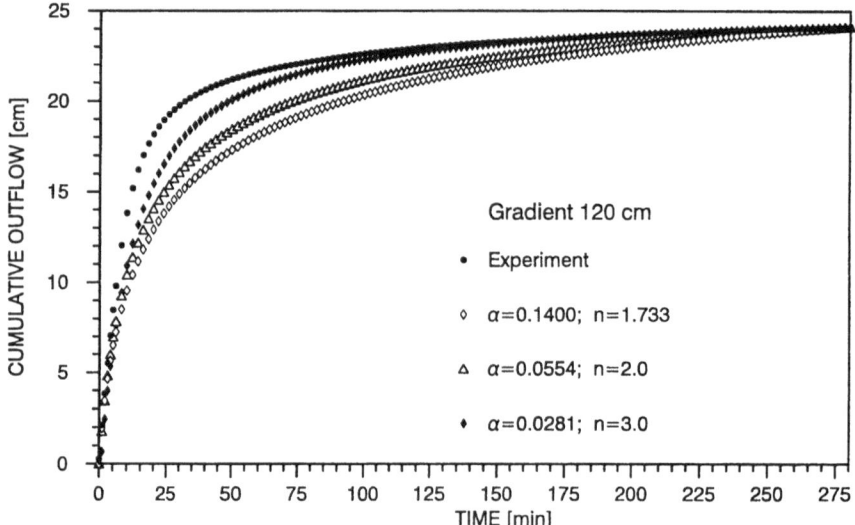

Fig. 4: Calculated and measued cumulative outflow of the 1.20 m experiment

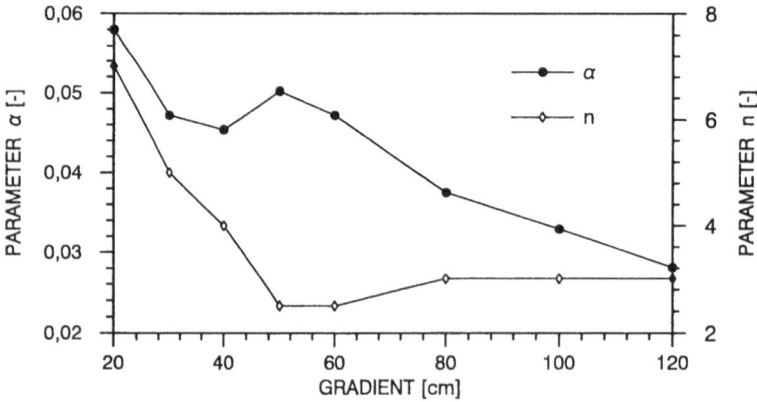

Fig. 5: Optimal parameters α and n

As shown in the Figures above, for each gradient-experiment the best fit is obtained for different pairs of values (α,n). In Fig. 5 for all selected experiments these pairs are given. The decrease of both parameters may be explained that for small gradient-experiments the range of values of the hydraulic function seems to be incomplete and with increasing gradients a more complet range of the function is inculded. That indicates, that parameters only fitted at experiments with small gradients (this means small pressure heads in the column) cannot be used to model unsaturated flow for large gradients. For processing an identification, the relevant range of pressure heads has to be considered at the design of the experiments (Dam van, 1992).

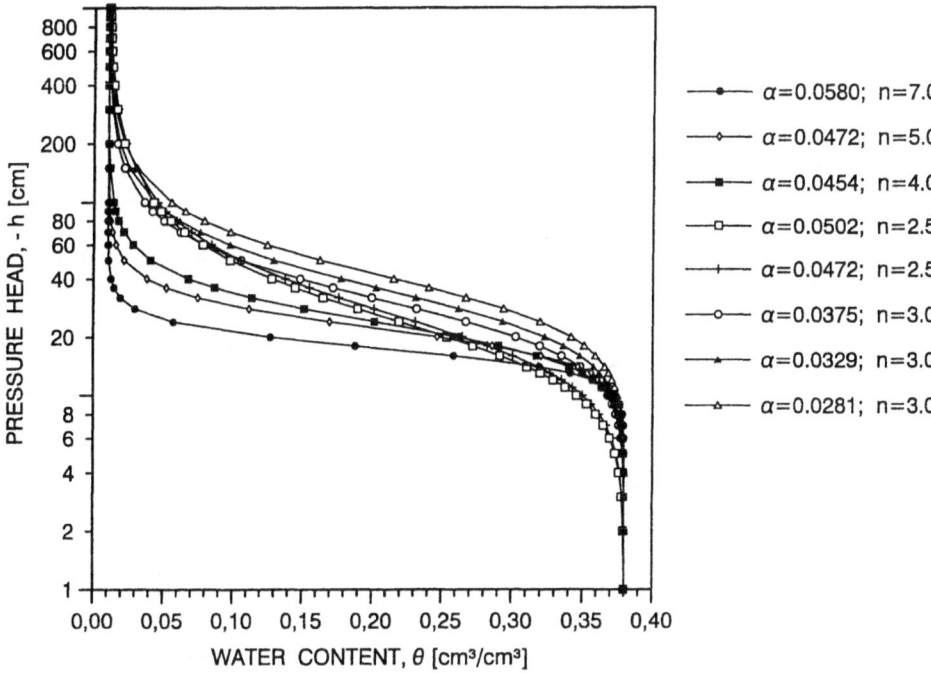

Fig. 6: Predicted water retention functions

In Fig. 6 the retention function θ (ψ) for the estimated best pairs of parameters are given. The parameter α causes a simple shift of the retention curve and does not influence the relative conductivity function K_r (ψ). The inverse of α represents in statistical terms the mean of the pore size distribution (Durner, 1991). The parameter n characterizes the width of the pore size distribution towards to large pores and thus the position of air entry value in the above mentioned function.

The slope of the function (see Fig. 6) decreaes for small parameters n. Considering this, the deceasing parameter values α and n, indicate that for small gradients the draining process is driven principally by active large pores. In addition to this, at increasing gradients more and more small pores are activated for the column outflow by the increasing pressure heads.

SUMMARY

The parameter α and n of the Mualem-Genuchten model are determined using cumulative ouflow data obtained of a series of drainage experiments with different groundwater tables combined with a numerical parameter estimation procedure. The identifiability of these parameters can be derived from the sensitivity functions of the outflow process. Qualitatively the fits show an expansion of the pore size range govering the gravity drainge process with increasing pressure head gradients. The meaning of an accurate formulation of the retention curve for the entire range is emphasized. The developed experimental set-up with constant water levels as a lower boundary is closer to natural unsaturated-saturated flow conditions as the commonly used small column experiments with artificial pressure heads as a lower boundary.

References

Dam, J. C. van, Stricker, J. N. M. and Droogers, P., 1992. Inverse method for determining soil hydraulic functions from one-step outflow experiments. Soil Sci. Soc. Am. J. 56: 1042-1050.

Durner, W., 1991. Vorhersage der hydraulischen Leitfähigkeit strukturierter Böden. Bayreuther Bodenkdl. Berichte, Bd. 20, Lehrstuhl f. Bodenkde. u. Bodengeographie der Univers. Bayreuth.

Genuchten, M. Th. van, 1980. A closed-form equation for predicting the hydraulic conductivity of unsaturated soils. Soil Sci. Soc. Am. J. 44: 892-898.

Hornung, U., 1983. Identification of nonlinear soil physical parameters from an input-output experiment. In: Deuflhardt, P. and Hairer, E. (Eds.). Workshop on numerical treatment of inverse problems in differential and integral equations. Birkhäuser, Boston, pp. 227-237.

Kool, J. B., Parker, J. C. and Genuchten, M. Th. van, 1985. Determining soil hydraulic properties from one-step outflow experiments by parameter estimation: I. Theory and numerical studies. Soil Sci. Soc. Am. J. 49: 1348-1354.

Kool, J. B., Parker, J. C. and Genuchten, M. Th. van, 1987. Parameter estimation for unsaturated flow and transport models - A review. J. Hydrol. 91: 255-293.

Kool, J. B. and Parker, J. C., 1988. Analysis of inverse problem for transient unsaturated flow. Water Resour. Res. 24: 817-830.

Mous, S. L. J., 1993. Identification of the movement of water in unsaturated soils: the problem of identifiability of the model. J. Hydrol. 143: 153-167.

Mualem, Y., 1976. A new model of predicting the hydraulic conductivity of unsaturated porous media. Water Resour. Res. 12: 513-522.

Nützmann, G., 1991. A simple finite element method for modeling one-dimensional water flow and solute transport in variably saturated soils. Acta hydrophys. 35: 33-59.

Application of Artificial Neural Networks for Site Characterization Using "Hard" and "Soft" Information

D. M. RIZZO and D. E. DOUGHERTY
Dept. of Civil & Environmental Engineering
University of Vermont
Burlington, VT 05405
U.S.A.

Site characterization of the subsurface materials is essential to any successful modeling or design effort. The physical parameters needed to accurately describe the groundwater system are often measured at a limited number of observation points (usually wells). The information gathered at these wells is often multivariate in nature. In this paper we will consider the problem of identifying hydraulic conductivity fields using both quantitative and qualitative information measured at the same spatial locations. Examples are presented in which the quantitative data, often referred to as "hard" data, will be the conductivity values obtained from pumping tests. The more qualitative data or "soft" data will consist of soil descriptions from driller well logs.

A number of methods are used throughout the geohydraulic community to realize maps of parameters such as hydraulic conductivity (three such methods include ordinary kriging [Journel and Huijbregts, 1978], indicator kriging [Journel, 1984] and probability kriging [Sullivan, 1984]). More traditional multivariate classification techniques (i.e. discriminant analysis, logistic regression) may also be used to determine a "best estimate" of the conductivity field.

An alternative method to produce estimates of hydraulic conductivity based on the application of artificial neural networks, *neural kriging* (NK), has been developed [*Rizzo and Dougherty*, 1994]. This method possesses many operational objectives of the kriging methods. However, the method is data driven and requires no estimate of a covariance function.

It is not possible to obtain hard or soft information from field measurements at the resolution needed by most numerical models. As a result, a means of mapping the hydraulic parameters that control subsurface flow from sparse field data to the scale of the model is the key to any reliable modeling effort.

The next section briefly describes the neural kriging network used to develop maps of spatially–distributed conductivity fields. Two example applications which incorporate mixed data into the characterization scheme are then presented. The first application involves a synthetic test case developed by the United States Geological Survey (USGS); the second application applies the characterization scheme to the

A. Peters et al. (eds.), Computational Methods in Water Resources X, 793–799.
© *1994 Kluwer Academic Publishers. Printed in the Netherlands.*

Lawrence Livermore National Laboratory (LLNL) site. The paper concludes with some summary remarks.

BRIEF REVIEW OF THE NEURAL KRIGING NETWORK

Artificial neural networks (ANNs) are biologically inspired computational systems that interconnect simple processing elements in highly parallel configurations. Examples of the ANN used in this work are shown in Figure 1. The network used in each of the following applications is a feed-forward network that will be trained using the counterpropagation training algorithm developed by *Hecht–Nielsen* [1987, 1988]. A brief review of its operation will be presented below. For more details see *Rizzo and Dougherty* [1994].

The counterpropagation algorithm consists of two phases—a training phase and an interpolation phase. During training, examples of some mapping (vectors x, and Y) are presented to the network. The weights are adjusted over a series of iterations until the network has satisfactorily mapped the training set inputs to the training set outputs. Once a satisfactory mapping has been obtained, training ends and the weights are fixed. These fixed weights are then used during the interpolation phase to map new inputs, x, to points at which we would like to predict the output variable, Y'.

In order to use a NK network that incorporates mixed data, a sequential process is employed. Step one consists of training vectors which contain both hard and soft data. Input vectors, x, consist of three componenets (see Figure 1 (a)). The first two components contain the two–dimensional spatial information for each well site. The third component is a real number ranging between 0.0 and 1.0 which corresponds to soil descriptions indicated in the driller well logs (see for example Table 1(b)). The corresponding output training vectors, Y, contain the hydraulic conductivity value measured at the respective well site encoded in a binary form (see Table 1 (a)). After training, the network is used to estimate conductivity at additional sites that contain only soft information (*i.e.* soil borings). Once the conductivity at these additional sites has been determined, the network may be retrained using all sites (wells and soil borings) as the new training patterns. Step two of the sequential training process uses the new training vectors to estimate values over the entire grid where hard and soft data do not exist. The new input training vectors contain only spatial information and the corresponding output vectors contain the conductivity values measured at those locations (see Figure 1 (b)).

APPLICATION 1: USGS SYNTHETIC TEST CASE

Hydraulic conductivity fields are constructed using a synthetic test case which was developed by the USGS as a controlled model calibration experiment using nonlinear regression [*Hill et al.*, 1993]. The aquifer material for this synthetic site is vertically homogeneous. As a result, only aerial distributions of the hydraulic conductivity field are constructed. Data are collected from the "site" over the course of four

Hard Data		
Binary Form	ft/d	Conductivity
0001	< 130	Low Conductivity
0010	$130 \leq 255$	Medium Low
0100	$255 \leq 380$	Medium High
1000	> 380	High

Soft Data	
	Soil description from driller's well logs
0.00	Fine to medium grained sand with some interbedded silts
0.33	Medium sand with some interbedded fine sand
0.66	Coarse sand with occasional stringers of fine sand
1.00	Very coarse sand with occasional thin gravel layers

Table 1: Hard data used to represent (a) the four discrete hydraulic conductivity classes, and (b) soft data representing the soil descriptions.

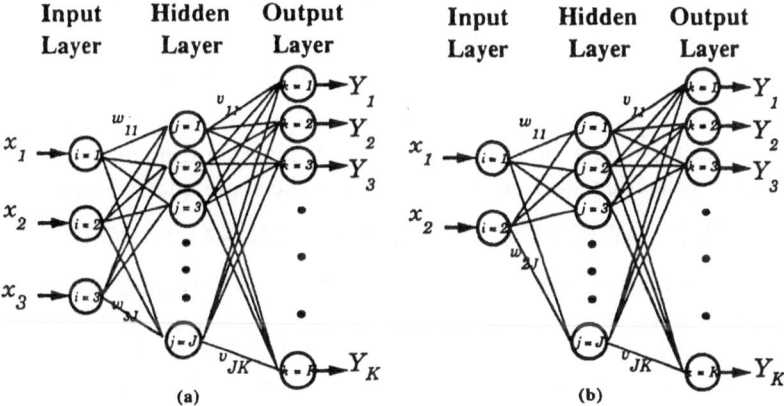

Figure 1: General schematics of the sequential neural kriging network used to incorporate mixed data. The network in (a) uses mixed data; the network of (b) uses hard data only.

Figures 2 (a) field seasons 1 through 4 show the results of constructing hydraulic conductivity fields using hard information from the corresponding four field seasons.

The conductivity fields shown in Figures 2 (b) field seasons 1 through 4 were created using a combination of both hard and soft data. The procedure for creating these realizations consists of the sequential training process described earlier. For example, to produce the conductivity field of Figure 2 (b) field season 1, the network was first trained using the hard and soft data of the original seven wells of Figure 2 (a) field season 1. The training patterns consist of input vectors $\mathbf{x} = (x_1, x_2, x_3)$. The first two components contain the x and y coordinates of each well site. the third component is a real number between 0.0 and 1.0 corresponding to soil descriptions indicated in the Table 1(b). The output vectors correspond to the binary vectors of Table 1(a) encoding the hydraulic conductivity measured at the corresponding well location.

After training, the network is used to predict the hydraulic conductivity at the locations which correspond to the ten wells (of the original 17) that contain only soft information (indicated by the circles of Figure 2 (b) field season 1). Once the conductivity values have been predicted for these ten locations, the second of the two ANNs used for mixed data assimilation is retrained using all 17 wells as training patterns to produce the map shown in Figures 2 (b) field season 1. Figures 2 (b) field season 2 through 4 were produced using the same sequential training process. Compare the map produced using hard information from all the wells of field season four (Figure 2 (a)) to that produced using mixed data from the wells of the second field season (see Figure 2 (b) field season two). The two maps are almost identical.

These preliminary results suggest that a method of pattern completion that incorporates mixed data types may be used within an optimization framework to reduce the number of data collection sites needed to produce the same quality map.

APPLICATION 2: LAWRENCE LIVERMORE NATIONAL LABORATORY

An optimal groundwater management model has been developed to assist in the design of groundwater remediation systems at LLNL. An important element of the management model consists of a numerical model which is used to simulate water movement in the Livermore aquifer and the spread of contaminants in response to various pumping scenarios. Figure 3 illustrates the mesh used in the numerical flow and transport model. A superimposed map of the site features outlines the LLNL Main Site property. The heterogeneity of the hydraulic parameters in aquifers predominantly controls the flow field and hence the spread of the contaminant plume. Hard data (hydraulic conductivity measurements obtained from pump tests) are known at 195 well locations throughout LLNL's Main Site. Soft data (soil sample descriptions at approximately 2 ft vertical intervals from driller well logs) are available at 170 of these well locations. The hydraulic conductivity values have been quantized into one of the eight discrete conductivity classes pre–selected by LLNL personnel.

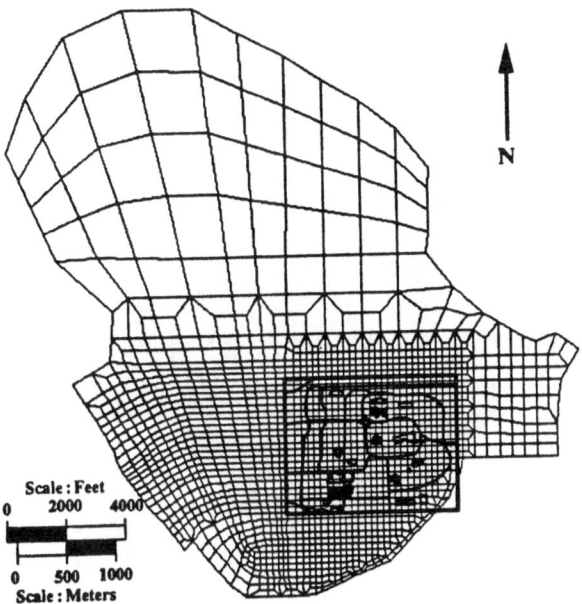

Figure 3: Map view of LLNL Main Site with site boundary and key surface features superimposed on the finite element mesh.

Figure 4 presents a map view of hydraulic conductivities estimated by the NK network using hard information only. The map extends well beyond the LLNL Main Site boundary in order to produce an estimated value of hydraulic conductivity at every nodal subdomain in the mesh of Figure 3. The same 170 well locations that contain mixed data were used as training vectors. The input training vectors $\mathbf{x}^m = (x_1, x_2)$, $m = 1, 2, \cdots 170$ correspond to the two–dimensional spatial coordinates of these wells locations. The output training vectors represent the hydraulic conductivity measured at each well site and quantized into one of the eight discrete conductivity classes. The hidden layer weights were initialized to the location of the measurement points plus a small random vector. Approximately 223 iterations were necessary to meet an allowable RMS error of 10^{-6}.

A plan view of hydraulic conductivity constructed by the NK network using mixed data is shown in Figure 5. The depth shown is at 130 feet, which was obtained by computing the geometric mean of the midpoints of all well screens. Network inputs $\mathbf{x}^m = (x_1, x_2, x_3)$, $m = 1, 2, \cdots, 170$ consist of two normalized spatial components and a third component ranging between 0 and 1 that encodes one of 16 soil types.

SUMMARY AND CONCLUSIONS

A new method of characterizing porous media which incorporates mixed data types has been developed using artificial neural networks. The method is data–driven and requires no estimate of a covariance function. The details of this method and how it

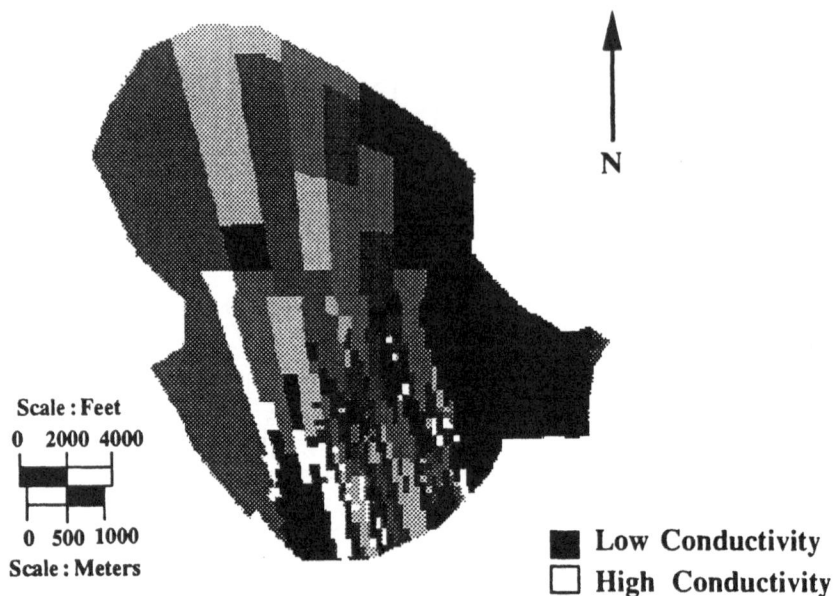

Figure 4: Hydraulic conductivity map (in plan view) produced by the NK network using hard data.

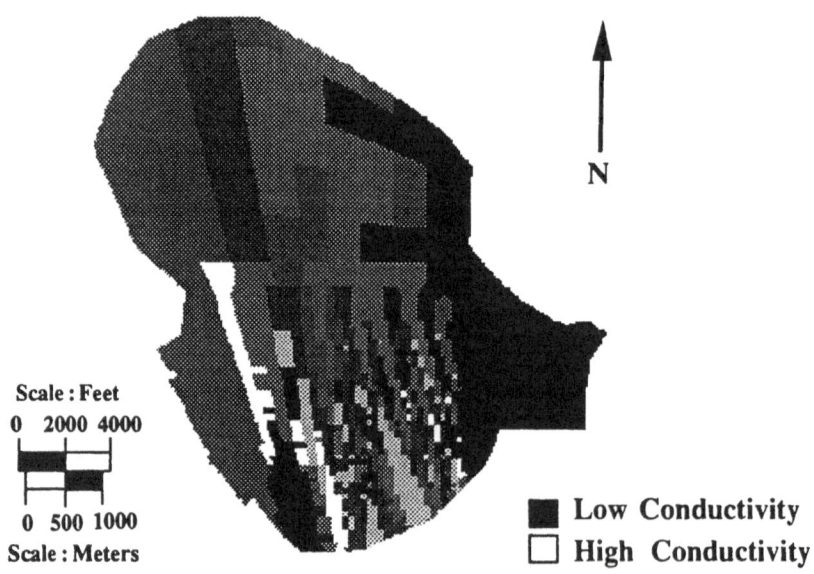

Figure 5: Hydraulic conductivity map (in plan view) produced by the NK network using mixed data.

can be extended to estimate parameters other than hydraulic conductivity and soil lithology are presented in *Rizzo* [1994].

Two examples of two–dimensional hydraulic conductivity estimates illustrate the advantages of using an approach which incorporates mixed data. These preliminary results suggest that using "soft" data in combination with "hard" data increases the degree of accuracy with which new spatial locations may be classified to discrete conductivity values. Moreover, the approach may be used to reduce the number of "hard" data observations needed to produce a map of given accuracy compared to using "hard" data only.

ACKNOWLEDGEMENTS

We are pleased to thank Mary Hill, Leah Rogers, Virginia Johnson, and Nai-Hsien Mao for their help. This work was partly supported by NSF Grant MSS–9214698 and by Subcontract B235648 with Lawrence Livermore National Laboratory.

REFERENCES

Hecht-Nielsen, R. (1987) "Counterpropagation networks", *IEEE International Conference on Neural Networks*, Vol II, pp. 19-32, July 24-27, San Diego, CA.

Hecht-Nielsen, R. (1988) "Applications of counterpropagation networks", *Neural Networks*, 1:131-139.

Hill, M. C., R. L. Cooley, and D. W. Pollock (1994) "A Controlled Experiment in Model Calibration Using Nonlinear Regression", *Water Resources Research*, in review.

Journel, A. G. (1984) *Geostatistics for Natural Resources Characterization, Part 1*, edited by Verly, M. David, A. G. Journel and A. Marechal, D. Reidel, Boston, 1984.

Journel, A. G. and C. J. Huijbregts (1978) *Mining Geostatistics* Academic Press.

Rizzo, D. M. (1994) "Optimal Groundwater Remediation Designs and Characterization of Aquifer Properties using Artificial Neural Networks", Ph.D. Dissertation, Dept. of Civil and Environmental Engineering, University of Vermont, Burlington.

Rizzo, D. M. and D. E. Dougherty (1994) "Characterization of aquifer properties using artificial neural networks, neural kriging", *Water Resources Research*, in press.

Sullivan, J. (1984) "Conditional recovery estimation through probability kriging–Theory and practice", *Geostatistics for Natural Resources Characterization, Part 1*, edited by Verly, M. David, A. G. Journel and A. Marechal, D. Reidel, Boston, 365-384.

Parameter Identification for Unsteady Heat Conduction Problem

Yasuhiko SANO, Akira ANJU and Mututo KAWAHARA
Department of Civil Engineering
Chuo University
1-13-27 Kasuga Bunkyou-Ku Tokyo 112
Japan

Abstract

In Japan, the outflow of pesticide scattered over the green in the golf course causes an environmental pollution problem. Aiming at the non-pesticide lawn management, we are trying to create the temperature control system for the ground. To simulate the heat conduction phenomenon, the parameter identification is one of the important problem. If the parameter is not accurately specified, a serious error between computed and observed results will be occurred. A method of the parameter identification is presented and the numerical examples are described using the method in this paper.

1 INTRODUCTION

For the purpose of creating the thermal management system of the ground, temperature near the ground surface should be kept constant. The method would be employed that the pipe is buried in the earth ground and that the temperature of water through the pipe is regulated.

The proper thermal conductivity should be selected to simulate the exact heat conduction phenomenon. Thus, the parameter identification is needed. The estimation of thermal conductivity of the ground is treated as an inverse analysis of parameter identification. To do this, the method based on the minimization principle is used. The cost function is expressed by the sum of square residual between the observed and the computed temperature. The thermal conductivity that minimizes the cost function can be searched by the Fretcher-Reeves method. The observed temperature data is obtained by the thermometer buried in the lawn field ground in the Chiba Prefectural Agricultural Laboratory.

2 BASIC EQUATION AND FINITE ELEMENT METHOD

Two dimensional unsteady diffusion equation is used as the governing equation as follows:

$$\rho C_p \frac{\partial T}{\partial t} - \beta \left(\frac{\partial^2 T}{\partial x^2} + \frac{\partial^2 T}{\partial y^2} \right) = Q \qquad in \ V \qquad (1)$$

and boundary conditions are;

$$T = \hat{T} \qquad on \ S_1, \qquad (2)$$

801

A. Peters et al. (eds.), Computational Methods in Water Resources X, 801–806.
© 1994 Kluwer Academic Publishers. Printed in the Netherlands.

$$\beta \frac{\partial T}{\partial n}(x, y) = \hat{q} \qquad\qquad on\ S_2. \qquad (3)$$

where ρ, C_p, β, Q are density, specific heat, thermal conductivity and inside heat radiation quantity respectively. Analytical domain is denoted by V, which is surrounded by the boundaries S_1 and S_2. Temperature and flux are prescribed on the boundary S_1 and S_2 respectively. The finite element equation is expressed in the following form discretizing equation (1) in space.

$$\overline{[M_{\alpha\beta}(\gamma)]}\{\dot{T}_\beta\} + [S_{\alpha\beta}(\beta)]\{T_\beta\} = \{\hat{\Omega}\} \qquad (4)$$

The γ means ρC_p (density times specific heat). The equation (4) is discretized in time applying the explicit Euler method. The resulted equation is as follows;

$$\overline{[M_{\alpha\beta}(\gamma)]}\frac{\{T_\beta\}^{n+1} - \{T_\beta\}^n}{\Delta t} + [S_{\alpha\beta}(\beta)]\{T_\beta\} = \{\hat{\Omega}\} \qquad (5)$$

where $\overline{[M_{\alpha\beta}(\gamma)]}$, $[S_{\alpha\beta}(\beta)]$, $\{\hat{\Omega}\}$ indicates lumped mass matrix, diffusion coefficient matrix and flux vector on natural boundary respectively. The time increment is denoted by Δt. The unsteady calculation can be performed using equation (5). This procedure is called as the forward analysis.

3 CONJUGATE GRADIENT METHOD

The cost function expressed by the sum of square residual between calculated and observed temperatures as follows:

$$J = \frac{1}{2}\int_{t_0}^{t_f}\{T(\beta) - T^*\}^T\{T(\beta) - T^*\}dt \qquad (6)$$

where $T(\beta)$, T^* mean calculated and observed temperature respectively and t_0, t_f denote starting and final times respectively. The parameter β that minimizes J can be searched by the iterative calculation. There are many minimization methods, the Fretcher-Reeves method that is a sort of the conjugate gradient method is used in this analysis. The gradient is computed as follows;

$$\{d\} = -\left\{\frac{\partial J}{\partial \beta}\right\} = -\int_{t_0}^{t_f}\left[\frac{\partial T}{\partial \beta}\right]^T\{T(\beta) - T^*\}dt \qquad (7)$$

Where $\left[\frac{\partial T}{\partial \beta}\right]$ is sensitivity matrix which is obtained by partially differentiating equation (5) with respect to parameter β, i.e.,

$$\overline{[M_{\alpha\beta}(\gamma)]}\frac{1}{\Delta t}\left(\left[\frac{\partial T}{\partial \beta}\right]^{n+1} - \left[\frac{\partial T}{\partial \beta}\right]^n\right)$$

$$= -\left(\frac{\partial}{\partial \beta}[S_{\alpha\beta}(\beta)]\{T_\beta\}^n + [S_{\alpha\beta}(\beta)]\left[\frac{\partial T}{\partial \beta}\right]^n - \left[\frac{\partial \hat{\Omega}}{\partial \beta}\right]\right) \qquad (8)$$

The equation (8) is calculated applying the explicit Euler method. Initial condition is $\left[\frac{\partial T}{\partial \beta}\right]_{t=t_0} = 0$. As the boundary condition is:

$$\left[\frac{\partial T}{\partial \beta}\right] = 0 \qquad\qquad on \; S_1. \tag{9}$$

Consider the scalar of $J(\{\beta\} + \alpha\{d\})$.

$$J(\{\beta\} + \alpha\{d\}) = J + \alpha\left\{\frac{\partial J}{\partial \beta}\right\}^T \{d\}$$

$$= \frac{1}{2}\int_{t_0}^{t_f}\left(\{T(\beta)\} + \alpha\left[\frac{\partial T}{\partial \beta}\right]\{d\} - \{T^*\}\right)^T\left(\{T(\beta)\} + \alpha\left[\frac{\partial T}{\partial \beta}\right]\{d\} - \{T^*\}\right) dt \tag{10}$$

The equation (10) is the Taylor expansion of the cost function. The α that minimizes $J(\{\beta\} + \alpha\{d\})$ is the scalar which gives the minimum value of the cost function at this stage. This α is obtained by partially differentiating $J(\{\beta\} + \alpha\{d\})$ with respect to α.

$$\alpha = -\frac{\{d\}^T\{\frac{\partial J}{\partial \beta}\}}{\{d\}^T \int_{t_0}^{t_f}\left[\frac{\partial T}{\partial \beta}\right]\left[\frac{\partial T}{\partial \beta}\right]^T dt\{d\}} \tag{11}$$

The parameter is renewed using $\{d\}$ and α, which are obtained by equations (7) and (11) respectively. The new parameter $\{\beta\}^{(i+1)}$ is expressed as:

$$\{\beta\}^{(i+1)} = \{\beta\}^{(i)} + \alpha\{d\}^{(i)} \tag{12}$$

The gradient is calculated using the Fretcher-Reeves method. The gradient is renewed as follows;

$$\varphi = \frac{\left(\{\frac{\partial J}{\partial \beta}\}^{(i+1)}, \{\frac{\partial J}{\partial \beta}\}^{(i+1)}\right)}{\left(\{\frac{\partial J}{\partial \beta}\}^{(i)}, \{\frac{\partial J}{\partial \beta}\}^{(i)}\right)} \tag{13}$$

$$\{d\}^{(i+1)} = -\left\{\frac{\partial J}{\partial \beta}\right\}^{(i+1)} + \varphi\{d\}^{(i)} \tag{14}$$

The gradient computed by equation (14) is used for the next cycle gradient. The calculation algorithm is summarized as follows;

1. Assume initial value $\{\beta\}^0$, and calculate $\{T\}$, $\left[\frac{\partial T}{\partial \beta}\right]$ by equation (5) and (8).
2. Compute $\{d\}^{(i)} = -\{\frac{\partial J}{\partial \beta}\}^{(i)}$ by equation (7).
3. Determine α that minimizes $J(\{\beta\}^{(i)} + \alpha\{d\}^{(i)})$ by equation (11).
4. Compute $\{\beta\}^{(i+1)} = \{\beta\}^{(i)} + \alpha\{d\}^{(i)}$ by equation (12).
5. Calculate $\{T\}$, $\left[\frac{\partial T}{\partial \beta}\right]$ by equation (5) and (8).
6. Compute $\{d\}^{(i+1)}$ by equation (7).   IF$|\{d\}^{(i+1)}| < \epsilon$ Then stop.
7. Compute $\varphi = \frac{(\{\frac{\partial J}{\partial \beta}\}^{(i+1)}, \{\frac{\partial J}{\partial \beta}\}^{(i+1)})}{(\{\frac{\partial J}{\partial \beta}\}^{(i)}, \{\frac{\partial J}{\partial \beta}\}^{(i)})}$ by equation (13).
8. Compute $\{d\}^{(i+1)} = -\{\frac{\partial J}{\partial \beta}\}^{(i+1)} + \varphi\{d\}^{(i)}$ by equation (14).
9. Set $i = i + 1$ and   Go To 3.

4 NUMERICAL EXAMPLE

To observe the thermal conductivity in the ground, thermometer was buried at the lawn field in the Chiba Prefectural Agricultural Laboratory. The cross-sectional view of lawn field is shown in Fig.1. Thermometer (A) is set at the depth of 3-cm from the ground surface. Thermometer (B) is stuck to the surface of pipe. The site is composed of three layers. Those are fine sand, coarse sand and gravel as shown in Fig.1. Thermometers are set more than two in every layer. Temperature was measured every thirty minute. The finite element mesh is shown in Fig.2. It is assumed that the flux on the boundary A-B, B-C and C-D are zero(S_2 boundary). On A-D boundary and pipe nodes, measured temperature is specified as the Dirichlet boundary condition(S_1 boundary). Temperature data from September.17 to October.7 and from Decsember.7 to Decsember.24, in 1993 are used for the numerical analysis. It is possible to control the temperature of the water through the pipe. The data obtained by the thermometers (A) and (B) are shown in Fig.3. These data are used as boundary conditions for the finite element analysis. It is assumed that ρC_p(density × specific heat) is 2.0×10^6 $[Kg/K \cdot m \cdot s^2]$ constant, and identification is performed for the parameter β only. The computed cost function is shown in Fig.4, and thermal conductivity of every layer is shown in Fig.5. Using obtained thermal conductivity, forward analysis is performed. The comparisons between the calculated and observed temperatures are shown in Fig.6.

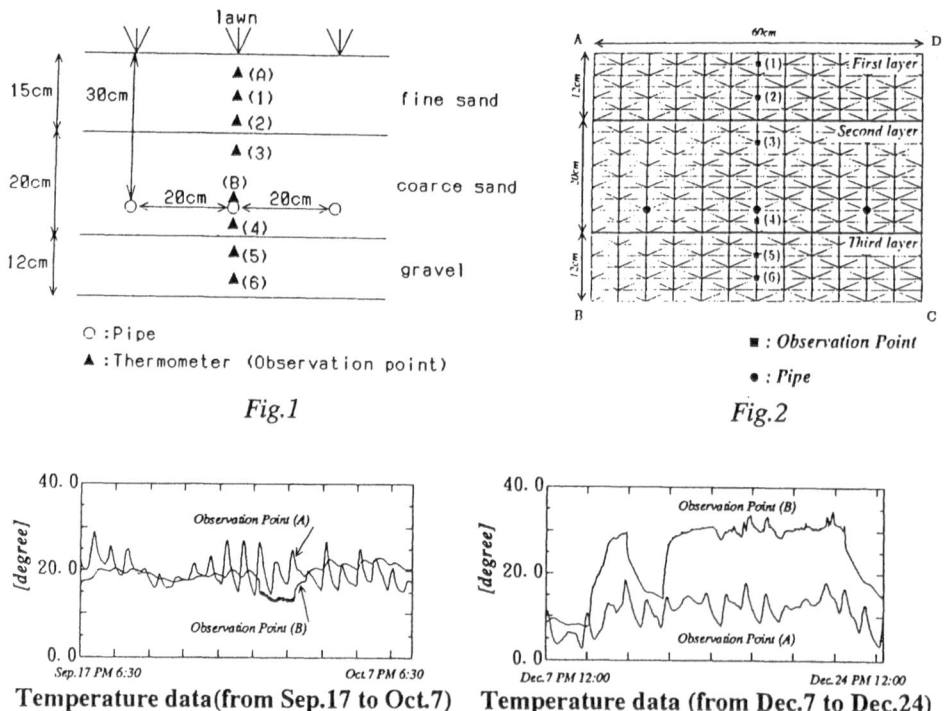

Fig.1

Fig.2

Temperature data(from Sep.17 to Oct.7)   Temperature data (from Dec.7 to Dec.24)

Fig.3

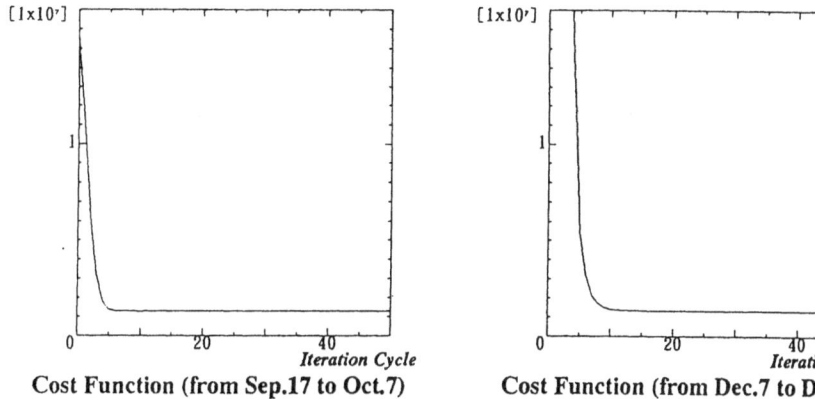

Cost Function (from Sep.17 to Oct.7) Cost Function (from Dec.7 to Dec.24)

Fig.4

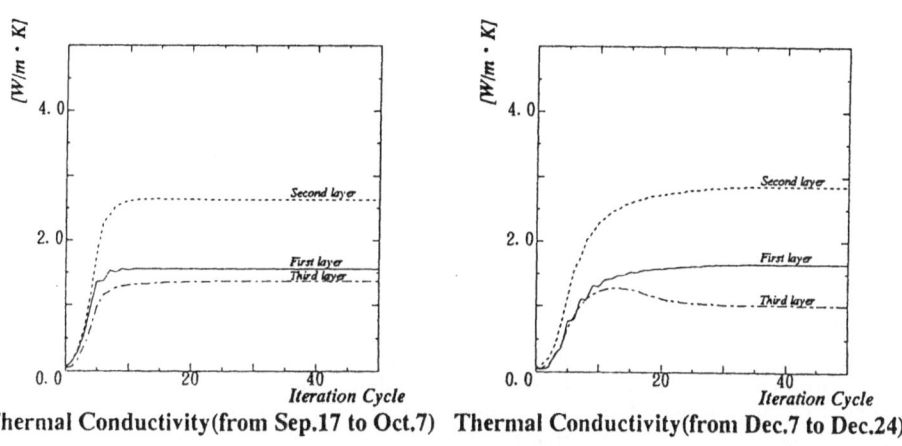

Thermal Conductivity(from Sep.17 to Oct.7) Thermal Conductivity(from Dec.7 to Dec.24)

Fig.5

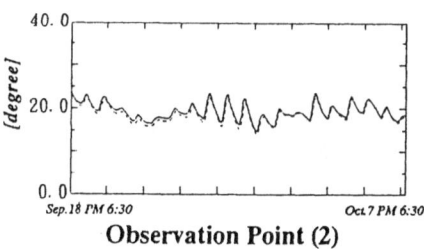

Observation Point (2)

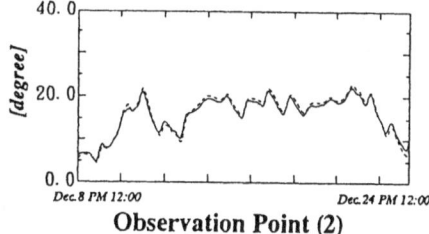

Observation Point (2)

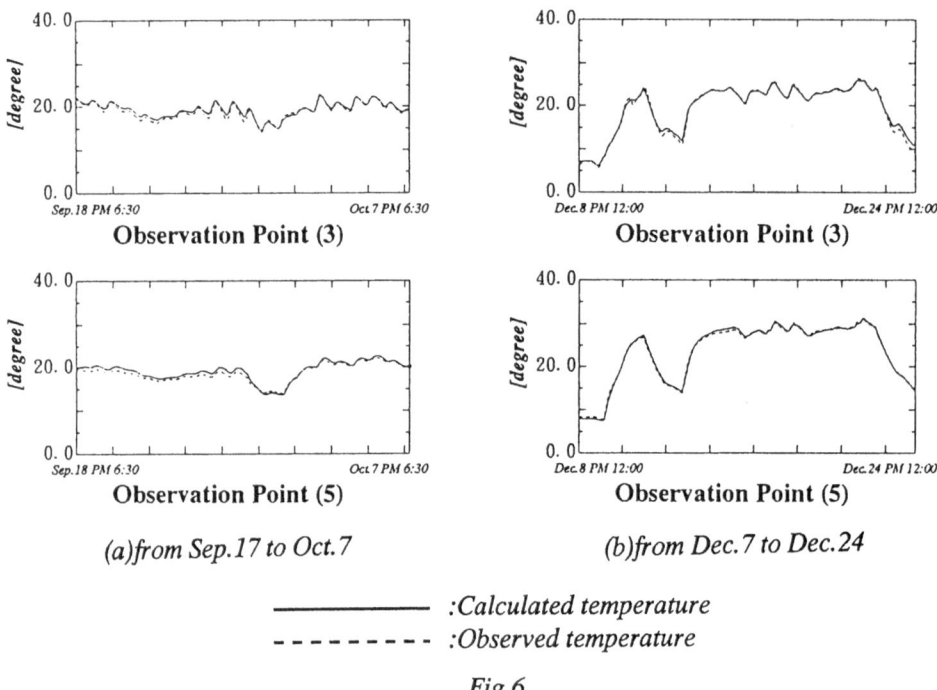

(a)from Sep.17 to Oct.7   *(b)from Dec.7 to Dec.24*

———————— :*Calculated temperature*
– – – – – – – – :*Observed temperature*

Fig.6

5 CONCLUSION

Using the data for relatively long term observation, the stable calculation is carried out. The numerical results show that the obtained parameter based on the data in fall is almost the same as that in winter. The results of the forward analysis using the obtained parameter are well in agreement with the observed value. The thermal conductivity seems independent of the seasonal change of temperature. If the calculation is performed using this thermal conductivity, it is possible to predict exact heat conduction phenomenon in soil. The results can be used for the creation of the thermal management system. However, the computation is sometimes overflowed. There are some computed results which are not independent of the observed data and of the position of the observation point. This investigation will be the future subject.

References

[1] K.Hatanaka, "A Basic Sutudy on Identification of Aquifer Parameters in Groundwater Hydorogy" *Finite Element in Fluid*, part.2 pp.901-908, 1993

Parameter Identification and Optimal Control of Underground Temperature

Seiichi SUZUKI and Akira ANJU and Mutsuto KAWAHARA
Department of Civil Engineering
Chuo University
1-13-27 Kasuga Bunkyou-ku Tokyo 112
Japan

1 INTRODUCTION

Pesticide pollution in the golf links becomes serious hazard in the environment. Its intensification in the field slowly contaminates the groundwater. Thus, the pesticide pollution control in the golf links is important for the environmental improvement. Generally, a lawn dies from too hot or cold. An agricultural specialist who is studying at an agricultural experimental site in Chiba-Prefecture in Japan illustrated that it is better for a lawn if the underground temperature is lowered in summer. If the underground temperature is optimally controlled by the underground flow through pipes, a lawn at a green in the golf links would be maintained without pesticide. But before this study, the underground thermal conductivity should be decided with the observed temperature to perform the exact numerical simulations.

Namely, this paper presents the parameter identification for the underground thermal conductivity and the optimal control for the underground temperature. The methods of the parameter identification and the optimal control are the Direct-Iteration method[1] and the Sakawa-Shindo method[2]. The finite element method in space and the Crank-Nicolson method in time are used for the discretization of the basic equations. For this study, an agricultural experimental ground site at Chiba-Prefecture in Japan is chosen as a numerical example.

2 Parameter Identification by Direct-Iteration Method

2.1 COST FUNCTION

The parameter identification problem is formulated to determine $\{\beta\}$ which is thermal conductivity so as to minimize the cost function:

$$J = \frac{1}{2} \sum_{t=t_0}^{t_f} (\{\theta(\beta)\} - \{\theta^*\})^T (\{\theta(\beta)\} - \{\theta^*\}) \tag{1}$$

where t_0 and t_f mean starting and final times for the parameter identification, $\{\theta\}$ and $\{\theta^*\}$ mean the computed and observed temperature. The cost function consists of the

807

A. Peters et al. (eds.), Computational Methods in Water Resources X, 807–814.
© 1994 Kluwer Academic Publishers. Printed in the Netherlands.

error between the computed and observed temperature. Namely, when the cost function is minimized, the optimal parameters can be obtained.

2.2 BASIC EQUATIONS

To treat the unsteady heat conduction problems, the basic equation can be expressed as follows:

$$\rho C_p \frac{\partial \theta}{\partial t} - \beta \left(\frac{\partial^2 \theta}{\partial x^2} + \frac{\partial^2 \theta}{\partial y^2} \right) = 0 \qquad in \ \ \Omega \qquad (2)$$

where θ is temperature, t is time, x and y mean space, ρ is density, C_p is specific heat and β is thermal conductivity.

The initial and the boundary conditions can be expressed as follows:

$$\theta_0 = \hat{\theta}_0 \qquad in \ \ \Omega \qquad (3)$$

$$\theta = \hat{\theta} \qquad on \ \ \Gamma_1 \qquad (4)$$

$$q = \beta \left(\frac{\partial \theta}{\partial x} l + \frac{\partial \theta}{\partial y} m \right) = \hat{q} \qquad on \ \ \Gamma_2 \qquad (5)$$

where θ_0 is initial condition, symbol $\hat{}$ means known values, l and m denote the direction cosine of the unit outward normal of the boundary. The domain of the computation is denoted by Ω, which is surrounded by the boundaries Γ_1 and Γ_2 where Γ_1 and Γ_2 mean the Dirichlet and the Neumann boundaries respectively. The basic equation (2) is discretized by the finite element method in space and the Crank-Nicolson method in time. The finite element equation can be obtained as follows:

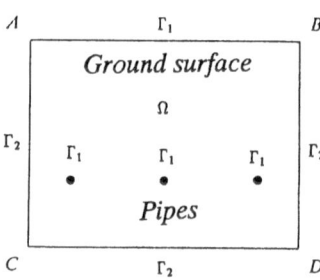

Fig.1 Computational domain

$$\left([M] + \frac{\Delta t}{2} [S(\beta)] \right) \{\theta^{n+1}\} = \left([M] - \frac{\Delta t}{2} [S(\beta)] \right) \{\theta^n\} + \Delta t \, \{\hat{F}^{n+1}\} \qquad (6)$$

where Δt is time increment, $\{\theta\}$ is state vector. The parameter n denotes time-step. The matrices $[M]$, $[S(\beta)]$ and the vector $\{\hat{F}\}$ are expressed as follows:

$$[M] = \int_\Omega \rho C_p \{\Phi\} \{\Phi\}^T d\Omega \qquad (7)$$

$$[S(\beta)] = \int_\Omega \beta \left(\{\Phi\}_{,x} \{\Phi\}_{,x}^T + \{\Phi\}_{,y} \{\Phi\}_{,y}^T \right) d\Omega \qquad (8)$$

$$\{\hat{F}\} = \int_{\Gamma_2} \{\Phi\} \hat{q} \, d\Gamma \qquad (9)$$

where $\{\Phi\}$ denotes the interpolation function. In this model,

$$\{\hat{F}\} = \{0\}. \qquad (10)$$

The reason is stated in the numerical example(1).

The finite element equation can be rewritten as:

$$\left([M] + \frac{\Delta t}{2} [S(\beta)] \right) \{\theta^{n+1}\} = \left([M] - \frac{\Delta t}{2} [S(\beta)] \right) \{\theta^n\}. \tag{11}$$

2.3   Direct-Iteration Method for minimizing Cost Function

There are a lot of methods for the non-linear minimization problems. In the Newton-Raphson method, Hessian matrix $\left[\frac{\partial^2 \theta}{\partial \beta^2} \right]$ has to be calculated. That is the main disadvantage of the method. But the Hessian matrix need not be computed in the Direct-Iteration method.

The optimal parameter $\{\beta^{(i+1)}\}$ is obtained with the iterative procedure as follows:

$$\{\beta^{(i+1)}\} = \{\beta^{(i)}\} + \{\Delta\beta^{(i)}\}. \tag{12}$$

The parameter increment $\{\Delta\beta^{(i)}\}$ is computed by the Direct-Iteration method. The state vector $\{\theta\}$ depends on the parameter $\{\beta\}$. The state vector can be expressed by the Taylor's theorem as follows:

$$\left\{ \theta(\beta^{(i+1)}) \right\} = \left\{ \theta(\beta^{(i)}) \right\} + \left[\frac{\partial \theta(\beta^{(i)})}{\partial \beta} \right] \{\Delta\beta^{(i)}\} + \{H.O.T.\}. \tag{13}$$

The following equation can be derived from the cost function(equation(1)) and the Taylor's theorem(equation(13)) as follows:

$$J = \frac{1}{2} \sum_{t=t_0}^{t_f} \left(\{\theta\} + \left[\frac{\partial \theta}{\partial \beta} \right] \{\Delta\beta\} - \{\theta^*\} \right)^T \left(\{\theta\} + \left[\frac{\partial \theta}{\partial \beta} \right] \{\Delta\beta\} - \{\theta^*\} \right). \tag{14}$$

The cost function J(equation(14)) should be minimized, i.e.,

$$\left\{ \frac{\partial J}{\partial \beta} \right\} = \sum_{t=t_0}^{t_f} \left[\frac{\partial \theta}{\partial \beta} \right]^T \left(\{\theta\} + \left[\frac{\partial \theta}{\partial \beta} \right] \{\Delta\beta\} - \{\theta^*\} \right) = \{0\}. \tag{15}$$

The parameter increment $\{\Delta\beta\}$ can be computed by equation(15) as follows:

$$\{\Delta\beta\} = -\left(\sum_{t=t_0}^{t_f} \left[\frac{\partial \theta}{\partial \beta} \right]^T \left[\frac{\partial \theta}{\partial \beta} \right] \right)^{-1} \left(\sum_{t=t_0}^{t_f} \left[\frac{\partial \theta}{\partial \beta} \right]^T \left(\{\theta\} - \{\theta^*\} \right) \right) \tag{16}$$

where $\left[\frac{\partial \theta}{\partial \beta} \right]$ is referred to the sensitive matrix.

2.4   SENSITIVE MATRIX

The sensitive matrix $\left[\frac{\partial \theta}{\partial \beta} \right]$ is utilized to compute the parameter increment $\{\Delta\beta\}$ (equation(16)). The sensitive matrix is obtained by differentiating the finite element equation(equation(11)) with respect to the parameter $\{\beta\}$, i.e.,

$$\left[\frac{\partial \theta}{\partial \beta} \right]^{n+1} = \left([M] + \frac{\Delta t}{2} [S(\beta)] \right)^{-1}$$
$$\left(-\frac{\Delta t}{2} \frac{\partial [S(\beta)]}{\partial \beta} \left\{ \{\theta^n\} + \{\theta^{n+1}\} \right\} + \left([M] - \frac{\Delta t}{2} [S(\beta)] \right) \left[\frac{\partial \theta}{\partial \beta} \right]^n \right). \tag{17}$$

To calculate the state vector $\{\theta\}$, the initial and the Dirichlet boundary conditions are assumed independent on the parameter $\{\beta\}$. Namely, the initial and the Dirichlet boundary conditions of the sensitive matrix can be written as:

$$\left[\frac{\partial\theta}{\partial\beta}\right]^{(0)} = 0 \qquad\qquad in \ \ \Omega \qquad\qquad (18)$$

$$\left[\frac{\partial\theta}{\partial\beta}\right] = \left[\frac{\hat{\partial\theta}}{\partial\beta}\right] = 0 \qquad\qquad on \ \ \Gamma_1 \qquad\qquad (19)$$

where $\left[\frac{\partial\theta}{\partial\beta}\right]^{(0)}$ denotes the initial condition, symbol $\hat{}$ means known values.

2.5 ALGORITHM (1)

1. Assume the initial parameters $\{\beta^{(0)}\}$.
2. Solve the state values $\{\theta^{(i)}(\beta^{(i)})\}$ by equations(3),(4),(5),(11).
3. Solve the sensitive matrix $\left[\frac{\partial\theta}{\partial\beta}(\theta^{(i)})\right]^{(i)}$ by equations(17),(18),(19).
4. Solve the cost function $J^{(i)}(\theta^{(i)})$ by equation(1).
5. Solve the parameter increment $\{\Delta\beta^{(i)}(\theta^{(i)}, \left[\frac{\partial\theta}{\partial\beta}\right]^{(i)})\}$ by equation(16).
6. Solve the next step parameters $\{\beta^{(i+1)}\} = \{\beta^{(i)}\} + \{\Delta\beta^{(i)}\}$ (equation(12)).
7. Judge the convergence: IF $(\frac{\partial J}{\partial\beta} < \varepsilon)$ THEN STOP ELSE $i = i + 1$ and GO TO 2.

where superscript (i) denotes the iteration cycle for the minimization, the parameter ε is a small number.

2.6 NUMERICAL EXAMPLE (1)

In this computation, three layers are assumed. The first, second and third layers are fine sand, coarse sand and gravel respectively. The temperature flux $\hat{q}$ (equation(5)) on CD (Fig.2) is assumed zero, because the observed temperature has no gradients concerned with the direction of y (Fig.4). The temperature flux $\hat{q}$ on AC and BD (Fig.2) are assumed zero because of symmetrical condition(Fig.1). Namely, the domain of the computation is shown in Fig.2. Equation(10) $\left(\{\hat{F}\} = \{0\}\right)$ is followed by $\hat{q} = 0$. The observed temperatures on the ground surface and the pipes are used as the Dirichlet boundary conditions $\hat{\theta}$ (equation(4),Fig.3) on AB and the pipes(Fig.2). When the cost function is minimized (Fig.5), the optimal parameter $\{\beta\}$ which is thermal conductivity is computed (Fig.6). The convergent condition is $\frac{\partial J}{\partial\beta} < \varepsilon = 10^{-4}$.

Table 1 Average error between the computed and observed temparature

Observational points	(1)	(2)	(3)	(4)	(5)	(6)	Total
Average error (degree)	0.029	0.261	0.179	0.005	0.028	0.025	0.044

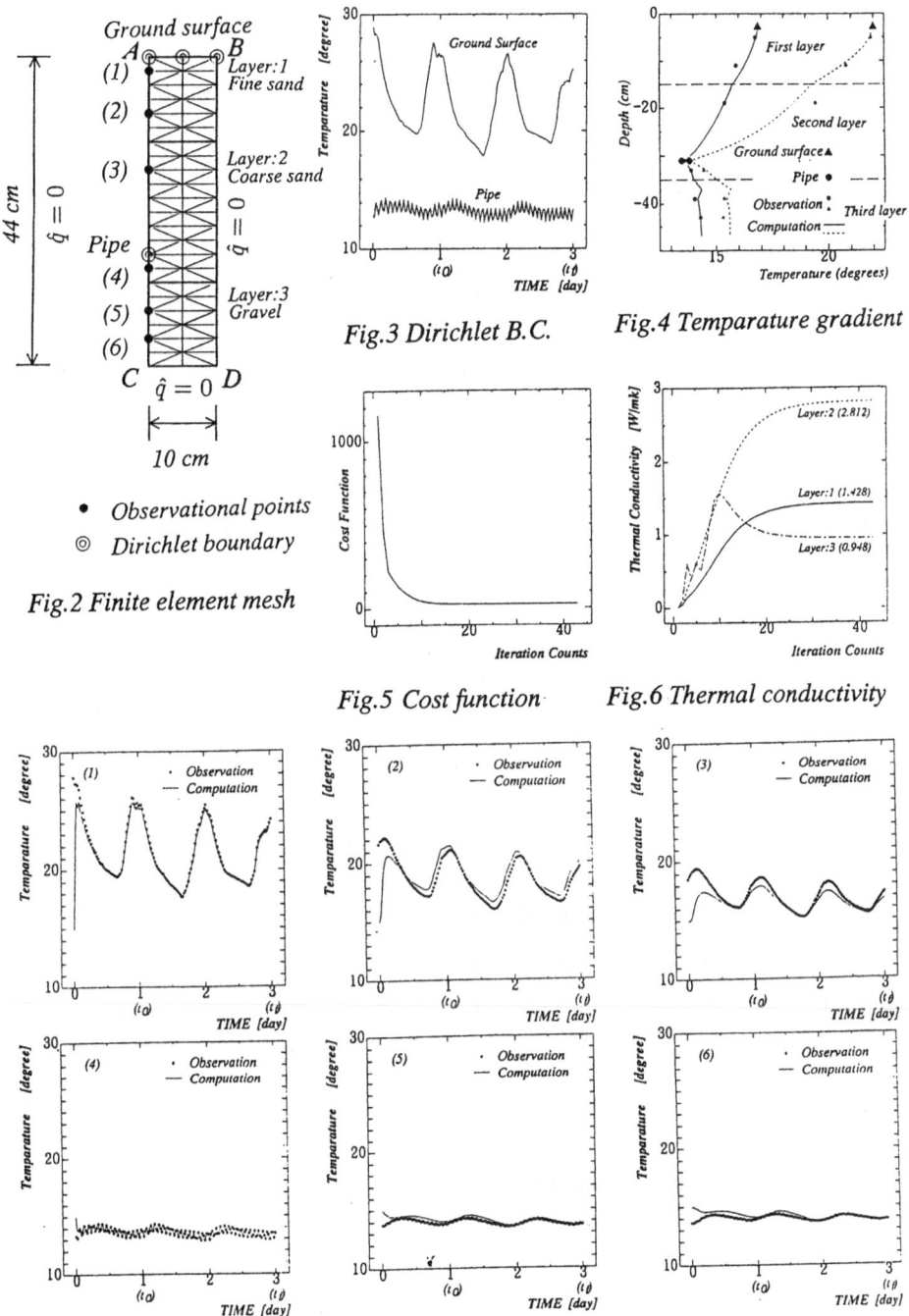

Fig.2 Finite element mesh

Fig.3 Dirichlet B.C.

Fig.4 Temparature gradient

Fig.5 Cost function

Fig.6 Thermal conductivity

Fig.7 Observed and Computed tempertature at observed points

3   Optimal Control by Sakawa-Shindo method

3.1   BASIC EQUATION

The basic equation(2) is discretized by the finite element method in consideration of equation(10) as follows:

$$\{\dot{\theta}\} = [M]^{-1}[S(\beta)]\,\{\theta\} \tag{20}$$

where thermal conductivity $\{\beta\}$ is obtained in the former section. Equation(20) is rewritten as follows:

$$\{\dot{\theta}\} = [A]\,\{\theta\} + [B]\,\{u\} + [C]\,\{f\} \tag{21}$$

where $\{\theta\}$, $\{u\}$ and $\{f\}$ are state vector, control vector and force vector. The matrices $[A]$, $[B]$ and $[C]$ are derived from $[M]^{-1}[S(\beta)]$ in equation(20).

3.2   COST FUNCTION

The optimal control problem is formulated to determine $\{u\}$ which is the control vector so as to minimize the cost function.

$$J = \frac{1}{2}\sum_{t=t_0}^{t_f}\Big((\{\theta\} - \{\theta^*\})^T[Q](\{\theta\} - \{\theta^*\}) + (\{u\} - \{u^*\})^T[R](\{u\} - \{u^*\})\Big) \tag{22}$$

where t_0 and t_f mean starting and final times for the optimal control. The vector $\{\theta\}$ and $\{\theta^*\}$ mean the computed and objective temperature at the reference points. The vector $\{u\}$ and $\{u^*\}$ mean the control temperature at the pipes and the ideal one for the cooling and heating machine. The vector $\{\theta^*\}$ and $\{u^*\}$ are known values. The matrices $[Q]$ and $[R]$ are weighting functions, which are assumed diagonal matrices with a constant coefficent q and r.

3.3   Sakawa-Shindo Method for Minimization Problems with Constraint

To apply the optimal control theory, it is necessary to introduce the Hamiltonian function as:

$$\begin{aligned} H = \quad & \frac{1}{2}\Big((\{\theta\} - \{\theta^*\})^T[Q](\{\theta\} - \{\theta^*\}) + (\{u\} - \{u^*\})^T[R](\{u\} - \{u^*\})\Big) \\ & + \{p\}^T\left([A]\,\{\theta\} + [B]\,\{u\} + [C]\,\{f\}\right) \end{aligned} \tag{23}$$

where $\{p\}$ denotes the Lagrange multiplier. The Euler-Lagrange equation and transversality condition can be described as follows:

$$\{\dot{p}\} = -\frac{\partial H}{\partial \theta} = -[Q](\{\theta\} - \{\theta^*\}) - [A]^T\{p\} \tag{24}$$

$$\{p(t_f)\} = \{0\}. \tag{25}$$

The Crank Nicolson method for the discretization in time is applied to equation(24) in the same method as the basic equation's.

$$\{p^{n-1}\} = \left([A]^T - \frac{2}{\Delta t}[I]\right)^{-1}\left(-[Q]\left(\{\theta - \theta^*\}^{n-1} + \{\theta - \theta^*\}^n\right) - \left([A]^T + \frac{2}{\Delta t}[I]\right)\{p^n\}\right) \tag{26}$$

where the parameter n expresses time-step, identity matrix is denoted by $[I]$. The Lagrange multiplier $\{p\}$ is computed by equations(25),(26).

To keep the stability of the computation, the Hamiltonian function is modified in the following form.

$$K^{(i)} = H^{(i)} + \left(\{u\}^{(i)} - \{u\}^{(i-1)} \right) [W]^{(i)} \left(\{u\}^{(i)} - \{u\}^{(i-1)} \right) \tag{27}$$

where superscript (i) means the iteration cycle for the minimization, the weighting function $[W]^{(i)}$ is assumed diagonal matrices with a constant coeffecient $w^{(i)}$ for ith iteration. The optimality condition can be written:

$$\frac{\partial K^{(i)}}{\partial u^{(i)}} = [R]\,\{u\}^{(i)} + [B]^T\,\{p\}^{(i-1)} + 2[W]^{(i)}\left(\{u\}^{(i)} - \{u\}^{(i-1)}\right) = 0 \tag{28}$$

which leads to the optimal control as:

$$\{u\}^{(i)} = -\left([R] + 2[W]^{(i)}\right)^{-1}\left([B]^T\,\{p\}^{(i-1)} - 2[W]^{(i)}\,\{u\}^{(i-1)}\right). \tag{29}$$

3.4 ALGORITHM (2)

1. Assume the initial control values $\{u\}^{(0)}$.
2. Solve the state values $\{\theta(u^{(0)})\}^{(0)}$ by equations(3),(4),(5),(11).
3. Solve the cost function $J(\theta^{(0)}, u^{(0)})$ by equation(22).
4. Solve the Lagrange multiplier $\{p(\theta^{(i-1)}, u^{(i-1)})\}^{(i-1)}$ by equations(25),(26).
5. Solve the control values $\{u\}^{(i)}$ by equation(29).
6. Compute $uu = \sum_{t=t_0}^{t_f}\left(u^{(i)} - u^{(i-1)}\right)^2$.
   IF $uu < \varepsilon$ THEN STOP ELSE GO TO 7.
7. Solve the state values $\{\theta(u^{(i)})\}^{(i)}$ by equations(3),(4),(5),(11).
8. Solve the cost function $J(\theta^{(i)}, u^{(i)})$ by equation(22).
9. Compute $JJ = J(\theta^{(i)}, u^{(i)}) - J(\theta^{(i-1)}, u^{(i-1)})$.
   IF $JJ < 0$ THEN $i = i + 1$ and GO TO 4.
   ELSE choose larger $W^{(i)}$ and GO TO 5.

where superscript (i) denotes the iteration cycle for the minimization, the parameter ε is a small number.

3.5 NUMERICAL EXAMPLE (2)

The model of optimal control is the same as the former section. When the cost function is minimized, the optimal control values are computed. Without consideration of a capacity of the machine which makes the cold and hot flow through pipes (weighting function $[R] = 0$), the underground temperature can be made almost constant (Fig.8). But this optimal control value would be impossible. In order to obtain the practical control values, the weighting function is changed in a various manner (Fig.9,10,11). The ideal temperature $\{u^*\}$ is assumed 10 degrees as a constant value. The objective temperature $\{\theta^*\}$ is assumed 20 degrees as a constant value at 4cm under the ground surface.

Ground surface _____
Objective point _ _ _ _ _
Control point __ _ __

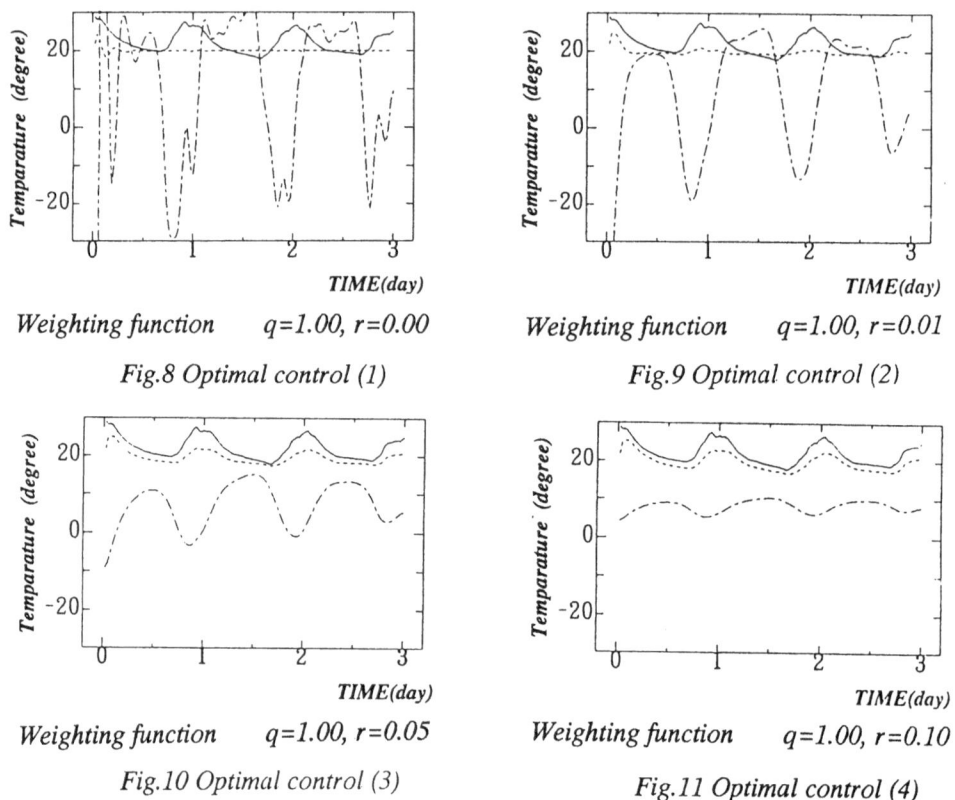

Weighting function $q=1.00, r=0.00$

Fig.8 Optimal control (1)

Weighting function $q=1.00, r=0.01$

Fig.9 Optimal control (2)

Weighting function $q=1.00, r=0.05$

Fig.10 Optimal control (3)

Weighting function $q=1.00, r=0.10$

Fig.11 Optimal control (4)

4   CONCLUSION

This paper is a first step study to maintain the golf links without pesticide for the groundwater's pollution control. This study succeeded to obtain the underground thermal conductivity(Fig.6) with the observed temperature. And then, the optimal control value can be decided (Fig.8) without any remitation. But this control seems out of realization range. It is possible to obtain the practical control(Fig.9,10,11) by adjusting the weighting function. To know the appropriate wighting function is another problem.

Bang-bang Control should be the feasible study, which is a turning on and off control. This study is now under investigation.

References

[1] K.Hatanaka, K.Kojima, M.Kawahara,"A Basic Study on Identification of Aquifer Parameters in Groundwater Hydrology", 8th International symposium on Finite Element in Fluids, vol.2,pp.901-908,(1993), eds. Oñate et.al, Barcelona.

[2] M.Kawahara, Y.Shimada, "Optimal Control with Constraint on The Control Value to Operate Water Gate of Dam", 8th International symposium on Finite Element in Fluids, vol.2,pp.1272-1280,(1993), eds. Oñate et.al, Barcelona.

PARAMETER ESTIMATION AND MODELING IN GROUNDWATER

J. TROESCH and U. KUHLMANN
Laboratory of Hydraulics, Hydrology and Glaciology
ETH-Zentrum
CH-8092 Zürich
Switzerland

Inverse numerical modelling of groundwater flow is a straightforward automatic method for estimating the model parameters using field data and prior information. With the model CASA the parameters such as permeability, storativity or storage coefficient, head and flow boundary conditions may be estimated. CASA is a transient Finite Element model with a combination of 1d, 2d and 3d elements. The Maximum Likelihood Method is used for the estimation of the optimal model parameters. The objective function is minimized with a choice of different optimization algorithms that may be used alternatively. The nonlinear optimization uses either the exact gradients or jacobians computed with the adjoint state method or a direct derivation of the FE-equations. Due to the resulting distribution of the residuals, the initial conceptual model may be improved iteratively.
The advantages of the method of inverse modeling are shown with a model of a complex transient fracture-flow system

INTRODUCTION

The poor knowledge of the aquifer properties is the main problem of groundwater modeling. Its spatial extension is uncertain, material properties vary in a wide range and the boundary conditions are quite often subject of philosophical discussions between hydrogeologists.

Inverse modelling is a method to estimate automatically optimal model parameters for a certain conceptual model. Therefore, time is not spent primarily for trial-and-error searching for parameters, but may be used for a better interpretation of the results and the

815

A. Peters et al. (eds.), Computational Methods in Water Resources X, 815–822.
© 1994 *Kluwer Academic Publishers. Printed in the Netherlands.*

statistical error structure of the inverse simulation. Thus, the conceptual model may be improved or new experiments may be designed to overcome uncertainties.

The model CASA is a transient 3d model based on an indirect approach to the inverse problem. Therefore, discrete performance measures of the modeled system, such as potential heads at specified observation points, are evaluated, solving the direct problem repeatedly. The appropriately weighted differences between computed and observed quantities are minimized. In the work presented here the Maximum Likelihood method is applied (Carrera and Neuman, 1986). In CASA, material properties, flux or head boundary conditions may be the unknown parameters. The prior information used is information on permeability, storage coefficient, flux, or head boundary conditions.

In a lot of cases, the minimum of the function is found with the Marquardt-Levenberg algorithm, but quite often a combined strategy is applied. At the beginning some steps of a gradient method are used to find the neighbourhood of the minimum. Then a Gauss-Newton method may be more efficient. Carrera et al (1989) and Kauffmann and Kinzelbach (1989) use also the Marquard-Levenberg method with good convergence. Odenwald and Herrling(1990) use a combined method. It is therefore important that the program offers the possibility to switch easily between different methods.

BASIC EQUATIONS

In saturated groundwater flow Darcy's law and the continuity equation give the equation for the potential head h

$$(k_{ij} h,_j),_i - q = S h,_t \qquad (i, j = 1 \ ... \ 3) \tag{1}$$

and subject to appropriate boundary conditions, where k_{ij} is the permeability tensor, q the flux and S the storativity or storage coefficient. Equation (1) is solved with a Galerkin Finite Element approximation. Quadratic isoparametric finite elements are used, 3d elements for the matrix, 2d elements for fractures and 1d elements for flow channels.

The parameter estimation is based on the Maximum Likelihood method, following the approach by Carrera and Neuman (1986). All types of model parameters may be estimated, such as permeability, storativity, areal recharge, or boundary conditions. The resulting negative Log-Likelihood-function is minimized with respect to the model parameters. If the error structure of the measurements is assumed to be known this is equivalent to minimize the objective function Z

$$Z = Z_h + Z_Q + \sum_i Z_i . \tag{2}$$

The function is a sum of a contribution of the head values Z_h, the flow rate values Z_Q and of the prior information Z_i, also called the plausibility criterion. These contributions can be written as

$$Z_h = (z_h^* - z_h)^T C_h^{-1} (z_h^* - z_h) \tag{3}$$

$$Z_Q = (z_Q^* - z_Q)^T C_Q^{-1} (z_Q^* - z_Q) \tag{4}$$

$$Z_i = (p_i^* - p_i)^T C_i^{-1} (p_i^* - p_i) \tag{5}$$

z_h^* :   measured head values

z_h :   calculated head

z_Q^* :   measured flux (eg. infiltration or exfiltration)

z_Q :   calculated flux

p_i^* :   prior information (permeability, storage coefficient, boundary conditions)

p_i :   calculated parameters

C_h, C_Q, C_i : Covariance matrix of the residuals. The inverse is a weight of the residuals

The potentials z_h and the fluxes z_Q are the result of a direct solution of equation (1). For the upscaling of local measurements interpolation or zonation has to be used. A well known method for interpolation is Kriging. From geological evidence, the aquifer may also be subdivided in different zones where the subzones are homogeneous unites.

IDENTIFIABILITY

The objective function Z should be a convex surface in a multidimensional parameter space. Unfortunately, this is often not the case and different parameters produce the same response of the model. As this surface may have several local minima's care has to be taken to check if the minimum found is global or not. This can be achieved with different starting points for the optimization runs.

If the function is insensitive to variations of the parameters, the function may have a banana-shaped form. In this case, the parameters can not be identified. Incorporation of prior information improves identifiably considerably, but in ill-posed problems the parameter estimates are close to their measurements. Better results are possible if more data, especially transient data and flow rate measurements are available. If flow rate data is incorporated in the objective function (2) then the convexity is improved considerably. In situ measurements of the flow rate are not always possible. In a lot of circumstances — in tunnels, drains or for river in- and exfiltration — the flow rate can be measured directly. In other cases it is preferable to compute good estimates — e.g. for lateral inflows — with hydrological methods.

APPLICATIONS

Several real world examples have been calculated until now. For a wide variety of problems the method proved to be useful.

Alluvial aquifers

For near surface alluvial aquifers with high permeability where a lot of measurements and good prior information of the material properties are available an optimal parameter set is found In most cases. If the convergence is slow or instabilities occur then the problem is ill-posed. In all cases, convergence could be improved with prior information on flow rates at constant head boundaries. Typically, these horizontal two-dimensional models have around 1000 nodes, the number of parameters to be estimated is 15-20. Simulation times for inverse runs on a SUN Sparc 10 for a stationary problem are approximately 0.5 to 2 hours.

Fracture zone model

In fractured rock models parameter estimation is more difficult, as even the hydrogeological concept is not straightforward. Fracture zones not detected during testing may influence the flow. It is therefore important to test different conceptual models and to interpret the results carefully. These concepts may be improved based on the error statistics of previous inverse runs. In the following such a fracture network model is presented.

A small scale local model with an extension of 100 m is discretized for the central BK-area at the Grimsel test site of the National Cooperative for the Storage of Radioactive Waste (Nagra). An important hydrogeological program with an intensive testing in numerous packed-off borehole sections gives a good data set. The conceptual model used in this case is a network of discrete fractures. From the drilling cores of the boreholes, three main groups of fractures can be distinguished. The K-fractures are perpendicular to the tunnel whereas the S^+ and the S^- fractures are sub parallel to the access tunnel. These fractures are discretized with two-dimensional elements, giving complicated intersections. The complete data set and a detailed description of the methodology is given in Kuhlmann and Correa (1994).

The intersections of the boreholes with the fractures are discretized with finite elements (modelled structure approach), the borehole itself having a fictitious high permeability (Fig. 1). The mesh size around the borehole is started with less than two times its radius. Due to the long packed-off intervals the boreholes are incorporated with 1d elements in order to establish the hydraulic connections between the fractures. The packers are simulated with separate impervious elements.

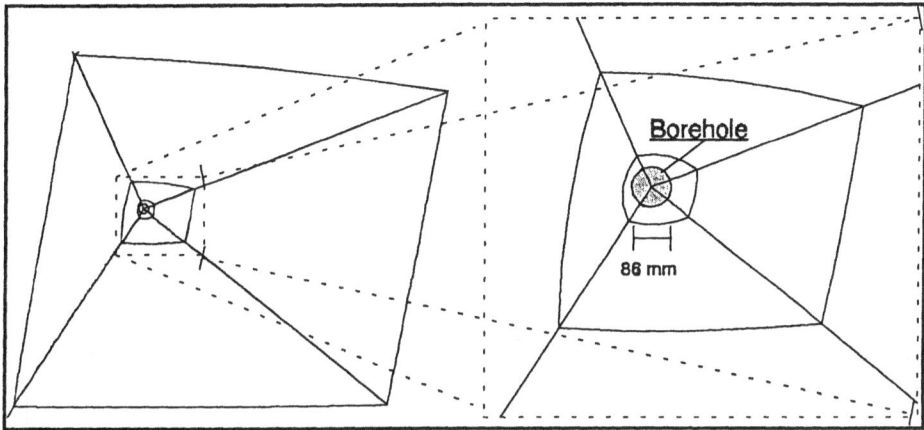

Fig. 1:   Borehole discretization and corresponding mesh refinement

At the outer boundary of the model no-flow boundaries are taken, as in this case the boundary conditions proved to be unimportant for the tested artificially induced flow field. In the injection point the flow rate is imposed.

For the time stepping a finite difference method is used. The initial time step of 0.001 s is enlarged by a factor of 1.2 until the maximum step of 120 s was reached. Thus, for the total simulation time of 20000 s, 300 time steps are needed.
Based on the hydrogeological concept two models are investigated with the inverse analysis. Model A consists of 5 fracture zones with 10 unknown parameters. The number of measurements to be interpreted are 32 values in 14 intervals i.e. 448 points. Model B contains one more fracture to account for a distinct connection between two intervals.

The prior information on transmissivities and storage coefficients are averaged data along the packed of sections. The transmissivity of individual fractures could not be tested. These values are used as best guess. However, the standard deviations are slightly increased to 0.5 related to the log-values. The errors of the head values are assumed to lack correlation in space and time, standard deviation is 0.1 m in the test sections.

The results of the inverse run are shown in Fig. 2 for model B, which showed some improvements compared to model A in model performance. The response of the model is in agreement with the measured system response. Significant deviations are observed in the injection interval, where the computed over pressure drops about 20 m compared to the measurements. This may be due to local inhomogeneities or matrix storage.

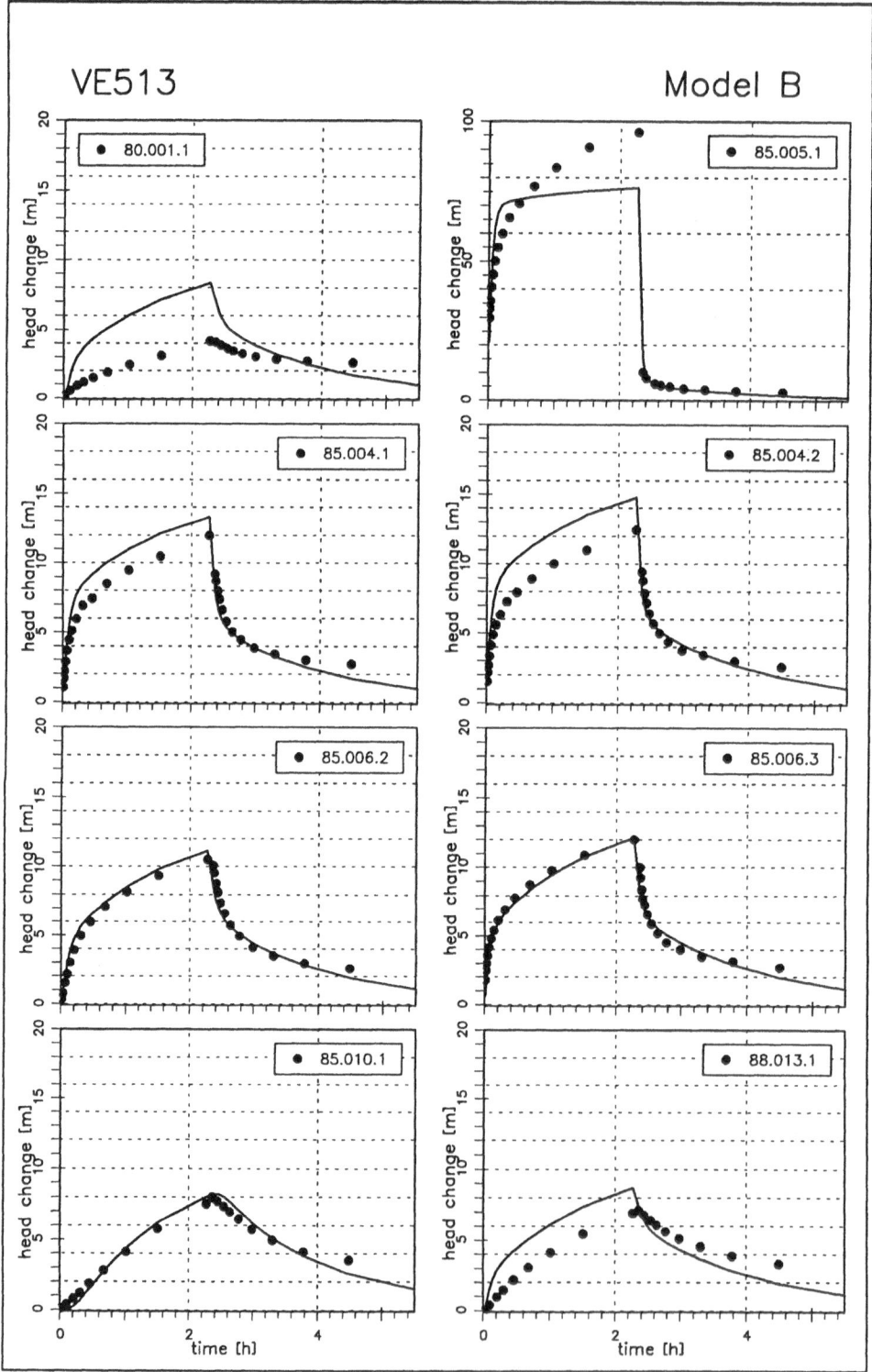

Fig. 2:   Comparison of computed and measured head variation for Model B

The residuals of the inverse simulation show the contributions of the individual test sections to the model performance (Fig. 3). It can be seen that two intervals make the largest contributions to the objective function. The interpretation of the sensitivities shows that the residuals of two distinct intervals are most sensitive to parameter variations (Fig. 4). To further improve the model an additional fracture intersecting these intervals may be necessary.

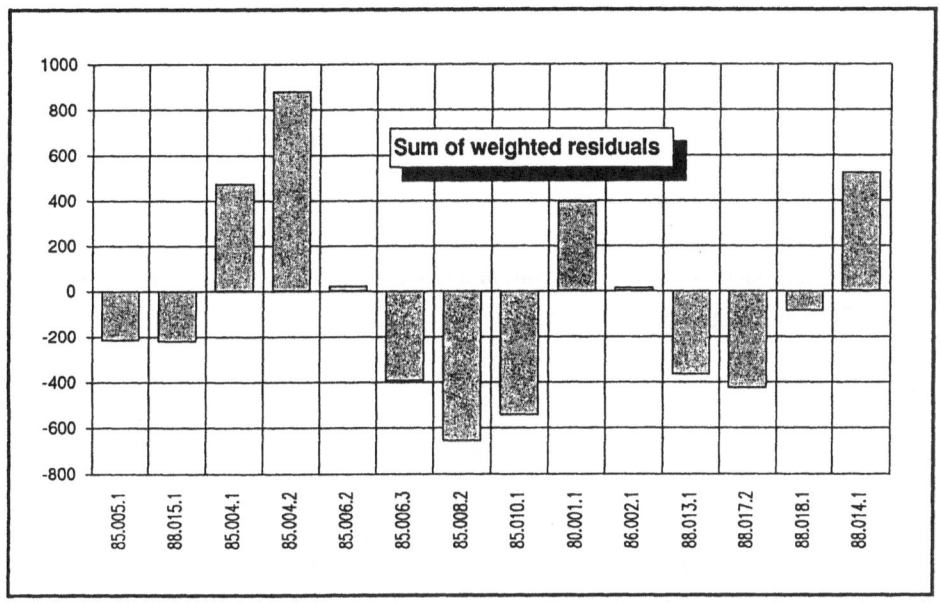

Fig. 3:   Sums (in time) of computed residuals for the individual test sections (Model B)

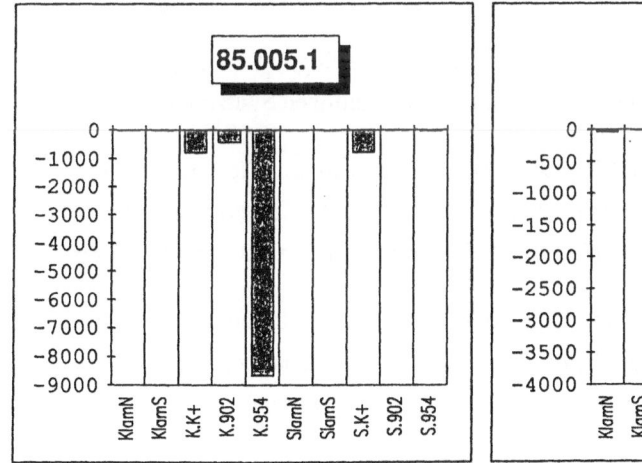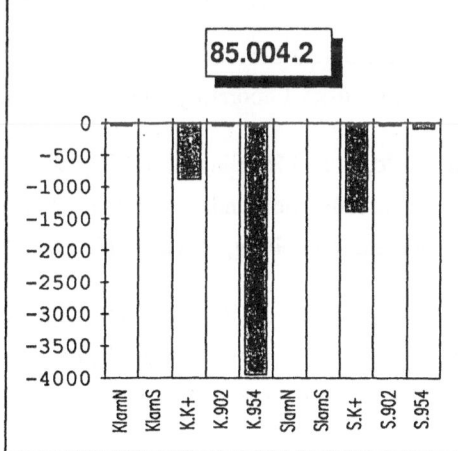

Fig. 4 Sums (in time) of sensitivities for Model B

CONCLUSIONS

Inverse modeling is a fast and straightforward method to compute optimal model parameters for 2d and 3d groundwater models. The a posteriori interpretation of the statistics of the residuals gives the possibility to improve the conceptual model, if zones with important deviations are detected and the sensitivity analysis gives information on relevant parameters.

In any case, it is useful to include measurements of the flow rates at fixed head boundaries. This improves the identifiability considerably. For transient problems computer time is high but especially in 3d problems no other reliable method exists.

ACKNOWLEDGEMENT

This work was funded in parts by the National Cooperation for storage of nuclear waste (Nagra), 5430 Wettingen, Switzerland.

REFERENCES

Carrera, J. and Neumann S.P., (1986) "Estimation of aquifer parameters under transient and steady state conditions", Water Resources Research, **22**,2, 199-242.

Kuhlmann, U. (1992) "Inverse Modellierung in geklüfteten Grundwasserträgern", Mitteilungen der Versuchsanstalt für Wasserbau, Hydrologie und Glaziologie, ETH Zürich, Nr. 120.

Kuhlmann U. and Correa N. (1994) "Fracture Zone Modelling in the central BK-Area at the Grimsel Rock Laboratory", Nagra Internal Report, Wettingen Switzerland.

Odenwald, B. and Herrling B. (1990) "A quick and stable algoritm for groundwater flow parameter estimation under uncertainty", in K. Kovar (ed.) Calibration and reliability in groundwater modelling, IAHS Publ. 195, IAHS Press, Wallingford, UK, pp 75-86

Carrera J., Samper J., Vives L., and Kuhlmann U. (1989) "Automatic inverse methods for the analysis of pulse tests: application to four pulse tests conducted at the Leuggern Borehole", Nagra Technical Report NTB 88-34, Baden, Switzerland.

Kauffmann C. and Kinzelbach W., (1989) "Parameter estimation in contaminant transport modelling", in H. E. Kobus, W. Kinzelbach (eds.) Contaminant transport in Groundwater, Balkema, Rotterdam, pp 355-362.

A REGULARIZATION APPROACH FOR DETERMINING HEAT SOURCE FROM OVERSPECIFIED BOUNDARY DATA

ZHENG KEWANG
Department of Applied Sciences
Hebei Institute of Light Industry and Chemical Technology
Shijiazhuang, Hebei 050018
China

In this paper we describe a regularization approach for determining an approximation to an unknown source term in a heat equation from overspecified initial boundary data. The method here is tested numerically.

INTRODUCTION

Consider the following problem[1]:

$$u_t = u_{xx} + f(t) \qquad 0 < x, \quad 0 < t \qquad (1)$$
$$u(x, 0) = g(x) \qquad 0 < x \qquad (2)$$
$$u(0, t) = \varphi(t) \qquad 0 < t, \quad \varphi(0) = g(0) = 0 \qquad (3)$$
$$u_x(0, t) = \psi(t) \qquad 0 < t \qquad (4)$$

here φ, ψ and g are assumed to be known data from which the pair $\{u, f\}$ is to be determined.

If the pair $\{u, f\}$ now solves this inverse problem then from (1), (2) and (4) we can first obtain the expression of u, in particular, it is convenient to write

$$K(x, t) = \frac{1}{2\sqrt{\pi t}} \exp(-\frac{1}{4} x^2 / t),$$

so that

$$u(x, t) = \int_0^t f(\tau) d\tau - 2 \int_0^t \psi(\tau) K(x, t - \tau) d\tau +$$
$$+ \int_0^\infty g(\xi) [K(x - \xi, t) + K(x + \xi, t)] d\xi. \qquad (5)$$

823

A. Peters et al. (eds.), Computational Methods in Water Resources X, 823–830.
© 1994 Kluwer Academic Publishers. Printed in the Netherlands.

Return to the observed condition (3) it is seen that the source f must satisfy the integral equation

$$Af(t) = \int_0^t f(\tau)\, d\tau = F(t), \tag{6}$$

$$F(t) = \varphi(t) + \frac{1}{\sqrt{\pi}} \int_0^t \frac{\psi(\tau)}{\sqrt{t-\tau}}\, d\tau -$$

$$\frac{1}{\sqrt{\pi}} \int_0^\infty g(x) \exp\left(-\frac{1}{4}\, x^2 / t\right) dx / \sqrt{t}. \tag{7}$$

For practical purposes it is may assumed that $\varphi, \psi \in C[0, T]$ and

$$g \in \bar{C}[0, X] = \{g: g \in C[0, \infty),\ \operatorname{supp}(g) \subset [0, X]\ \}.$$

And then we consider the integral operator A in (6) as a mapping of space $C[0, T]$ into itself. Unfortunately, the problem of finding the solution $f(\tau)$ of equation (6) in the space $C[0, T]$ from the initial data F in the same one is not well-posed on the pair of spaces (C, C) in the sense of Hadamard. It is mainly because the inverse operator A^{-1} is not continuous on AC. However, in practice (by physical measurements) we often know only an approximate right-hand member $F_\delta(t)$ in (6), so the ill-posedness of the problem leads to great difficulties in determining approximate solutions of equation (6).

REGULARIZATION APPROACH

The ultimate goal of this section is to construct a numerical procedure for finding an approximation to the unknown source $f(\tau)$. For this purpose we replace equation (6) with its finite-difference approximation

$$A^h f^h = F^h$$

where

$$A^h = (a_{ij}),\qquad\qquad a_{ij} = \begin{cases} 0 & i < j, \\ h & i \geqslant j, \end{cases}\qquad i, j = 1, 2, \cdots, n$$

$$t_i = ih,\qquad \tau_j = jh,\qquad h = T / (n+1),$$

f^h and F^h are difference functions:

$$f^h = (f_1^h, f_2^h, \cdots, f_n^h),\qquad F^h = (F_1^h, F_2^h, \cdots, F_n^h).$$

Now consider the following functional

$$M_h^\alpha[f^h, F^h] = \sum_{i=1}^{n} h\,[\sum_{j=1}^{i} a_{ij}\, f_j^h - F_i^h]^2 + \alpha \sum_{j=1}^{n} h\,[f_j^h]^2 +$$

$$+ \alpha \sum_{j=0}^{n} h\,[\frac{f_{j+1}^h - f_j^h}{h}]^2 \qquad (8)$$

$(f_0^h = f_1^h, \qquad f_n^h = f_{n+1}^h)$.

Theorem 1. For every difference function F^h and every positive number α, there exists a unique function f_α^h for which the functional M_h^α attains its greatest lower bound:

$$M_h^\alpha [f_\alpha^h, \ F^h] = \inf M_h^\alpha [f^h, F^h].$$

Proof. By considering the first variation of the functional M_h^α, a straightforward calculation shows that the desired function f_α^h should be determined by the Euler equation

$$\alpha L^h[f^h] = \sum_{i=j}^{n} a_{ij} \sum_{k=1}^{i} a_{ik} f_k^h - \sum_{i=j}^{n} a_{ij} F_i^h, \qquad j=1, 2, \cdots, n. \qquad (9)$$

$$L^h[f^h] = (f_{j+1}^h - 2f_j^h + f_{j-1}^h) / h^2 - f_j^h$$

and the boundary conditions

$$f_0^h = f_1^h \ , \qquad f_n^h = f_{n+1}^h \qquad (10)$$

And further, it should be pointed out that under conditions (10) the associated homogeneous equation

$$\alpha L^h[f^h] = \sum_{i=j}^{n} a_{ij} \sum_{k=1}^{i} a_{ik} f_k^h \qquad j=1, 2, \cdots, n$$

cannot possess a nontrivial solution. For, were f^h such a solution, multiplying both sides by f^h, summing over j, we should get the following equality

$$-\alpha \{ \sum_{j=1}^{n} [f_j^h]^2 + \sum_{j=1}^{n} [\frac{f_{j+1}^h - f_j^h}{h}]^2 \} = \sum_{i=1}^{n} [\sum_{j=1}^{i} a_{ij} f_j^h]^2,$$

which would contradict the hypothesis that α is positive. Consequently, under conditions (10) the inhomogeneous equation (9) has one and only one solution f_α^h. Thus the theorem is proved.

The function f^h_α in the above theorem can be regarded as the result of applying to F^h an operator $f^h_\alpha = R(F^h, \alpha)$ depending on a parameter α. And the following theorem will tell us that if the function $\alpha = \alpha(\delta) = \delta^\wedge$, $0 < \lambda \leq 2$, (δ denotes the error in the initial data) is selected as a regularization parameter, then the operator $R(F^h_\delta, \alpha(\delta))$ is a regularizing one and hence the function $f^h_{\alpha(\delta)}$ can be taken as an approximate solution of equation (6).

Theorem 2. Let $f_T(\tau) \in C^1[0, T]$ be the sought exact solution of equation (6) with the exact right-hand member $F_T(t)$:

$$Af_T(t) = \int_0^t f_T(\tau) d\tau = F_T(t), \quad F_T(0) = 0.$$

Then for every positive number ε there exist $\delta(\varepsilon) > 0$ and $h(\varepsilon) > 0$ such that for $\delta \leq \delta(\varepsilon)$ and $h \leq h(\varepsilon)$, the inequality

$$\sum_{i=1}^n h[E^h_{\delta \cdot i} - F^h_{T \cdot i}]^2 \leq \delta^2$$

implies that

$$|f^h_{\alpha(\delta) \cdot j} - f^h_{T \cdot j}| < \varepsilon, \quad\quad j = 1, 2, \cdots, n$$

where

$$F^h_{T \cdot i} = F_T(t_i), \quad\quad f^h_{T \cdot j} = f_T(t_j), \quad\quad i, j = 1, 2, \cdots, n$$

and $f^h_{\alpha \cdot \delta} = R(F^h_\delta, \alpha(\delta))$ is the minimizer of functional $M^{\alpha(\delta)}_h[f^h, F^h_\delta]$.

Proof. By the definition of $f^h_{\alpha(\delta)}$, we have

$$M^{\alpha(\delta)}_h[f^h_{\alpha(\delta)}, E^h_\delta] \leq M^{\alpha(\delta)}_h[f^h_T, F^h_\delta].$$

That is,

$$\sum_{i=1}^n h[\sum_{j=1}^i a_{ij} f^h_{\alpha(\delta) \cdot j} - E^h_{\delta \cdot i}]^2 + \alpha(\delta) \{ \sum_{j=1}^n h[f^h_{\alpha(\delta) \cdot j}]^2 +$$

$$\sum_{j=0}^n h[\frac{f^h_{\alpha(\delta) \cdot j+1} - f^h_{\alpha(\delta) \cdot j}}{h}]^2 \}$$

$$\leq \sum_{i=1}^n h[\sum_{j=1}^i a_{ij} f^h_{T \cdot j} - E^h_{\delta \cdot i}]^2 + \alpha(\delta) \{\sum_{j=1}^n h[f^h_{T \cdot j}]^2 +$$

$$\sum_{j=0}^n h[\frac{f^h_{T \cdot j+1} - f^h_{T \cdot j}}{h}]^2\}$$

$$= \sum_{i=1}^n h[F^h_{T \cdot i} - E^h_{\delta \cdot i}]^2 + \alpha(\delta) \int_0^T [f^2_T(\tau) + f'_T(\tau)^2] d\tau + \xi(h)$$

$$(\xi(h) \to 0 \text{ as } h \to 0)$$

$$\leq \delta^2 + \delta^\wedge \int_0^T [f^2_T(\tau) + f'_T(\tau)^2] d\tau + \delta^2 \leq D\delta^\wedge, \quad (h \leq h(\varepsilon), \ 0 < \delta < 1)$$

$$D = 2 + \int_0^T [f_\tau^2(\tau) + f_\tau'(\tau)^2] \, d\tau,$$

and hence

$$\sum_{j=1}^n h [f_{a(\delta),j}^h]^2 + \sum_{j=0}^n h \left[\frac{f_{a(\delta),j+1}^h - f_{a(\delta),j}^h}{h} \right]^2 \leqslant D,$$

$$\sum_{i=1}^n h \left[\sum_{j=1}^i a_{ij} f_{a(\delta),j}^h - E_{..i}^h \right]^2 \leqslant D \delta^{\wedge}.$$

We define now a function with continuous variable:

$$f_{a(\delta)}^h(\tau) = f_{a(\delta),j}^h + \frac{f_{a(\delta),j+1}^h - f_{a(\delta),j}^h}{h} (\tau - \tau_j)$$

$$\tau \in [\tau_j, \tau_{j+1}], \quad j = 0, 1, \cdots, n.$$

Then both $f_\tau(\tau)$ and $f_{a(\delta)}^h(\tau)$ belong to the set

$$E = \{ f(\tau) : \int_0^T [f^2(\tau) + f'(\tau)^2] d\tau \leqslant 2D \},$$

which is a compact subset of space $C[0, T]$. And hence $Af_\tau(t)$, $Af_{a(\delta)}^h(t) \in AE$. Consequently, it follows by continuity of A^{-1} on AE that for every $\varepsilon > 0$ there exists a number $\eta(\varepsilon) > 0$ such that

$$| Af_\tau(t) - Af_{a(\delta)}^h(t) |_{L_2[0,T]} \leqslant \eta(\varepsilon)$$

implies $| f_\tau(\tau) - f_{a(\delta)}^h(\tau) |_{C[0,T]} \leqslant \varepsilon$

and hence

$$| f_{\tau,j}^h - f_{a(\delta),j}^h | \leqslant \varepsilon, \qquad j = 1, 2, \cdots, n.$$

Now since

$$| Af_\tau(t) - Af_{a(\delta)}^h(t) |_{L_2}^2 = \int_0^T [\int_0^t f_\tau(\tau) d\tau - \int_0^t f_{a(\delta)}^h(\tau) d\tau]^2 dt$$

$$= \sum_{i=1}^n h [\int_0^{t_i} f_\tau(\tau) d\tau - \sum_{j=1}^i a_{ij} f_{a(\delta),j}^h]^2 + \xi(h) \qquad (\xi(h) \to 0 \text{ as } h \to 0)$$

$$\leqslant 2 \sum_{i=1}^n h [F_{\tau,i}^h - F_{\delta,i}^h]^2 + 2 \sum_{i=1}^n h [F_{\delta,i}^h - \sum_{j=1}^i a_{ij} f_{a(\delta),j}^h]^2 + \delta^2$$

$$\leqslant 3 \delta^2 + 2D \delta^{\wedge} \leqslant \delta^{\wedge} (3 + 2D), \qquad\qquad (0 < \delta < 1)$$

so taking

$$\delta(\varepsilon) = [\eta(\varepsilon) / \sqrt{3 + 2D}]^{2/\wedge}.$$

justifies our conclusion at last.

After the determination of $f_{a(\delta)}^h$, of course, the numerical approximation to $u(x, t)$ can be determined directly by (5). But then in concern with (7) we also need the following

Theorem 3. Suppose that $\varphi_T(t)$, $\psi_T(t)$, $g_T(x)$ and $F_T(t)$ are connected by (7), i. e.,

$$F_T(t) = \varphi_T(t) + \frac{1}{\sqrt{\pi}} \int_0^t \frac{\psi_T(\tau)}{\sqrt{t-\tau}}\, d\tau -$$

$$\frac{1}{\sqrt{\pi}} \int_0^X g_T(x) \exp\left(-\frac{1}{4} x^2 / t\right) dx / \sqrt{t}, \qquad (X=T)$$

and that

$$F^h_{\delta \cdot i} = \varphi^h_{\delta \cdot i} + \frac{1}{\sqrt{\pi}} \sum_{j=1}^i b_{ij}\, \psi^h_{\delta \cdot j} - \sum_{j=1}^n C_{ij} g^h_{\delta \cdot j} \qquad i = 1, 2, \cdots, n$$

where

$$b_{ij} = \begin{cases} 2\sqrt{h}\,(\sqrt{i-j+1} - \sqrt{i-j}) & j \leqslant i, \\ 0 & j > i, \end{cases}$$

$$C_{ij} = \mathrm{erf}\left(\frac{j+1}{2} \sqrt{h/i}\right) - \mathrm{erf}\left(\frac{j}{2} \sqrt{h/i}\right).$$

$$i, j = 1, 2, \cdots, n.$$

Then for every positive number δ there exist $\triangle(\delta) > 0$ and $h(\delta) > 0$ such that for $\triangle \leqslant \triangle(\delta)$ and $h \leqslant h(\delta)$, the three inequalities

$$|\varphi^h_{\delta \cdot i} - \varphi^h_{T \cdot i}| < \triangle, \qquad |\psi^h_{\delta \cdot j} - \psi^h_{T \cdot j}| < \triangle \quad \text{and} \quad |g^h_{\delta \cdot j} - g^h_{T \cdot j}| < \triangle$$

$$(i, j = 1, 2, \cdots, n)$$

imply together that

$$\sum_{i=1}^n h\,[E^h_{\delta \cdot i} - F^h_{T \cdot i}]^2 \leqslant \delta^2,$$

where

$$\varphi^h_{T \cdot i} = \varphi_T(t_i), \qquad \psi^h_{T \cdot j} = \psi_T(\tau_j) \quad \text{and} \quad g^h_{T \cdot j} = g_T(x_j).$$

Proof. Since

$$\sum_{i=1}^n h\,[E^h_{\delta \cdot i} - F^h_{T \cdot i}]^2 = \sum_{i=1}^n h\,\Big\{ \varphi^h_{\delta \cdot i} - \varphi^h_{T \cdot i} + \frac{1}{\sqrt{\pi}}\Big[\sum_{j=1}^i b_{ij}\, \psi^h_{\delta \cdot j} - \int_0^{t_i} \frac{\psi_T(\tau)}{\sqrt{t_i - \tau}}\, d\tau\Big]$$

$$- \Big[\sum_{j=1}^n C_{ij} g^h_{\delta \cdot j} - \frac{1}{\sqrt{\pi}} \int_0^X g_T(x) \exp\left(-\frac{1}{4} x^2 / t_i\right) dx / \sqrt{t_i}\,\Big]\Big\}^2$$

$$= \sum_{i=1}^n h\,\Big[\,(\varphi^h_{\delta \cdot i} - \varphi^h_{T \cdot i}) + \frac{1}{\sqrt{\pi}}\sum_{j=1}^i b_{ij}\,(\psi^h_{\delta \cdot j} - \psi^h_{T \cdot j}) - \sum_{j=1}^n C_{ij}\,(g^h_{\delta \cdot j} - g^h_{T \cdot j})\Big]^2 + \xi\,(h)$$

$$\leqslant 3\triangle^2\,\Big\{\sum_{i=1}^n h + \frac{1}{\pi}\sum_{i=1}^n h\,[\sum_{j=1}^i b_{ij}]^2 + \sum_{i=1}^n h\,[\sum_{j=1}^n C_{ij}]^2\,\Big\} + \xi\,(h)$$

$$\leqslant 3\triangle^2\,\Big(T + \frac{4}{\pi}\sum_{i=1}^n ih^2 + \sum_{i=1}^n h\,\Big) + \xi\,(h) \qquad (\xi\,(h) \to 0 \text{ as } h \to 0)$$

$$\leqslant 3\angle^2 \left(2T + \frac{2}{\pi} T^2 \right) + \angle^2 T = \angle^2 T \left(7 + 6T / \pi \right), \qquad\qquad (h \leqslant h(\delta))$$

this suggests that we put

$$\angle(\delta) = \delta / \sqrt{T(7+6T / \pi)}.$$

A NUMERICAL EXAMPLE

The method presented above is applied to an example $(T=X=1)$.
Take

$$\varphi_T(t) = t^3/3 - t^4/2 + 0.0002\sqrt{t / \pi} \left\{ 1 + 4t \left[\exp\left(-\frac{1}{4} / t \right) - 1 \right] \right\},$$

$$\psi_T(t) = 256 t^{9/2} / (315\sqrt{\pi}), \qquad g_T(x) = 0.0001 x (1-x^2),$$

and hence

$$F_T(t) = t^3/3 - t^4/2 + t^5/5,$$

which has yielded the true solution of equation (6) $f_T(t) = t^2 (t-1)^2$.
Now we try to recover it from the initial data

$$\varphi_\delta(t) = \varphi_T(t) + \delta \cdot \sin(50 \pi t), \qquad \psi_\delta(t) = \psi_T(t) + \delta \cdot \sin(50 \pi t)$$

and

$$g_\delta(x) = g_T(x) + \delta (1-x) [1 - (1-x)^2].$$

The actual calculation proceeds as shown in the following table. (Note symmetry of f_T with respect to $t = 0.5$)

Table. The reconstruction of the source term $f_r(t)$

t	0.025	0.05	0.075	0.1	0.125
$f_T(t)$	0.000594	0.002256	0.004813	0.0081	0.011963
$f_\alpha(t)$	0.003830	0.005317	0.007679	0.010750	0.014383

t	0.15	0.175	0.2	0.225	0.25
$f_T(t)$	0.016256	0.020844	0.0256	0.030407	0.035156
$f_\alpha(t)$	0.018434	0.022773	0.027280	0.031841	0.036356

t	0.275	0.3	0.325	0.35	0.375
$f_T(t)$	0.03975	0.0441	0.048125	0.051756	0.054932
$f_\alpha(t)$	0.040728	0.044872	0.048713	0.052179	0.055214

t	0.4	0.425	0.45	0.475	0.5
$f_T(t)$	0.0576	0.059719	0.061256	0.062188	0.0625
$f_\alpha(t)$	0.057767	0.059794	0.061266	0.062159	0.062457

$n = 159$, $h = 1/160$, $\delta = 0.0001$, $\lambda = 1.2$

REFERENCES

[1] Cannon, J. R. and Zachmann, D. (1982) "Parameter determination in parabolic partial differential equations from overspecified boundary data", Int. J. Engng Sci. 20, No. 6, 779-788.

[2] Tikhonov, A. N. and Arsenin, V. Y. (1977) Solutions of ill-posed problems, John Wiley & Sons. New York, Toronto, London, Sydney.